Handbook of Aflatoxins

Handbook of Aflatoxins

Edited by **Debbie Harms**

R Callisto Reference

New York

Published by Callisto Reference,
106 Park Avenue, Suite 200,
New York, NY 10016, USA
www.callistoreference.com

Handbook of Aflatoxins
Edited by Debbie Harms

International Standard Book Number: 978-1-63239-367-8 (Hardback)

Printed in the United States of America.

Contents

Preface

It is often said that books are a boon to mankind. They document every progress and pass on the knowledge from one generation to the other. They play a crucial role in our lives. Thus I was both excited and nervous while editing this book. I was pleased by the thought of being able to make a mark but I was also nervous to do it right because the future of students depends upon it. Hence, I took a few months to research further into the discipline, revise my knowledge and also explore some more aspects. Post this process, I begun with the editing of this book.

Researches show that aflatoxins are responsible for deteriorating around 25% of the world's food crops. This book talks about several topics related to aflatoxin contamination, its measurement & analysis and approaches for prevention & control of aflatoxins on crops and different foods. It talks about the essence that these subjects have for a country, taking the case of China as an example and mentions instances that illustrate the ubiquity of aflatoxins in several commodities. This book further discusses the concept of measurement and examination of aflatoxins from historical facets, legal challenges, and the state of the art procedures and techniques. It concludes with discussions on actions to avoid and mitigate the genotoxic effect of one of the most discernible aflatoxins, AFB1. Also, it marks the interventions to decrease known aflatoxin-induced diseases at agricultural and dietary levels and approaches that can monitor aflatoxin levels. Along with the precautionary management, many approaches have been employed, comprising of physical, chemical biological cures and solvent extraction to detoxify AF in contaminated feeds and feedstuffs.

I thank my publisher with all my heart for considering me worthy of this unparalleled opportunity and for showing unwavering faith in my skills. I would also like to thank the editorial team who worked closely with me at every step and contributed immensely towards the successful completion of this book. Last but not the least, I wish to thank my friends and colleagues for their support.

Editor

Part 1

Aflatoxin Contamination

Aflatoxin Contamination and Research in China

Huili Zhang[1,2], Jianwei He[2], Bing Li[3],
Hui Xiong[3], Wenjie Xu[3] and Xianjun Meng[1]
[1]*Academy of Food Science Shenyang Agricultural University, Shenyang,*
[2]*School of Life Science, Liaoning University, Shenyang,*
[3]*College of Light Industry, Liaoning University, Shenyang,*
China

1. Introduction

Aflatoxins(AF) are highly poisonous secondary metabolites produced by *Aspergillus flavus* and *Aspergillus parasiticus*. They have been found in moldy human food and animal feeds and have been implicated in numerous animal disorders. *A. parasiticus* produces four major aflatoxins: B1 , B2 , G1 and G2 , while AFB1 is the most toxic in the group and the toxicity is in the order of B1 > G1 > B2 > G2. Since the 1960 outbreak of Turkey X disease, when more than 10,000 turkeys died after being fed with aflatoxin contaminated peanut meal, scientists in China have paid more attention to the studies on aflatoxins including its distribution, pollution, health hazards, testing, monitoring, detection technology, managing, microbiology, ecology, toxicology, and policies in controlling aflatoxins. In this review, we present a brief report on the situation of aflatoxin contamination and research progress in China.

2. Distribution of aflatoxin contamination

2.1 The distribution in cereals, oils and foodstuffs

In general, the nationwide aflatoxin contamination was mainly in cereals, oils and foodstuffs. The aflatoxin B1 (AFB1) content detected in vegetable oil products was far higher than in food products. Based on 1,000 investigations of susceptible aflatoxin contamination from nearly 20 provinces between 2002 and 2008, contamination was reported in almost every province. The majority of the samples tested shows the presence of aflatoxins. The overall level of contamination in southern part of China is higher than in the northern region. The most severe province is Guangxi. The main reason is due to the hot and humid southern climate. Climatic condition significant influences the level of aflatoxin contamination. When in serious drought and/or high temperature conditions, or when the soil humidity is below the normal level before harvest, it increase in the number of *A. flavus* spores in the air resulting more fungal infection, and thus high level aflatoxin accumulation. During end processing and packaging, storage of animal-derived food, "cold chain" transfer or pollution of the packing material could also lead to the *A. flavus* infection and aflatoxin contamination (Duan *et al.*, 2009).

Studies on the level of aflatoxin B1 in 486 foodstuff and 146 oil samples collected from 18 cities in 2008 (Zhang, 2008) demonstrated that the levels of aflatoxin contamination were between 0.02 and 54.20 µg/kg in foodstuff and 0.41 and 36.54 µg/kg in oil products respectively. While the detection rate ranged from 0.41% to 2.06%, respectively. A similar study in 2004 reported that the aflatoxin B1 detection rate was as high as 58% in 17 grain samples, which was the most severe incidence in Guangxi province (Wang, 2004). Samples collected from 5 provinces including Sichuan (mainly in Chongqing), Guangdong, Guangxi, Hubei and Zhejiang showed that the aflatoxin B1 detection rates were 70.27% and 24.24% in corn and peanut respectively (Liu, 2006). The aflatoxin B1 detection rates in peanut oil, peanut and corn samples collected from Yunnan province were 100%, 24.32% and 5.26% respectively. The aflatoxin levels in samples were 16.05% above the legally allowed limit, which was similar to the level in peanut samples from Beijing (Wei, 2002; Gao et al., 2007) .

2.2 Distribution in dairy products

The investigation of aflatoxin contamination in dairy products indicated that aflatoxin M1 (AFM1), a hydroxylated metabolite of AFB1 secreted in milk, was commonly detectable in most of the dairy samples tested. This phenomenon is correlated well with the distribution of AFB1. From a survey of more than 1,000 samples in 17 provinces between 1991 and 2005, the following reported detection rate ranged from 4.0% to 73.7%.

The scientists of Guangxi Anti-Epidemic and Basic Course Section monitored the AFM1 contamination in 100 samples of milk and dairy products in 15 provinces between 1991 and 1999, the AFM1 contamination levels in milk were from 0.2 µg/kg to 1.9 µg/kg (Tang, 1999a). In 1991, a study using the HPLC on 57 milk samples and 15 milk powder samples from Shanghai, the detection rate of AFM1 were 26.7% and 73.7%, respectively (concentration ranges between 0.025 and 0.95 µg/kg (Zhu, 1991). A similar survey was performed using TLC in 1995 on 59 milk and 53 milk powder samples from Fuzhou, reported that the detection rates were 4.0% and 13.19% respectively (concentration range between 0.06 and 0.20g/kg (Lin, 1995).

The data indicated that the detection ratio of AFM1 correlate with the high content of AFB1 in animal feed. The amount of AFB1 consumption by animals influences the amount of AFM1 secreted in milk in a dose-dependent manner. Again, the highest level was detected in milk and dairy products from Guangxi province.

2.3 Distribution in feed

The investigation of aflatoxin contamination in animal feed demonstrated the wide distribution of aflatoxins. A survey done in more than 1,000 samples in 20 provinces from 2003 to 2008 showed that aflatoxins present in most of the samples, as stated below. This analysis indicated that the general level of serious contamination in southern region is similar to that in northern region. Irradiation has been suggested as a possible means of controlling insects and microbial populations in stored food under moist storage condition (Xiao et al., 2007).

Aflatoxins in feeds has long been a problem in Huanan, Huabei and Huazhong large geographic regions. Detections of AFB1 in 109 samples showed that the aflatoxin detectable rate and the average content were 83.9% and 24.6µg/kg in corn, 100% and 8.27µg/kg in complete feed, 100% and 6.81µg/kg in animal and plant protein, 100% and 13.3µg/kg in mycoprotein, respectively (Wang *et al.*, 2003). The data showed the relevant ratio, average content and above limit ratio of aflatoxins in feedstuffs were 92.1%, 8.15g/kg and 6.6%, respectively. These values were 100%and 5.95µg/kg in dairy cattle mix feed as studied during 2006 and 2007 (Ao & Chen, 2008).

2.4 Distribution in fermented flavoring

The investigation demonstrated that the safety of fermented flavoring food products such as soy sauce is very optimistic in China. A latest survey of 203 samples of national brand soy sauce samples in 2010 showed that the aflatoxin level is below the maximum allowed level set forth by European Commission (Qi & Che, 2010). This may be contributed by the fact that soybean, row material of fermentation, is not susceptible to infection of aflatoxin-producing fungi preharvest, eventhough the growth condition of *Aspergillus oryzae* and *Aspergillus niger*, are similar to that of the aflatoxin-producing fungi.

The maximum amount AFB1 allowed in brewed soy sauce in China was set by law at 5µg/kg. In order to understand the AFB1 contamination of the brewed soy sauce in China, 203 soy sauce samples from different provinces in China were tested for the establishment of emergency response and early warning systems of AFB1 (Sun *et al.*, 2010). The study concluded that the soy sauce is safe for consumption. The average AFB1 content in the brewed soy sauce from the five provinces in China were 0.3560µg/kg, 0.4636µg/kg, 0.5273µg/kg, 0.3143µg/kg and 0.2083µg/kg respectively. All of the tested samples were bellow the maximum allowed level set forth in China and the EU countries, which is 2µg/kg.

2.5 The distribution in traditional chinese medicine

Due to a great variety of traditional Chinese medicines and the wide area of planting regions, the traditional Chinese medicinal herbs can be infected with aflatoxin-producing fungus, *A. flavus*, in the process of processing, storage and transportation. Aflatoxin-producing fungus exists in soil and air and Chinese medicinal herbs can be infected by *A. flavus* directly. Studies demonstrated previously that the aflatoxin contamination in Chinese herbal medicine is another issue of concern in preventing aflatoxin contamination in the food.

The investigation of AFB1 in regular Chinese herbal medicine and Chinese traditional patent medicines using the method of ELISA has been reported (Ren & Ma, 1997). It was demonstrated that the presence of AFB1 in traditional Chinese medicinal materials was common. Results suggested that the positive rate and contents of AFB1 were serious enough to alert our concern. Studies in 20 different provinces during 1997~2001 period showed that the AFB1 content in 83%~100% samples was over the limits allowed, with several samples seriously over limits. The indirectly competitive enzyme-linked immunosorbent assay on seven Chinese medicines showed that the aflatoxin content in severe cases reached as high as 200~229 ng/g in Shenqu and 1,056 ng/g in Yueju baohe pellet, respectively (Liu, 2001). Other report showed 85% of samples detected the presence of aflatoxin at a concentration less than 1ng/g (Tang, 1999b).

3. The toxicity of aflatoxin

A series following surveys of nearly 10,000 people from 2006 to 2009 show that the toxicity has a positive correlation with the distribution of AFB1. The total morbidity in the southern region is more serious than in the northern region. Guangxi and Zhejiang provinces are the high incidence regions, which is significantly higher than the already reported AFB1 contamination area, Guangdong, Hunan and Singapore. It is widely accepted that aflatoxin contamination in the region is correlated well with the onset of liver cancer in human. Studies showed that AFB1 causes the p53 gene mutation in human cancer cell. P53 is a tumor suppressor, a transcription factor involved in the regulation of the cell cycle.

3.1 Harm to human
3.1.1 Carcinogenecity

Aflatoxins are the most notorious within mycotoxins. These toxins target liver of human and animal and can lead to hepatic cancer and even death. Among them, the AFB_1 was categorized as No. 1 carcinogen by International Agency for Research on Cancer (IARC) in 1988. A significant negative relationship between the amount of aflatoxin in food with incidence of liver cancer was observed (Gao, 1998).

3.1.1.1 Hazards

To explore the epidemiological feature, as well as the changing rule of the morbidities of malignant diseases, especially, liver cancer, in the population of Fushui county, Guangxi province in the period of 1997~2003 (Huang & Wei, 2006), morbidity data of all malignant diseases in Fushui county, Guangxi province, were collected. Population data were collected as well. The population construction, by ages and sexes, was calculated, referring to the data of overall survey in 1990. They calculated statistically the yearly rates of liver cancer, in order to produce the changing trends comparing the data with history data. Results showed the mean morbidity rate of liver cancer in Fushui county was 52.79/105 (or 50.50/105, if adjusted to the Chinese population, 1964). Liver cancer morbidity rate was the highest in all of the malignant diseases occupying 57.01% of the morbidity rate of overall malignant diseases. Male is more susceptible than female. The ratio of morbidity rate between men and women was 4.93:1. Morbidity rates of liver cancer rose by ages, with the mid-age of 47.58. Morbidity rates of liver cancer in these years remained relatively stable. Comparing with 1970's data, these rates seemed already slightly reduced. Considering that mortality rates (a replacement of cancer morbidity rates) of liver cancer already rise obviously in Guangxi, as well as in the rest area of China. The trend that morbidity of liver cancer in Fusui county reduces a little comparing with historical data. This could be considered as a reflection of the effect of cancer control, which had already been being carried out in Fusui county. The facts that mid-age group of morbidity is decreased and that morbidities in younger age group is significantly reduced are the important evidences of the effect of the field cancer control.

In 2008, polymorphism studies on CYP3A5 genes in 210 patients from high aflatoxin contamination area showed that about 60% of the total individuals are those with high level CYP3A5 expression[25]. This percentage is far higher than that has been reported in Guangdong, Hunan provinces and Singapore, which are considered low aflatoxin B1 contamination. Consequently, the contamination of aflatoxin is the main reason of the occurrence of liver cancer in this area (Lu *et al.*, 2008). In the same year, studies on the relationships of the aflatoxin exposure, glutathione transferase gene polymorphism and high risk group with primary hepatocellular carcinoma shown that the exposure of aflartoxin is the main risk factor of the occurrence of liver cancer in this area (Tang *et al.*, 2008).

In 1970~1999, there were 4,215 new liver cancer cases in Zhongshan. Its crude incidence rate, China and world standardized rates were 13.0/105, 12.5/105, 16.8/105, respectively. There is no increasing or decreasing trend for its incidence rates in 1975~1994. However, a declining tendency between 1995 and 1999 was observed. The liver cancer incidence rate during this period in Zhongshan was moderate comparing with the worldwide statistics, but at middle-high level and at low level compared with urban and rural pilot areas in China at the same time period.

The crude and standardized incidences of liver cancer were analyzed by collecting the disease information from the rural area in Ningbo from 2006 to 2008 (Cui et al., 2009). The results show that the crude incidence of hepatoma of the rural residents in Ningbo from 2006 to 2008 is 38.66/105. The age standardized incidence of this disease is 32.14/105. The incidence of hepatoma increased with age. Its incidence in male is 2.77 times of that in female. As to the diagnosis technology, imageology is the most persuasive method to make a definite diagnosis with a ratio of 58.93%. Next effective method is the pathological examination with a ratio of 36.72%. Hepatoma incidence of rural residents in Ningbo is above the average ratio of that in Zhejiang province and China.

3.1.1.2 Pathogenesis

The aflactoxin can result in cancer by a variety of molecular mechanisms.

Aflatoxin exposures can begin in utero and continue through childhood. A mutation in the P53 tumor suppressor gene from AGG to AGT (arginine to serine) transversion at codon 249 (Ser249 mutation) has been reported for hepatocellular carcinoma and matched plasma DNA found in plasma of young children from a region of high aflatoxin exposure (Xu, 2009). This gene mutation in tumor-derived DNA has recently been detected in plasma or serum DNA from adult hepatocellular carcinoma patients. The presence of this mutation before hepatocellular carcinoma onset (e.g., in patients with cirrhosis and patients without clinically diagnosed liver disease) may indicate that the mutation is a marker of chronic exposure to aflatoxin (Kirk et al., 2005). This mutation has been detected in areas with high aflatoxin exposure while it is rare in the low aflatoxin exposure regions (Duan et al., 2005).

A close relationship between the expression of survivin, a newly founded inhibitor of apoptosis protein (IAP), and the abnormality of Wnt signal transduction pathway, was revealed (Ban & Cao, 2005), (Jiao et al., 2007). The HBV is prevalent in high hepatic cancer risk areas, so there is a synergetic effect between the two risk factors. Using population-based case-control study to find the main risk factors of hepatocellular carcinoma (HCC), when people exposed with three main environmental factors (HBsAg, intake of moldy food and drinking raw water), the ORs of hepatic cancer were increased by several times suggesting a conjugated effect between HbsAg and AFB1 albumin adduct.

Concerning the coordinate cancergenic mechanism between AFB1 and HBV, it is concluded that: i. both of the risk factors can reduce the gene expression level of drug-metabolizing enzyme; ii. Chronic inflammatory reaction increased possibility of p53 mutation induced by AFTB1. iii. Chronic infection of HBV changes AFTB1 to an active form. iv. HBX inhibits the nucleus excision repair of DNA, hindering the repair of AFB1-DNA adduct and similar DNA damage and accelerating the process of carcinomatous change of hepatic cells. Besides, the sensibility of host to AFB1 and fatty degeneration of liver could also come to be carcinogens (Xu, 2009).

3.1.2 Chronic intoxication

3.1.2.1 Correlated event

It is widely reported that the aflatoxins cause many human acute intoxication events. For example, farmers in three families ate mildewed rice (the aflatoxin content reached 225.9 µg/kg) in Taiwan province. This event led to 25 persons poisoned and the deaths of three children, among 39 persons involved. There was an explosion of Toxic Hepatitis caused by aflatoxin in 200 villages, 397 persons got the disease and 106 persons dead (Wu, 2007).

3.1.2.2 Symptoms

After a person ate aflatoxin contaminated food, it may cause fever, abdominal pain, vomiting, more seriously splenohepatomegalia, hepatalgia, skin mucous membrane stained yellow, ascites, edema of lower limbs and dysfunction of liver after 2~3 weeks. The cardiac dilatation, pulmonary edema, coma, spasm may also occur (Xiao & Xing, 2003).

3.2 Harm to animals

The aflatoxin can cause damage to the liver, and the sensitivity of aflatoxin is closely related to animal size, species, gender, age, and nutrition. The aflatoxin can damage to the animal embryo, decrease the liver function, cause a decline in milk and egg production, and decrease animal immunity if infection of micro organism happens repeatedly. During the growth period, animals at young stage are more likely to be infected. The clinical manifestations poisoning include low reproductive capacity, gastrointestinal dysfunction, decline in feed utilization. Moreover, dairy cattle could produce AFM1 and M2.

3.3 Financial loss

According to statistics, the aflatoxins contamination of animal feed in USA led to about 10% financial loss. Besides, the death of livestock results in a severe loss to the agriculture (Zhang, 2008). At the same time, aflatoxins can reduce the production of the food and fiber crops.

4. The biological research

The microbiology research of the aflatoxins in China started fairly late. Reviews on biosynthesis of aflatoxins have not been reported until 2003 and researches on aflatoxin resistance were started only since 2001. Molecular biological methods have been carried out in aflatoxin research during the last two years, such as gene chip (microarray) technology used for gene expression studies.

4.1 Biosynthesis of aflatoxins
4.1.1 Process of synthesis

Based on the improvement in the analysis of the aflatoxins biosynthesis, the aflatoxin biological synthesis process was summarized (Xu & Luo, 2003). In the initial period, with Acetyl-CoA as the original unit and malonyl-CoA as the elongation unit, the reaction is catalysized by polyketide synthase to form the aflatoxin backbone, polyketone. In general, the specific process of the synthesis scheme of AFB1 and AFG1 is from Acetyl-CoA to caproyl-CoA, norsolorinic acid, averantin, averufin (AVF), versiconal hemiacetal acetate, versiconal, versicolorin B, versicolorin A, versicolorin, O-methylsterigmatocystin, and finally to AFB1 and

AFG1; the synthesis of AFB2 and AFG2 is: The front part of the process is the same to AFGB1 and AFG1, the difference is versicolorin B changed to dihydro versicolorin, then to dihydro-O-methylsterigmatocystin, to AFB2 and AFG2.

4.1.2 Factors involved in the synthesis

The factors closely related to the synthesis of aflatoxins include the genes, enzymes and the environmental conditions. The genes related to the aflatoxin biosythesis were analyzed by the technology of gene chip as well as RT-PCR method (Hu & Xu, 2009). Six abnormally expressed genes were detected. The six genes are *aflA*, *aflE*, *aflF*, *aflR*, *aflT* and *aflX*. According to the result, the different expression level of *aflR* has close correlation with the production of aflatoxin.

Some related factors in the synthesis of aflatoxins were studied (Lu *et al.*, 2010). The results show that several dehydrogenases, peroxidases, cyclases, methyltransferases and oxidoreductase have a key role in the biosynthesis of aflatoxins. The activity of those enzymes affected the yield of aflatoxins directly. On the other hand, the most important environmental factors are carbon and nitrogen source, power of hydrogen, temperature, water activity and plant metabolites.

4.2 The resistant research

Studies on resistance to aflatoxigenic fungi through molecular biology in China include: the synthesis of artificial antigens, the aflatoxin resistant microorganism and catabolic enzymes, screening of important resistance genes and molecular markers. Aflatoxins are small molecules, thus the immunization of aflatoxins was achieved through coupling with large proteins. With the m-chloroperbenzoic acid (MCPBA) as oxygenant turning the aflatoxin G1 to 8.9- epoxide, a compound AFG1-BSA was obtained after the epoxide coupling with BSA in a two-phase reaction system (Zhang & Li, 2008). Ultraviolet scanning of the compound showed a significant difference comparing with the scanning result of aflatoxin G1 and a different fluorescence intensity between them, which indicated the coupling of BSA and AFG1. This analytical method promoted the study on the preparation of monoclonal antibody and immunoaffinity column.

The mixture of broad bean and wheat flour during fermentation was used to screen antagonistic bacteria against aflatoxigenic *A. flavus* (Gao & Ding, 2010). A strain L4 with strong antifungal activity against the aflatoxin-producing fungus *A. flavus* was selected using agar medium (BAM). According to its morphological, physiological and biochemical characteristics and 16S rRNA gene sequence homology analysis, L4 was identified as *Bacillus subtilis*. When L4 and *A. flavus* were co-cultured for 15 days, the weight of the mycelium and the production of aflatoxin B1 were both significantly lower than those of *A. flavus* cultured without L4. The accumulation of AFB1 was greatly inhibited, the suppression effective ratio was 93.7%. When L4 culture supernatant was mixed with the spore suspension of *A. flavus* at ratio of 1:1 and then inoculated on corn, the germination and growth of *A. flavus* was completely inhibited.

Using the method of filter paper diffusion. a strain of marine microorganism which exhibited highly inhibitory effect on *Aspergillus flavus* was screened (Kong & Liu, 2010). With the aid of 16S rDNA gene sequence, this marine strain was finally identified as a marine strain of *Bacillus megaterium*. Then, its inhibitory effects on mycelium extending, spore germination and aflatoxin biosynthesis of *A. flavus* were further studied. Quantitative analysis kit for aflatoxins (Beacon) was used to determine the concentration of aflatoxin. The results showed that this marine strain exhibited good inhibition to the mycelium growth, spore germination and aflatoxin biosynthesis in *A. flavus*. Eighty-seven percent spore (1×10^9CFU mL^{-1} *B. megaterium*) and 50.75% aflatoxin (1×10^8CFU mL^{-1} *B. magaterium*) were inhibited, compared with control group. The possible mechanism is that some kinds of metabolites secreted by this marine strain can inhibit the mycelium growth and spore germination of *A. flavus*.

Aflatoxin-detoxifizme (ADTZ), being from *Amillariella tabescens*, can effectively decompose aflatoxins. To secretively express ADTZ in *Pichia pastoris* with higher performance, through optimizing the 5′coding region of its cDNA according to the preferred codons of *P. pastoris* (Zuo & Liu, 2007). Two-step DNA synthesis was used to synthesize the cDNA sequence being optimized of ADTZ (OPT-ADTZ). OPT-ADTZ was inserted in the constitutive plasmid pGAPZαA to construct the recombinant plasmid pNOA. pNOA was linearized and then transformed into *P. pastoris* GS115. Then code-optimized ADTZ was constitutively and secretively expressed in *P. pastoris*. In seed of Balsampear Fruit, the antifungal activity of ribosome inactivating proteins (RIPs) were examined (Liu, 2001). In the research aimed at developing a rapid and reliable screening method for selecting *A. flavus* infection resistance in peanut, two DNA markers closely linked with the resistance to *A. flavus* infection were identified using BSA technique. The two specific fragments were about 440bp and 520bp, respectively. They were named as marker E45M53-440 and E44M5-520 (Lei, 2009). The potential usage of the two markers can be in determining or selecting the resistance to the infection by *A. flavus*.

5. Main methods of detection and screening

Monitoring programs have been established to reduce the risk of aflatoxin consumption by human and animals. Analytical testing methods of large numbers of samples of foodstuffs have been developed for rapid detection of Aflatoxins. Current analytical techniques are more accurate in characterization and quantitation of aflatoxins. These include high pressure liquid chromatography (HPLC), Gas chromatography (GC) and serum assay (ELISA), which are much better than the early thin layer chromatography (TLC) technique (Zhang *et al.*, 2008).

5.1 Thin Layer Chromatography (TLC)

Thin Layer Chromatography is a chromatography technique used to separate mixtures, which is performed on a sheet of glass, plastic, or aluminum foil coated with a thin layer of adsorbent material usually silica gel, aluminium oxide, or cellulose. It can be used to monitor the progress of molecule migration to identify compounds present in a given substance and to determine the purity of a substance. TLC can also be used on a small semi-preparative scale to separate mixtures of up to a few hundred milligrams. The mixture is not "spotted" on the TLC plate as dots, but rather applied to the plate as a thin even layer horizontally to and just above the solvent level. For small-scale analysis, TLC can be far more efficient in term of time and cost than chromatography. To analyze the amount of aflatoxin in samples by TLC, the small-scale target can be visible at UV light under 365 nm wavelength. According to the intensity, size and color of the spots on TLC plates, the type and its exact form of the compounds can be determined (Xie, 2007). As the TLC analysis is often affected by many factors, the accuracy of this method is poor. With the improvement of extraction and isolation method as well as the application of new reagents, the TLC detection becomes a simple and widely used analysis method. It is still used in China today.

5.2 High Pressure Liquid Chromatography (HPLC)

Using the liquid chromatography (HPLC) method with immuno-affinity column cleanup through post-column derivatization system, aflatoxins can be adsorbed in the immunaffinity column and eluted with organic solvent. The HPLC method with fluorescence detector using post-colummn derivatization system is a commonly used method in different countries (Wang, 2004). This method is more sensitive and accurate. Furthermore, this method is one of the best method for determining aflatoxin in traditional Chinese medicines (Ma, 2007).

5.3 Micro-column method

Employing the micro-column method to analyze aflatoxin is first to build up the micro-column chromatography tube using sample-extracted solvent and then the aflatoxin would be adsorbed by the florisil adsorbent as the alumina absorb the foreign matter. Under 365 nm UV light, the amount of aflatoxin can be calculated by the intensity of the blue-violet light reflected from the compound. This method is accurate, simple, rapid and reproducible. However, the micro-column is considered a qualitative method (Xie, 2007).

5.4 Enzyme-linked Immunosorbent assay (ELISA)

ELISA is widely used in food and feed industries to determine the content of aflatoxin in food products. Though the ELISA is accurate stable and reproducible, the analysis sometimes shows false positive, or false negative due to enzyme instability and variations of enzymes reaction conditions. So the application of this method in analysis of aflatoxin remains to be improved in future (Ma, 2007).

5.5 Immunochromatography

Immunochromatography is a kind of immunoassay technique developed in recent years. It is simple, rapid and is suitable for prescreening a large number of samples and for analyzing on the spot (Ma, 2007).

6. Prophylactico-therapeutic measures

Efforts to minimize adverse effects of aflatoxins include monitoring, managing and controlling their levels in agricultural products from farm to market and to table. While an association between aflatoxin contamination and inadequate storage conditions has long been recognized, studies have been focused on developing commercial crop cultivars that are resistant to *Aspergillus flavus*, such as peanut varieties Guihua 22 and Yueyou 58. Meantime, selecting the rational planting techniques and harvesting method, reasonable storage conditions and inhibitor are equally important.

6.1 Control measures in oils and foods
6.1.1 Selecting the crop cultivar with high level of resistance to aflatoxins

To date, many countries paid much attention to researches on the development of this method. In China, the new peanut cultivars such as Guihua 22 and Yueyou 58 have been cultivated (Wang, 2004). Since the Vitamin E is an essential factor in the synthesis of aflatoxins, it is important to select a cultivar with low Vitamin E content in seed coat to reduce aflatoxin contamination.

6.1.2 Using reasonable planting techniques and harvesting methods
It is very important to use suitable planting technique and harvesting method. The unsuitable handling process will cause damage to kernal, which will result in fungal infection and aflatoxin contamination. So during the harvest and storage, any measures that can reduce physical damage to the kernal of crops including insect pest and rats will surely reduce fungal growth and aflatoxin contamination (Wu, 2007).

6.1.3 Using reasonable storage condition
The storage condition is also an important determinant for reducing aflatoxin contamination. The AFB1 content of dry hot peppers stored under different storage conditions was analyzed to provide the theoretical basis for improving quality of dry hot pepper during storage. The results showed that the storage conditions of low temperature, low moisture content, low relative humidity and sealed package could significantly reduce the occurrence and accumulation of AFB1 in dry hot pepper.

6.1.4 Reasonable adoption of antiseptic
Utilizing antiseptic agents is effective in preventing aflatoxin contamination. The most commonly used antiseptics are sodium benzoate, sorbitol, propionic acid and propionates. The antiseptic compounds that contain propionic acid, propionates and sorbitol are in high demand. Studies showed that the removing rate of AFG B1 was more than 90% when using ozone to treat AFG B1 in contaminated crops (Luo et al., 2003).

6.2 Control measures in feed
Except low temperature, low moisture, control of oxygen and using antiseptics, adequate dilution may also prevent the aflatoxin contamination in feed (Wang & Zhang, 2006). However, this method is only applied to the least aflatoxin contaminated situation. The specific dilution methods are to analyze the exact content of aflatoxins and mix those feed for which their toxin content is in the borderline with the unmolded feed materials according to the normal feeding amounts. However, if the diluted feed hasn't been used up soon, it will extend aflatoxin contaminations. The regulating strategy include: based on the fact, adding methionine and electrolyte to improve the hepatic function and increase the natural concentration, especially the level of non-contaminated proteins.

6.3 Control measures in fermented condiment
Control of aflatoxin in fermented condiment should start from the raw materials. Controlling the key factors such as water activity in the soy bean and the humidity in environment is important. The best parameter should be controlled to keep the water content in soy bean below 13% and the humidity no less than 65% in depositories (Wang & Z., 2009). Combined with the application of antiseptic and ozone treatment the quality of the finished products and semi-finished products can be enhanced.

6.4 Control measure in chinese medicines
Similar to fermented condiment, the control of aflatoxin in Chinese medicine should also start from the resources. Thus, control measures of the herbal medicine can be taken to

enhance their resistance based on a developed management practice (Tang, 1999b). In recent years, a new technique of utilizing antagonistic microorganisms and the change from wild fermentation to pure fermentation are becoming effective measures to reduce aflatoxin contamination. Similar to crops, the control methods in storage also include the temperature., humidity, and oxygen etc.

7. Methods of degradation and removing

The practical methods applied for reducing aflatoxin production are: sorting, processing, and the sun light (ultraviolet) sterilization. These traditional meathods have been used since 1995 in China. Significant emphasis has been placed on detoxification of contaminated lots by irradiation, ammonia fumigation, chemical method, oxidants, microbial and enzymatic methods commonly used in treating corn and peanuts.

7.1 Adsorption techniques

Adsorption is the most common method to reduce aflatoxin contamination. For example, the harmful damaging effect of aflatoxins to animals can be reduced by adding several nutrition unactive adsorbents to the feed. Additionally, the adsorbents also can remove a portion of the aflatoxins (Xu et al., 2001). Among which the activated charcoal, mannan oligosarccharide (MOS), aluminosilicate, hydrated aluminosilicate and bentonite have been best studied. However, the positive effect has been observed only in the laboratory. Commercial utilization of the absorption meathod is rarely used in practical production in China.

With nanomaterial silicate adsorbent added to the feed contaminated by aflatoxins, it can significantly reduce the residual toxin in chicken muscle and liver. This is promising for producing safe animal products that meets international standards (Feng, 2004). The new adsorbent can also effectively reduce its harmful effect on growth, visceral function and the immune system of boilers. In the study on the adsorption of Silicate structure adsorbent NSP in feed of pigs, it was discovered that the absorption function in three forms: adsorption inside layers, adsorption between layers, and adsorption at the edges (Qi, 2002).

Absorption of several organic absorbents(KGM, Detoxification substance, Sorbent C) detoxification of AFB1 in animal was studied (Yu, 2007). In vitro experiment to study the absorption characteristics through different absorption, different content, different pH, and different temperature indicated that in high temperature, absorption capabilities of the three sorbents are worse than that in low temperature. Absorption capability of KGM is very weak in high temperature, while detoxification substance and sorbent C are obviously better. The three sorbents adsorb better in alkaline than in acidic conditions. But in acidic pH, sorbent C is worse than the other two.

Qingdao Agricultural University tested glucomannan to adsorb aflatoxins, the EGM at the concentration of 78.54%, 83.71% and 0.11% showed the best AFB1 adsorption ability, when the concentration of glucomannan is 0.11% (Yu, 2007).

7.2 Aflatoxins detoxification by ammonia gas

Aflatoxin detoxification in peanut and peanut meal by ammonia gas was tested (Liang,2009). Single factor test showed that ammonia temperature, time and water content of samples greatly affected AFB_1 degradation. The optimal conditions for best result are 10% amonia by volume, 24% peanuts meal moisture, which gave 100% AFB1 degradation. There is no detectable AFB1 after ammonia fumigation (Liang,2009).

7.3 Alkali refining
Studies by Liuzhou Health and Epidemic Prevention Station and Food Bureau suggested that one part contaminated feed immersed in two parts NaOH should be boiled for 1~2h before feeding to animals. In addition, use lime cream, pure potash and kali to soak aflatoxin contaminated corns for 2~3h, followed by washing in clean water, and drying. The detoxication efficiency can reach 60%~90% (Fan, 2003).

7.4 Oxidants
The treatment with 5% sodium hypochlorite for several seconds could reduce aflatoxin by 98%~100% (Zhang et al., 2004). After analyzing the difference among different time and the products of different places, the ClO_2 was effective in detoxicating aflatoxin when the aflatoxin contaminated corns were infused in the 250ug/mL ClO2 for 30~60 min (Zhang & Zhu, 2001). The treatment with 2% sodium bisulfite for 3 days showed best effect on aflatoxin detoxication (Feng, 2002).

7.5 Micro - organisms
The aflatoxin degradation ability of some food micro-organisms such as the lactic acid bacteria and yeasts was investigated (Zhu & Lin, 2001), (Li et al., 2003). The concentration of aflatoxin, the quantity of fungus and the temperature have a combined effect on the toxin binding ability by lactic acid bacteria. In yeasts, in exponential phase, it showed highest toxin binding ability and the higher concentration of aflatoxin, the higher the binding ability. The enzymatic detoxification of aflatoxin is an effective and safe method, highly selective, no harmful effect on nutrition value and no adverse effect to the treated products (Gong et al., 2004). A new technology on aflatoxin detoxification was developed in recent year. Thoroughly enzymatical hydrolyzation of the peanut meal to achieve full ionization of slightly dissolved aflatoxins from hydrophobic amino acid residues. Then retain the greater part of aflatoxin through successive filtration, thereby make markedly reduction of aflatoxin content (Xu & Luo, 2003).

8. The laws and regulations in controlling aflatoxins in China

Due to the risk of aflatoxin contamination of foods and feed on human health and livestock productivity, the Chinese Government has imposed laws and regulations limiting total amount of aflatoxins allowed in foodstuffs and feedstuffs. This has minimized potential exposure to aflatoxins. The maximum level of aflatoxins allowed in many commodities has been established. "Food Hygiene Law of the People's Republic of China" specifically prevents the sale of aflatoxin contaminated commodities and has set limits in food no more than 20 ppb total aflatoxins, and 10 ppb in rice and 0.5 ppb AFM1 in milk, butter and fresh pork. There is a zero tolerance in infant formula that no trace amount of aflatoxins shall be detected.

8.1 Related laws about aflatoxins
At present, Chinese laws about aflatoxin contamination are greatly improved. The Ministry of Health of the People's Republic of China has established a number of hygiene control measures to prevent aflatoxin contamination.
Food hygiene law of the People's Republic of China warns that the food, which is mould or mixed with foreign matters or those with abnormal flavour properties, may be harmful to

human health. "Food Hygiene Control Regulations Article IV" stated clearly that rural and state-owned farms should be organized and guided to harvest in time, threshing, dry, removing impurities, to prevent food mildew pollution during harvesting process. Article VI points out that we should actively carry out the "no worms, no mildew, no rat, and no sparrow" activities. ARTICLE II and III of Prevention Aflatoxin Contamination on Food Hygiene Regulations make clear that we should prevent food mildew and deterioration to achieve the objectives of mould proof and poison removal. Article IV provides that when using grain and oil whose aflatoxin content is higher than allowed level, effective measures must be taken to remove the toxins through technical procedures. The products can only be consumed when the product meets the food safety criterion. Article VI requires that, to ensure infant food safety, a zero tolerance policy should be adopted and food sector should provide non-aflatoxin detectable grain, as materials of infant milk replacer. For aflatoxin monitoring and management, Chinese Health and Quarantine law also established relevant regulations.

8.2 The organizations involved in aflatoxin control supervision

Not one or two departments can accomplish aflatoxin control supervision in the process of strengthening food safety supervision system. Team work may play an important and positive role. Management of aflatoxin control mainly involves the following departments.

8.2.1 Hygiene management department

Due to the problem with aflatoxin contamination during food processing, transportation and marketing process, especially peanuts. The hygiene administrative departments are required to perform some relative control measures on preventing aflatoxin contamination, e.g , Food hygiene law of the People's Republic of China.

8.2.2 Health and quarantine departments

Aflatoxin contamination of food is difficult to prevent, therefore, the aspect of food quarantine is particularly important. China has made specific provisions on the highest aflatoxin tolerance amount in all kinds of food. The health and quarantine departments must adopt the advanced science and technology in aflatoxin testing, strictly implement supervision, to reduce aflatoxin hazard to human health.

8.2.3 Disease control department

Because aflatoxin is extremely poisonous substances, it has an aneretic role on human and animal's liver tissue, accompanied with stem cell degeneration and necrosis, eventually result in serious organ damage or even death. Aflatoxins not only damage liver organ in animals, but also affect embryo development in animal. Due to immuno-suppression and recurrent infections aflatoxin contamination in animal feed will reduce milk and eggs production. Experimental results show that aflatoxin toxicities are different depending on animal species, age, and gender. In general, the younger the animals the higher the sensitivity to aflatoxins. Aflatoxins can also pass through food chain to human body through consumption and accumulation in animals. Disease control department should create a healthy environment, maintaining the social stability and national security, improve people's health through the prevention and control of diseases resulted from aflatoxin contamination. Under the leadership of the ministry of health, technological management and technical service will be enhanced.

8.3 Aflatoxin quarantine requirements

Chinese government has strict regulations on the maximum amount aflatoxin allowed in different foodstuffs. In corn, peanuts, peanut oil, nuts and dried fruit (walnut, almond) the maximum amount allowed is 20µg/kg(AflatoxinB1);. While in rice and oils (sesame oil, rapeseed oil, soybean oil, sunflower oil, oil, tea oil, sesame oil flax, corn germ oil, rice bran oil, cottonseed oil) is 10µg/kg (aflatoxin B1). In milk, milk products and butter (disinfection, fresh raw milk, whole milk powder, and evaporated milk, sweet condensed milk, butter) is 0.5 µg/kg (Aflatoxin M1). No aflatoxins shall be detected in any infant formula.

9. References

Ao, Z. & Chen, D. (2008). Recent Trends of Mycotoxin contamination in Animal Feeds and Raw Materials in ChinaT, *China Animal Husbandry & Veterinary Medicine* 35, 23-26.

Ban, K. & Cao, J. (2005). The Expression of Survivin mRNA and Its Clinical Significance in Hepatocellular Carcinoma of High Aflatoxin B1 Risk Area, *Chinese Journal of Clinical Oncology* 32, 614-617.

Cai, F. & Gao, W. (2010). Pollution status of Aflatoxin in Chinese materia medicia and its control techniques, *China Journal of Chinese Materia Medica* 35, 2503.

Cui, J., Zhang, T. & Yang, X. (2009). An analysis on morbidity of liver cancer between 2006 and 2008 in Ningbo, *Chinese Rural Health Service Administration* 29, 917-918.

Duan, W., Lu, H. & Pang, J. (2009). Studies on flatoxin contamination in Animal Derived Food and it's control, *Xiandai Nongye Keji* 206-208.

Duan, X., Ou, S. & Li, H. (2005). Experimental induced tree east-zhejiang process of liver cancer cell apoptosis and related gene research on P53 mutations loci., *Journal of Medical Reaserch* 34, 28-31.

Fan, H. (2003). Research Advances in integrated control of Aflatoxin in Feed, *Hebei Animal & Veterinary Sciences* 19, 40-41.

Feng, D. (2002). Research on Detoxification effect of sodium hypochlorite to Aflatoxin in peanut cake *Journal of South China Agricultural University* 118, 65-69.

Feng, J. (2004), Research on adsorption and detoxication effect of ann to Aflatoxin in chicken Daily diet, Animal science institute Zhejiang university

Gao, H. (1998). Introduce some international material of food additives, *China Food Additives* 51-54.

Gao, X., Ji, R. & Li, Y. (2007). Pollution Survey of Aflatoxin B1 in Grain, Oils and Food on Beijing City, *Journal of Hygiene Research* 36, 237-239.

Gao, Y. & Ding, W. (2010). The Screening of Produce Toxic Aflatoxin Efficient Connect Antibacterial Intraditional Ferment of Watercress., *Microbiology* 37, 369-374.

Gong, C., Jiang, L. & Zhang, Y. (2004). Hazard of Aflatoxin in Food and Its Methods for Removing, *Food Research and Developent* 25, 120-123.

Hu, N. & Xu, Y. (2009). Research on the Application of Differentially Expressed Genes Related Analysis in the Process of Aflatoxin Biosynthesis, *Food Science* 30, 208-211.

Huang, T. & Wei, Z. (2006). An analysis on morbidity of liver cancer between 1997 and 2003 in Fushui *Guangxi Medical Journal* 28, 1337-1338.

Jiao, Y., Ban, K. & Cao, J. (2007). Expression and Exon 3 Mutation of β-Catenin in Human Hepatocellular Carcinoma, *Chinese Journal of Cancer* 26, 1085-1089.

Kirk, G., Turner, P. & Gong, Y. (2005). Absence of TP53 codon 249 mutations in young Gninean childrenwith high aflatoxin exposure, *Cancer Epidemiol Biomarkers Prev* 14, 2053-2055.

Kong, Q. & Liu, Q. (2010). Inhibitory effect on the growth and aflatoxin biosynthesis of Aspergillus flavus by a strain of marine Bacillus, *Journal of Zhejiang University (Agriculture & Life Sciences)* 36, 387-392.

Lei, Y. (2009), The molecular markers of resistance of peanut resistance of Aflatoxin infection chinese agricultural science research institute. Master degree theses.

Li, Z., Yang, B. & Yao, J. (2003). Surface of binding of Aflatoxin B1 by Lactic acid bacteria, *Chinese Journal of Food Hygiene* 15, 212-215.

Liang, J. (2009). Study on the Effective for Inactivating the Aflatoxins in Peanut and Peanut Meal by Ammonia Gas, Sichuan Agricultural University.

Liang, Y. & Huang, R. (2000). Detection of the aflatoxin B1 in Chinese meteria medica, *Chinese Journal of Modern Applied Pharmacy* 17, 224-226.

Lin, J. (1995). Preliminary survey of AFM1 in milk and milk powder on Fuzhou, *Shanxi Medical Journal* 17, 32-35.

Liu, F. (2001). The Inhibitory Effect Of Plant Ribosome Inactivating Proteins (RIPs) On Aspergillus Flavus, *Acta Scientiarum Naturalium Universitatis Nankaiensis(Natural Science Edition)* 34, 78-81.

Liu, P. (1999). Determination of Aflatoxin in Eighteen lots of Chinese Herbs, *Chinese Pharmaceutical Journal* 24, 287-290.

Liu, X. (2006). Strengthen research work of exposure and control of mycotoxin, *Chinese Journal of Preventive Medicine* 40, 307-308.

Lu, H., Feng, Z. & Feng, X. (2008). Polymorphisms of CYP3A7 Gene of 210 Persons in Guangxi with Heavy Aflatoxin B1 Contamination, *Carcinogenesis, Teratogenesis & Mutagenesis* 20, 294-298.

Lu, Z., Wu, S. & Sun, C. (2010). The Relationship of Aflatoxin Biosynthetic Gene Expression and Environmental Factors, *Biotechnology Bulletin* 56-61.

Luo, J., Li, R. & Chen, L. (2003). Research On O_3 To Degrade AFB1 In Cereals, *Grain Storage* 23-28.

Ma, L. (2007), Research in high sensitivity testing technology of Aflatoxin B1, Master Dissertation in Chinese Academy of Sciences.

Qi, L. (2002), Research on adsorption effect of ann to Aflatoxin in duck Daily diet, Animal science institute Zhejiang university.

Qi, X. & Che, Z. (2010). Introduction of the detection and coercion of aflatoxin in fermentation products, *China Condiment* 22-28.

Ren, F. & Ma, H. (1997). Determination of Aflatoxin B1 in Traditional Chinese Medicine by special agent kit, *Chinese Journal of Pharmaceutical Analysis* 17, 280-284.

Sun, X., Yan, l., Sun, T. & Zhu, Y. (2010). Risk assessment of aflatoxin B_1 in soy sauce[*Chinese Journal of Microbiology* 22, 748-752.

Tang, J. (1999a). Research progress of aflatoxin in Guangxi, *Journal of Traditional Chinese Veterinary Medicine* 33-35.

Tang, Y. (1999b). Experimental survey research of aflatoxin B1 in traditional Chinese materia medica, *Chinese Pharmaceutical Journal* 24, 287-291.

Tang, Y., Huang, T. & Wang, J. (2008). Research on Aflatoxin exposure and GSH shift enzyme gene polymorphism in primary liver cancer high-risk groups, *Journal of Guangxi Medical University* 25, 694-698.

Wang, R., Miao, C. & Zhang, Z. (2003). Investigation of mould toxin pollution of feed and feedstuff, *Feed Industry* 24, 11-15.

Wang, S. & Zhang, L. (2006). Research on aflatoxins in feed, *Hebei Xumu Shouyi* 48.-51.

Wang, T. & Z., C. (2009). A tentative discussion on aflatoxins pollution in condiment fermenting, *China Condiment* 34, 35-38.

Wang, X. (2004). Aflatoxin contamination in oil and grain product and it's prevention *Science and Technology of Food Industry* 141-142.

Wei, Y. (2002). Pollution Survey of Aflatoxin B1,in Grain, Oil and Food on Ten Counties of the Yunnan Province, *Resources and Production.*

Wu, D. (2007). Hazard of Aflatoxin in Oil and Grain Product and Its Prevention, *Grain Processing* 32, 93-96.

Xiao, C., Feng, G., Wei, J. & Zheng, J. (2007). A Review: Aflatoxins In Feeds, *Livestock And Poultry Industry* 217, 4-8.

Xiao, L. & Xing, W. (2003). The Damage of Aflatoxins and Its Control, *World Agriculture* 287, 40-43.

Xie, G. (2007). Research in Determination Methods of Aflatoxins, *Feed Industry.*

Xu, J., Ji, R. & Luo, X. (2001). Removing of adsorbent and mycotoxin, *Chinese Journal of Food Hygiene* 13, 35-37.

Xu, J. & Luo, X. (2003). Molecular biology of aflatoxin biosynthesis, *Journal of Hygiene Research* 32, 628-631.

Xu, M. (2009). Progress in the possible mechanism of Aflatoxin Hepatocarcinogenesis *International Journal of Laboratory Medicine* 3, 547-548.

Yu, Z. (2007), Resesrch of Absorption Capacity of Esterified Glucomannan in Vitro to AFB1. Qingdao Agricultural University.

Zhang, C. (2008), Investigation and risk assessment of aflatoxin B_1 in main foods of China, Master Dissertation in Northwest University of Farming and Forestry Technology.

Zhang, C., Yue, T., Gao, Z. & Wang, J. (2008). Research Advance on the Determination Methods of Aflatoxin B1 in Food, *Academic Periodical Of Farm Products Processing* 18-22.

Zhang, G., He, R. & Qi, D. (2004). Research advance in detoxification of aflatoxin in feedstuff, *China Feed* 36-40.

Zhang, J. & Li, P. (2008). Synthesis of Aflatoxin G1 Artificial Antigen, *Food Science* 29, 194-197.

Zhang, Y. & Zhu, B. (2001). Research on the Effect of ClO2 Detoxication of the AflatoxinB1 in Corns, *Food Science* 68-70.

Zhu, X. (1991). Preliminary survey of AFM1 in ate milk and milk powder on inhabitant in urban area of Shanghai, *Tumor* 11, 175-176.

Zhu, X. & Lin, J. (2001). Study on Aflatoxin Degradation Ability of Some Food Micro - Organisms, *Food Science* 65-68.

Zuo, Z. & Liu, D. (2007). Studies on Constitutive and Secretive Expression of Codon-Optimized Recombinant Aflatoxin-Detoxifizyme (rADTZ) in Pichia pastoris, *Journal of Agricultural Science and Technology* 9, 87-94.

2

Occurrence of Aflatoxin M1 in Dairy Products

Laura Anfossi, Claudio Baggiani,
Cristina Giovannoli and Gianfranco Giraudi
Department of Analytical Chemistry, University of Turin,
Italy

1. Introduction

Aflatoxin M1 (AFM1) is a major metabolite of aflatoxin B1 (AFB1), which is formed when animals ingest feed contaminated with aflatoxin B1. The AFB1, once ingested by the animal, is rapidly absorbed by the gastrointestinal tract and is transformed into the metabolite AFM1, which appears in the blood after 15 minutes and is then secreted in the milk by the mammary gland (Van Egmond, 1989; Battacone, et al. 2003). The amount of AFM1 which is found in milk depends on several factors, such as animal breed, lactation period, mammary infections etc… It has, anyway, been demonstrated that up to 6% of the ingested AFB1 is secreted into the milk as aflatoxin M1 (Van Egmond & Dragacci, 2001) and, because AFM1 is relatively resistant to heat treatments (Yousef & Marth, 1989; Galvano et al., 1996), it is almost entirely retained in pasteurized milk, powdered milk, and infant formula. Moreover, only a limited decrease of AFM1 content has been verified in UHT milk after long storage (Galvano et al., 1996; Martins & Martins, 2000; Tekinsen & Eken, 2008). The hepatotoxicity and carcinogenic effects of AFB1 have been clearly demonstrated, thus it has long been classified as a group 1 human carcinogen by the International Agency on Research on Cancer (IARC, 2002). Initially, the IARC classified AFM1 as a possible carcinogen for humans (group 2b) since toxicological data was limited (IARC, 1993). However, genotoxicity and cancerogenity of AFM1 have been observed in vivo, although lower than those of AFB1, and its cytotoxicity has been definitively demonstrated (Caloni et al., 2006). As a result of these and other further investigations, the IARC moved aflatoxin M1 from group 2B to group 1 human carcinogen (IARC, 2002).

Considering that milk and milk derivatives are consumed daily and, moreover, that they are of primary importance in the diet of children, most countries have set up maximum admissible levels of AFB1 in feed (European Commission, EC, 2003a) and of AFM1 in milk, which vary from the 50 ng/kg established by the EU, to the 500 ng/kg established by US FDA (EC, 2003b; U.S. Food and Drug Administration, FDA, 2011). More restrictive MRLs have been implemented by the EU for the presence of AFM1 in baby food (EC, 2004) Regulations for aflatoxin M1 existed in 60 countries by the end of 2003, most of them being EU, and candidate EU countries, but some other countries in Africa, Asia and Latin America also apply the limit of 50 ng/kg. The higher regulatory level (500 ng/kg) is applied in the United States and in several countries in Asia and in Latin America, where it is also established as a harmonized MERCOSUR limit (FAO, 2011).

Based on admissible levels, on measured values in milk obtained in various monitoring programs and on typical diets, the intake of aflatoxin M1 from milk has been calculated to vary between 0.1 ng/person per day in Africa to 12 ng/person per day in the Far East (Europe: 6.8 ng/person per day, Latin America: 3.5 ng/person per day, the Middle East: 0.7 ng/person per day) (Creppy, 2002). The level of attention in the control of AFM1 contamination in milk is high all over the world, as attested by the number of scientific papers dealing with development and validation of analytical methods for measuring such a contaminant, the published survey studies on this argument, and by the attention paid by various international organisms.

However, the stability of AFM1 determines the persistence of such toxic compound in a number of other foodstuffs of wide human consumption, which are subject to less scrutiny, except in some geographical regions such as in the Middle East. In particular, the resistance to heat treatment and mild acidic conditions used in the production of cheese or other dairy products (such as, for example, yogurt, butter, cream and ice cream) has been accounted for the contamination of such products (Oruc et al., 2006; Colak, 2007). In addition, several authors have demonstrated that AFM1 is bound to milk proteins (Kamkar et al., 2008; Mendonca & Venancio, 2005; Prandini et al., 2009), mainly casein, and that therefore the toxin is more concentrated in cheese than in the milk used to produce it. As a result of the affinity of AFM1 for milk proteins, the toxin is distributed unevenly between whey and curd. In 2001, Govaris et al. (Govaris et al., 2001) first discussed the contrasting results reported until then, which regarded the distribution of AFM1 between whey and curd during cheese manufacturing. Differences in published results were attributed both to the variability of cheese-making processes investigated by the various authors and to the method of analysis employed to measure AFM1. More recent papers report results in greater agreement among themselves and demonstrates that the highest concentration of the toxin is found in the curd, regardless of the procedures applied in cheese-making and the method of analysis employed (Colak, 2007; Kamkar et al., 2008; Motawee et McMahon, 2009; Deveci, 2007; Manetta et al., 2009). According to Motawee et al and Deveci et al. approximately 60% of the AFM1 is found in the curd. Kamkar et al. found an even greater amount of AFM1 in the curd (3-times the content of whey). Accordingly, about half of the AFM1 from contaminated milk is found in cheese (Oruc et al., 2006; Colak, 2007), which means that levels of contamination could be very high, given that a kilogram of cheese is produced from several litres of milk, depending on type and maturity level of cheese (for example, 4.5 l of milk give 1 kg of mozzarella cheese, while as much as 16 l of milk are needed to obtain 1 kg of parmesan). As a matter of fact, AFM1 has been found in dairy products at levels which are 2-5 times higher than in the milk (Kamkar et al., 2008; Govaris et al., 2001, Motawee & McMahon, 2009; Deveci, 2007; Manetta et al., 2009). Moreover, substantially all authors who investigated the fate of AFM1 during cheese-making and cheese maturation agree to conclude that AFM1 content does not change significantly during these steps. These findings have also been confirmed by recent survey studies (Table 2) regarding the incidence of AFM1 contamination in cheese, which demonstrate the presence of AFM1 at various levels with a relevant incidence of positive samples (> 50 ng/kg), and in some case of highly contaminated samples (> 250 ng/kg). Occurrence of AFM1 in dairy products other than cheese has also been assessed (Kim et al., 2000; Maqbool et al., 2009; Lin et al., 2004; Martins & Martins, 2004) and demonstrates the potential risk for consumer health due to the widespread contamination of milk-derived products.

Despite this evidence, an adequate regulation about admissible limits of AFM1 in dairy products is still lacking in most countries. The strategy applied by several countries (i.e: EU and USA) is based on the assumption that a strict control of milk would prevent contamination of derived products. Therefore, the establishment of admissible limits for aflatoxin B1 in feed and of very severe MRLs in milk are judged to be sufficient to protect consumers from risk due to aflatoxin intake. On the other hand, specific maximum admissible levels for AFM1 in cheese have been set up in some countries and are summarised in Table 1 (Creppy, 2002; Italian Health Department, 2004; Dashti et al., 2009; Amer & Ibrahim, 2010; Sarımehmetoglu et al., 2004). The majority of countries which established a limit fixed it at 250 ng/kg, which corresponds to the assumption that cheese is made with milk which complies to regulations (i.e: contaminated at a level below 50 ng/kg) and that AFM1 concentration could rise up to 5-fold due to dehydration. However, some countries have decided on a zero tolerance strategy (Rumania and Egypt), to give the maximum consumer health protection at the expense of milk and cheese producers. Contrarily, in 2004, Italy raised the limit applicable to hard cheese to 450 ng/kg to protect parmesan production, which was generally highly contaminated in that year as the result of a foregoing peak of AFB1 contamination in feed. Interestingly, the vast majority of surveys on the occurrence of AFM1 in cheese have been carried out in those countries that in fact set up an admissible level in cheese and not just in milk (Table 2). Particularly noteworthy is that most studies have been carried out on Turkish cheese or on cheese consumed in Turkey.

Country	MRL (ng/kg)	Ref
Argentina	500	Dashti et al., 2009
Austria	250	Dashti et al., 2009
Switzerland	250	Creppy, 2002, Dashti et al., 2009
Egypt	0	Amer & Ibrahim, 2010
Honduras	250	Dashti et al., 2009
Italy	250 (450[a])	Italian Health Department, 2004
Rumania	0	Dashti et al., 2009
The Netherlands	200	Creppy, 2002
Turkey	250	Sarımehmetoglu et al., 2004

[a] limited to hard cheese

Table 1. International admissible levels for aflatoxin M1 in cheese

2. Methods of analysis of AFM1 in dairy products

Several methods for aflatoxin M1 determination have been developed, including high-performance liquid chromatography associated with fluorescence or mass spectrometric detection. Immunochemical methods have also been described and are employed as screening methods in routine analysis, mainly because of their simplicity and rapidity. However, the rate-determining step and the major source of errors in the analysis of cheese is the extraction of AFM1, which, in fact, strongly limits the number of samples to be analysed, with its being the most time-consuming, tedious and costly step of the entire analytical protocol.

2.1 Confirmatory and validated methods of analysis

Analytical methods for measuring aflatoxin M1 in milk have been widely described and a lot of HPLC-based methods are available. Some of them have been validated both in interlaboratory trials (Dragacci, & Grosso, 2001; Gallo et al., 2006; Gilbert & Anklam, 2002) and according to latest EU rules (Muscarella et al., 2007). In past years, TLC methods have also been widely used and, even more recently, a TLC protocol to determine AFM1 in milk has been reported and validated (Grosso et al., 2004). As regards cheese and other dairy products, some instrumental analysis methods have been described (Oruc et al., 2006; Kamkar et al., 2008; Mendonca & Venancio, 2005; Govaris et al., 2001; Deveci, 2007; Hisada et al., 1984; Pietria et al., 1997; Manetta et al., 2005). Validation according to EU regulation has been reported for an LC-FLD method applied to yogurt (Tabari et al., 2011). Schematically, confirmatory analytical protocols consist of: (i) extraction of the toxin with some organic solvent (dichloromethane, chloroform, methanol, acetonitrile); (ii) clean-up, which usually exploits the affinity and selectivity of antibodies immobilized in a solid-phase extraction (SPE) column (Immuno Affinity Chromatography) to reduce matrix interfering components and to strongly concentrate the target compound; alternatively C18-SPE is used for the purpose; (iii) chromatographic separation by reverse-phase HPLC; (iv) detection of the native fluorescence of AFM1. In 2005, Manetta and co-workers described a particularly sensitive method of analysis (LOD as low as 1 ng/kg in cheese) which used post-column derivatisation to enhance AFM1 fluorescence (Manetta et al., 2005). Mass spectrometric detection has also been successfully applied for the determination of AFM1 in different types of cheese samples (Cavaliere et al., 2006) and for the simultaneous detection of the toxin with other eight mycotoxins (Kokkonen et al., 2005). The exploitation of a very selective detection, such as tandem mass spectrometry, moreover permitted the application of simplified extraction procedures (Cavaliere et al., 2006).

2.2 Rapid techniques for measuring AFM1 in cheese

Historically, the first visual and rapid methods for the detection of AFM1 in milk were TLC methods. TLC-based analytical methods were developed for the measurement of the toxin present in cheese and dairy products too and were recognized as reference methods (see for example: AOAC 980.21 and 947.17 visual methods and Bijil et al., 1987). Nevertheless, immunoassays nowadays play a major role in the monitoring of AFM1 as a first level screening analysis. A number of commercial immunoassay kits (mainly ELISA methods) (International Standards Organisation, ISO, 2002) are available, which state their applicability not only in milk, but also in yogurt, cheese and any other sort of dairy products. However, since the lack of specific regulations in most countries, ELISA kits are principally intended for milk analysis. Therefore, their performances are valued for this purpose, as for example in the work of Rubio et al. who compared five commercial immunoassay kits aimed at the measurement of the target toxin in milk (Rubio et al., 2009). Each of the five kits was singularly evaluated and compared with the other, emphasising strong limitations in some of them. Immunoassay techniques which regard AFM1 determination in milk have been also reported in literature (Pestka et al., 1981; Tihrumala-Devi et al., 2002; Magiulo et al., 2005), while few papers report results aimed at demonstrating that immunoassays are reliably applicable for measuring AFM1 in dairy products: examples are represented by the work of Kim et al. who demonstrated the applicability of the developed ELISA in yogurt samples (Kim et al., 2000) and of a previously published work of our group where the modification of a commercial ELISA intended for milk analysis for measuring AFM1 in cheese was described (Anfossi et al., 2008). On the other hand, commercial ELISA kits have been widely used to study the fate of AFM1 during cheese-making or the occurrence of the toxin in various cheeses by several authors. Lopez et al. evaluate the performance of one of these commercial kits (Ridascreen Aflatoxin M1, R-Biopharm, Darmstadt, Germany) in the determination of the target compound in cheese and validate it by comparison with a thin layer chromatographic reference method, according to AOAC (Lopez et al., 2001). A nice approach for the rapid detection of AFM1 in milk, which exploits components of an immunoassay, carried out "on-column" instead of in a microtitre plate, was proposed by Sibanda et al. (1999). This visual assay has been extended and applied to yogurt and kefir by Goryacheva et al. (2009). Briefly, a specific antibody is immobilized on a gel-support, which is packed into a cartridge to form an immune-layer. A solution containing the toxin is mixed with an antigen-peroxidase conjugate and passed through the immune-layer, thus, a competition between the toxin and the antigen-peroxidase is established for binding to the immobilized antibody to take place. After washing, a chromogenic substrate solution of the peroxidase is added to the column, and the developed colour is observed. In the absence of the toxin, the antigen-peroxidase conjugate present is bound by antibodies and remains in the immune-layer; therefore intense colour development is observed. In the presence of the toxin, the binding of the antigen-peroxidase conjugate is inhibited, and, consequently, colour intensity would be lower or completely absent. The on-column assay coupled with the pre-concentration obtained by the same immune-layer allowed AFM1 detection at a level low enough to raise regulatory concern. The latest goal of researchers in the development of new rapid techniques in mycotoxin analysis is the exploitation of the immunochromatographic assay, also called lateral flow immunoassay (LFIA) or gold-colloid-based immunoassay, to produce fully-portable devices, which require no laboratory equipment, minimum skilled personnel, minimum sample preparation, and no hazardous chemicals (Krska & Molinelli, 2009). The assay can be typically concluded in few minutes and results can be both visually estimated or read by an appropriate reader. A commercial LFIA for the quantitative detection of AFM1 in milk is available (Rosa Aflatoxin M1 SL, Charm) and has been validated in an interlaboratory trial, confirming its reliability in the 300-550 ng/kg range. A more sensitive one-step device is also available from the same supplier (Rosa aflatoxin M1 MRL, Charm). Very recently, Wang et al. published the first LFIA for the assessment of AFM1 in milk. Nevertheless, AFM1 could only be detected at levels higher than 1 µg/kg. The requirement of extracting the toxin in a liquid medium from cheese samples, which would involve the use of organic solvent and laboratory equipment, together with the lack of specific regulations, has, until now, discouraged researchers from developing LFIA for measuring AFM1 which could be applicable to cheese.

2.2.1 Extraction of AFM1 from cheese samples to be analysed by rapid techniques

As discussed above, often a rapid and simple analytical method of measurement loses part or all of its advantages because for the need of time-consuming and laborious sample treatments. In addition, sample manipulation often involves the use of hazardous chemicals and laboratory equipment (centrifuge, evaporation systems, etc). As a typical example, the extraction protocol required before measuring the target toxin by means of the Ridascreen ELISA kit (R-Biopharm, Germany), which is the most widely used in AFM1 monitoring in cheese (Tekinsen & Eken, 2008; Colak, 2007; Dashti et al., 2009; Amer & Ibrahim, 2010; Sarimehmetoglu et al., 2004; Lopez et al., 2001; Virdis et al., 2008; Yapar et al., 2008; Ardic et al., 2009; Gurbay et al., 2006; Fallah et al., 2009), consists of the following procedures:

(i) a 2g-portion of cheese is homogenised and extracted with 40 ml of dichloromethane for 15 min; (ii) 10 ml of the extract is filtered and the solvent is evaporated under a nitrogen flux at 60°C; (iii) the residue is re-dissolved in a methanol-phosphate buffer mixture (50/50); (iv) the fat components are removed by adding an equal volume of hexane to the methanol-phosphate solution, shaking for 1 min, and (v) separating from the organic layer by centrifugation (15 min); (vi) finally, after discarding the upper organic layer, the lower aqueous-methanolic layer is diluted with a buffer and used in the assay.

Some other authors used a different commercial ELISA kit (Tecna srl, Trieste, Italy), whose extraction procedure is almost identical to that described above, except for the volume of dichloromethane used in the first step. These procedures evidence three major drawbacks: (i) large volumes of organic solvent are used particularly chlorinated ones, which means too that samples should be small (to limit the volumes of hazardous solvents being used) thus limiting representativeness; (ii) analysis should be conducted in an equipped laboratory; (iii) the procedure is long and laborious, with several steps, therefore increasing the sources of possible errors. Recently, we described a very simple and fast procedure for the extraction of AFM1 from dairy products, which uses an aqueous extracting medium and which allows the processing of several samples at the same time (Anfossi et al., 2008). The proposed method is based on the observation that AFM1 is bound to milk proteins, thus a protocol aimed at re-dissolving proteins from cheese (routinely employed in cheese analysis with the purpose of measuring total protein content) has been applied. The procedure involves: sample homogenisation and addition of a citrate solution; 15-min heating (50°C) under stirring; followed by centrifugation (15 min). The upper fat layer is discarded and the underlying layer is directly used in the ELISA. The validity of the approach has been verified on yogurt samples and different types of cheese: fresh, cream, soft, semi-hard, hard, blue, and elastic cheese. Validation of the described extraction has been made by comparing results on naturally contaminated cheeses with those obtained through a HPLC-FLD reference method. The extraction method is simple, relatively rapid and does not involve the use of any hazardous chemicals. Noteworthy is, the extraction medium, being completely aqueous and buffered at pH 8, makes it easy to combine with immunoassays.

3. Incidence of contamination of AFM1 in cheese

Since the late nineties of the last century, when the toxicity of aflatoxin M1 was brought to light and global regulations regarding aflatoxins started to be defined, monitoring of aflatoxin M1 in milk has been carried out. Some authors also investigated the occurrence of AFM1 in dairy products, although to a much lesser extent. These works have been already reviewed elsewhere (Govaris et al., 2002), therefore the latest five-years results have been summarised here.

3.1 Survey studies from 2006 to date

Several surveys have been conducted over the last five years on the occurrence of AFM1 in dairy products and, in particular, in cheese. The latter have been summarised in Table 2.

The first self-evident observation is that the problem of AFM1 contamination in cheese is mostly perceived in a specific geographical area, as almost all investigations have been carried out in the Middle East, except from the study conducted by Oliveira et al. (2011) in

Brasil and by Virdis et al. (2008) and Montagna et al. (2008) in Italy. In part, this may be explained by the fact that admissible levels in cheese have been set up by countries of the same region. Another factor which could be accounted for is the typical diet and the commercial relevance of cheese within various countries. Habits and the typical diet are difficult to quantify. However, there is a mismatch between the attention paid to the risk of aflatoxin contamination in cheese and geographical distribution of cheese production and consumption. More than 99% of the global production of cheese in 2010 was attributable to only 11 countries; in details, 47% in European countries (EU together with Switzerland) and 32% in the USA. Within the European Union, France (13.3%), Germany (8.3%), and Italy (7.8%), play the major role as cheese producers. In parallel, data on cheese consumption confirms the prominence of European countries (45% of global consumption of cheese in 2010) and the USA (32%); and specifically of France (10.7%), Germany (8.5%) and Italy (9.6%) within the European Union. Interestingly, Brasil and New Zealand are strong cheese consumers (5% of the cheese globally consumed is attributable to each of these countries) and Brasil is also a producer of a certain relevance (4% of total cheese produced annually in the world) (United States Department of Agriculture Foreign Agricultural Service, 2011a; 2011b). In this context, from one point of view, the interest in monitoring cheese safety in Italy and Brasil is not surprising, on the other hand a lack of data regarding other countries (for example United States, France and Germany) is evident.

Considering which analytical methods have been employed to conduct survey studies which have been published in the last five years, almost all use ELISA immunoassays to measure AFM1, thus confirming that the availability of simple and cost-effective techniques allows large monitoring programs to be carried out. The exception established by the work of Oliveira et al. who carried out a survey program by exploiting an HPLC method for measuring AFM1 in 48 samples, further confirms that the use of instrumental techniques limits the number of samples to be considered. In conclusion, there is a strong consistency in the analysis methods and a certain territorial homogeneity in considered samples, although this does not mean that samples are similar to each other concerning cheese-making, maturation and composition. In contrast, results on the level and incidence of AFM1 contamination are highly variable. Some authors found very low contamination levels and a great incidence of negative samples (Amer & Ibrahim, 2010; Dashti et al., 2009; Montagna et al., 2008; Er et al., 2010). On the contrary, other authors, who use the same analytical method and even analysed samples coming from the same country found a much larger incidence of positive samples and generally a much higher level of contamination (Tekinsen & Eken, 2008; Yapar et al., 2008; Ardic et al., 2009). A partial explanation of the discrepancy of results on Turkish cheese is the number of samples analysed which is, in some cases, too limited to be really representative. According to Govaris et al. (2001), the type of cheese-making could also influence toxin amount and, in fact, works have been done of different types of cheese.

However, the most populated level is the one which corresponds to AFM1 < 50 ng/kg in most works; some noticeable exceptions are represented by the level of AFM1 occurrence in Iran in 2008-2009 (Fallah et al., 2009; Rahimi et al., 2009) and in Turkey in 2008, according to Teckinsen & Eken (2008) and Ardic et al. (2009). Finally, we can observe that samples with AFM1 contamination beyond the admissible limits (where they exist) have been found in not insignificant percentages and that very high AFM1 concentrations (> 450 ng/kg) have been measured in 58 samples (4.6% of the total), both of which highlight the need for further and continuous control to preserve consumer health.

	Analysed samples (TOT)	Types of cheese	N of samples with AFM1 at level/TOT (%)				Ref
			<50 ng/kg	51-250 ng/kg	251-450 ng/kg	>450 ng/kg	
Turkey 2006	39	1	71.8	28.1	0	0	Gurbay et al. 2006
Turkey 2008	105	5	28.6	33.3	35.2 [a]	2.8 [a]	Yapar et al., 2008
Italy 2008	265	15	83.3	16.6	0	0	Montagna et al., 2008
Iran 2008	210	2	23.3	52.2	14.7	9.4	Fallah et al., 2009
Turkey 2008	132	1	17.4	55.3	19.7 [b]	7.6 [b]	Tekinsen & Eken, 2008
Italy 2008	41	1	9.8% positives, range 79.5-389 ng/kg				Virdis et al., 2008
Kuwait 2009	40	28	70.0	27.5	0	2.5	Dashti et al., 2009
Turkey 2009	193	1	17.6	56.0	14.0 [b]	12.5 [b]	Ardic et al., 2009
Iran 2009	88	1	53.4% positives, range 82-1254 ng/kg				Rahimi et al., 2009
Turkey 2010	70	1	92.9	7.1	0	0	Er et al., 2010
Egypt 2010	150	3	66.7	33.3	0	0	Amer & Ibrahim, 2010
Brasil 2011	48	2	77	18.8 [c]	4.2 [c]	0	Oliveira et al., 2011
Iran 2010	80	2	Average contamination: 22.3 (creamy cheese) and 43.3 ng/kg (feta cheese)				Mohamadi & Alizedeh, 2010

[a] reported contamination levels : 250-400 and >400 ng/kg
[b] reported contamination levels : 250-500 and >500 ng/kg
[c] reported contamination levels : 51-200 and 200-400 ng/kg

Table 2. Survey of AFM1 contamination in cheese from 2006 to date.

3.2 Occurrence of AFM1 in Italian cheese: results of a survey study conducted in 2010
The occurrence of AFM1 in Italian cheese was investigated during a one-year monitoring program in 2010. More than a hundred samples, belonging to different milking animal (cow, sheep, goat), manufacturing (industrial or traditional), feeding of dairy cattle (grazing or composite feed), and cheese maturation (long maturation, medium maturation, fresh) have been collected and analysed. Samples were extracted by using the above described aqueous approach and were analysed by a commercial ELISA kit. The complete method of analysis – extraction and quantification - had been validated in a previous work through comparison with a HPLC-FLD reference method on various classes of Italian cheese (Anfossi et al., 2008). In these conditions, the ELISA was demonstrated to have a limit of detection of 25 ng/kg, a dynamic range of 30-500 ng/kg

and relative standard deviations lower than 20%. It should be noted that the described method contemplates a corrective factor in the AFM1 quantification which makes results independent from the water content of samples. Indeed, different cheese could have very variable water content (usually indicated by the humidity percentage), depending on the preparation process and ripening, however this parameter has been included in the calculation of the amount of the target toxin, as discussed. The first aim of the work was the assessment of the occurrence of the aflatoxin M1 in Italian cheese. Italy is a producer of cheese of global importance and, in the meantime, Italians are strong consumers of both national and imported cheeses. In addition, there is a countless variety of the types of cheese that can be found on the Italian market; several of them originate from small producers who follow ancient recipes and traditional cheese-making methods. The complexity of this situation makes it difficult to generalize and classify samples so as to find exhaustive information regarding samples. Besides this first purpose of snapshotting the amplitude of the risk associated to AFM1 contamination of Italian cheese, the main objective of the work has been the identification of correlations between levels of contamination and some external factors which were identified as potentially influencing the presence and the concentration of the toxin. For this purpose, samples were divided into four categories according to: the animal which supplied the milk used to produce the cheese, the type of manufacturing, the season of production, and the maturation of the cheese. Within each category, samples were further sub-divided into groups (Table 3), which were compared with each other by statistic tests to evidence significant differences between groups.

3.2.1 Materials and methods
Samples classified as industrial were obtained from local supermarkets, while samples classified as small-scale were kindly provided by the Slow Food association (Cuneo, Italy) and by Eataly Distribuzione srl (Cuneo, Italy). Hard and medium maturing cheese samples were stored at -18°C until analysed. Soft cheese samples were immediately analysed without freezing. All samples were analysed before their expiry dates. A portion of sample (100 g *ca*) was roughly cut and then thoroughly minced and homogenized in a kitchen mixer. Aflatoxin M1 extraction was performed as previously described. In details, 5 g of homogenised cheese sample was weighed in a 50-mL conic tube, 20 mL of the extraction solution was added and the combination was maintained at 50°C for 15 min under vigorous stirring. The slurry was then centrifuged in a refrigerated centrifuge (25°C) for 15 min at 3200 x g. The fatty semi-solid upper layer was discarded and the liquid serum was withdrawn and directly analysed. Samples were extracted in single and analysed in triplicate. ELISA analyses were carried out as previously described (Anfossi et al., 2008). Briefly, 60 µL of AFM1 standard solutions or sample extracts was added to the same amount of the diluted antiserum and incubated in non-coated wells for 50 min. One hundred microliters of the mixture were transferred into coated wells and incubated for 15 min. After washes, 100 µL of the diluted anti-rabbit antibody labelled with the peroxidase was incubated in wells for 15 min. Colour development was obtained by a 20 min incubation with the TMB solution, followed by the addition of the stop solution. Finally, absorbance was recorded at 450 nm. Aflatoxin M1 concentrations were determined by interpolation on a linear calibration curve. Linearization of the calibration curve was performed by the logit-log transformation, by plotting the logit of the ratio (in percent) between the absorbance at each concentration of analyte (B) and the absorbance in the absence of analyte (B_0) against

the log of analyte concentration. The best data fit was obtained by linear regression of the standard points. Statistical analysis of data was carried out by the SigmaPlot 11.0 software (Systat Software Inc., CA, USA). First, the Shapiro-Wilk test on distribution of data was carried out. To be able to include undetectable samples (AFM1 concentration below 25 ng/kg, which is the detection limit of the method) in the statistical analysis, they were randomly ordered and a concentration value comprises between 0 and 25 ng/kg was attributed to each of them, by random number generation. Statistical differences between groups were evaluated by means of the Mann- Withney test on ranks (for the comparison between two groups) and of the extended Kruskall-Wallis ANOVA test on ranks (for the comparison between more than two groups) (Massart et al., 1988).

3.2.2 Correlation of aflatoxin M1 contamination with type of manufacturing, season of production, species of the animal that produced the milk, and cheese maturation

More than 83% of analysed samples showed detectable levels of toxin (> 25 ng/kg); most of the positive samples were measured to contain AFM1 between 50 and 150 ng/kg, with the exception of fresh cheese and of cheese made with goats' milk alone or mixed with other types of milk (Table 3). These groups generally showed a lower AFM1 content. Cheeses made with sheeps' milk have an equal distribution between the contamination levels below 50 ng/kg and between 50 and 150 ng/kg.

Statistical data analysis brought in light that the only factor which determined significant differences among groups was the origin of the milk. More specifically, cheese made with cows' milk showed itself to be more contaminated than cheese made with goat or sheep (or mixed goat/sheep, mixed goat/cow and sheep/cow) milk. This result agrees with the previous observation that milk from goats and sheep is less contaminated than cows' milk, both because of the different digestive apparatuses and mechanism of AFB1 assimilation of animals, and for the different feeding used in cow's breeding compared to ovine and caprine (Barbiroli et al., 2007; Hussain et al., 2010; Fallah et al., 2011). In fact, cattle fodders are more likely to be contaminated with AFB1 than those used to feed sheep and goats. This finding also confirms previous observations of other authors (Montagna et al., 2008), who also reported that cow's cheeses are more contaminated than others.

As a consequence of this first observation, samples made with cow's milk (82 samples) were isolated from the rest and the statistical analysis was repeated on them for the other three identified categories: manufacturing, cheese ripening, and production season. In this way, a further significant difference could be emphasized; industrial products were discovered to be less contaminated than small-scale products, probably because checks conducted on milk to be used in cheese production are more stringent in industrial scale production than in artisanal contexts. In addition, artisans often makes use of only one milk source, which can occasionally be contaminated with high AFM1 levels (although within the legal limit) thus determining a peak of contamination which would be found also in the derived cheese, while industrial production uses dilution of milk from various sources. This finding is in contrast to that recently obtained by Fallah et al. (2011). On the other hand, contrary to what appears at first sight from the data shown in Table 3, maturation does not influence AFM1 content in cheese. Several other authors observed that maturation does not significantly alter the AFM1 concentration, as would be reasonable to expect, given an appropriate correction of concentration values for the water content of the cheese analysed. A decrease of aflatoxin M1 concentration rather than an increase during maturation could be assumed, because of degradation of the toxin with time. Nevertheless, this degradation has not been pointed out in any previous works aimed at assessing the fate of the toxin (Oruc et al., 2006; Colak, 2007; Kamkar et al., 2008; Mendonca & Venencio, 2005; Prandini et al., 2009; Govaris et al., 2001; Motawee & McMahon, 2009; Deveci, 2007).

Category	Group	Analysed samples (TOT)	N of samples contaminated at a level/TOT (%)			
			<50 ng/kg	50-150 ng/kg	150-250 ng/kg	>250 ng/kg
Maturation	Long (>3 months)	29	31.0	62.1	3.4	3.4
	Medium (>45 days; <3 months)	46	43.5	54.3	2.2	0.0
	Fresh (<45 days)	27	55.6	40.8	3.7	0.0
Manufacturing	Big brands	38	47.4	51.4	0.0	0.0
	Small-scale	64	52.6	53.1	4.7	1.6
Production season	Winter-spring	65	38.4	56.9	3.1	1.5
	Summer-autumn	37	51.4	45.9	2.7	0.0
Milk from	Cow	82	39.0	56.1	3.7	1.2
	Sheep	6	50.0	50.0	0.0	0.0
	Goat	6	83.3	16.7	0.0	0.0
	Buffalo	3	33.3	66.7	0.0	0.0
	Mix [a]	5	60.0	40.0	0.0	0.0
TOTAL		102	44	54	3	1

Table 3. Number of cheese samples analysed for the various groups identified as potentially influencing the level of AFM1 contamination and distribution of samples between these groups as a function of the level of AFM1 contamination.

The production season is also irrelevant according to statistical analysis. The factor "season of production" was defined to evaluate the influence of animals feeding (grazing or composite feed) on the assumption that animals fed on pasture would be less exposed to AFB1 ingestion and, consequently, would produce less AFM1 contaminated milk. Accordingly, cheeses made during summer and autumn, which belong to milk from grazing animals, would be less contaminated than cheese made during winter and spring, which belong to milk from animals fed with composite and stored fodder. Actually, according to information (when available) provided by producers of samples analysed in our work, animals were fed in pastures during summer and autumn, whereas they consumed stored feed during most of the spring. Therefore, groups to be compared were defined as reported in Table 3. The irrelevance of the period of production (and consequently, or partially consequently, of animal feeding) observed on Italian cheese samples could be explained by the fact that aflatoxin producing fungi also affects crops in the field. Nevertheless, the main limitation in making this analysis is the uncertainty of attribution of samples. In fact, some samples were accompanied by exhaustive information (period of production, animal feeding), however for most of them information was incomplete or unavailable. In these cases, attribution to groups was assumed on the basis of generic information regarding the type of cheese, the expiry date and the similarity to other samples. Therefore, results on this factors cannot be considered as conclusive and would need further investigation. In fact, Taikarimi and co-workers observed that the season of production is relevant in determining aflatoxin M1 in cheese and demonstrated that cheese produced in winter are more contaminated than those produced in summer (Tajkarimi et al., 2008). Accordingly, Fallah et al. (2011) observed that samples produced in winter-spring are more contaminated than those produced in summer and autumn.

As in the case of samples from cows' milk, a further statistical analysis should be conducted by separately isolating the two categories of industrial and small-scale manufacturing samples and re-run statistical tests on the remaining categories (season of production and maturation) to highlight eventual significant differences which may have been hidden by the non-random distribution of samples between groups. However, the number of samples in each category and groups would become non-representative, therefore it would be interesting to increase the number of analysed samples to achieve more conclusive results.

Despite the high incidence of AFM1 at detectable concentrations all samples were contaminated beyond the admissible limit (250 ng/kg), except for 1 hard cheese, which still complied with legal limits (because MRL for hard cheese has been raised to 450 ng/kg in Italy since 2004). It is likely that the screening of milk (by control organisms or, most likely, by internal audit) is in general adequate to also secure the safety of cheeses, as undertaken by those countries that established admissible limits in milk and not in other dairy products.

4. Conclusions

Some of the inconsistencies highlighted by surveys conducted over the past five years could be clarified in light of these results, namely by separate samples according to the origin of the milk and to the type of manufacturing. For example, Virdis and co-workers found low positivity in Italian cheeses in 2008 compared to our survey, however it is justifiable since their study regarded specifically goats' cheese, which showed itself to be less contaminated than that of cows also in this study. The same is true for the work carried out by Gurbay et al (2006). On the other hand, authors who found high levels of contamination analysed cheese samples exclusively from cows' milk (Tekinsen & Eken, 2008; Dashti, et al., 2009; Yapar et al., 2008; Fallah et al. 2009) or samples produced at least partially from cows' milk (Ardic et al., 2009). Oliveira and co-workers reported a distribution of contamination levels which is in good accordance with that observed in the present study. More controversial are the results shown by Er et al. (2010) and Amer & Ardic (2009). The latter reported low contamination levels; however, few details regarding the type of samples are stated in the text. Er et al. showed very low incidences of AFM1 contamination in cheeses made from cows' milk (Er et al., 2010), which is in contradiction with all other published studies.

In general, most works were limited to reporting the occurrence of the toxin and the level of contamination, without correlating this information with any characteristics of the analysed samples. Therefore, conclusions were partial and related to specific circumstances and did not permit authors to generalise their observations. The reported findings of the study conducted in a one-year survey on various types of cheese in Italy and their correlation to some of the factors which could influence aflatoxin M1 presence in cheese allowed the identification of some relevant factors (milk origin, manufacturing type) and to rationalise the results of the study and also preceding observations. The statistical approach is promising; however, further investigations on already identified factors, together with attempts to widen the number of considered factors, would occur.

From the point of view of the risk to consumers posed by AFM1 intake with cheese, the assumption seems verified that control strategies to limit AFB1 in feed and AFM1 in milk are an adequate protection for consumer health. Nevertheless, data representing the occurrence of aflatoxin M1 in cheese belonging to those countries which represent the principal cheese producers and consumers (United States, France, Germany) would be of great interest to further support this conclusion and to procure reliable suggestions to those who have legislative responsibility on this matter. Finally, it has been demonstrated once more that immunochemical methods of analysis, associated with rapid and simple treatments of samples, allow large screening surveys to be completed, thus providing researchers with a lot of information. The advantage of having readily available data should be, however, counterbalanced by appropriate methods of data management to achieve meaningful conclusions.

5. Acknowledgment

Authors wish to thank Dr. F. Baldereschi (Slow Food, Italy) and Dr. S. Del Treppo (Eataly, Italy) for kindly providing most of the small-scale cheese samples and related information.

6. References

Amer, A.A.; Ibrahim, M.A.E. (2010) Determination of aflatoxin M1 in raw milk and traditional cheeses retailed in Egyptian markets. *Journal of Toxicology and Environmental Health Sciences* Vol. 2, No. 4, (September 2010) pp. 50-53, , ISSN 1528-7394

Anfossi, L.; Calderara, M.; Baggiani, C.; Giovannoli, C.; Arletti, E.; Giraudi, G. (2008) Development and Application of Solvent-free Extraction for the Detection of Aflatoxin M_1 in Dairy Products by Enzyme Immunoassay. *Journal of Agricultural and Food Chemistry*, Vol. 56, No. 6, pp. 1852-1857, ISSN 0021-8561

Ardic, M.; Karakaya, Y.; Atasever, M.; Adiguzel, G. (2009) Aflatoxin M1 levels of Turkish white brined cheese *Food Control* Vol. 20, pp. 196–199, ISSN 0956-7135

Barbiroli, A.; Bonomi, F.; Benedetti, S.; Mannino, S.; Monti, L.; Cattaneo, T.; Iametti, S. (2007) Binding of Aflatoxin M1 to Different Protein Fractions in Ovine and Caprine Milk *Journal of Dairy Science*, Vol. 90, pp. 532–540 ISSN 1525-3198

Battacone, G.; Nudda, A.; Cannas, A.; Cappio Borlino, A.; Bomboi, G.; Pulina, G. (2003) Excretion of Aflatoxin M1 in Milk of Dairy Ewes Treated with Different Doses of Aflatoxin B1 *Journal of Dairy Science*, Vol.86, No.8 (December 2003) pp. 2667–2675 ISSN 0022-0302

Bijl, J.P.; Van Peteghem, C.H.; Dekeyser, D.A. (1987) Fluorimetric determination of aflatoxin M1 in cheese. *Journal of AOAC* Vol. 70, No. 3, pp. 472-5. ISSN 1060-3271

Caloni, F.; Stammati, A.; Friggé, G.; De Angelis, I. (2006) Aflatoxin M1 absorption and cytotoxicity on human intestinal in vitro model *Toxicon* Vol. 47, pp. 409–415, ISSN 0041-0101

Cavaliere, C.; Foglia, P.; Guarino, C.; Marzioni, F.; Nazzari, M.; Samperi, R.; Lagana, A. (2006) Aflatoxin M1 determination in cheese by liquid chromatography–tandem mass spectrometry *Journal of Chromatography A*, Vol. 1135, No. 2, pp. 135–141 ISSN 0021-9673

Charm Science Inc (Lawrence, MA, USA) Charm SL Aflatoxin M1 Quantitative Test for Milk 21/04/2011 Available from:
http://www.charm.com/en/products/rosa-milk/aflatoxin-m1.html

Colak, H. (2007) Determination of Aflatoxin M1 Levels in Turkish White and Kashar Cheeses Made of Experimentally Contaminated Raw Milk *Journal of Food and Drug Analysis*, Vol. 15, No. 2, pp. 163-168 ISSN 10219498

Creppy, E.E. (2002) Update of survey, regulation and toxic effects of mycotoxins in Europe *Toxicology Letters* Vol. 27, pp. 19–28, ISSN 0378-4274

Dashti, B.; Al-Hamli, S.; Alomirah, H.; Al-Zenki, S.; Bu Abbas, A.; Sawaya W. (2009) Levels of aflatoxin M1 in milk, cheese consumed in Kuwait and occurrence of total aflatoxin in local and imported animal feed. *Food Control* Vol. 20, pp. 686–690, ISSN 0956-7135

Deveci, O. (2007) Changes in the concentration of aflatoxin M1 during manufacture and storage of White Pickled cheese. *Food Control* Vol. 18, pp. 1103–1107, ISSN 0956-7135

Dragacci, S.; Grosso, F. (2001) Immunoaffinity Column Cleanup with Liquid Chromatography for Determination of AflatoxinM1 in Liquid Milk: Collaborative Study *Journal of AOAC International* Vol. 84, No. 2, ISSN 1060–3271

Er, B.; Demirhan, B.; Onurdag, F.K.; Yentur, G. (2010) Determination of Aflatoxin M1 in milk and white cheese consumed in Ankara region, Turkey. *Journal of Animal Veterinary Advances* Vol. 9, No. 12, pp. 1780-1784, ISSN 160-5593

European Commission (2003a) Commission Regulation (EC) No 100/2003 *Official Journal of the European Community* Vol. L 285, pp. 33-37

European Commission (2003b) Commission Regulation (EC) No 2174/2003, *Official Journal of the European Community* Vol. L326, pp. 12-15

European Commission (2004) Commission Regulation (EC) No 683, 2004, *Official Journal of the European Community* Vol. L106, pp. 3-5

Fallah, A.A.; Jafari, T.; Fallah, A.; Rahnama, M. (2009) Determination of aflatoxin M1 levels in Iranian white and cream cheese. *Food and Chemical Toxicology* Vol. 47, pp. 1872–1875, ISSN 0278-6915

Fallah, A.A.; Rahnama, M.; Jafari, T.; Saei-Dehkordi, S.S. (2011) Seasonal variation of aflatoxin M1 contamination in industrial and traditional Iranian dairy products *Food Control*, in press, ISSN 0956-7135

FAO Corporate Document Repository. (April 2011) Worldwide regulations for mycotoxins in food and feed in 2003. 20/04/2011. Available from:
http://www.fao.org/docrep/007/y5499e/y5499e0n.htm#TopOfPage

FDA U.S. Food and Drug Administration (April 2011) Guidance for Industry: Action Levels for Poisonous or Deleterious Substances in Human Food and Animal Feed. 20/04/2011. Available from:
http://www.fda.gov/Food/GuidanceComplianceRegulatoryInformation/Guidan ceDocuments/ChemicalContaminantsandPesticides/ucm077969.htm

Gallo, P.; Salzillo, A.; Rossini, C.; Urbani, V.; Serpe, L. (2006) Aflatoxin M_1 determination in milk : Method validation and contamination levels in samples from Southern Italy. *Italian Journal of Food Science* Vol. 18, No. 3, pp. 251-259, ISSN 1120-1770

Galvano, F.; Galofaro, V.; Galvano, G. (1996) Occurrence and stability of aflatoxin M1 in milk and milk products: A Worldwide Review. *Journal of Food Protection* Vol.59, pp. 1079-1090, ISSN 0362-028X

Gilbert, J.; Anklam, E. (2002) Validation of analytical methods for determining mycotoxins in foodstuffs *Science* Vol. 21, No. 6, pp. 468-486, ISSN 01659936

Goryacheva, I.Y.; Karagusheva, M.A.; Van Peteghem, C.; Sibanda, L.; De Saeger, S. (2009) Immunoaffinity pre-concentration combined with on-column visual detection as a tool for rapid aflatoxin M1 screening in milk. *Food Control* Vol. 20, pp. 802–806, ISSN 0956-7135

Govaris, A.; Roussi, V.;. Koidis, P.A.; Botsoglou, N. A. (2001) Distribution and stability of aflatoxin M1 during processing, ripening and storage of Telemes cheese. *Food Additives and Contaminants*, Vol. 18, No. 5, pp. 437- 443, ISSN 0265± 203X

Grosso, F.; Fremy, J.M.; Bevis, S.; Dragacci, S. (2004) Joint IDF-IUPAC-IAEA(FAO) interlaboratory validation for determining aflatoxin M1 in milk by using immunoaffinity clean-up before thin-layer chromatography *Food Additives and Contaminants* Vol. 21, No. 4, pp. 348-57, ISSN 0265–203X

Gurbay, A.; Engin, A.B.; Çaglayan, A.; Sahin, G. (2006) Aflatoxin M1 levels in commonly consumed cheese and yogurt samples in Ankara, Turkey *Ecology of Food and Nutrition*, Vol. 45, pp. 449–459, ISSN 0367-0244

Hisada, K.; Terada, H.; Yamamoto, K.; Tsubouchi, H.; Sakabe, Y. (1984) Reverse phase liquid chromatographic determination and confirmation of aflatoxin M1 in cheese. *Journal of AOAC* Vol.67, No. 3, pp. 601-606. ISSN 1060-3271

Hussain, I.; Anwar, J.; Asi; M.R.; Munawar, M.A.; Kashif, M. (2010) Aflatoxin M1 contamination in milk from five dairy species in Pakistan. *Food Control* Vol. 21, pp. 122–124, ISSN 0956-7135

International Agency for Research on Cancer (1993) *Monographs on the evaluation of the carcinogenic risk of chemicals to humans: some naturally occurring substances, food items and constituents, heterocyclic aromatic amines and mycotoxins*; Vol. 56, 397-344 IARC, Lyon, France 1993,

International Agency for Research on Cancer (2002) *Monograph on the Evaluation of carcinogenic risks in humans: Some Traditional Herbal Medicines, Some Mycotoxins,. Naphthalene and Styrene*; Vol. 82, pp. 171–274 IARC, Lyon, France

International Standards Organisation, ISO (2002). Milk and milk products. Guidelines for a standardized description of competitive enzyme immunoassays-determination of AFM1 content. Standard 14675. Geneva, Switzerland.

Italian Health Department (2004) Sampling and analytical methods for aflatoxin detection in cheese (24 August 2004) D.G.V.A/IX/25664/f.5.b.b.2/P

Kamkar, A.; Karim, G.; Aliabadi, F.S.; Khaksar, R. (2008) Fate of aflatoxin M1 in Iranian white cheese processing. *Food and Chemical Toxicology* Vol. 46, pp. 2236–2238 ISSN 0278-6915

Kim, E.K.; Shon, D.H.; Ryu, D.; Park, J. W.; Hwang, H. J.; Kim, Y. B. (2000) Occurrence of aflatoxin M1 in Korean dairy products determined by ELISA and HPLC. *Food Additives and Contaminants* Vol. 17, pp. 59 – 64, ISSN 0265–203X

Kokkonen, M.; Jestoi, M.; Rizzo, A. (2005) Determination of selected mycotoxins in mould cheeses with liquid chromatography coupled to tandem with mass spectrometry. *Food Additives and Contaminants* Vol. 22, No. 5, pp. 449-56, ISSN 0265–203X

Krska, R.; Molinelli, A. (2009) Rapid test strips for analysis of mycotoxins in food and feed. *Analytical and Bioanalytical Chemistry*, Vol. 393, No.1, pp. 67-71, ISSN 1618-2642

Lin,L.C.; Liu, F.M.; Fu, Y.M.; Shih, D.Y. (2004) Survey of Aflatoxin M1 Contamination of Dairy Products in Taiwan. *Journal of Food and Drug Analysis,* Vol. 12, No. 2, pp. 154-160, ISSN 10219498

Lopez, C.; Ramos, L.; Ramadan, S.; Bulacio, L.; Perez, J. (2001) Distribution of aflatoxin M1 in cheese obtained from milk artificially contaminated *International Journal of Food Microbiology* Vol. 64, pp. 211–215, ISSN 0168-1605

Magliulo, M.; Mirasoli, M.; Simoni, P.; Lelli, R.; Portanti, O.; Roda, A. (2005) Development and Validation of an Ultrasensitive Chemiluminescent Enzyme Immunoassay for Aflatoxin M$_1$ in Milk. *Journal of Agricultural and Food Chemistry*, Vol. 53, No. 9, pp. 3300-3305, ISSN *0021-8561*

Manetta, A.C.; Di Giuseppe, L.; Giammarco, M.; Fusaro, I; Simonella, A.; Gramenzi, A.; Formigoni, A. (2005) High-performance liquid chromatography with post-column derivatisation and fluorescence detection for sensitive determination of aflatoxin M1 in milk and cheese *Journal of Chromatography A*, Vol. 1083, No. 1-2, pp. 219–222 ISSN 0021-9673

Manetta, A.C.; Giammarco, M.; Di Giuseppe, L.; Fusaro, I.; Gramenzi, A.; Formigoni, A.; Vignola, G.; Lambertini, L. (2009) Distribution of aflatoxin M1 during Grana Padano cheese production from naturally contaminated milk. *Food Chemistry* Vol. 113, pp. 595–599, ISSN 0308-8146

Maqbool, U.; Anwar-Ul-Haq; Ahmad, M. (2009) ELISA determination of Aflatoxin M1 in milk and dairy products in Pakistan *Toxicological and Environmental Chemistry* Vol. 91, No. 2 (March 2009), pp. 241–249, ISSN 0277–2248

Martins, M.L.; Martins, H.M. (2000) Aflatoxin M1 in raw and ultra high temperature-treated milk commercialized in Portugal. *Food Additives & Contaminants: Part A* Vol. 17, No. 10, pp. 871- 874, ISSN 0265-203X

Martins, M.L.; Martins, H.M. (2004) Aflatoxin M1 in yogurts in Portugal. *International Journal of Food Microbiology* Vol. 91, No. 3 (March 2004), pp. 315-317, ISSN 0168-1605

Massart, D.L.; Vandenginste, B.G.M.; Deming, S.N.; Michotte, Y.; Kaufman, L. (1988) Non-parametric tests for comparison of methods. In: Data handling in science and technology, Vol. 2 Chemometrics: a textbook. Vandenginste, B.G.M.; Kaufman, L. (Eds.) 48-51 Elsevier Science Publishing Company Inc, ISBN 0-444-42660-4, New York, NY, USA

Mendonca, C.; Venancio, A. (2005) Fate of aflatoxin M1 in cheese whey processing *Journal of the Science of Food and Agriculture* Vol. 85, pp. 2067–2070, ISSN 0022–5142

Mohamadi, H.; Alizadeh, M. (2010) A Study of the Occurrence of Aflatoxin M1 in Dairy Products Marketed in Urmia, Iran. *Journal of Agricultural, Science and Technology* Vol. 12, pp. 579-583, ISSN 1561-7645

Montagna, M.T.; Napoli, C.; De Giglio, O.; Iatta, R.; Barbuti, G. (2008) Occurrence of Aflatoxin M1 in Dairy Products in Southern Italy. *International Journal of Molecular Science* Vol.9, pp. 2614-2621, ISSN 1422-0067

Motawee, M.M.; McMahon, D.J. (2009) Fate of aflatoxinM1 during Manufacture and Storage of Feta Cheese. *Toxicology and Chemical Food Safety* Vol 74, No. 5, pp. 42-45 ISSN 0278-6915

Muscarella, M.; Lo Magro, S.; Palermo, C.; Centonze, D. (2007) Validation according to European Commission Decision 2002/657/EC of a confirmatory method for aflatoxin M1 in milk based on immunoaffinity columns and high performance liquid chromatography with fluorescence detection. Analytica Chimica Acta Vol. 594, No. 2 (July 2007), pp. 257-264, ISSN 0003-2670

Oliveira, C.A.F.; Franco, R.C.; Rosim, R.E.; Fernandes, A.M. (2011) Survey of aflatoxin M1 in cheese from the North-east region of Sao Paulo, Brazil. *Food Additives and Contaminants: Part B* Vol. 2011, pp. 1–4, ISSN 1939–3210

Oruc, H.H.; Cibik, R.; Yilmaz, E.; Kalkanli O. (2006) Distribution and stability of Aflatoxin M1 during processing and ripening of traditional white pickled cheese. *Food Additives and Contaminants*, Vol. 23, No. 2, pp. 190–195, ISSN 0265–203X

Pestka, J.J.; Li, Y.; Harder, W.O.; Chu, F.S. (1981) Comparison of radio immunoassay and enzyme linked immunosorbent assay for determining aflatoxin M_1 in milk. *Journal of AOAC*, Vol. 64, pp. 294-301, ISSN 1060-3271

Pietria, A.; Bertuzzia, T.; Bertuzzia, P.; Piva, G. (1997) Aflatoxin M1 occurrence in samples of Grana Padano cheese. *Food Additives and Contaminants* Vol. 14, No. 4, pp. 341-344, ISSN 0265–203X

Prandini, A.; Tansini, G.; Sigolo, S.; Filippi, L.; Laporta, M.; Piva G. (2009) On the occurrence of aflatoxin M1 in milk and dairy products. *Food and Chemical Toxicology* Vol. 47, pp. 984–991 ISSN 0278-6915

Rahimi, E.; Karim, G.; Shakerian A. (2009) Occurrence of aflatoxin M_1 in traditional cheese consumed in Esfahan, Iran *World Mycotoxin Journal* Vol. 2, No 1 (February 2009), pp.91-94, ISSN1875-0710

Rubio, R.; Berruga, M.I.; Román, M.; Molina, A. (2009) Evaluation of immunoenzymatic methods for the detection of aflatoxin M1 in ewe's milk. *Food Control* Vol. 20, pp. 1049–1052, ISSN 0956-7135

Sarımehmetoglu, B.; Kuplulu, O.; Celik, T.H. (2004) Detection of aflatoxin M1 in cheese samples by ELISA. *Food Control* Vol. 15, pp. 45–49, ISSN 0956-7135

Sibanda, L.; De Saeger, S.; Van Peteghem, C. (1999) Development of a portable field immunoassay for the detection of aflatoxin M in milk. *International Journal of Food Microbiology* Vol. 48, pp. 203–209, ISSN 0168-1605

Tabari, M.; Karim, G.; Ghavami, M.; Chamani, M. (2011) Method validation for aflatoxin M1 determination in yogurt using immunoaffinity column clean-up prior to high-performance liquid chromatography *Toxicology and Industrial Health* Epub 8 March 2011, ISSN 0748-2337

Tajkarimi, M.; Aliabadi-Sh, F.; Salah Nejad, A.; Poursoltani, H.; Motallebi, A.A.; Mahdavi, H. (2008) Aflatoxin M1 contamination in winter and summer milk in 14 states in Iran. *Food Control* Vol. 19, pp. 1033–1036, ISSN 0956-7135

Tekinsen, K.K; Eken, H.S. (2008) Aflatoxin M1 levels in UHT milk and kashar cheese consumed in Turkey. *Food and Chemical Toxicology* Vol. 46, pp. 3287–3289, ISSN 0278-6915

Tihrumala-Devi, K.; Mayo, M.A.; Hall, A.J.; Craufurd, P.Q.; Wheeler, T.R.; Waliyar, F.; Subrahmanyam, A.; Reddy, D.V. (2002) Development and application of an indirect competitive enzyme-linked immunoassay for aflatoxin m(1) in milk and milk-based confectionery *Journal of Agricultural and Food Chemistry*, Vol. 50, No. 4, pp. 933-937, ISSN 0021-8561

United States Department of Agriculture Foreign Agricultural Service (2011) Cheese Production Selected Countries. 21/04/2011. Available from: (http://www.fas.usda.gov/dlp2/circular/1998/98-01Dairy/cheese.pdf)

United States Department of Agriculture Foreign Agricultural Service. Cheese Production and Consumption: Summary For Selected Countries 21/04/2011. Available from: http://www.fas.usda.gov/psdonline/psdReport.aspx?hidReportRetrievalName= Cheese+Production+and+Consumption%3a+Summary+For+Selected+Countries& hidReportRetrievalID=1233&hidReportRetrievalTemplateID=7

Van Egmond, H.P. (1989) Aflatoxin M1: Occurrence, toxicity, regulation. In *Mycotoxins in Dairy products*. Van Egmond H.P. (Ed.) 11-59 Elsevier Applied Science Publisher, ISBN 1-85166-369-X, London, United Kingdom

Van Egmond, H.P.; Dragacci, S .(2001) Liquid Chromatographic Method for Aflatoxin M1 in Milk. In: *Methods in Molecular Biology, Vol 157 Mycotoxin Protocols II*. M.W. Trucksess; A.E. Pohland (Ed) 59-69 Humana Press, ISBN 0-89603-623-5, Totowa (NJ), United States

Virdis, S.; Corgiolu, G.; Scarano, C.; Pilo, A.L.; De Santis, E.P.L. (2008) Occurrence of Aflatoxin M1 in tank bulk goat milk and ripened goat cheese. *Food Control* Vol. 19, pp. 44–49, ISSN 0956-7135

Wang, J. J.; Liu, B.H.; Hsu, Y.T.; Yu, F.Y (2011) Sensitive competitive direct enzyme-linked immunosorbent assay and gold nanoparticle immunochromatographic strip for detecting aflatoxin M1 in milk *Food Control* Vol. 22, pp. 964-969, ISSN 0956-7135

Yapar, K.; Elmali, M.; Kart, A.; Yaman, H. (2008) Aflatoxin M1 levels in different type of cheese products produced in Turkey *Medycyna Weterynaryina* Vol. 64, No. 1, pp. 53-55 ISSN 00258628

Yousef, A. E.; Marth, E. H. (1989) Stability and degradation of aflatoxin M_1. In: *Mycotoxins in dairy products*. Egmond, H. P. van (Ed) 127-161 Elsevier Applied Science Publisher, ISBN 1-85166-369-X, London, United Kingdom

3

Silage Contribution to Aflatoxin B_1 Contamination of Dairy Cattle Feed

Alonso V.A.[1,2], González Pereyra M.L.[2], Armando M.R.[2], Dogi C.A.[2],
Dalcero A.M.[1,4], Rosa C.A.R.[5,6], Chiacchiera S.M.[3,4] and Cavaglieri L.R.[1,4]

[1]*Departamento de Microbiología e Inmunología, Facultad de Ciencias Exactas,*
Físico-Químicas y Naturales, Universidad Nacional de Río Cuarto, Cuarto, Córdoba,
[2]*Consejo Nacional de Investigaciones Científicas y Técnicas (CONICET),*
[3]*Departamento de Química, Facultad de Ciencias Exactas,*
Físico-Químicas y Naturales, Universidad Nacional de Río Cuarto, Cuarto, Córdoba,
[4]*Consejo Nacional de Investigaciones Científicas y Técnicas (CONICET),*
[5]*Departamento de Microbiologia e Imunología Veterinaria,*
Universidade Federal Rural do Rio de Janeiro, Instituto de Veterinaria, Rio de Janeiro,
[6]*Conselho Nacional de Pesquisas Científicas (CNPq),*
[1,2,3,4]*Argentina*
[5,6]*Brazil*

1. Introduction

Dairy production systems have traditionally relied on direct utilization of pastures and annual soiling crop. This feeding strategy is complemented by the use of other feeds such as grains, balanced feed, silage, hay and industrial products, the level of use was variable and it defined in any way the degree of intensification of each dairy production systems.

Over recent decades, this intensification has been increasing at an accelerated rate, partly because the farms that remain, integrated into general agricultural-livestock mixed models, increasing land for agriculture, as a result of best price-cost and simplicity of production. This change in management practices in dairy cattle breeding, from the extended to semi-intensive or intensive form, has meant a change in the way animals are fed.

The change from grazing over large areas of land to cowshed feeding with grain-based concentrates and silage has greatly improved productivity increase on the number of animals per hectare and, in turn, improved performance and milk production per cow due to the nutritional advantages afforded by the new way of eating. The dairy industry has been driven to higher levels of efficiency and competitiveness. This management system makes storing feed necessary as it is used throughout the year whether it is produced in the same establishment or not. This raises the concern to protect these products from damage by insects, pests and fungal contamination in order to maintain an appropriate level of feed security. Storage systems for feed, both silage and whole grains are a man-made ecosystem in which quality and nutritive changes occur because of interactions between physical, chemical and biological factors.

The deterioration by fungi and mycotoxin contamination is one of the greatest risks of stored feed. Apart from reducing palatability and feed consumption, fungal growth leads to loss of nutrients and dry matter causing in animal performance (O´Brien et al., 2005). Fodder, cereals and seeds used in feed for dairy cattle are naturally in contact with yeasts and filamentous fungi, the contamination of raw materials occurs frequently in the field, because of the infection of plant symbiotic fungi as phytopathogens. This contamination can also occur during harvesting, transport and storage of these products and post harvest mishandling can lead to rapid spoilage. In well-preserved forages fungal growth depends on moisture conditions of the plant during harvest. Stored feed, moisture, temperature and oxygen availability are key conditions that determine risk degree of fungal contamination. The critical water activity for safe storage is 0.7 to 0.8 (Magan & Aldred 2007; Scott, 1957). When this level is exceeded, large degrading ability fungi as *Eurotium sp.*, and species of *Aspergillus* and *Penicillium* can grow. Increase in respiratory activity, due to the development of these fungi, leads to an increase in the temperature of feed that can lead to the contamination by other fungi especially thermophilic fungi and, therefore, to further deterioration.

Silage is one of the main constituents in the diets of dairy cattle and its deterioration and aflatoxin contamination can lead to considerable production losses and a major impact on human health.

2. Breeding and feeding systems on dairy farms

In many systems of milk production mainly in the northern hemisphere, the dairy cows are housed in stockyard due to extreme weather conditions, either high or low temperatures. These intensive production systems use a minimal proportion of grass per cow. In other systems, where climates are more benign and temperate, the production system is typically extensive grazing.

In general, worldwide, the diversity of soils, climates and production scales do not allowa single production system; it is clear that there has been a gradual shift from purely pastoral models to semi-intensive systems (López 2008). In the first instance, the producers began to incorporate ration, preferably, corn grain or commercial feed and for this, they took the shackles of milking, where feeders are installed. Simultaneously, the corn silage began to spread, both as a reserve fodder as well as balanced diet. At this point, producers required new ways of providing meals.

This intensification is necessarily accompanied by a significant increase of the scale, this fact causes many people to use new technologies to keep the cows in confinement.

The development in milk production in recent years has followed an intensification which has resulted in a change in the use of feed, evolving from simple grazing feeding systems based on mixed feed formulation combining grains and forages.

Although the current systems of feeding in major milk producing areas in Argentina have particular differences in the degree of intensification, they can be considered supplement grazing systems (or semi-intensive). Through this enhancement, production level was able to grow extensively. The levels of milk production increased from 12 L to 20-30 L. However, animal numbers by hectare did not increase. That supplementation can not only avoid the seasonality of production due to the availability of pastures in different seasons, but also allow to balance the dietary components optimizing milk production per cow (West, 2003). However, many authors argue not to forget grass, which remains the staple feed "of ruminant herbivores" as well as the cheapest cost of production.

The composition of feed rations for dairy cows consists of:

• Pastures (including small grain winter and summer)
• Conserved forage (silage, hay)
• Concentrate

Figure 1 shows estimated components proportion, which may vary slightly according to season and geographical area. Perennial pastures are usually based on alfalfa pasture. Forages are used both for direct consumption of pasture (winter and summer soiling) and as conserved forage in the form of rolls or bales of hay. Typically, 10% of forage is intended for these purposes and often rye, oats, moha, wheat and sorghum are selected in dairy farms according to acreage and selected pasture. As concentrates, grain corn, grain sorghum, cotton seed, wheat bran, dregs of malt, peanut shells, and sunflower expeller, are used among others. It is also common to use commercial pelleted feed.

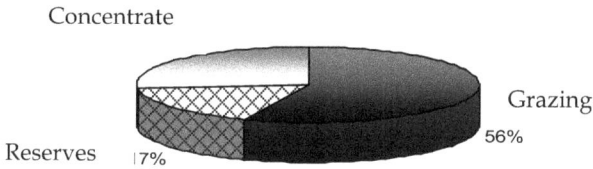

Concentrate

Grazing
56%

Reserves 17%

Fig. 1. Typical diet for milking cows (Chimicz & Gambuzzi 2007).

3. Corn silage

Corn (*Zea mays* L.) is the most widely grown crop in the Americas, extensively used for animal feeding and human consumption due to its nutritional value. A large percentage of the world corn production is destined to animal feeding. Silage is a widespread practice to preserve forages during extended time periods. The production of corn silage entails incorporation of the whole plant and its storage is based on the principle of preservation under anaerobic conditions with growth of lactic acid bacteria which promote a natural fermentation that lowers the pH to a level at which clostridia and most fungal growth are inhibited. In dairy cows, silage is a preferred food by the vast majority of producers.

As corn silage consists of grinding and storing the whole corn plant, it includes not just grain but a high percentage of stalks and stover and represents a new important bulky feed source for dairy and beef cattle. Nutritionally, corn silage, for example, has a balance between the energy density of the grain and fibber and digestibility of the green plant that makes it suitable for feeding ruminants in the phases of maximum nutritional needs (Molina et al., 2004).

4. Ensiling and storage conditions

Silage is a method of forage preservation based on lactic acid fermentation, usually spontaneous under anaerobic conditions, where the pH reaches values of 2-3 being an important indicator of forage conservation (Johnson et al., 2002). Air must be removed as much as possible from the silo in order to obtain good silage quality. To achieve this goal, certain management aspects must be emphasized. Forage should be harvested, chopped, packed well and covered in the silo as fast as possible. Air and rain infiltration can cause poor fermentation and spoilage in the silo. Rain will increase moisture/seepage, favour growth of undesirable bacteria (for example *Clostridium sp.*), and wash nutrients away. The resulting silage will have low nutritional value and will likely be avoided by cows (low dry matter intake). Intake is directly related to milk production in lactating dairy cows, therefore low intake equals low milk yield.

Maize, sorghum and barley malt are the main forages used for silage (Driehius & Oude Elferink, 2001). Ideal fermentation is dependent upon decisions and management practices implemented before and during the ensiling process. The primary management factors that are under the control of the producer are:

1. Stage of maturity of the forage at harvest.
2. The type of fermentation that occurs in the silo or bunker.
3. Type of storage structure used and methods of harvesting and feeding.

During the ensiling process, some bacteria are able to break down cellulose and hemicellulose to various simple sugars. Other bacteria break down simple sugars to smaller end products (acetic, lactic and butyric acids). The most desirable end products are acetic and lactic acid. As the bacteria degrade starches and sugars to acidic and lactic acids, dry matter is lost.

Quality silage is achieved when lactic acid is the predominant acid produced, as it is the most efficient acid fermentation and will drop the silage pH quickly. The faster the fermentation is completed, the more nutrients will be retained in the silage.

At least six phases can be described during the ensiling process (Table 1), in a first phase the aerobic bacteria predominant on the forage surface continue respiring within the silo structure. This phase is undesirable since the aerobic bacteria consume soluble carbohydrates that might otherwise be available for the beneficial lactic acid bacteria or for the animal consuming the forage. Phase I ends once the oxygen has been eliminated from the silage mass. Under ideal crop and storage conditions, this phase will last only a few hours.

After the oxygen in the ensiled forage has been used by the aerobic bacteria, Phase II begins. This is an anaerobic fermentation where the growth and development of acetic acid-producing bacteria occur. These bacteria ferment soluble carbohydrates and produce acetic acid as an end product. Acetic acid production is desirable as it can be utilized by ruminants in addition it initiates the pH drop necessary to set up fermentation phases. As the pH of the ensiled mass falls below 5.0, the acetic bacteria decline in numbers as this pH level inhibits their growth. This signals the end of Phase II. In forage fermentation, Phase II lasts no longer than 24 to 72 h. Phase III begins when the increasing acid inhibits acetic bacteria. The lower pH enhances the growth and development of another anaerobic group of bacteria, those producing lactic acid.

Phase IV is a continuation of Phase III as lactic-acid bacteria start to increase in number, ferment soluble carbohydrates and produce lactic acid. Lactic acid is the most desirable of the fermentation acids and for efficient preservation, should comprise greater than 60 percent of the total silage organic acids produced. When silage is consumed, lactic acid will also be utilized by cattle as an energy source. Phase IV is the longest phase in the ensiling process as it continues until the pH of the forage is low enough to inhibit the growth of all bacteria. When this pH is reached, the forage is in a preserved state. No further destructive processes will occur as long as oxygen is kept from the silage.

Phase V is the storage time when the final pH is reached, and the good conditions of anaerobiosis are supported.

Phase VI refers to the silage when it is cut to be used as feed. The Phase VI occurs on any surface of the silage that is exposed to oxygen during storage and in the feed bunk.

	Phase I	Phase II	Phase III	Phase IV	Phase V	Phase VI
Age of silage	0-2 days	2-3 days	3-4 days	4-21 days	21 days	
Activity	Cell respiration; production of CO_2, heat and water	Production of acetic acid and lactic acid ethanol	Lactic acid formation	Lactic acid formation	Material storage	Aerobic decomposition on re-exposure to oxygen
Temperature change	20-32 °C	32-29 °C	29 °C	29 °C	29 °C	29 °C
pH change	6.5-6.0	6.0-5.0	5.0-4.0	4.0	4.0	4.0-7.0
Produced by		Acetic acid and lactic acid bacteria	Lactic acid bacteria	Lactic acid bacteria		Mold and yeast activity

Table 1. Silage fermentation phases and storage

5. Influence of pH and water activity on the silage contamination

The current system of dairy animal production requires a thorough knowledge of production, processing and quality of all feed used. Contamination of feed intended for animal consumption usually reflects the incidence of fungal infection in the original crop.

Temperature, humidity, oxygen availability and pH conditions vary during the silage process and microbiota may also change from one stage to another. However, poor storage conditions - including excessive moisture or dryness, condensation, heating, leakage of rainwater and insect infestation - can lead to undesirable fungal contamination, mycotoxin production and the reduction of nutritional value.

The forage quality is evaluated through physicochemical and fermentative conditions such as pH, water activity (a_W), percentages of ammonium / total nitrogen (Teimouri Yansari et al. 2004). The water content of a substrate does not give a direct index of a_W for microbial growth. The availability of water in hygroscopic materials such as grains is measured as equilibrium relative humidity (ERH), a_W or water potential (ψ). The last two measures are most appropriate for situations where the availability of water in the substrate is the factor that controls growth.

The pH in the silage provides an indication of the type and range of the fermentation process. The acid pH resulting from fermentation prevents proper development of viable cells. Only a few yeasts, other microorganisms tolerant to this pH and spores as Clostridia and *Bacillus* can survive in dormant state (Driehius & Oude Elferink, 2001).

The silage can be contaminated and damaged by fungi from the soil and essentially can contaminate forages in various stages and plant management. The process of preservation by acidification, dehydration and exclusion of O_2 in the early stages of storage does effectively restrict the development of these microorganisms. Moreover, the improper extraction of silage (straight cut, little waste, little oxygenation) and the mixing of different sections of the silo before being incorporated into the mixer, could enhance the final feed contamination with aflatoxigenic fungi and aflatoxins (Borreani & Tabacco, 2010).

Comparative multivariate statistical studies on the influence of pH and a_W on the fungal count and on the incidence of AFB_1 in dairy cattle feedstuff, were performed using principal component analysis. In Figure 2, the "biplot" graphic in which the variables: total fungal count, *Aspergillus* count, *A. flavus* count and incidence of aflatoxin B_1 (AFB_1), depending on the type of food and a_W, are shown.

Corn silage at a_W 0.97 is closely related to total fungal count and *Aspergillus* spp. So is for that same feed at a_W 0.98, 0.99, 0.96, and 0.93. This positive relationship shows that at a_W 0.93 or higher; the corn silage contributes to finished feed contamination by fungi such as aflatoxicogenic fungi.

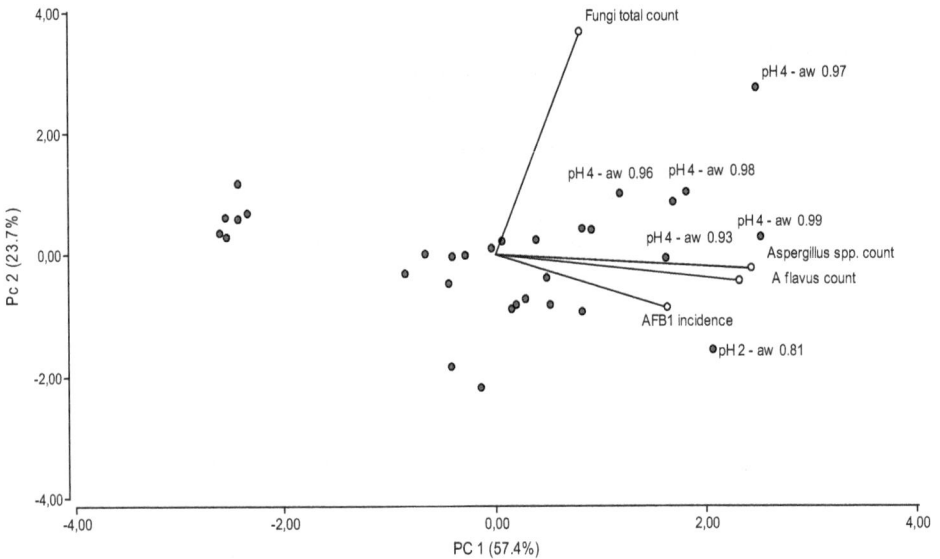

Fig. 2. Graph "biplot" principal component analysis to study variables (total fungal count, *Aspergillus* spp count of, *A. flavus* count, incidence of AFB_1) depending on the type of food and a_W.

A multivariate statistical comparative study in terms of the type of feedstuff and pH among the variables total fungal count, *Aspergillus spp* count, *A. flavus* count, and AFB_1 incidence are shown in Figure 3.

According to the principal component analysis, the contribution of total fungi to finished feed is mainly given by the silage at pH 4 and 5.

The contribution of *Aspergillus* spp. and *A. flavus* corresponds mainly to the silage at pH 4.5. These studies allow to highlight that silage, when reaches these pH values, will be affected by contamination with *Aspergillus* spp.s and *A. flavus*. This fact will determine the contribution of fungal contamination from silage to finished feed that will be consumed by dairy cattle.

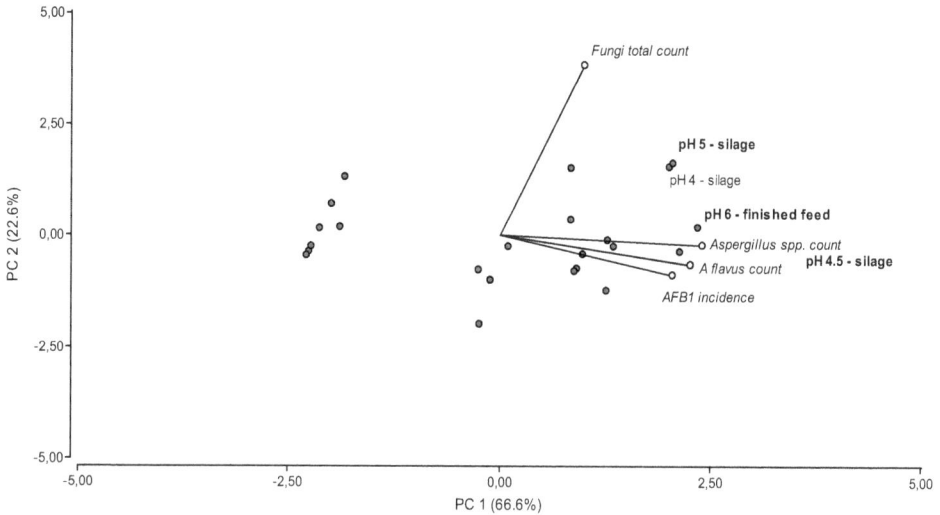

Fig. 3. Graph biplot of principal component analysis to study variables (total fungal count, *Aspergillus* spp count, *A. flavus* count, AFB₁ incidence) depending on the type of food and pH.

6. Silage mould pathogens

Fungi growing in silage expose animals to respiratory problems, abnormal rumen fermentation, decreased reproductive function, kidney damage, skin and eye irritation (Akande, 2006; Scudamore & Livesey, 1998). Fungal concentrations in forage above 1×10^4 CFU g^{-1} may be the reason for these problems. Thus, the fungal colony count is an indicator of forage quality (Di Costanzo et al., 1995). Currently, the Good Manufacturing Practises International (GMP 2008) recommends a limit set as 1×10^4 CFU g^{-1} in feedstuff.

The major fungal species isolated from feed for dairy cattle, belong to *Aspergillus*, *Penicillium* and *Fusarium* genera (El-Shanawany et al., 2005, Garon et al., 2006; Gonzalez Pereyra et al., 2008; Rosa et al., 2008; Simas et al., 2007).

Several species within these genera are capable of producing mycotoxins, in exposed animals or humans.

Strains of *A. flavus* and aflatoxins are the main grains and corn plant contaminants (Chulze 2010). *A. flavus* can infect pre-and post-harvest corn and a significant increase in the content of aflatoxins may occur if the drying and storage phases do not perform correctly.

In Argentina, in studies on the fungal contamination in dairy cattle feed it can be seen how corn silage influences the degree of contamination of the ration supplied to livestock (Gonzalez Pereyra et al., 2008).

The multivariate analysis through principal component analysis (PCA) allows biplot graph (Figure 4) expressing the associations between finished feed contamination and raw materials that mainly contribute with fungal contamination. It can be seen that the obtained silage fungal counts are strongly correlated with the finished feed contamination. A similar correlation was observed between the finished feed and cotton seed. Raw materials such as corn and brewer's grains, do not have correlation with finished feed, in other words, do not contribute to the increase of fungal contamination.

In relation to sampling periods, it can be seen in the same figure that December is associated with a higher contamination in silage and in finished feed. The prediction ellipse confirms that during December all feed adds high fungal contamination to finished feed (98% confidence ellipse).

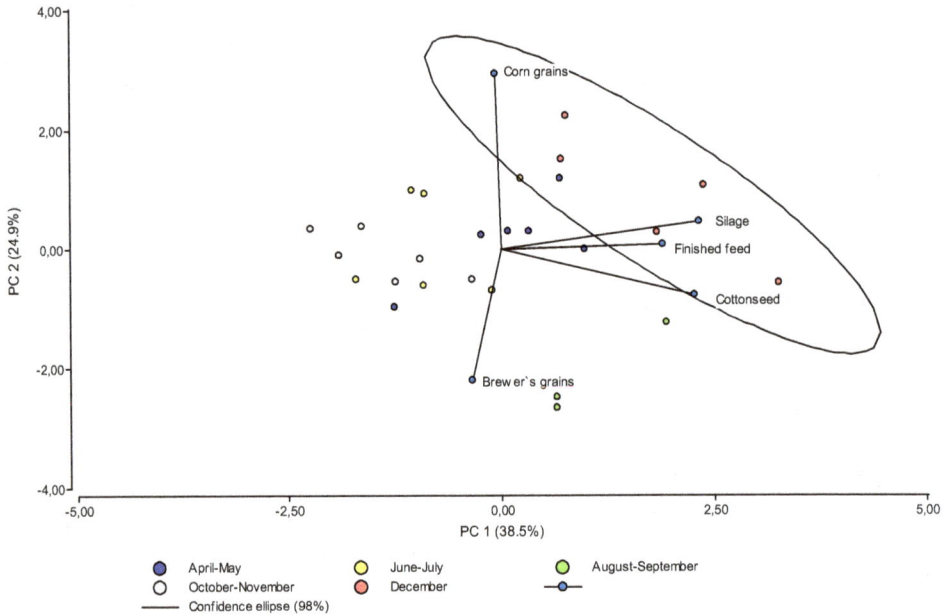

Fig. 4. Graph "biplot" principal component analysis of the feed materials and sampling periods in terms of total fungal counts.

7. Corn silage sections

Corn silage can be divided into three main sections corresponding to the upper (generally more exposed to fungal contamination) middle (best-preserved portion) and lower portions. Statistical analysis based on fungal counts were performed on the three sections of the silo showed that levels of found contamination were significantly different in each studied section (p <0.05). Table 2 shows the associations between different levels of silage for fungal contamination. The upper and lower sections had the same levels of fungal contamination whereas pollution in the middle section did not show association with low levels. This was an expected result since the anaerobic environment and low pH silage allow good conservation in half portions of the silage that do not have contact with air or ground as in upper and lower section.

Proper storage is related the state of compaction. The most compact the silo, the teast possibility of losing reduced pH and anaerobic conditions. The extraction method is also highly important. Even if the upper section is in contact with air, it is less affected when the silo is firmly packed. In silos visibly unarmed, the upper and lower sections (more than 10 cm) are visibly contaminated and altered in colour and smell.

Silage	Total fungal count (log $_{10}$CFU g^{-1})	
	Mean ± Standard Error	LSD (p < 0.05)
Upper	7.13 ± 0.58	b
Middle	4.83 ± 0.57	a
Lower	6.99 ± 0.58	b

Different letters indicate significant differences (p <0.05) according to the test of Fisher's least significant difference (LSD). The count data were transformed to log$_{10}$ (x +1) to achieve homogeneity of variance

Table 2. Total fungal counts (CFU g^{-1}) present in samples of silages at different sampling sections of the silo.

Table 3 details the total fungal counts, expressed in CFU g^{-1}, obtained for each raw material and finished feed at different sampling periods. The silage was considered by averaging the counts obtained in the three sections for each sampling period.

In corn, the average values in total fungal counts during all sampling periods ranged from a minimum of 2.36 x 10^3 to a maximum of 7.00 x 10^5 CFU g^{-1}, corresponding to the periods August-September and October-November, respectively. In general, the fungal counts showed low variability throughout the year, finding associations in contamination levels during the first three bimonthly sampling. Table 4 shows the percentage of samples whose total fungal counts exceeded hygienic limit of 1 x 10^4 CFU g^{-1} established by Good Manufacturing Practices (GMP, 2008) for feedstuff. The levels of fungal contamination in silage were among the highest. All tested samples were positive for low levels of fungi, with a maximum of 2.10 x 10^5 CFU g^{-1} and 80% had values of total fungal counts greater than 1 x 10^4 CFU g^{-1} (Table 4).

Sampling period	Total fungal count (CFU g^{-1})				Finished feed
	Dairy cows feedstuff				
	Ingredients				
	Corn grains	Cotton seed	Brewer's grains	Corn Silage	
April-May	5,50 x 10^5 c	2,10 x 10^5 ab	1,70 x 10^6 b	5,13 x 10^7 a	9,17 x 10^6 b
June-July	3,65 x 10^5 c	5,00 x 10^4 bc	2,12 x 10^4 a	1,33 x 10^7 a	2,49 x 10^5 b
August-September	2,36 x 10^3 c	2,40 x 10^5 c	8,90 x 10^6 a	4,22 x 10^7 a	2,27 x 10^7 b
October-November	7,00 x 10^5 a	3,00 x 10^3 a	4,12 x 10^6 a	2,64 x 10^6 a	4,80 x 10^5 a
December	2,30 x 10^5 b	5,75 x 10^5 b	3,80 x 10^5 a	2,18 x 10^8 b	3,08 x 10^6 b

Different letters indicate significantly different values according to Fisher's least significant difference test (LSD) p=0.05. The count data were transformed to log$_{10}$ (x +1) to achieve homogeneity of variance. Statistical results should be read vertically to each food type separately.

Table 3. Total fungal colony count (CFU g^{-1}) from food samples for dairy cows during different sampling periods.

The mean values of total fungal counts in the finished feed, varied between 10^5 and 10^7 CFU g^{-1}. During October and November the count levels found were low, while during other periods of the year the pollution levels were consistent with each other. As it can be seen, 100% of the silage samples exceeded the limit of hygienic quality. It is important to observe that silage samples exceeded 100% the GMP recommendation in lower and upper sample sections.

Dairy cattle feedstuff		Total fungal count (CFU g⁻¹)	
		Contaminated samples (%)	Simples exceeding HLQ (%)[b]
Ingredients	Corn grains	80	90
	Cotton seed	100	80
	Brewer's grains	100	100
	Corn silage Upper	92	100
	Corn silage Middle	85	81
	Corn silage Lower	100	100
Finished feed		90	100

[b] LCH: Hygienic limit quality by GMP (2008) 1x10⁴ CFU g⁻¹.

Table 4. Samples percentage that exceeded hygienic limit according to good manufacturing practices (GMP, 2008).

Figure 5 shows the $\log_{10}$ CFU g⁻¹ for each type of food. Silage was the substrate with higher levels of pollution, followed by cotton seed. These substrates have a difference of almost two $\log_{10}$ units in relation to the other contaminated foods. The principal component analysis indicates that both components made the greatest contribution of fungal contamination to finished feed (Figure 4).

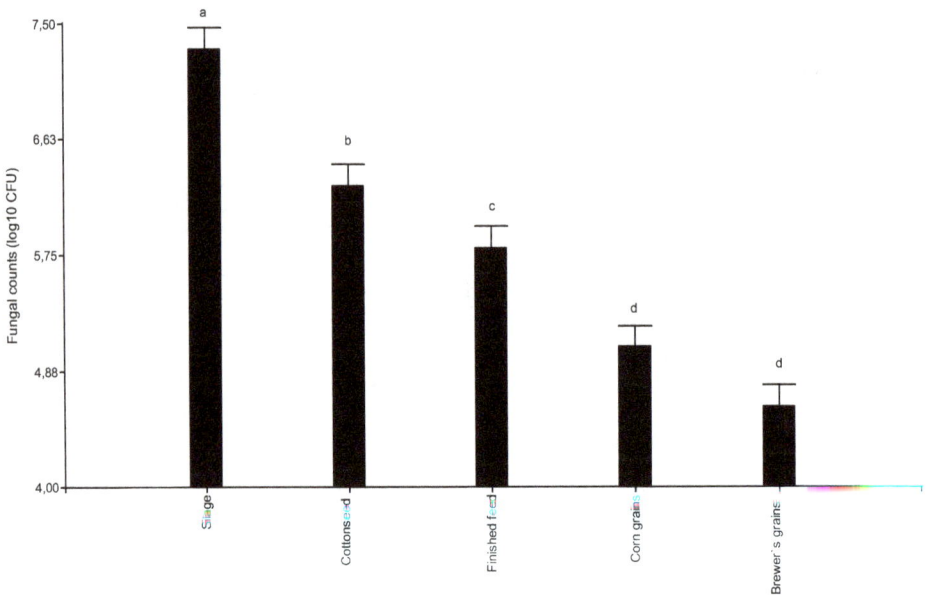

Fig. 5. total fungal count ($\log_{10}$CFU g⁻¹) of ingredients and finished feed for dairy cattle consumption.

8. Fungal genera distribution

Mycological survey of the strains isolated from different feeds, showed that the main toxigenic genera were present at high levels and in all types of feed samples.

Finished feed samples showed a high variability in their isolation, finding a high frequency of *Aspergillus spp*, in all sampling periods. They were isolated in 100% of the samples during the periods April-May, June-July and December (Figure 6).

Fusarium spp. were also one of the most frequent followed by yeasts. Penicillium spp were isolated throughout the sampling, although less frequent. They were isolated from 50% samples during June to July and August-September. December had further fungal diversity.

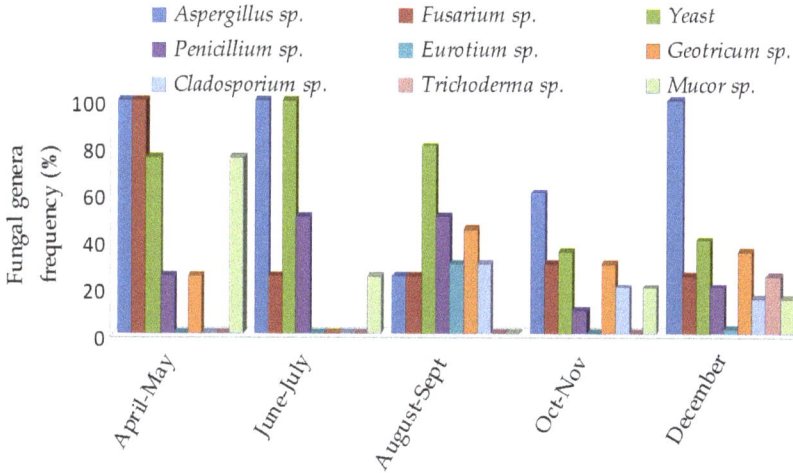

Fig. 6. Frequency of isolation of fungal genera (%) in finished feed during different sampling periods.

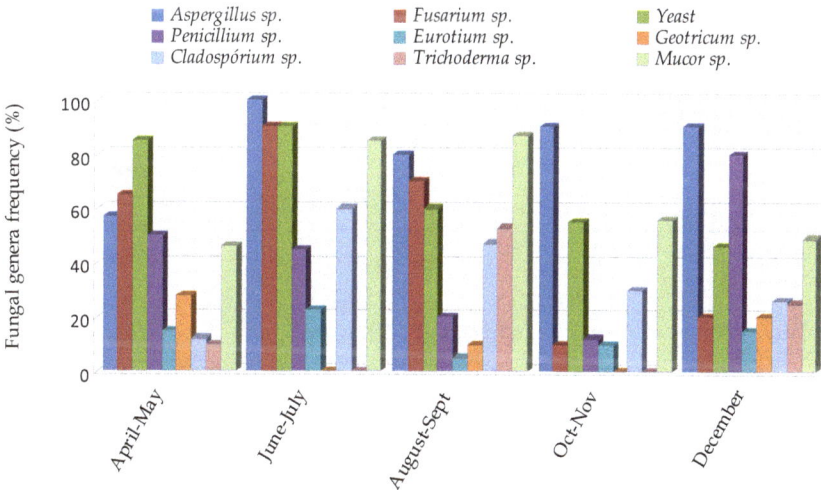

Fig. 7. Frequency of isolation of fungal genera (%) present in silage maize at different sampling periods.

The distribution of fungal genera in corn silage is presented in Figure 7. The incidence of important toxigenic genera was very high throughout the period. The rates of isolation of *Aspergillus spp* ranged between 90 and 100% in all sampling months, except for April-May, when the incidence, although lower, was also important (60%). For *Penicillium* spp. the isolation frequency was from 12 to 80% in December.

The incidence of *Fusarium spp* was high during the first three bimonthly sampling. They were isolated at 90% during the period June-July.

9. Incidence and toxigenic potential of *Aspergillus* section *Flavi*

It is of particular interest to describe the behaviour of the population of *Aspergillus* Section *Flavi*, its ability to produce AFs in silage for dairy cows, as it gives the possibility of contamination with aflatoxin B_1 in feedstuff. A widespread population of aflaoxicogenic *Aspergillus* has been described in raw materials and especially in silage samples intended for dairy cattle.

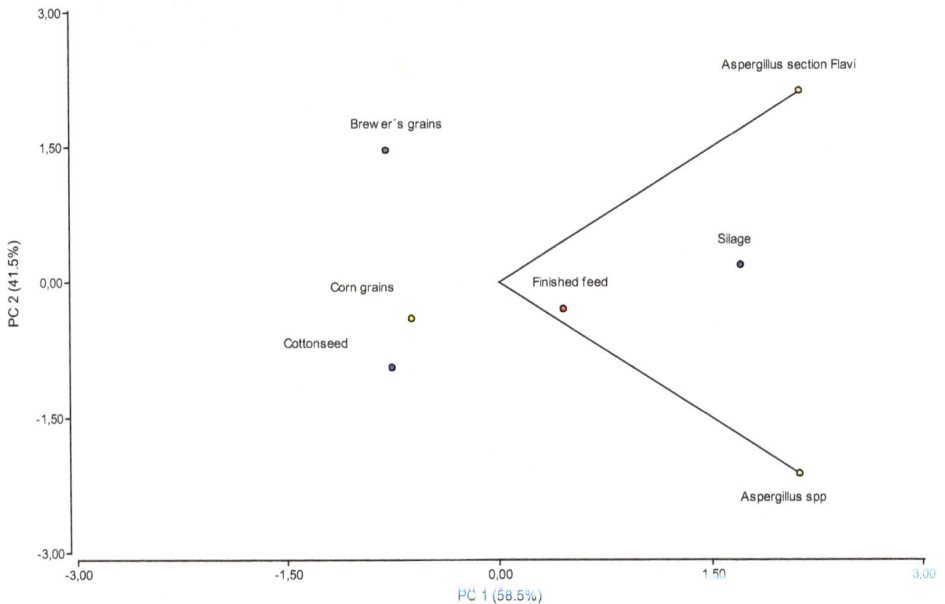

Fig. 8. Relative density of *Aspergillus* spp isolated from feed

Aspergillus flavus was also predominant in the other studied raw materials: 74% in cottonseed, 60.5% in corn and 39.7 in finished feed. Aspergillus parasiticus, although less frequent, it was present in all the substrates with rates of 5.6% in silage and 10% in cotton seed.

The finished feed showed a wide diversity of species, and *A. fumigatus* (19.7%) followed *A. flavus*. Analyzing the obtained results, it is estimated that the major contribution of this fungus to the finished feed comes from corn silage.

Dairy cattle feedstuff		Total fungal count (CFU g⁻¹)	
		Aspergillus genera	*Aspergillus* section *Flavi*
Ingredients	Corn grains	3,20 x 10⁴ **b**	7,07 x 10³ **b**
	Cotton seed	1,28 x 10⁵ **bc**	3,03 x 10⁴ **bc**
	Brewer's grains	0 **a**	0 **a**
	Corn silage	1,95 x 10⁷ **d**	1,20 x 10⁶ **d**
Finished feed		6,06 x 10⁵ **c**	3,72 x 10⁴ **c**

Table 5. Total fungal counts of *Aspergillus* section *Flavi* and *Aspergillus sp* (CFU g⁻¹) from dairy cattle feedstuff during different sampling periods

Figure 9 shows the principal component analysis for *Aspergillus* spp variables depending on the kind of the studied feedstuff. This analysis shows that pattern of behaviour in relation to the kind of feed between the species of *Aspergillus* and *Aspergillus* section *Flavi*. There was a positive correlation between the presence of *Aspergillus* section *Flavi* and *Aspergillus* genera, according to the kind of feed (Figure 9). It is important to emphasize that silage was the ingredient with a greater presence of these fungi. Thus, this is ingredient that contributes with the greatest contamination to finished feed.

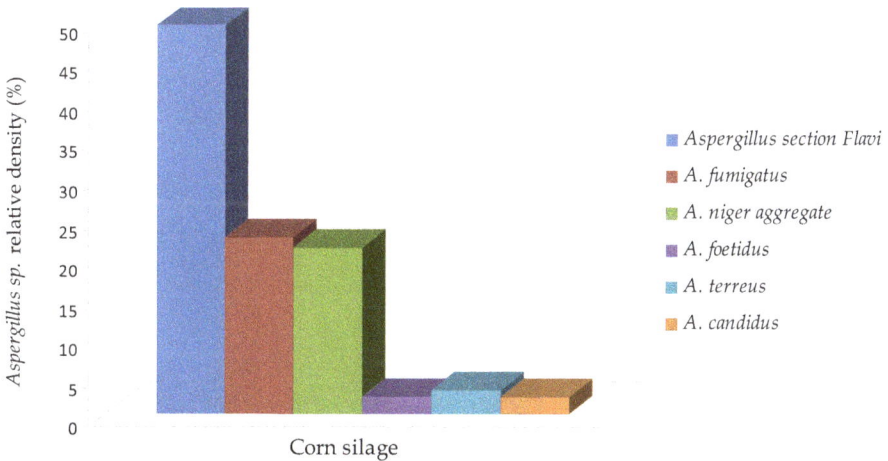

Fig. 9. Graph "biplot" principal analysis component of variable *Aspergilllus spp.* and *Aspergillus* section *Flavi* in relation to the kinds of feed.

10. Aflatoxin in silage

In cattle, chronic ingestion of mycotoxins causes various adverse effects such as increased susceptibility to disease, loss of reproductive performance, and in case of dairy cattle, a decrease in yield and quality of milk production. These effects are caused because the exposure of animals to mycotoxins causes a decrease in consumption or feed refusal, a reduction of nutrient absorption, an impairment of metabolism, and changes in the endocrine and immune system suppression. Exposure of cattle to mycotoxins generally occurs through consumption of contaminated feed. Nelson et al., (1993) described as "mycotoxicosis" to diseases caused by exposure to food mycotoxin-contaminated rations.

Aflatoxins, particularly AFB_1 have been described both acute and chronic (Meggs, 2009). In June 2004, in Kenya there was an outbreak of acute aflatoxicosis, high levels of AFB_1 in stored corn at high humidity conditions were found (Lewis, 2005). Aflatoxin B_1 has been found in different countries as a contaminant in feed of dairy, cottonseed, barley, soy bran, pellet wheat, peanut shells, corn silage and sorghum silage (Decastelli et al., 2007; Sassahara et al., 2005).

For dairy cattle the problem does not end in animal disease or production losses since the mycotoxins in feed can lead to their presence or their metabolic products in dairy products which will be eventually affecting human health.

In the case of AFB_1, its presence in the food of dairy cattle leads to the emergence of AFM_1 in milk and dairy products (Boudra et al., 2007; Veldman et al., 1992).

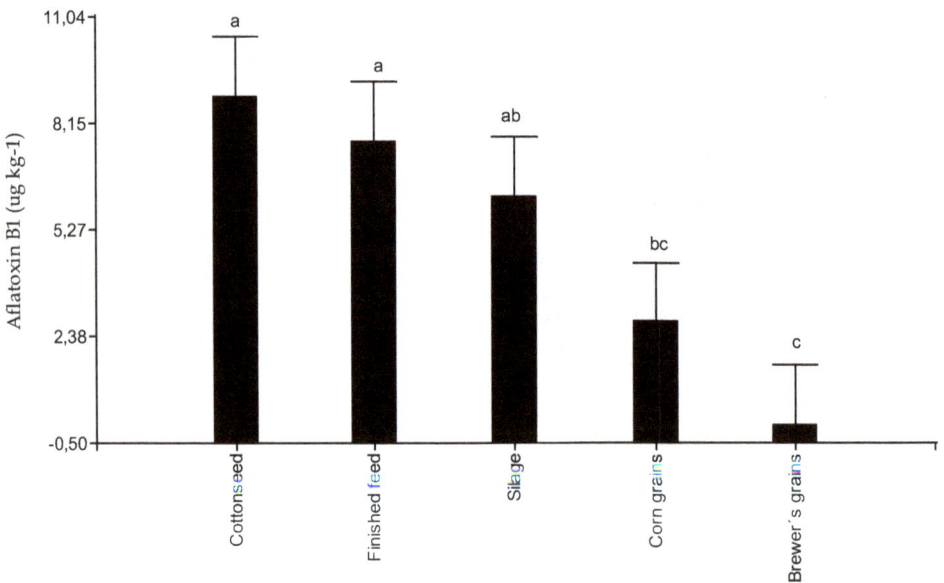

Fig. 10. AFB_1 levels in raw materials and finished feed intended for dairy cattle.

The natural occurrence of AFB_1 in feeds for dairy cows has shown that, in many cases aflatoxin levels exceeded regulation limits. Multivariate statistical studies show that silage makes the main contribution of AFB_1 to finished feed (Figure 11).

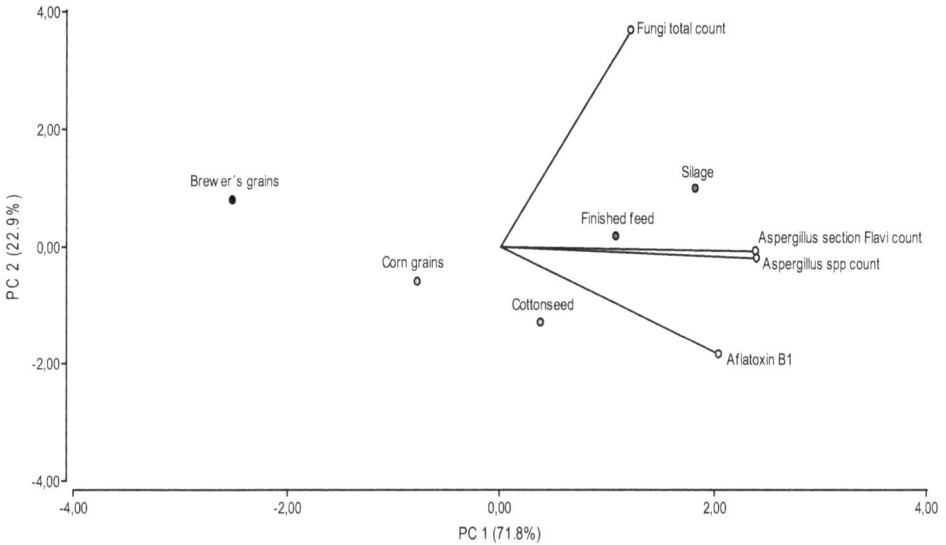

Fig. 11. Principal component analysis for variables: total fungal count, *Aspergillus spp* count, *Aspergillus* Section *Flavi* count and AFB₁ incidence.

11. References

Akande KE, Abubakar MM, Adegbola TA, Bogoro SE. (2006). "Nutricional and health implications of mycotoxins in animal feed: a review". Pakist J Nutrit. 5, 398-403.

Borreani G, Tabacco E. (2010). "The relationship of silage temperature with the microbiological status of the face of corn silage bunkers". J Dairy Sci. 93, 2620-2629.

Boudra H, Barnouin J, Dragacci S, Morgavi DP. (2007). "Aflatoxin M₁ and ochratoxin A in raw bulk milk from French dairy herds." J. Dairy Sci. 90, 3197-3201.

Chulze SN. (2010). "Strategies to reduce mycotoxin levels in maize during storage: a review". Food Additives and Contaminants" 27, 651–657

Decastelli L, Lai J, Gramaglia M, Monaco A, Nachtmann C, Oldano F, RuYer M, Sezian A, Bandirola C. (2007). "Aflatoxins occurrence in milk and feed in Northern Italy during 2004–2005". Food Control 18, 1263–1266.

Di Costanzo A, Johnston L, Felice L. Murphy M. (1995). "Effect of molds on nutrient content of feeds reviewed". Feed Stuffs. 16, 17-20.

Driehuis F, Oude Elferink SJ. (2001). "The impact of the quality of silage on animal health and food safety: a review". Vet Q. 22, 212-216.

El-Shanawany A, Eman A, Mostafa M. Barakat A. (2005). "Fungal populations and mycotoxins in silage in Assiut and Sohag governorates in Egypt, with a special reference to characteristic Aspergilli toxins". Mycopathologia. 159, 281–289.

Garon D, Richard E, Sage L, Bouchart V, Pottier D, Lebaill P. (2006). "Mycoflora and multimycotoxin detection in corn silage: experimental study." J Agricult. Food Chemist. 54, 3479-3484.

GMP (2008). Certification Scheme Animal Feed Sector, 2006. Version Marzo 2008. Appendix 1: Product standards (including residue standards). The Hague, the Netherlands: Productschap Diervoeder. pp 1 – 39.

Gonzalez Pereyra ML, Alonso VA, Sager R, Morlaco MB, Magnoli CE, Astoreca AL, Rosa CA, Chiacchiera SM, Dalcero AM, Cavaglieri LR. (2008a). "Fungi and selected mycotoxins from pre- and postfermented corn silage". J of App Microb. 104, 1034–1041.

Johnson LM, Harrison JH, Davidson D, Mahanna WC, Shinners K, Linder D. (2002). "Corn silage management: effects of maturity, inoculation, and mechanical processing on pack density and aerobic stability". J Dairy Sci. 85, 434-44

Lewis L, Onsongo M, Njapau H, Schurz-Rogers H, Luber G, Kieszak S, Nyamongo J, Backer L, Dahiye AM, Misore A, DeCock K, Rubin C. (2005). "Aflatoxin Contamination of Commercial Maize Products during an Outbreak of Acute Aflatoxicosis in Eastern and Central Kenya". Environ Health Perspect. 113, 1763-1767.

López A. (2008). Argentina Milk production systems. Red ICAARG. Cattle Milk Production Area. Veterinary Science Faculty of Buenos Aires, UBA.

Magan N, Aldred D. (2007). "Post-Harvest Control Strategies: Minimizing Mycotoxins In The Food Chain". Int J of Food Microbiol. 119, 131-139.

Meggs W. (2009). "Epidemics of mold poisoning past and present". Toxicol and Indust Health 25, 9-12.

Molina AM, Roa LB, Alzate SR, Serna de León JG, Arango AF. (2004). "Silage as a livestock feed source". Rev Lasall Investig. 1, 66-71.

O'Brien M, O'Kiely P, Forristal P, Fuller HT. (2005). "Fungi isolated from contaminated baled grass silage on farms in the Irish Midlands". Microbiology Letters. 247, 131–135.

Rosa CA, Cavaglieri LR, Ribeiro JMM, Keller KM, Alonso VA, Chiacchiera SM, Dalcero AM, Lopes CWG. (2008). "Mycobiota and naturally-occurring ochratoxin A in dairy cattle feed from the Río de Janeiro State, Brazil." World Mycot J. 1, 195-201.

Sassahara M, Pontes D, Yanaka K. (2005). "Aflatoxin occurrence in foodstuff supplied to dairy cattle and aflatoxin M1 in raw milk in the north of Paraná state". Food and Chem Toxicol. 43, 981-984.

Scott WJ. (1957) "Water relations of food spoilage micro-organisms". Advan Food Res 7, 83-127.

Scudamore K A, Livesey CT. (1998). "Occurrence and significance of mycotoxins in forage crops and silage: A review". J. Sci. Food Agric. 77, 1–17.

Simas MS, Botura MB, Correa B, Sabino M, Mallmann CA, Bitencourt T, Batatinha M. (2007). "Determination of fungal microbiota and mycotoxins in brewers grain used in dairy cattle feeding in the State of Bahia, Brazil". Food Control. 18, 404-408.

Teimouri Yansari A, Valizadeh R, Naserian A, Christensen DA, Yu P, Eftekhari Shahroodi F. (2004). "Effects of alfalfa particle size and specific gravity on chewing activity, digestibility, and performance of Holstein dairy cows". J Dairy Sci. 87, 3912-24.

Veldman AJ, Meijs AC, Borggreve GJ, Heeres van der Tol JJ. (1992). "Carry-over of aflatoxina from cows' food to milk." Anim Prod. 55, 163–168.

West JW. 2003. "Effects of Heat-Stress on Production in Dairy Cattle". Dairy Science, 86:2131-2144.

4

Aflatoxin Contamination Distribution Among Grains and Nuts

Eduardo Micotti da Gloria
University of Sao Paulo – ESALQ,
Brazil

1. Introduction

Grains (cereals and oilseeds) and nuts in general are subject to mold attack, in preharvest and postharvest. Among molds that can attack these foods *A. flavus*, and *A. parasiticus* are important because they can produce aflatoxins that are considered a potent natural toxin (Wild & Gong, 2010). Aflatoxin can be produced mainly by different *Aspergillus* species, but *Emiricella* and *Petromyces* have been reported as aflatoxin producers (Frisvad et al., 2005).

Aflatoxin contamination has been reported for grains as corn, soya, wheat, rice, and cottonseed, and nuts such as peanuts, almonds, Brazil nuts, hazelnuts, walnuts, cashew nuts, pecans, and pistachio nuts (Fuller et al., 1977; Ayres, 1977; Moss, 2002; CAST, 2003; Gürses, 2006). Despite aflatoxin contamination having been observed in several foodstuffs, the contamination of maize, peanuts, and oilseeds can be considered, in terms of diet exposure, the most important worldwide (Benford et al. (2010).

Based on deleterious problems that aflatoxin can cause to human and animal health, some countries established a maximum concentration for aflatoxins in specific products. According to published data (Van Egmond, 2007), until 2003 one hundred countries had established legal limits for mycotoxins, and most of them regulated the aflatoxins presence in food and feeds.

Several biotic and abiotic factors can determine fungal infection and growth, as well as aflatoxin production in preharvest. Temperature, water availability, plant nutrition, infestation of weeds, birds, and insects, plant density, crop rotation, drought stress, presence of antifungal compounds, fungal load, microbial competition, substrate composition, and mold strain capacity to produce aflatoxin are some important factors. The incidence of these factors is different in preharvest among plants and production areas of the same farm, among different farms of the same region and among different producer regions. Even among grains of the same ear or peanuts of the same pod the differences can occur. In postharvest, factors such as temperature, availability of water, oxygen, and carbon dioxide, insect and rodents infestation, incidence of broken grains or nuts, the cleaning of the product, toxigenic fungal load, microbial competition, antifungal compound presence, and substrate composition are important too. Transport, waiting time for drying, drying system (temperature and drying rate), and storage conditions can affect these factors during the postharvest period (Dorner, 2008; Diener et al., 1987; Campbell et al., 2006; Molyneux et al., 2007).

As a result of variable conditions that can occur during pre and postharvest, the aflatoxin contamination level among grains and nuts within the same lot can have an extremely uneven distribution. The uneven distribution of aflatoxin contamination was observed in different foodstuffs, such as peanuts, maize, almonds, Brazil nuts, and pistachios (Cucullu et al., 1966; Whitaker et al., 1994; Shotwell et al., 1974; Schatzki & Pan, 1996; Steiner et al., 1992; Shade et al. 1975; Ozay et al., 2007). In a contaminated lot, just a few grains and nut kernels can have quite high concentration levels of aflatoxin, and most of them do not have detectable contamination. Table 1 shows some high individual concentrations detected in a peanut, a maize grain, a Brazil nut, in a pistachio, and a cottonseed. The high concentration observed in an individual grain or kernel can result, for example in maize, in a contamination level of 136 µg/kg, when just one grain is contaminated, considering 0.34 g as the average weight of maize grain, and the high concentration showed in the table 1.

The not uniform distribution of contamination within a lot represents a great challenge to measure the true contamination level of the lot. If several samples are collected from the same lot of a commodity, completely different contamination results can be obtained, as shown in table 2. Several theoretical distribution models have been investigated as possible models to describe the observed distribution of aflatoxin test results. Among them, are the negative binomial (Whitaker et al., 1972; Whitaker & Wiser, 1969; Knutti & Schlatter, 1978; Knutti & Schlatter, 1982), compound gamma (Knutti & Schlatter, 1978; Knutti & Schlatter, 1982; Giesbrecht & Whitaker, 1998), log normal (Giesbrecht & Whitaker, 1998; Brown, 1984), truncated normal (Giesbrecht & Whitaker, 1998), Waibel (Waibel, 1977), 3-parameter Weibull (Sharkey et al., 1994; Schatzki, 1995), exponential, chi-square (Tiemstra, 1969), logistic, and Neiman-A (Whitaker et al.,1972). Additionally, evaluations of several sampling plans to detect aflatoxin contamination have been done, and they have shown, with some differences due to plan characteristics and product to be sampled, that results obtained by sampling plans always involve a certain degree of uncertainty (Whitaker et al., 2005b).

Product	Aflatoxin b1 concentration reported (µg/kg)	Reference
Brazil nuts	4,000	Steiner et al. (1992)
Brazil nuts	25,000	Stoloff et al. (1969)
Pistachio nuts	1,400,000	Steiner et al. (1992)
Peanuts	1,100,00	Cucullu et al. (1966)
Maize	400,000	Shotwell et al. (1974)
Cottonseed	5,750,000	Cucullu et al. (1977)

Table 1. Concentration reported for individual grain or nut

Lot number	Aflatoxin analysis (µg/Kg)										Average
1	0	0	0	0	8	8	15	16	16	125	19
2	0	0	0	0	0	0	0	8	22	198	22
3	0	0	0	0	0	0	0	9	12	285	31
4	5	12	56	66	70	92	98	132	141	164	84
5	18	50	53	72	82	108	112	127	182	191	100
6	29	37	41	71	95	117	168	174	183	197	111

Source: Dickens and Whitaker (1986)

Table 2. Example of variation that can be observed among sample results when a peanut lot is sampled

Despite uneven contamination representing a problem for the task of sampling, it consists in an opportunity to segregate aflatoxin contaminated grains and nuts from an entire lot. As contamination is concentrated in few grains or nuts the removal of those material can to reduce the aflatoxin levels.

The fungal growth in grains and nuts is normally related to some changes in their bio-chemical and sometimes in the visual characteristics (Pomenranz, 1992; Wacowicz, 1991). Discoloration or staining of skin or kernel material, appearance of fluorescent material, changes in the standard of reflectance and transmittance spectroscopy, density and size changes in relation to sound grains and nuts are some characteristics that have been observed as consequence of fungal growth (Kumar & Agarwal, 1997; Pomeranz, 1992).

Some technologies able to detect and remove grains and nuts with the previously mentioned differences in their characteristics have been studied and used to improve the overall quality of commodities, but their efficiency to be used as a way to reach a reduction of aflatoxin levels in specific commodities must be evaluated. Table 3 shows some technologies which have been studied and used to segregate aflatoxin contamination in lots of commodities.

Technology	Product
Electronic color sorting	Grains and nuts
Hand picking	Nuts
Blanching and electronic color sorting	Peanuts
Gravimetric table	Grains and nuts
Size separation	Grains and nuts
Flotation	Maize and peanuts

Table 3. Examples of technologies studied to improve the overall quality of commodities or to reduce aflatoxin contamination levels of an entire lot

2. Segregation by appearance features

Fungal growth can cause chemical changes in grains or nuts, which can result in some modifications in color or form. The modifications are not always visible to the naked eye, some of them can be visible just with the aid of specific techniques or equipment. Color changes in grains or nuts can appear as a result of biochemical reactions or due to the fungal mycelium itself. According to Robin et al. (1995), hydrolysis of the macromolecules, e.g., proteins, lipids, and polysaccharides, occurs during mold infection, resulting in the release of free amino acids, free fatty acids, and simple sugars. These breakdown products contribute to color development in, e.g., peanut kernels during roasting of blanching before electronic color sorting.

The detection of fungal changes in grains and nuts makes it possible to know where fungal growth, and probable mycotoxin production, has occurred. As the presence of a fungus does not assure mycotoxin presence (Gloria et al., 2006), some researchers have tried to show correlations between changes in grains or nuts and their mycotoxin concentration. The correlation between poorly graded categories of grains and nuts and aflatoxin concentration has been shown for peanuts, maize, and almonds (Whitaker et al., 1998; Johansson et al., 2006; Whitaker et al., 2010).

The optical detection of faulty grains, nuts, kernel of nuts (blemished, discolored, and misshapen), and gross contaminants (glass, stones, insects, rotten product, extraneous vegetable material, etc.) has been carried out by visual color sorting (hand picking) or by an electronic color sorting (automatic sorting). Sorting of food products using the human eye and hand is still widely practiced where labour rates remain low. However, where the cost of labour has increased, automated techniques have been introduced (Bee and Honeywood, 2002).

There are several possible characteristics in the appearance which have been studied as indicators of aflatoxin presence. The BGYF (Bright Greenish-Yellow Fluorescence) was studied as an aflatoxin contamination indicator to maize (Shotwell & Hesseltine, 1981), pecans (Tyson and Clark, 1974), pistachio nuts (Dickens and Welty, 1975), dried figs (Steiner et al., 1988), and Brazil nuts (Steiner et al., 1992), as shown in Figure 1. The BGYF is produced by the oxidative action of heat-labile enzymes (peroxidases) in living plant tissue on kojic acid, which is produced by *A. flavus*. The method is not a definitive indicator of aflatoxin because it can produce false positive or negative results. False negative occurs when the aflatoxin contaminated maize grain does not present the fluorescent compound because peroxidase or kojic acid were not present to produce it. False positive occurs when contaminated maize sometimes does not exhibit BGYF, while kernels infected with *A. flavus* strains that produce kojic acid but do not produce aflatoxin exhibit BGYF, and thus are aflatoxin "false positives" when a maize grain is examined with a black light (Wilson, 1989; Wiclow, 1999). Hadavi (2005) studied the application of BGYF to segregate contaminated pistachio nuts, and concluded that the BGYF can be used to remove nuts with high aflatoxin level. Nowadays, BGYF is not currently used as a technique of decontamination of aflatoxin contaminated maize, it has been used as a technique for analyzing samples to detect aflatoxin contamination.

Fig. 1. Brazil nut kernels with Blue Greenish-Yellow Fluorescen (BGYF)

Other types of fluorescence have been studied as a way to indicate contaminated peanuts, almonds, and maize (Pelletier & Reizner, 1992; Shade and King, 1984; Yao et al., 2010). A device capable of measuring fluorescence intensities from peanut surfaces and physically rejected peanuts having undesired fluorescence properties was described (Pelletier et al., 1991), however a comparison of the efficiency between it and the color sorting process in peanuts lots showed that it was not effective as an aflatoxin decontamination technique (Pelletier & Reizner, 1992). Farsaie et al. (1981) developed an automatic sorter to remove fluorescent in-shell pistachio nuts, and an aflatoxin reduction by ca. 50% was reported. Steiner et al (1992) showed that fluorescence (yellow fluorescence) was a good indicator for aflatoxin contamination in kernels of Brazil nuts, but it was not good for in-shell pistachio nuts or kernels of pistachio nuts. For Brazil nuts, the hand picking segregation based on segregation of kernels with fluorescence has been used in Bolívia as an aflatoxin decontamination technique. Yao et al. (2010) reported good correlation between single kernel fluorescence hyperspectral data and aflatoxin concentration in maize.

Despite the fluorescent characteristic of grains and nuts being a possibility to segregate contaminated material, nowadays other color characteristics are used more often as an aflatoxin reduction technology. Color changes can be detected by the naked human eye or by optical systems using different technologies (Bee & Honeywood, 2002). Color sorting by the human eye and hand picking has been used as a feasible process to improve overall quality of nuts, mainly in some world regions where the cost of labour is sufficiently low to justify the economic feasibility of the process. For grains such as cereals, even in regions where the cost of labour is low, hand picking is not a feasible process. In spite of its higher cost in developed countries, hand picking is still used in certain cases to achieve a better removal of contaminated material and aflatoxin reduction, as happens to peanuts in the USA (Kabak et al., 2006).

The efficiency of color sorting to improve overall quality and also to reduce aflatoxin contamination depends on the product and the characteristics of the hand picking process or electronic sorter used. Electronic color sorting segregates grains or nuts with color off-standard in relation to a defined standard for sound grains and nuts which present low probability of aflatoxin contamination (Bee & Honeywood, 2002). Color sorting can be used alone or together with other processes such as blanching used for peanuts. Some reports on the performance of the electronic color sorting to reduce aflatoxin contamination have been published. Dickens and Whitaker (1975) showed that hand picking was more efficient to segregate aflatoxin contamination than electronic color sorting, as the latter also showed variable performance in aflatoxin reduction depending on the lot processed, however a great improvement in the optical technology occurred in the last thirty-five years, therefore nowadays it is correct to believe that color sorters have a better performance than before. Shade et al. (1975) also reported a better efficiency of the hand picking than the electronic color sorting to segregate aflatoxin contamination in almonds. Escher (1974) observed that color sorting was not successful in pecans because inherent intense fluorescence in the kernels. They investigated electronic color sorting and hand picking finish almonds products and they found contamination just in the electronic finish product. However, a great improvement in the optical technology occurred in the last thirty-five years, therefore nowadays is correct to believe that color sorters have a better performance than that time. Whitaker (1997) reported an evaluation of the performance of blanching and electronic color sorting process applied to 8911 contaminated peanut lots during the years of 1990 to 1994, as shown in table 4. The average reduction of aflatoxin contamination reported was of 89.9% and weight loss of 16.8%. Pearson (1996) reported a machine vision system to automatically segregate stained pistachio nuts which presented hulls with abnormal coloration, which is an indication of nuts with early splitting hulls. The early splitting pistachio nuts present higher probability to be contaminated with aflatoxin than the not stained nuts or nuts with closed hulls (Sommer et al., 1986). Two years later, Pearson & Shatzki (1998) reported an evaluation of this system and concluded that the sorter could be applied in the product recovery, and in the preparation of the product for very stringent markets. Visual sorting with hand picking based on color characteristics has been used for improvement of the overall quality of nuts, e.g. peanuts and shelled Brazil nuts in some processing plants in Brazil (Figures 2 and 3). Galvez et al. (2002) proposed a method to reduce aflatoxin in raw peanuts based on roasting, manual de-skimming and human sorting. The method was able to reduced aflatoxin of high and low contaminated samples. Campbell et al. (2003) observed that for walnuts the main commercial sorting used in the USA was based in color sorting to separate light colored shells (high value) from darker shells, and darker shells contained some shriveled and darkened kernels but until that time there was not information about the correlation of those types and aflatoxin content. De Mello & Scussel (2009) evaluated different types of sorting processes and concluded that color sorting for in-shell Brazil nuts did not show a safe segregation of contaminated nuts.

Crop Year	Lots processed	Aflatoxin contamination (µg/Kg)		Reduction (%)	Weight loss
		Before	After		
1990	5479	56.3	3.6	90.7	16.8
1991	669	36.6	2.5	92.0	14.7
1992	311	33.0	2.5	90.4	13.9
1993	1861	35.8	3.6	88.0	18.9
1994	591	31.0	3.4	86.6	14.1
Average/Total	8911	48.1	3.5	89.9	16.8

Source: Whitaker (1997)

Table 4. Aflatoxin reduction in contaminated peanut lots after blanching and electronic color sorting

Color sorting technology has shown several innovations over the last years which have improved the efficiency to remove poor quality grains, nuts and extraneous material (Bee & Honeywood, 2002). According to Wicklow & Pearson (2006), sorters used in the past had limited capacity to separate molded products because their optical system was based on mono-chromatic red-filters. However, the near-infrared is nowadays a feasible technology to be used in sorters, thus bi-chromatic color sorters have had their capability of detection extended beyond visible light, which made it possible to detect color and bio-chemical changes due to fungal growth. Sorters have used near-infrared transmittance (NIRT) and near-infrared reflectance (NIRR) spectroscopy to evaluate internal quality in many whole nuts.

Fig. 2. Hand picking of peanuts based on color and other characteristics of kernel

Fig. 3. Sorting of Brazil nuts kernels based on color and other characteristics of during hand shelling step

Some color sorters using those innovations were checked to evaluate the performance of aflatoxin segregation. Hirano et al. (1998) evaluated a method of transmittance near infra red to detected mouldy peanuts and could distinguished moudy from sound nuts by transmittance ration of 700 nm to 1100 nm. According to authors the trigrycerides hydrolysis caused by fungal growth was responsible for spectral differences. Pearson et al. (2001) evaluated transmittance spectra (500 to 950 nm) and reflectance spectra (550 to 1700 nm) to distinguish aflatoxin contamination in a single whole maize grain. More than 95% of maize grains were correctly classified as containing either high (>100 ppb) or low (< 10 ppb) levels of aflatoxin. Classification accuracy for kernels between 10 and 100 ppb was only about 25%, but according to researches these grains do not usually affect sample concentrations and are not as important. Pearson et al. (2004) evaluated a commercial sorter based on that technology and observed a reduction of 81% and 85% for aflatoxin and fumonisin B1, respectively.

3. Segregation by size features

Aflatoxin contamination has been related to smaller grains and nuts in commodities lots (Dorner et al. 1989; Whitaker et al., 2005; Schatzki & Pan, 1996). Besides the high correlation between aflatoxin and size, infected grains and nuts can be more friable than not infected ones (Shotwell et al., 1974), therefore the handling of the product can generate fragments, as shown in figure 3, of infected material, and it can contribute to the total aflatoxin level of the lot (Meinders & Hurburg, 1993; Piedade et al., 2002). Therefore, segregation by size has been studied as a way to remove aflatoxin contamination in commodities lots. Generally, size sorting is carried out by using sieves with holes that allow small grains or nuts to pass through, while retaining larger ones. The size sorting process can involve different sieves with decreasing hole sizes. The process is primarily used to categorize commodities by grains and nuts in size, where the largest categories are more valued in the market, and to clean the product to improve the overall quality of the lot. Thus, aflatoxin reduction by size sorting is normally a secondary result.

In spite of this, some data have been reported about the correlation between size and aflatoxin levels, and the effect of processes based on size segregation in the aflatoxin levels of processed lots of commodities. Brekke et al. (1975) evaluated cleaning procedures which remove broken kernels and foreign materials in white and yellow maize lots and they could not observe satisfactory aflatoxin reduction. Cole et al. (1995) reported that using a farmer stock peanut the sizing and electronic color sorting process were responsible by 29 and 70% of the aflatoxin reduction. Piedade et al. (2002) investigated the aflatoxin segregation when a sieve of 4.5 mm of round-holes was used to sieve maize samples and they observed that the largest grain fraction had lower average levels (84.8 µg/Kg) than the smallest one (204.0 µg/Kg). However, due to weight participation of each fraction in the total sample weight, the contribution of the smallest fraction was lower than the largest, so the segregation by size would not be able to reduce the aflatoxin leves in the whole sample. Meinders & Hurburg (1993) also detected a concentration in the aflatoxin levels as decreasing maize fractions from 6.3 to 1.8 mm were analyzed. Schatzki & Pan (1996) showed a positive relation between small pistachio nuts and aflatoxin levels. Whitaker et al. (2005) evaluated the aflatoxin distribution among peanut size categories using 46 peanut mini-lots. A negative correlation between size and aflatoxin content was observed. The shelled peanuts showed an average contamination of 75.3 µg/Kg, before the sorting, and after the sorting the six categories showed average contaminations of 42.5, 66.2, 93.6, 116.7, 105.1 and 133.6 µg/Kg. Only the two largest categories showed aflatoxin levels lower than the initial level.

Dowell et al. (1990) reported data about aflatoxin reduction when belt screen was used to screen unshelled peanuts to separate loose kernels and small pods. An average of 35% of reduction in the aflatoxin levels was observed when 17 lots were processed with belt screen. According to Dorner (2008), that type of device has been widely used by the USA peanut industry. Whitaker reported that the initial mean aflatoxin concentration of 73.7 µg/Kg was reduced to means of 42.5 and 66.2 µg/Kg in the large (named jumbo) and medium size categories of peanut, respectively, but was increased to 93.6, 105.1, and 133.6 µg/Kg in the smaller categories number one, sound split, and oil stock categories, respectively.

Fig. 4. Broken maize grains that can be remove by size sorting

4. Segregation by density

Grains and nuts, in which fungal growth and insect attack occurred, can present lower density than sound ones (Kabak et al., 2006). This characteristic has been used to separate poor quality material in commodities. In addition, the possibility to remove poor quality material brings the possibility to reduce aflatoxin contamination in food lots, because normally, the aflatoxin contamination is concentrated in poor quality material. Research on aflatoxin segregation by differences in density has been carried out, e.g., in maize, peanuts, and Brazil nuts.

Huff (1980) obtained 60% of aflatoxin levels when buoyant maize in water was removed. Sucrose solutions could improve the aflatoxin reduction to 90%, as the concentration of sucrose was increased up to 40%, but in this case 53% of maize was removed. Huff et al. (1982) also observed that flotation of maize in water and in 30% sucrose solution were efficient to segregate the aflatoxin contamination. Kirksey et al. (1989) studied the aflatoxin distribution in relation to peanut kernel density. They put 500 g of peanuts in 2000 mL of tapwater, and 15-30% of the kernel rose to the surface as buoyant kernels and they contained an average of 95% of total aflatoxin present in the samples. It was observed that kernels floated due to a hollow space inside them between cotyledons, which consisted in a reservoir of air to flotation, fungal growth, and aflatoxin production. Henderson et al. (1989) patented a procedure based on flotation of contaminated peanuts, but this procedure has not been widely used due to an additional drying step necessary after the flotation process. Gnanasekharan et al. (1992) found a negative correlation between aflatoxin content and density of peanut kernels, showing that kernels of low density have high probability to be contaminated. Steiner et al. (1992) reported that the weight of kernels in Brazil nuts evaluated was not a good indicator of aflatoxin contamination. Clavero et al. (1993) evaluated a method of flotation based on maize grain immersion in hydrogen peroxide. They observed a segregation of 90% in the initial aflatoxin contamination. The method was based on the catalase reaction with hydrogen peroxide. Clavero et al. (1993) demonstrated that *A. parasiticus* can produce catalase in peanut milk. Then, it was hypothesized that catalase produced by *A. parasiticus* would react with hydrogen peroxide and promote the formation of oxygen bubbles on the surface of the mold-infected kernels, causing their flotation.

5. Sampling procedures based on grain and nut types with high contamination probability

Grains and nuts, in which fungal growth and insect attack occurred, can present lower density than sound ones (Kabak et al., 2006). This characteristic has been used to separate poor quality material in commodities. In addition, the possibility to remove poor quality material brings the possibility to reduce aflatoxin contamination in food lots, because normally, the aflatoxin contamination is concentrated in poor quality material. Research on aflatoxin segregation by differences in density has been carried out, e.g., in maize, peanuts, and Brazil nuts.

Huff (1980) obtained 60% of aflatoxin levels when buoyant maize in water was removed. Sucrose solutions could improve the aflatoxin reduction to 90%, as the concentration of sucrose was increased up to 40%, but in this case 53% of maize was removed. Huff et al. (1982) also observed that flotation of maize in water and in 30% sucrose solution were efficient to segregate the aflatoxin contamination. Kirksey et al. (1989) studied the aflatoxin distribution in relation to peanut kernel density. They put 500 g of peanuts in 2000 mL of tapwater, and 15-30% of the kernel rose to the surface as buoyant kernels and they contained an average of 95% of total aflatoxin present in the samples. It was observed that kernels floated due to a hollow space inside them between cotyledons, which consisted in a reservoir of air to flotation, fungal growth, and aflatoxin production. Henderson et al. (1989) patented a procedure based on flotation of contaminated peanuts, but this procedure has not been widely used due to an additional drying step to be necessary after the flotation process. Gnanasekharan et al. (1992) found a negative correlation between aflatoxin content and density of peanut kernels, showing that kernels of low density have high probability to be contaminated. Steiner et al. (1992) reported that the weight of kernels in Brazil nuts evaluated was not a good indicator of aflatoxin contamination. Clavero et al. (1993) evaluated a method of flotation based on maize grain immersion in hydrogen peroxide. They observed a segregation of 90% in the initial aflatoxin contamination. The method was based on the catalase reaction with hydrogen peroxide. Clavero et al. (1993) demonstrated that *A. parasiticus* can produce catalase in peanut milk. Then, it was hypothesized that catalase produced by *A. parasiticus* would react with hydrogen peroxide and promote the formation of oxygen bubbles on the surface of the mold-infected kernels, causing their flotation.

In Brazil, the animal production industry, mainly poultry sector, has used gravimetric tables, a machine that segregate maize grains in high and low density fractions, to obtain mycotoxin segregation in maize. The high density fractions, which contain grains with low probability of mycotoxin contamination, is intended to make feeds for younger poultry, which are more susceptible to mycotoxin effects. The low density fraction, which has high probability of mycotoxin contamination, is intended to make feed for other poultry.

6. Sampling procedures based on grain and nut types with high contamination probability

The uneven distribution of aflatoxin contaminated grains and nut inside a lot normally represents a problem for measuring the true average level of aflatoxin contamination. However, some researchers have tried to take advantage of the distribution concentrated in few grains and nuts which can present different visual, optical, or physical characteristics in relation to sound ones that are not contaminated. From the analysis of samples containing only poor quality material, they have tried to improve the sampling plans efficiency to indicate lots which are above or under an established limit of acceptance for aflatoxin contamination.

Whitaker et al. (1998) studied the possibility to measure the aflatoxin contamination of farmers' stock peanuts by measuring the contamination in various peanut-grade categories. It was observed that best indicator for the aflatoxin concentration in the lot was the aflatoxin mass combined in the Loose Shelled Kernels (LSK), Damaged Kernels (DAM), and Other Kernels (OK). Whitaker et al. (1999) evaluated the performance of sampling plans based on the measurement of aflatoxin contamination in peanut-grade categories collected from a 2 Kg sample of the farmers' stock peanut lots, and establishing an acceptance limit of 50 µg/Kg. They observed that sampling plans based on combined mass of aflatoxin in LSK, DAM, and OK gave the best operating curve. Johansson et al. (2006) studied the possibility to predict aflatoxin in maize lots using poor-quality grade components. The aflatoxin mass combined in Damaged Kernels (DAM), and in Broken Kernel and Foreign Material was highly correlated with aflatoxin contamination in the lot, so they suggested that the measured aflatoxin mass combined with grade components could be used as a screening method to predict aflatoxin in maize lots.

Otherwise, Gloria et al. (2010) compare the performance of a sampling plan based on measuring the aflatoxin contamination in combined types of damaged grain maize, which was withdrawn from an 1 Kg sample of maize, with a sampling plan based on measuring aflatoxin in all types of grain (sound and damaged) in a sample test of ca. 5 Kg. The best operating curve was obtained by the sampling plan based on a 5 Kg test sample.

7. Conclusions

Several technologies for aflatoxin contamination segregation have been proposed in the scientific literature, but just some are currently used by the industry. Some processes have been used to improve the overall quality of commodities, and the reduction of aflatoxin is just a consequence and not the objective. The electronic color, in the visible or near infra-red wavelenghts, alone or combined with other technology of sorting, is the technology most widely used by the industry and which has shown a great improvement of modern optical possibilities and consequently improve aflatoxin remotion.

8. References

Ayres, J.L. (1977). Aflatoxins in Pecans: Problems and Solutions. *Journal of the American Oil Chemists' Society*, Vol. 54, No 3, pp A229-A230

Benford, D.; Leblanc, J.C. & Setzer, R.W. (2010). Application of the margen of exposure (MoE) approach to substances in food that are genotoxic and carcinogenic. Example: Aflatoxin B1 (AFB1). *Food Chemical Toxicology*, Vol. 48, pp 534-541

Brown, G.H. (1984). The distribution of total aflatoxin levels in composited samples of peanuts. *Food Technology in Australia*, Vol. 36, pp 128-130

Campbell, B.C.; Molyneux, R.J. & Shatzki, T. (2003). Current Research on Reducing Pre-and Post-harvested Aflatoxin Contamination of U.S. Almond, Pistachio, and Walnut. *Journal of Toxicology TOXIN REVIEWS*, Vol. 22, No.2 & 3, pp 225-266

Campbell, B.C.; Molyneus, R.J. & Shatzki, T. (2006). Advances in Reducing aflatoxin Contamination of U.S. Tree Nuts, In: *Aflatoxin and Food Safety*, Abbas, H.K. (Ed.), 483-516, CRC Press, Boca Raton, FL,USA

Clavero, M.R.S.; Hung, Y.; Beuchat, L.R & Nakayama, T. (1992). Catalase content in peanut milk as affected by growth by *Aspergillus parasiticus*. Journal of Food Protection, Vol 56, pp 55-57

Clavero, M.R.S; Hung, Y.C.; Beuchat, L.R.; Nakayama, T. (1993). Separation of aflatoxi-contaminated kernels from sound kernels by hydrogen peroxide treatment. *Journal of Food Protection*, Vol. 56, No.2, pp 130-136, 156

Cole, R.J.; Dorner, J.W. & Holbrook, C.C. (1995). Advances in mycotoxin elimination and resistance. In: Advances in Peanut Science, Pattee, H.E. & Stalker, H.T. (Eds.), 456-474, American Peanut Research Educational Society, Stillwater, OK, USA

Cucullu, A.F.; Lee, L.S.; Mayne, R.Y. & Goldblatt, L.A. (1966). Determination of aflatoxin in individual peanuts and peanut sections. *Journal of the American Oil Chemists' Society*, Vol. 43, No. 2, pp 89-92

Cucullu, A.F.; Lee, L.S. & Pons, W.A. Relationship of physical appearance of individual damaged cottonseed to aflatoxin content. *Journal of the American Oil Chemists' Society*, Vol. 54, No.3, pp A235-A237

De Mello, F.R. and Scussel, V. Development of Physical and Optical Methods for in-shell Brazil nuts Sorting and Aflatoxin. *Journal of Agricultural Science*, Vol. 1, No. 2, pp 3-14

Dickens, J. & Welty, R.J. (1975). Fluorescence in pistachio nuts contaminated with aflatoxin. *Journal of American Oil Chemists' Society*, Vol.52, pp 448-450

Dickens, J. & Whitaker, T.B. (1975). Efficacy of Electroni Color Sortin and Hand Picking to Remove Aflatoxin Contaminated Kernels from Comercial Lots of Shelled Peanuts. *Peanut Science*, Vol. 2, No.2, pp 45-50

Dickens, J.W. & Whitaker, T.B. (1986). Sampling and Sample Preparation Methods for Mycotoxin Analysis, In: *Modern Methods in the Analysis and Structural Elucidation of Mycotoxins*, Cole, J. (Ed.), 29-49, Academic Press, Inc., Oakland, FI, USA

Diener, U.L.; Cole, R.J.; Sanders, T.H.; Payne, G.; Lee, L.S. & Klich, M.A. (1987). Epidemiology of Aflatoxin formation by *Aspergillus flavus*. *Annual Review of Phytopathology*, Vol.25, No. , pp 249-270

Dorner, J.W. (2008). Management and prevention of mycotoxins in peanuts. *Food Additives and Contaminants*, Vol. 25, No. 2, pp 203-208.

Dowell, F.E.; Dorner, J.W.; Cole, R.J. & Davidson, J.I. (1990). Aflatoxin reduction by screening farmers stock peanuts. *Peanut Science*, Vol. 17, pp 6-8

Escher, F. (1974). Mycotoxin problems in the production and processing of peanuts and pecans in the USA. *Lebensmitt Wiss Technol*, Vol. 7, pp 255

Farsaie, A.; McClure, W.F. & Monroe, R.J. (1981). Design and development of an automatic electro-optical sorter for removing BGY fluorescent pistachio nuts. *Transactions of the ASAE*, Vol. 24, No.5, pp 1372-1375

Frisvald, J.C.; Skoube, P. & Samson, R.A. (2005). Taxonomic comparison of three different groups of aflatoxin producers and a new efficient producer of aflatoxin B1, sterigmatocystin and 3-O-methylsterigmatocystin, *Aspergillus rambellii* sp. nov.*Systematic and Applied Microbiology*, Vol. 28, pp 442-453

Fuller, G.; Spooncer, W.W.; King, A.D.; Shade, J. and Mackey, B. (1977). Survey of aflatoxins in California tree nuts. *Journal of the American Oil Chemists' Society*, Vol. 54, No.3, pp A231-A234

Galvez, F.C.; Francisco, M.L.D.L.; Lustre, A.O. & Resurrecion, A.V.A. (2002). *Controlo f Aflatoxin in raw peanuts through proper manual sorting*. United States Agency for International Development, Peanut Collaborative Research Support Program, Project 04 (USA and Philippines), Monograph Series No. 3, Retrieved from http://168.29.148.65/images/pdfs/reports/monograph3.PDF

Gloria, E.M.; Ciacco, C.F.; Lopes-Filho, J.F.; Ericsson, C. & Zocchi, S.S. (2006). Distribution of aflatoxin contamination in maize samples. *Ciência e Tecnologia de Alimentos*, Vol. 24, No. 1, PP 71-75

Gloria, E.M.; Janeiro, V.; Abdallah, M.F.I.; Borges, F.C.; Bertelli, M.; Gasparotto, B.; Tomazella, T.; Calori-Domingues, M.A. & Zocchi, S.S. (2010). Aflatoxin contamination estimate of corn truck loads based on different quality kernels. *Proceedings of 124th AOAC Annual Meeting & Exposition*, Orlando, FL, USA, September, 2010

Giesbrecht, F.G. & Whitaker, T.B. (1998). Investigations of the Problems of Assessing Aflatoxin Levels in Peanuts. *Biometrics,*Vol. 54, June, pp 739-753

Gnanasekharan, M.S.; Chinnan, M.S. & Dorner, J.W. (1992). Methods for Characterization of Kernel Density and Aflatoxin Levels of Individual Peanuts. *Peanut Science*, Vol.19, pp 24-28

Gürses, M. Mycoflora and Aflatoxin Content of Hazelnuts, Walnuts, Peanuts, Almonds and Roasted Chickpeas(LEBLEBI) Sold in Turkey. (2006). *International Journal of Food Properties*, Vol. 9, No. 3, pp 395-399

Hadavi, E. (2005). Several physical properties of aflatoxin contaminated pistachio nuts: Application of BGY fluorescence for separation of aflatoxin contaminated nuts. *Food Additives and Contaminants*, Vol. 22, No. 1, pp 1144-1153

Hirano, S.; Okawara,N. & Narazaki, S. (1998). Near Infra red Detection of Internally Mold Nuts. *Bioscince, Biochemistry, and Biotechnology*, Vol. 62, No.1, pp 102-107

Henderson, J.C.; Kreutcher, S.H.; Schmidt,A.A. & Hagen, W.R. (1989). Flotation separation of aflatoxin contaminated grain or nuts. U.S. Patent No. 4.795.651.

Huff., W.E. (1980). A physical method for the segregation of aflatoxin-contaminated corn. *Cereal Chemistry*, Vol. 57, No. 4, pp 236-238

Johansson, A.S.; Whitaker, T.B.; Hagler, W.M.; Jr, Bowman, D.T.; Slate, A.B. & Payne, G. (2006). Predictin Aflatoxin and Fumonisin in Shelled Corn Lots Using Poor-Quality Grade Components. *Journal of AOAC International*, Vol. 89, No.2, pp 433-440

Kabak, B.; Dobson, A.D.W. & Var, I. (2006) Strategies to Prevent Mycotoxin Contamination of Food and Animal Feed: A Review. *Critical Rewies in Food Science and Nutrition*, Vol. 46, pp 593-619

Knutti, R. & Schlatter, C. (1978). Problems of assessing aflatoxin in peanut-Proposal for a sampling and analysis plan for control of imports. *Mitteilungen aus dem Gebiete der Lebensmitteluntersuchung und Hygiene*

Knutti, R. & Schlatter, C. (1982). Distribution of aflatoxin in whole peanut kernels, sampling plans for small sample. *Zeitschrift für Lebensmitteluntersuchung und-Forschung A*, Vol. 174, No. 2, pp 122-128

Kumar, M. & Agarwal, V.K. (1997). Fungi detected from different types of seeds discoloration in maize. *Seed Research*, Vol. 25, pp 88-91

Meinders, B.L. & Hurburg Jr., C.R. (1993). Properties of corn screenings. *Transaction of the ASAE*, Vol. 36, pp 811-819

Molyneux, R.J.; Mahoney, N. & Campbell, B.C. (2007). Mycotoxins in edible tree nuts. *International Journal of Food Microbiology*, Vol. 119, No. , pp 72-78

Moss, M. Risk assessment for aflatoxins in foodstuffs. (2002). *International Biodeterioration & Biodegradation*, Vol. 50, No.3-4, pp 137-142

Van Egmond, H.P.; Schothorst, R.C. & Jonker, M.A. (2007). Regulations relating to mycotoxins in food.*Analytical and Bioanalytical Chemistry*, Vol. 389, pp 147-157

Ozay, G.; Seyhan, F.; Yilmaz, A. Whitaker, T.B.; Slate, A.B. & Giesbrecht, F. (2007). Sampling Hazelnuts for Aflatoxin: Effect of Sample Size and Accept/Reject Limit on Reducing the Risk of Misclassifying Lots. *Journal of AOAC International*, Vol. 90, No. 4, pp 1028-1035

Pearson, T. (1996). Machine vision system for automated detection of stained pistachio nuts. *Lebensmittel-Wissenschaft und-Technologie*, Vol. 28, No.6, pp 203-209

Pearson, T.C. & Schatzki, T.F. (1998). Machine Vision System for Automated Detection of Aflatoxin-Contaminated Pistachios. *Journal of Agricultural Food Chemistry*, Vol. 46, No. 6, pp 2248-2252

Pearson, T.C.; Wicklow, D.T.; Maghirang, E.B.; Xie, F. & Dowell, F.E. (2001). Detecting Aflatoxin in Single Corn Kernels by Transmittance and Reflectance Spectroscopy. *Transaction of the ASAE*, Vol. 44, No. 5, pp 1247-1254

Pearson, T.C.; Wicklow, D.T. and Pasikatan, M.C.(2004). Reduction of aflatoxin and fumonisin contamination in yellow corn by high-speed dual-wavelenght sorting. *Cereal Chemistry*, Vol. 81, pp 490-498

Pelletier, M. J.; Spetz, W. L. and Aultz, T. R. (1991).Fluorescence sorting instrument for the removal of aflatoxin from large numbers of peanuts. *Review of Scientific Instruments* , Vol.62, No.8, pp1926-1931

Piedade, F.; Fonseca, H.; Gloria, E.M.; Calori-Domingues, M.A.; Piedade, S.M.S. & Barbin, D. (2002). Distribution of Aflatoxins in contaminated corn fractions segregated by size. *Brazilian Journal of Microbiology*, Vol. 33, pp 12-16

Pomenranz, Y. (1992). Biochemical, funcitional, and nutritive changes during storage. In: *Storage of Cereal Grains and Their Products*, Sauer, D.B. (Ed.), 55-141, 4th ed., American Association of Cereal Chemists, St. Paul, Minesota, USA

Robin, Y; Chiou, Y; Wu, P.Y. & Yen, Y.H. (1995). Color sorting of lightly roasted and deskinned peanut kernels to diminish aflatoxin and retain the processing potency. *Developments in Food Science*, Vol. 37, pp 1533-1546

Schade, J. & King, Jr. A.D. (1984). Fluorescence and Aflatoxin Content of Individual Almond Kernels Naturally Contaminated with Aflatoxin. *Journal of Food Science*, Vol. 49, No. ,pp 493-497

Schade, J.E.; McGreevy, K.; King, Jr. A.D.; Mackey, B. & Fuller, G. (1975). Incidence of Aflatoxin in California Almonds. *Applied Microbiology*, Vol. 29, No. 1, pp 48-53.

Sharkey, A.J.; Roch, O.G. & Coker, R.D. (1994). A case-study on the development of a sampling and testing protocol for aflatoxin levels in edible nuts and oil-seeds.*The Statistician*, Vol. 43, No. 2, pp 267-275

Shatzki, T.F. (1995). Distribution of Aflatoxin in Pistachios. 2. Distribution in Freshly Harvested Pistachios. *Journal of Agricultural and Food Chemistry*, Vol.43, No.6, pp 1566-1569

Shatzki, T. & Pan, J. (1996). Distribution of Aflatoxin in Pistachios. 3. Distribution in Pistachio Process Streams. *Journal of Agricultural and Food Chemistry*, Vol. 44, No. 4, pp1076-1084

Shotwell, O.L.; Goulden, M.L. and Hesseltine, C.W. (1974). Aflatoxin: Distribution in contaminated corn. *Cereal Chemistry*, Vol. 51, pp 492-499

Shotwell, O.L. &Hesseltine, C.W. (1981). Use of bright greenish yellow fluorescence as a presumptive test for aflatoxin in corn. *Cereal Chemistry*, Vol.58, No.2, pp124-127

Sommer, N.F.; Buchanan, J.R. & Fortlage, R.J. (1986). Relation of early splitting and tattering of pistachio nuts to aflatoxin in the orchard. *Phytopathology*, Vol. 76, pp 692-694

Steiner, W.; Rieker, H. & Battaglia, R. (1988). Aflatoxin contamination in dried figs: distribution and association with fluorescence. *Journal of Agricultural and Food Chemistry*, Vol.36, No., pp 88-91

Steiner, W.E.; Brunschweiler, K.; Leimbacher, E. and Schneider, R. (1992). Aflatoxins and Fluorescence in Brazil Nuts and Pistachio Nut. *Journal of Agricultural and Food Chemistry*, Vol. 40, No.12, pp 2453-2457

Stoloff, L.; Campbell, A.D.; Becwith, A.C.; Neshiem, S.; Winbrush, J.S.Jr.; & Fordham, A.M. (1969). Sample preparation for aflatoxin assay. The nature of the problem and approaches to a solution. *Journal of the American Oil Chemists' Society*, Vol. 46, pp 678-684

Tiemstra, P.J. (1969). A study of the variability associated with sampling peanuts for aflatoxin. *Journal of the American Oil Chemists' Society*, Vol. 46, pp 667-672

Tyson, T.W. & Clark, R.L. (1974). An investigation of the fluorescent properties of aflatoxin infected pecans. *Transaction of the ASAE*, Vol. 17, No. 5, pp 942-945

Wacowicz, E. (1991). Changes of chemical grain components, especially lipids, during their deterioration by fungi. In: *Cereal Grain:Mycotoxins, Fungi and Quality in Drying and Storage*, Chelkowski, J. (Ed.), 259-280, Elsevier, Amsterdam, Netherlands

Waibel, Von J. (1977). Stichprobengrösse für die Bestimmung von Aflatoxin in Erdnüssen. *Deutsche Lebensmittel-Rundsschaw*, Vol 73, No. 11, pp 353-357

Whitaker, T.B. & Wiser, E.H. (1969). Theoretical Investigations Into the Accuracy of Sampling Shelled Peanuts for Aflatoxin. *Journal of the American Oil Chemists' Society*, Vol. 46, July , pp 377-379

Whitaker, T.B.; Dickens, J.W.; Monroe, R.J. & Wiser, E.H. (1972). Comparasion of the Observed Distribution of Aflatoxin in Shelled Peanuts to the Negative Binomial Distribution. *Journal of the American Oil Chemists' Society*, Vol. 9, No. 10, pp 590-593

Whitaker, T.B.; Giesbrecht, F.G.; Hagler, Wu,J.; Hagler, W.M.& Dowell,F.E. (1994). Predicting the Distribution of Aflatoxin Test Results from Farmers' Stock Peanuts. *Journal of Association of Official Analytical Chemists*, Vol.77, pp 659-666

Whitaker, T.B.; Giesbrecht, F.G. & Wu, J. (1996). Suitability of Several Statistical Models to Simulate Observed Distribution of Sample Test Results in Inspection of Aflatoxin-Contaminated Peanut Lots. *Journal of AOAC International*, Vol. 79, No. 4, pp 981-988

Whitaker, T.B. (1997). Efficiency of the Blanching and Electronic Color Sorting Process for Reducing Aflatoxin in Raw Shelled Peanuts. *Peanut Science*, Vol. 24, pp 62-66

Whitaker, T.B.; Hagler, W.M.Jr.; Giesbrech, F.G.; Domer, J.W.; Dowell, F.E. & Cole, R.F. (1998). Estimating Aflatoxin in Farmers' Stock Peanut Lots by Measuring Aflatoxin in Various Peanut-Grade Components.*Journal of AOAC International*, Vol 81, No. 1, pp 61-67

Whitaker, T.B.; Hagler, Jr. W.M. & Giesbrecht, F. (1999). Performance of Sampling Plans To Determine Aflatoxin in Farmers' Stock Peanut Lots by Measuring Aflatoxin in High-Risk-Grade Components. *Journal of AOAC International*, Vol 82, No.2, pp 264-270

Whitaker, T.B.; Dorner, J.W.; Lamb, M. & Slate, A.B. (2005).The effect of sortin farmers' Stock Peanut by Size and Color on Partitioning Aflatoxin int Various Shelled Peanut Grade Sizes. *Peanut Science*, Vol. 32, pp 103-118

Whitaker, T.B.; Slate, A.B. & Johansson, A.S. (2005b). Sampling feeds for mycotoxin analysis. In: *The Mycotoxin Blue Book*, Diaz, D.E. (Ed.), 1-23, Nottingham University Press, Thrumpton, Nottinghan, United Kingdon

Whitaker, T.B.; Slate, A.; Adams, T.B.J.; Jacobs, M. & Gray, G. (2010). Correlation Between Aflatoxin Contamination and Various USDA Grade Categories of Shelled Almonds. *Journal of AOAC International*, Vol 93, No. 3, pp 943

Wicklow, D.T. & Pearson, T.C. (2006). Detection and removal of single mycotoxin contaminated maize grains following harvest. *Proocedings of 9th International Working Conference on Stored Product Protection*, Campinas-SP, Brasil, October, 2006

Wicklow, D.T. (1999). Influence of Aspergillus flavus Strains on Aflatoxin and Bright Greenish Yellow Fluorescence of Corn Kernels. Vol. 83, No. 12, pp 1146-1148

Wild, C.P. & Gong, Y.Y. (2010). Mycotoxins and human disease: a largely ignored global health issue. *Carcinogenesis*, Vol. 31, No.1, pp 71-82

Wilson, D. (1989). Analytical method for aflatoxin in corn and peanuts. *Archives of Environmental Contamination and Toxicology*, Vol. 18, No. 3, 308-314

Yao, H.; Hruska, Z.; Kincaid, R.; Brown, R.; Cleveland, T. & Bhatnagar, D. (2010). Correlation and classification of single kernel fluorescen hyperspectral data with aflatoxin concentration in corn kernels inoculated with *Aspergillus flavus* spores. *Food Additives and Contaminants*, Vol. 27, No. 5, pp 701-709

5

Aflatoxins in Pet Foods:
A Risk to Special Consumers

Simone Aquino[1] and Benedito Corrêa[2]
[1]Universidade Nove de Julho/ UNINOVE, São Paulo
[2]Instituto de Ciências Biomédicas/ USP, São Paulo
Brazil

1. Introduction

Mycotoxin contamination in pet food poses a serious health threat to pets. Cereal grains and nuts are used as ingredients in commercial pet food for companion animals such as cats, dogs, birds, fish and rodents. Cereal by-products may be diverted to animal feed even though they can contain mycotoxins at concentrations greater than raw cereals due to processing (Moss, 1996; Brera et al., 2006). Several mycotoxin outbreaks in commercial pet food have been reported in the past few years (Garland and Reagor, 2001; Stenske et al., 2006).

Most outbreaks of pet mycotoxicosis, however, remain unpublished and may involve the death of hundreds of animals (MSNBC News Services, 2006). The term "companion animal" implies the existence of a strong human–animal bond between pets and their owners (Adams et al., 2004). A pet is often regarded as a family member by its owner and a person may develop strong relationships with animals throughout his or her lifetime. Pet interactions and ownership have been associated with both emotional and physical health benefits (Milani, 1996; Adams et al., 2004). The human–animal bond has resulted in over sixty four million American households in owning one or more pets in 2006, thereby creating a huge market for the pet food industry (APPMA, 2006). Dogs and cats continue to be the most popular pet to own, found in at least one out of three US households. The breakdown of pet ownership in the USA according to the 2009-2010 National Pet Owners Survey is above of a hundred millions of dogs and cats (Table 1).

Bird	15.0
Cat	93.6
Dog	77.5
Freshwater Fish	171.7
Saltwater Fish	11.2
Reptile	13.6
Small Animal	15.9

APPA's 2009/2010

Table 1. Total Number of Pets Owned in the U.S. (millions)

The health problems of pets, are therefore more of an emotional concern as compared to a mainly financial concern in farm animals (Dunn et al., 2005; Milani, 1996). In 2009, forty five billion was spent on pets in the U.S. Seventeen billion went to Pet food (Table 2).

Food	$17.56 billion
Supplies/OTC Medicine	$10.41 billion
Vet Care	$12.04 billion
Live animal purchases	$2.16 billion
Pet Services: grooming & boarding	$3.36 billion

APPA's 2009/2010

Table 2. Expenses on Pet care in 2009 in U.S. (billion)

1.1 Aflatoxins

Mycotoxins are secondary fungal metabolites that exert toxic effects on animals and human beings. Secondary fungal metabolites are not necessary for the growth or reproduction of the fungus. Not all fungi are capable of producing mycotoxins; those that can are referred to as toxigenic. The major aflatoxins (AFs) consist of aflatoxins B_1, B_2, G_1 and G_2 produced by certain toxigenic strains of *Aspergillus flavus, Aspergillus parasiticus,* and *Aspergillus nominus* (Richard, 2007; Puschner, 2002). Aflatoxin M_1 (AFM$_1$), a hydroxylated metabolite, found primarily in animal tissues and fluids (milk and urine) is a metabolic product of aflatoxin B_1 (AFB$_1$) and this mycotoxin is not found in feed grains. Aflatoxins can be present in milk of dairy cows, meat of swine or chicken eggs if the animals consume sufficient amounts in their feed (Robens and Richard, 1992).

Many toxigenic fungi produce mycotoxins only under specific environmental conditions. Grains stored under high moisture/humidity (>14%) at warm temperatures (>20°C) and/or inadequately dried can potentially become contaminated. Warm (air temperature of 24°C–35 °C) and humid (moisture content of substrate between 25% and 35%) conditions lead to extensive mold growth and aflatoxin production (Ominski et al., 1994 Puschner, 2002). The amount of water activity (a_w) is a measure of the amount of water available for bacterial and fungal growth. The pure water value is 1.0 and the decrease of a_w confers a protective result against toxigenic molds, since the minimum a_w permitting fungal germination and growth ranged from 0.80 to 0.82, according to Pitt and Miscamble (1995). Hunter (1969) proposed the value of 0.87 as the minimum required for aflatoxin production. Grains must be kept dry, free of damage and free of insects. Initial growth of fungi in grains can form sufficient moisture from metabolism to allow for further growth and mycotoxin formation. These conditions allow mold "hot spots" to occur in the stored grain. Traditionally, mycotoxin-producing fungi have been divided into two groups: "field" (plant pathogenic) and "storage" (saprophytic) fungi. Even though production can occur after harvest under inadequate storage conditions, large-scale contamination typically occurs in the field (Puschner, 2002).

Toxic secondary fungal metabolites may pose a significant risk to human and animal health if cereal grains and animal feed become colonized by toxigenic fungi. Aflatoxins have been found in many agricultural commodities but most commonly in corn, cottonseed, ground nuts, and tree nuts. The occurrence of a toxigenic fungus on a suitable substrate does not necessarily mean that a mycotoxin is also present (CAST, 1989).

Mycotoxicoses are reported in small animals. However, it is evident that there is little in the scientific literature on mycotoxicoses in pets (Puschner, 2002). An attempt is made to compile additional information on mycotoxins that have caused disease in small animals after experimental exposure. Aflatoxins, tricothecenes, tremorgens, and other mycotoxins are discussed in view of particular hazards and concerns for small animals, but undoubtedly, aflatoxins are the most documented of all mycotoxins (CAST, 1989).

Since the 1950s, there have been many reports and studies on aflatoxin metabolism, toxicity, residues, and species susceptibilities in domestic animals (mainly swine, cattle, and poultry). Research on the toxicity of aflatoxins in dogs began in the 1960s (Armbrecht et al., 1971; Chaffee et al., 1969; Newberne et al., 1966). Compared with other species, the effects of aflatoxins in dogs are less well documented, yet there are reports of aflatoxicosis in dogs after eating moldy food (Bailly et al., 1997, Ketterer et al., 1975) or contaminated grain (Bastianello et al., 1987).

1.2 Contaminated ingredients

Aflatoxins have elicited great public health concern because of their widespread occurrence in several dietary staples such as peanuts, tree nuts, corn, dried fruits, silage, and forages, all of which are used as animal feed ingredients. Monitoring these substrates for mycotoxins, especially AFB_1, is crucial to prevent outbreaks of acute mycotoxicosis and to diminish exposure risk of animals and humans to these harmful toxins (CAST, 2003, Pereyra et al., 2008).

Sharma and Márquez reported that the samples of cat and dog foods which had high amount of AFB_1, AFM_1 and AFP_1 and the ingredients were cheese, dry milk powder, oil seed meal, soya, cereals and rice, but maize was the main ingredient in all contaminated samples. Siame et al. (1998) reported that aflatoxin was the most common toxin detected in foods and feeds samples containing sorghum and maize. Scudamore et al. (1997) has presented that aflatoxin B_1 was the mycotoxin found most frequently in rice bran, maize products, palm kernels and cottonseeds. Aflatoxins are also commonly found in peanuts, raw milk and tree nuts (Haschek et al., 2002).

A total of 35 samples of pet food of 12 different trademarks, out of which, 19 samples were of dog and 16 were of cat foods with different ingredients and flavours were analysed by Sharma and Márquez (2001) in Mexico and amounts of aflatoxins (B_1, B_2, G_1, G_2, M_1, M_2, P_1 and aflatoxicol) were determinated in these samples and the presence of aflatoxins and aflatoxicol were observed in most of the samples. Aflatoxin B_1 was the mycotoxin found with higher frequency (0.885) and its level was higher in 17% of samples (of both dog and cat foods).

The authors described also that the highest level of AFB_1 were found in cat food of three different trademarks with concentration of 46.1, 30.8 and 22.2 ng/g. In case of dog food, two samples contained 39.7 ng/g and 27.0 ng/g of AFB_1. A higher incidence of AFM_1 was observed in three samples (21.37, 19.37 and 10.8 ng/g). AP_1 was found in one sample (12.52 ng/g) of dog food and other aflatoxins were found in traces. Two samples (one each cat food and dog food) contained a high concentration of total aflatoxins (72,4 and 59.7 ng/g).

According to Joint FAO/WHO Expert Committee on Food Additives (JEFCA) the tolerance limit is 20 ng/g to AFB_1, 50 ng/g of total aflatoxins in food and 0.5 ng/g of AFM_1 in milky (Bhat, 1999).

1.3 Pet food: How is it made?

Commercially prepared pet foods are an easy and economical way to fulfill the nutrient requirements in pets. These types of foods provide more than 90% of the calories consumed by pets in North America, Japan, Northern Europe, Australia, and New Zealand. Dogs, cats, hamsters, rabbits, birds, chinchillas and fishes are the main focus to pet food industry. There are three basic forms of commercial pet foods: dry, semi-moist, and moist or canned. The main difference in this categorization scheme is based on the water content of the food with dry foods containing usually less than 11% water, semi-moist foods containing 25 to 35% water, and moist or canned food containing 60 to 87% water (Zicker, 2008).

Most manufactured pet foods are formulated to meet specific nutrient goals to support growth, maintenance, or gestation/ lactation as recommended by the Association of American Feed Control Officials (AAFCO, 2007). The nutrients that are targeted include the calories, protein, fat, carbohydrate, vitamins, and minerals required to sustain life and, where possible, optimize performance (Zicker, 2008). Sorghum, maize, soya, rice, cereals, meal of meat and bones, by products of birds, fish, chicken, derived product of egg and milk were the main ingredients of pet food (Sharma and Márquez, 2001).

1.3.1 Dry food

Dry food is by far the major segment of the pet food industry attributable to its convenience to store and feed. Dry food particles are usually formed through a process called extrusion, which utilizes the same technology as that to produce breakfast cereals for people. Other methods include baking, flaking, pelleting, and crumbling of foods to achieve a dry form. Dry foods are protected against spoilage due to their low water content. To produce extruded foods, ingredients determined by the formulation are compounded and mixed homogeneously and then passed through an extruder. The extruder uses a combination of steam, pressure, and temperature to rapidly cook foods, then pushes the mixture through a faceplate where a revolving knife slices the extruded mix into the final kibble product. The extrusion process puts the ingredients through a temperature between 100 to 200°C and 34 to 37 atm pressure, which is high enough to effectively achieve a food sterilization process that meets industry standards The resultant extruded material has a moisture of approximately 25% before drying, where the final moisture content of 8 to 10% is attained. At this level of moisture mold formation is inhibited (Zicker, 2008; Crane et al., 2000; Miller and Cullor, 2000).

1.3.2 Canned food

Moist or canned foods historically comprised a much greater segment of the manufactured pet foods market but they have decreased in use. Moist foods are high in water content, usually 60 to 87%, and require the presence of gelling agents such as starch or gums to achieve their final consistency. Moist foods go through a process that results in a well sterilized final product similar to canned products for human consumption. Ingredients are mixed, ground together, and then cooked into a hot mixture for transfer to the can. The slurry is allotted into the cans and the top is sealed under steam, which displaces any air, resulting in an anaerobic environment. Finally the cans are sterilized in a machine called a retort where temperatures of 121°C are maintained for a minimum of 3 minutes (Zicker, 2008).

1.3.3 Semi-moist food

Semi-moist foods are a smaller but significant portion of the manufactured pet food market. Semi-moist foods require the use of humectants and acidification to control water content and inhibit mold growth. Semi-moist foods also have a low fiber content and relatively high sugar content, which make them highly palatable but also not an ideal choice to deliver weight control applications based on fiber. Semi-moist foods are manufactured in a way similar to extruded food but the water content is maintained at a higher level because of the added humectants. The final moisture content of 25 to 35% is more prone to mold and spoilage, which is mitigated by mold or bacterial inhibitors as well as managing the a_w component of the food. The addition of humectants helps to keep this at a low level of a_w, which effectively inhibits their growth despite higher total water content. It is apparent that much effort is put toward producing products that not only meet nutrient targets but that are also safe for their intended purposes. In addition to the care paid to details during the formulation and manufacturing process, companies maintain a quality control programs that further ensure safety and adequacy or products (Zicker, 2008).

Thermal inactivation is a good alternative for products that are usually heat processed. The processes described use high temperature (100 to 200 °C) that is an important physical factor to fungal control, because when heated to high temperatures, bacteria and fungi can be killed. However, the temperatures applied in the pet food processes are not enough to control the pre-formed aflatoxins in the ingredients. Mycotoxins are, in general, chemically and thermally stable, rendering them unsusceptible to commonly used feed manufacturing techniques (Kabak et al., 2006; Leung et al., 2006).

By-products commonly used in animal feeds (e.g. dried distillers grains and solubles) may also contain concentrated (i.e. higher) levels of mycotoxins relative to the grains (corn) they are derived from (Schaafsma et al., 2009). Aflatoxins are stable to moderately stable in most food processes. Aflatoxins are stable up to their melting point of around 250 °C and are not destroyed completely by boiling water, autoclaving, or a variety of food and feed processing procedures (Feuell 1966; Van Der Zijden et al., 1962).

1.4 Outbreaks of mycotoxicoses in pets

The effects of mycotoxins on companion animals are severe and can lead to death. As early as 1952, a case of hepatitis in dogs was directly linked to consumption of moldy food (Devegowda and Castaldo, 2000). A careful survey of the early outbreaks showed that they were associated with Brazilian peanut meal and the mycotoxin contaminated feed was groundnut cake. This outbreak occurred in the 1960 when more than 100,000 young turkeys, ducklings and young pheasants on poultry farms in England died in the course of a few months from an apparently new disease that was termed "Turkey X disease", because the cause of the disease was unknown. An intensive investigation of the suspect peanut meal was undertaken and it was quickly found that this peanut meal was highly toxic to poultry and ducklings with symptoms typical of Turkey X disease. Speculations made during 1960 regarding the nature of the toxin suggested that it might be of fungal origin. In fact, the toxin-producing fungus was identified as *Aspergillus flavus* and the toxin was given the name Aflatoxin. This discovery has led to a growing awareness of the potential hazards of these substances as contaminants of food and feed causing illness and even death in humans and other mammals (Bradburn and Blunden, 1994; Asao et al., 1963). Following the discovery of AF the agent responsible for the 1952 case was identified as AFB_1 (Newberne et al., 1966) and the symptoms of aflatoxicoses in dogs were also elucidated (Newberne et al., 1966; Ketterer et al., 1975).

Mycotoxins were detected in food for dogs, cats, birds, rodents and fishes with different prevalences across regions. Wild bird seed, for instance, has been found to be most contaminated among different pet food products (Henke et al., 2001). In the case study realized by Ketterer et al. (1975), three dogs on a farm in Queensland became ill (severe depression, anorexia, and weakness) and died at different times within a month following consumption of a commercial dog food mixed with AF-contaminated bread.

In 1998, 55 dogs died in Texas after eating dog food containing levels of aflatoxin that varied between 150 and 300 ppb (parts per billion). The corn in the diets was contaminated with aflatoxin (Bingham et al., 2004). Aflatoxins have been the most common cause of acute mycotoxin outbreaks in commercial dog food because corn is the usual source of aflatoxins in these cases. A commercial dog food with a high aflatoxin level was responsible for the acute deaths of 23 dogs in the United States in 2005 (Lightfoot and Yeager, 2008).

Pereyra et al. (2008) described an acute aflatoxicosis case on a chinchilla farm in Argentina. Chinchillas (*Chinchilla lanigera*) are rabbit-sized crepuscular rodents native to the Andes Mountains in South America. Chinchillas are farm raised and are currently used by the fur industry and as pets. Chinchillas are known to be very sensitive to mycotoxins, and a large number of animals often die if acute aflatoxicosis occurs. Clinical signs that may indicate mycotoxicosis on a farm include low feed intake, diarrhea, weight loss, poor condition of the skin, fur discoloration, sudden death, and a predisposition to secondary infections (Pereyra et al., 2008). In this case of chinchilla's farm the feed samples had undergone a pelleting process by an expander at 90°C for 60 min. This oat-based commercial feed was suspected to have caused the death of 200 animals.

The available reports of acute mycotoxicosis, however, cannot provide the whole picture of the mycotoxin problem associated with pet foods since only a small number of food poisoning cases are published. Veterinarians, furthermore, often overlooked mycotoxins as the cause of chronic diseases such as liver and kidney fibrosis, infections resulting from immunosuppression and cancer. These findings suggest that mycotoxin contamination in pet food poses a serious health threat to pet species. The public has recently begun a shift to organic pet foods. The public perception is that organic foods are safer due to the lack of pesticide residues. In the case of mycotoxins, however, the avoidance of insecticides and fungicides may result in increased crop pest damage, fungal growth and mycotoxin production (Boermans and Leung, 2007).

2. Mycotoxin risks assessment

"Risk assessment" is the systematic scientific characterization of potential adverse effects resulting from exposure to hazardous agents (NRC, 1993; Faustman and Omenn, 2001). Risk is the probability that a substance will produce a toxic effect. Risk involves two components: toxicity and exposure. Thus mycotoxins of relatively low toxicity may pose significant risks if exposure is great, frequent, and long. Conversely, mycotoxins of high toxicity, such as aflatoxins, may pose virtually no risk if exposure can be substantially reduced. "Exposure assessment" determines what type, levels, and duration of exposures are expected. Although the exposure of pet animals to mycotoxins in grain-based pet food is generally low, it is unavoidable and occurs throughout the entire life of the animal. The toxicity of a substance is dependent on its chemical, physical and biological properties. Often referred to as "hazard", toxicity is an inherent property of the compound and the animal being exposed (Faustman and Omenn, 2001).

The objectives of classical mammalian toxicity studies developed for risk assessment are as follows:

1. Hazard identification — determine the kinds of adverse effects
2. Dose–response assessment — determine the potency or sensitivity of effects
3. No observable adverse effect level (NOAEL) and lowest observed adverse effect level (LOAEL).

Toxicological information on mycotoxins is important because it allows us to judge the relative risk that may result from exposure to these toxic substances. Today's short-term repeat dose toxicity testing is used to derive symptoms as the main objective with mortality as a secondary objective (Paine and Marrs, 2000).

From this data, LD50 can be calculated as an indicator of acute response but LD50 alone gives little information on chronic response. Subchronic toxicity studies of 90 day exposures are used to determine the chemical dose an animal can consume daily without any demonstrable effect (NOAEL), and to characterize the effects of the chemical when administered at doses above the NOAEL. Chronic toxicity studies measure the effect of doses below the NOAEL on the normal life span of the animal. Chronic studies are often used to determine if the substance causes "delayed" effects on reproduction, development or cancer. One dose in chronic studies should cause subtle signs of toxicity such as reduced weight gain or a minor physiological response (LOAEL) (Boermans and Leung, 2007).

These classical toxicity studies using dosages in both effects and no effect range are designed to derive data to be applied to risk determination. Toxicity testing has made great strides, ever increasing our ability to detect sensitive toxic endpoints. Routine haematology, blood biochemistry, histology, and cytology are being supplemented by sophisticated diagnostic equipment including ultrasound imaging, magnetic resonance imaging (MRI), and electron microscopy. New technologies are extremely sensitive at detecting effects at sub-clinical dosages. Molecular techniques (e.g. DNA Microarray) detect alterations at the molecular level and help elucidate modes of action. Toxic endpoints should, however, have a level of clinical significance. What should be the most sensitive toxicological parameter may soon have to be answered (Boermans and Leung, 2007).

Most mycotoxin research has been designed to investigate toxic effects and therefore dosages used are in the toxic range. In such experiments the lowest experimental dose causing a toxic effect may be far greater than the threshold of the adverse effect and therefore overestimates the true LOAEL. Furthermore, if an experimental dose falls into the no-effect range, it may be far below the threshold dose and therefore greatly underestimates the true NOAEL. These factors introduce variability and uncertainty in the estimation of NOAEL and LOAEL (USEPA, 1995).

As for all risk assessments, pet health risk assessment requires data on toxicity and exposure. Pet species are seldom used for toxicity studies and therefore data obtained from other species are used in the risk assessment for pets. This results in a level of uncertainty when extrapolating toxicity data from experimental animals to pet species (Faustman and Omenn, 2001). The process of human health risk assessment (Covello and Merkhofer, 1993) can be applied to the risk determination in animals. In order to estimate the risk associated with mycotoxin exposure, we need to determine the dose a pet can consume in the food on a daily basis for their entire life with no adverse effect (i.e. NOAEL). This level can then be divided by an appropriate safety factor. Safety Factors (SF) (i.e. uncertainty factors) are numerical values applied to the NOAEL or other effect levels to account for any uncertainty in the data (Boermans and Leung, 2007).

Uncertainty includes species extrapolation, the nature and severity of effect, differences in pet breeds, and variability in LOAEL estimation. SF's could be adjusted according to epidemiological data on pets. Human SF numbers are often selected as a factor of 10, 100, or 1000 and set by the World Health Organization/Food and Agriculture Organization (WHO/FAO). Pet food SF's could be set by the pet food industry to apply standard safety guidelines. This process of combining the qualitative and quantitative aspects of toxicity and exposure to derive a quantitative level of risk is called "risk characterization" (Faustman and Omenn, 2001).

To complete the process a safe pet dietary level (SPDL) can be determined using a pet specific food factor (FF). The food factor is the amount of food consumed daily and accounts for the differences in the quantity of food consumed by different animal species. A safe pet dietary level (SPDL) would be equivalent to the human Maximal Permissible (tolerance) Level in foods (MPL). Risk management refers to the process by which policy actions are chosen to control hazards identified in the risk assessment/risk characterization processes (Faustman and Omenn, 2001; Covello and Merkhofer, 1993).

Calculation of SPDL would provide producers with a pet specific maximal permissible (tolerance) level of mycotoxin in food. Managers would consider scientific evidence and risk estimates along with processing, engineering, economic, social and political factors in evaluating alternative options (Boermans and Leung, 2007).

3. Mechanisms of toxicity in pets

The aflatoxins are primarily hepatotoxic or cause liver damage in animals; aflatoxin B_1 is the most toxic, followed by aflatoxins G_1, B_2, and G_2. Susceptibility varies with breed, species, age, dose, length of exposure and nutritional status. Aflatoxins may cause decreased production (milk, eggs, weight gains, etc.), are immunosuppressive, carcinogenic, teratogenic and mutagenic (Miller and Wilson, 1994). Aflatoxins are acutely toxic, carcinogenic, teratogenic, mutagenic, and immunosuppressive to most mammalian species. Animal species display differing degrees of susceptibility to aflatoxins, however, and it is now recognized that young animals are more susceptible. The primary clinical effects in aflatoxicosis are related to hepatic damage in all species studied (Boermans and Leung, 2007; Puschner, 2002; Plumlee, 2004).

After ingestion, aflatoxins are absorbed into the circulatory system, from which they are largely sequestered into the liver. Aflatoxins are then metabolized in the liver by microsomal mixed-function oxidases and cytosolic enzymes (Eaton et al., 1994a). The toxicity of aflatoxins is a result of the formation of the reactive aflatoxin B_1 8,9-epoxide, which binds covalently to cellular macromolecules such as DNA, RNA, and protein enzymes resulting in damage to liver cells (Cullen and Newberne, 1994).

Binding to these macromolecules results in adduct formation and is thought to ultimately result in damage to and necrosis of hepatocytes and other metabolically active cells. Typically, hepatocellular damage leads to impaired liver function, bile duct proliferation, bile stasis, and liver fibrosis. Epoxide formation may also occur in other tissues such as renal proximal tubular epithelium. Primary metabolites are further detoxified by conjugation with glutathione, glucuronic acid, amino acids, sulfate, or bile salts, and they are eliminated via feces and urine (Cullen and Newberne, 1994; Eaton et al., 1994b).

In addition to their hepatotoxic properties, aflatoxins are also carcinogenic. The binding of DNA causes genotoxicity and mutation in cells. Aflatoxin B_1 (AFB_1) has become an important model agent in the fields of experimental mutagenesis, carcinogenesis and biochemical and molecular epidemiology (Groopman et al., 1988). AFB_1 is metabolized by the microsomal mixed function oxygenase enzyme system (localized mainly on the endoplasmic reticulum of liver cells, but also present in kidney, lungs, skin, and other organs) to a variety of reduced and oxidized derivatives including an unstable reactive epoxide (Busby and Wogan, 1984). Epoxidation of the double bond of the terminal furan ring of AFB_1 results in AFBi-8,9-epoxide, which can form adducts with nucleophilic sites in DNA, primarily at the N [7] atom guanine, as well as reacting with RNA and protein (Croy et al., 1978; Neal et al., 1986).

In dogs, the most prominent clinical signs of aflatoxicosis are related to the impairment of liver function. In most reported cases, dogs either died suddenly or after a short clinical course. In addition to hepatitis and sudden death in dogs, symptoms of acute aflatoxicoses in both dogs and cats include vomiting, depression, polydipsea, and polyuria, weakness, anorexia, diarrhea, icterus, epistaxis, and petechiae on mucous membranes. It is thought that hemorrhagic diathesis secondary to protein synthesis inhibition and clotting factor deficiency is the cause of death in affected dogs (Hussein and Brasel, 2001; Puschner, 2002).

Necropsy observations revealed enlarged livers, disseminated intravascular coagulation, and internal hemorrhaging. In subacute aflatoxicosis (at 0.5–1 mg/kg of pet food over 2–3 week), dogs and cats become lethargic, anorexic, and jaundiced (Newberne et al., 1966). This can be followed by disseminated intravascular coagulation and death. The vomitus specimens from one dog contained high levels of AF (100 $\mu g/g$ of AFB_1 and 40 $\mu g/g$ of AFG_1). Death usually occurs in 3 days with LD50 levels ranging from 0.5 to 1.0 mg/kg in dogs and 0.3 to 0.6 mg/kg in cats depending on the age of the animal (Newberne et al., 1966).

The oral median lethal dose (LD50) of purified aflatoxin found by Cullen and Newberne (1994) in dogs was 0.80 mg/kg of body weight (BW). The authors reported the LD50 for cats as 0.55 mg/kg of BW. Dogs have developed acute, subacute, and chronic aflatoxicosis from batches of commercial dog food containing 0.1 to 0.3 mg/kg (ppm) of aflatoxin and after eating moldy bread with aflatoxin concentrations of 6.7 and 15 mg/kg (ppm), respectively (Bailly et al., 1997; Bastianello et al., 1987; Ketterer et al., 1975).

In acute aflatoxicosis, dogs exposed to > 0.5–1 mg aflatoxin/kg body weight (BW) typically die within days, showing enlarged livers, disseminated intravascular coagulation and internal hemorrhaging (Bohn and Razzai–Fazeli, 2005). Sub-acute aflatoxicosis (0.5–1 mg aflatoxin/kg pet food) is characterized by anorexia, lethargy, jaundice, intravascular coagulation and death in 2–3 weeks. Similar hepatotoxic effects can also be produced by chronic aflatoxin exposure with 0.05– 0.3 mg aflatoxin/kg pet food over 6–8 weeks. The chronic carcinogenic dose of aflatoxins is much lower than the acute dose. Newberne and Wogan (1968) have experimentally induced malignant tumors in rats with a continual exposure of <1 mg aflatoxin B_1/kg feed. Since aflatoxins are both acute and chronic hepatotoxins and carcinogens, the actual number of dogs affected by aflatoxins would be far more than the total number reported in acute poisoning cases (Boermans and Leung, 2007).

A 2-kg feed sample taken from a chinchilla farm located in the province of Córdoba, in the central region of Argentina, was analyzed. This study evaluated macroscopic and histologic changes in the livers of dead chinchillas. The authors reported that all chinchillas were kept under the same husbandry conditions on the farm. The hatchery had 200 animals that all received the same feed. All of these animals died naturally after the consumption of feed by an acute aflatoxicosis. Analyses of the pelletized feed for AFs by TLC revealed that the feed sample was contaminated at a mean level of 212 ppb ± 8.48 ppb of AFB_1 (Pereyra et al., 2008). Macroscopic inspection of the livers revealed general enlargement, pale-yellowish coloration, hypertrophy, rounded hepatic borders, and increased friability. Livers from chinchillas with aflatoxicosis were 38–71% larger than those from control animals. The color of the livers from chinchillas with acute aflatoxicosis was yellowish gray to pale yellow with gray spots (8 of 9 affected livers). Histopathology revealed severe, diffuse cytoplasmic vacuolation, with the appearance of many large and fewer small cytoplasmic vacuoles in hepatocytes in HE-stained tissue sections. Frozen sections of liver stained confirmed the presence of lipid within the cytoplasmic vacuoles (Pereyra et al., 2008).

The frequency of chromosomal aberrations in bone marrow cells, after a single i.p. aflatoxin B_1 (AFB$_1$) dose, was examined in male Chinese hamsters *(Cricetulus griseus)*. There was a significant increase in aberrant cells within 5 days of administration of a dose of 0.1 µg-5 mg AFB$_1$/kg, and on the 36th day. After a single dose of 5 mg AFB$_1$/kg the enhanced frequency of aberrant cells was monitored up to day 104 with no sign of a decrease to control level. The results indicate that the minimum mutagenic effect of an AFB$_1$ dose in this system is 0.1 µg/kg. Attention is drawn to the long-term presence of chromosomal aberrations even after a single i.p. exposure to AFB$_1$ (Bárta et al., 1990). According to Schmidt and Panciera (1980) aflatoxin caused primarily foetal growth retardation in hamsters and hepatic and renal necrosis occurred in the pregnant females.

Rabbit is considered as one of the most suitable and sensitive animal model for studying the teratogenic potential of a chemical (World Health Organization, 1993). Wangikar et al. (2005) showed that AFB$_1$ was found to be teratogenic in rabbits when given by oral route during gestation days 6–18 and the dose of 0.1 mg/kg could be considered as the minimum oral teratogenic dose. In this study the mean fetal weights were significantly reduced and the gross anomalies observed included wrist drop and enlarged eye socket whereas, skeletal anomalies were agenesis of caudal vertebrae, incomplete ossification of skull bones and bent metacarpals. The visceral anomalies of microphthalmia and cardiac defects were observed. The characteristic histological findings of fetal tissues were distortion of normal hepatic cord pattern and reduced megakaryocytes in liver, fusion of auriculo-ventricular valves, mild degenerative changes in myocardial fibers, microphthalmic eyes and lenticular degeneration. There was no dead fetus in any group.

Avian species are more susceptible than other affected species, such as dogs, cattle, swine, and humans, to aflatoxicosis (Robens and Richard, 1992). Aflatoxin and fusariotoxin are often responsible for avian mycotoxicosis. Clinical signs of chronic afalotoxicosis often include lethargy, weight loss, anorexia, regurgitation, and polydipsia (Degernes, 1995; Rauber, 2007). Mycotoxins are hepatotoxic and histologic changes include increased content of hepatic glycogen, portal infiltrate of monocytes, increased lipid droplet accumulation, hepatic necrosis and bile duct hyperplasia (Degernes, 1995; Ergün et al., 2006).

Changes in levels of specific neurotransmitters in the pons and brain stem have also been noted in some species (Yegani et al., 2006). The commercial product to birds are presented as a mixed of grains, that are more susceptible to fungal attack. Hepatic changes have been shown to occur in turkeys at levels as low as 100 to 400 ppb (Schweitzer et al. 2001). In the United States, the acceptable level of total aflatoxins in food for human consumption is less than 20 mg/kg, except for Aflatoxin M$_1$ in milk, which should be less than 0.5 mg/kg (Lightfoot and Yeager, 2008).

There are no reports about aflatoxicosis in aquarium small fishes. Aflatoxin contamination has been generally detected in fish farmed widely in the tropical and subtropical regions. Shi et al. (2010) studied tilapias that were fed six diets containing different levels of AFB$_1$ (19, 85, 245, 638, 793 and 1641 µg/kg), which were prepared with AFB$_1$-contaminated peanut meal. The results indicated that dietary AFB$_1$ led to aflatoxicosis effects in tilapia in a dose- and duration-dependent manner. No toxic effects of AFB$_1$ were found during the first 10 weeks, but by 20 weeks, the diet with 245 µg AFB$_1$/kg or higher doses reduced the growth and induced hepatic disorder, resulting in decreased lipid content, hepatosomatic index, cytochrome P450 A1 activity, elevated plasma alanine aminotransferase activity and abnormal hepatic morphology in these fishes.

The aflatoxin-treated Indian major carp (1.25 mg/kg body weight) revealed a reduction of total protein, globulin levels, bacterial agglutination titre, NBT and serum bactericidal activities, as well as an enhanced A:G ratio without change in albumin concentration. Thus, AFB_1 proved to be immunosuppressive to fishes even at the lowest dose of toxin treatment (Sahoo and Mukherjee, 2001).

4. Pet food regulation and recall

To ensure safety, pet foods and individual pet food ingredients are regulated by several governmental agencies in addition to meeting manufacturer's quality control and storage standards (Miller et al., 2000). Considering that the intrinsic toxicological properties of a chemical cannot be altered, regulatory agencies consider exposure mitigation the only meaningful opportunity for risk reduction (NRC, 1993). Government regulations of mycotoxin contamination, however, are often compromised by the analytical detection limits, regional prevalence, as well as trade relationships amongst different countries instead of fulfilling the scientific approach of risk assessment and safety determination (Leung et al., 2006).

Scientifically based regulations for the acceptable limit of mycotoxins in pet food would be beneficial. Strict regulations, however, would create greater competition with the human food chain resulting in increased pet food costs and decreased industry profits. It is also possible that the avoidance of severe regulations will promote mycotoxin outbreaks (Boermans and Leung, 2007). Safety and efficacy of foods intended for cats and dogs are of prime interest to manufacturers. Long-lived, healthy consumers (pets) contribute to greater sales, so breakdowns in product quality can have catastrophic effect on profits or even company viability. Recent problems with contamination, while affecting only a small percentage of commercial pet foods, impacted the entire pet food industry (Williard et al., 1994; Anonymous: FDA, 2005; Anonymous: FDA, 2007).

Such experiences have reaffirmed the need for manufacturers to devote extensive resources to documenting product quality. In many cases the processes already in place exceed the recognized standards within the industry. Nonetheless, most companies have increased the screening and sourcing control on ingredients used in pet foods. Regulatory standards are provided at several levels to ensure safety and adequacy of commercial products. In addition, the manufacture and regulation of pet foods is continually progressing forward, which should result in even more veterinary and consumer confidence in commercially manufactured foods (Zicker, 2008).

The FDA has action levels for aflatoxins regulating the levels and species to which contaminated feeds may be fed (CAST, 2003). The European Community levels are more restrictive (FAO Food and Nutrition Paper No. 81, 2004). In the United States, the FDA regulates foods and ingredients that are shipped across state or international boundaries under the authority of the Federal Food, Drug and Cosmetic Act (FFDCA) (Price et al. 1993; Van Houweling et al., 1977).

The FDA regulates enclose cat food, bag of dog food, box of dog treats or snacks. The FDA's regulation of pet food is similar to that for other animal feeds (Table 3). The FFDCA requires that pet foods, like human foods, be safe to eat, produced under sanitary conditions, contain no harmful substances, and be truthfully labeled. AFB_1 is the most toxic type and is regarded as the "sentinel" substance for all other aflatoxins. Aflatoxin control limit adopted in the US is 20 parts per billion for aflatoxin B_1 (Phillips, 2007). Harmonized regulations for aflatoxins exist in MERCOSUL, a trading block consisting of Argentina, Brazil, Paraguay and Uruguay. Worldwide limits for total aflatoxins in feed may vary (from 0.01 to 50 µg/kg), depending on the destination of the feedstuff as for dairy cattle, for example, that is 50 µg/kg. A relatively flat distribution is apparent with the most occurring limits set at 20 mg/kg (FAO, 2011).

Mycotoxin	Grain for human food		Grain for animal feed	
	USA [a]	EU [b]	USA [a]	EU [b]
Aflatoxins	20 ppb	2-4 ppb [c]	20-300 ppb [d]	10-50 ppb [d]

[a] Munkvold, 2003a

[b] Commission Regulation (EC) N° 1126/2007

[c] Varies among specific food items

[d] Varies among livestock species

Adapted from Schmale and Munkvold, 2011

Table 3. Recommendations and regulations for safe limits on mycotoxin concentrations in grain in the United States and European Union, as of 2008.

FDA ensures that the ingredients used in pet food are safe and have an appropriate function in the pet food (FDA, 2011). There is no requirement that pet food products have pre-market approval by the FDA. However, depending on the ingredient, the quality control steps may include testing for nutrient content, aflatoxins, or other contaminants that may pose safety risks. Careful testing of susceptible commodities for aflatoxins is necessary and the contaminated lots are eliminated. These standards must meet regulatory requirements for the particular industry standard. However, in many cases company quality control standards exceed the minimal regulatory requirement to insure safety and efficacy of product for dogs and cats. Specifically, in the United States, the FDA monitors pet food and individual pet food ingredients for pesticides, mycotoxins, and heavy metals as part of its Feed Contaminants Program (Van Houweling et al., 1977). For contaminants not covered by a tolerance, an action level, or advisory level, the limit remains unknown, although it is assumed to be theoretically at zero. In the present day analytical methods have become so sensitive that minuscule amounts of contaminants can be detected (Zicker, 2008).

Recalls are actions taken by a firm to remove a product from the market. Recalls may be conducted on a firm's own initiative, by FDA request, or by FDA order under statutory authority (FDA, 2011).

Class I recall: A situation in which there is a reasonable probability that the use of or exposure to a violative product will cause serious adverse health consequences or death.

Class II recall: A situation in which use of or exposure to a violative product may cause temporary or medically reversible adverse health consequences or where the probability of serious adverse health consequences is remote.

Class III recall: A situation in which use of or exposure to a violative product is not likely to cause adverse health consequences.

Market withdrawal: Occurs when a product has a minor violation that would not be subject to FDA legal action. The firm removes the product from the market or corrects the violation. For example, a product removed from the market due to tampering, without evidence of manufacturing or distribution problem, would be a market withdrawal.

Medical device safety alert: Issued in situations where a medical device may present an unreasonable risk of substantial harm. In some case, these situations also are considered recalls.

Miller and Cullor (2000) compared commercial pet foods with other sources of poisoning in dogs and cats. Food ranked well below drugs, insecticides, plants, rodenticides and cleaning products, in terms of frequency of occurrence. Only 1.7% of reported poisonings of dogs and cats could be attributed to food of any type. Despite these statistics, adverse signs in pets are very frequently blamed on the pet's food. Incidents of pet food contamination and illness still occur. In 2005, more than 75 dogs died in the United States after consuming pet food contaminated with aflatoxins, and hundreds more experienced severe liver problems associated with the intoxication (FDA, 2005).

The contaminated pet food was shipped to 22 different states and at least 29 different countries. Diamond Pet Food has discovered aflatoxins in many products manufactured in South Carolina and the problem was associated with the growth of the fungus *Aspergillus flavus*, on corn and other crops. The U.S Food and Drug Administration posted a recall on December 20, 2005, and nineteen different types of pet food (cats and dogs) produced at a single facility in Gaston, South Carolina were removed from sale. Sixteen batches of pet food were found to be contaminated with aflatoxins at levels greater than or equal to 20 ppb. The veterinarians were alarmed because this outbreak caused 100 dog deaths in weeks. The widespread panic that followed this tragic event motivated many pet food companies to set-up routine testing services for aflatoxins (Schmale III, D.G. and Munkvold, 2011, FDA, 2005).

In the end of 2010, the Kroger Company recalled pet food packages that could be contaminated with aflatoxins distributed in stores around many states in USA (Alabama, Arkansas, Georgia, Indiana, Illinois, Kentucky, Texas, Louisiana, Mississippi, Michigan, Missouri, North Carolina, Ohio, South Carolina, Tennessee, Virginia and West Virginia). The recall involved five different kinds of cat and dog foods. The Kroger Company advised the customers to consult veterinarians if their animals showed any signs of sluggishness or lethargy combined with reluctance to eat. Yellowish tints to the eyes or gums, severe blood or diarrhea were also included in the alert of warning signs divulged by industry.

A product recall may be the most effective means of containing the risk in a swift manner. Most commonly, pet food recalls are limited in scope (eg, a single manufacturer) and involve a quickly identified and understood contaminant (eg, *Salmonella* spp., mycotoxins). While recalls may be expensive to conduct, the potential repercussions of failure to honor the request in terms of legal liability and company/brand reputation may be much more costly in the long term. Veterinarians who suspect a case of pet food-borne illness should collect as much information on the food in question as feasible. In fact, a record of the dietary history of a sick animal is always prudent and may become important later if a pattern emerges or a notice of a recall is announced at a later date. Pertinent information may include the manufacturer's or distributor's name and address, the product and variety names, a description of the type of product, and any lot or date codes on the packaging. Effort should be made to determine the place and date where the food was obtained (Dzanis, 2008).

Pet food companies report that even minor changes to color, odor or texture of a pet food that have no bearing on safety are frequently reported to increase complaints to the companies' consumer relations department. Except for overt moldiness, obvious rancidity, or visible inclusion of foreign materials, most incidents of pet food contamination are unlikely to be apparent on gross inspection. Thus, collection of samples for laboratory analysis may be indicated when the food is suspect. Proper handling of the sample as legal evidence may be critical if there is a possibility of a lawsuit at a later date (Miller and Cullor, 2000).

In submission of pet food samples suspected of contamination, effort should be made to improve the chances of detecting the possible contaminant. Vague references to "look for poison" on a sample submission form does not give much assistance. A tentative diagnosis, or at least a thorough description of clinical signs and laboratory findings, may give clues to the facility running the analysis on the suspected food as to which contaminants are likely and hence which analyses to conduct (Dzanis, 2008).

5. Diagnostic and treatment

The diagnosis of mycotoxicosis is a common challenge for veterinarians, because the mycotoxin-induced disease syndromes can easily be confused with other diseases caused by pathogenic microorganisms. The liver is the primary target organ of acute injury from AF ingestion in all species. Although it is difficult to prove that a particular disease outbreak was caused by a mycotoxin (CAST, 2003).

A diagnosis of mycotoxicosis is usually made by feed analysis and histopathology because clinical signs of aflatoxicosis can be nonspecific and confusing. Histologic evaluation of the livers of affected animals and analysis of the feed for mycotoxin content are crucial to confirm the clinical diagnoses. Histopathology signs as bile-duct hyperplasia, hepatocellular degeneration, fatty change of hepatocytes, and mononuclear-cell infiltration of the hepatic parenchyma were observed in broiler chickens fed 1 ppm AFs (Eraslan et al., 2006; Ortatali and Oguz, 2001).

In a 2005 research study, broilers were fed a combination of AFs and fumonisins. The livers of affected birds were enlarged, yellowish, friable, and had rounded borders (Miazzo et al., 2005).

The HE-stained tissue sections were characterized by multifocal cytoplasmatic vacuolation, with a variable location within hepatic lobes. Hepatocellular damage manifested by marked cytoplasmic vacuolation and pyknotic nuclei was reported in a 2006 study of rats administered 2 mg/kg body weight of AFB_1 (Sakr et al., 2006).

Testing for mycotoxins in food and in the patient can be difficult because of variation in toxic concentration and the inconsistent production of toxins (LaBonde, 1995). A complete blood cell count, serum chemistry panel, and analysis of bile acids, ammonia, and urine help to rule out other causes of acute or chronic liver disease (e.g., infectious, neoplastic, chemical, drug-induced, congenital). Serum activity of hepatic enzymes (alanine aminotransferase, aspartate aminotransferase, and alkaline phosphatase) is usually elevated. Serum ammonia and bilirubin concentrations are often increased. If bleeding disorders are found on clinical examination, determination of coagulation times may be helpful. If an animal has died, macroscopic findings may include generalized icterus, liver damage, ascites, widespread hemorrhage, and edema of the gallbladder (Bastianello et al., 1987).

Histologically, varying degrees of liver damage are observed depending on the length of exposure to aflatoxins and their concentrations in the diet. Typical lesions in chronic and subacute cases are bile duct proliferation, varying degrees of fibrosis, hepatocellular fatty degeneration, and megalocytosis. Acutely poisoned dogs show massive fatty degeneration and centrilobular necrosis of the liver as well as widespread hemorrhage. In addition to liver lesions, renal proximal tubular necrosis is often present in dogs poisoned by aflatoxins. Confirmation of aflatoxicosis should include testing of the suspect feed source for aflatoxins (Trucksees and Wood, 1994).

Even if the feed is not visibly moldy, mycotoxins may be present. It is recommended to contact a veterinary diagnostic laboratory for sampling and shipping instructions. Some laboratories also offer testing of fresh liver for aflatoxin B_1. Additionally, a liver biopsy may be useful in ruling out other etiologies of liver disease (Puschner, 2002).

Treatment for hepatic dysfunction is symptomatic and supportive (e.g., fluids, B-complex vitamins, glucose). In many cases, lactated Ringer's solution supplemented with potassium (20 mEq/L) is administered as a maintenance solution. In cases with hypoalbuminemia, administration of dextrose is recommended. Aflatoxicosis resulting in severe hepatic failure may lead to a hypocoagulable status, requiring correction with frozen plasma or whole blood. No antidote is available. The prognosis depends on the extent and severity of liver dysfunction. Monitoring serum biochemical parameters may help to evaluate the extent of liver damage. If liver damage is extensive, the prognosis is guarded to poor. Ammoniation and certain adsorbents are effective in reducing or eliminating the effects of aflatoxins in animals (Park et al., 1988; Puschner, 2002).

While there is no specific treatment for mycotoxicosis, birds that are at high risk of exposure may benefit from supplementation with glucomannans and organic selenium, which appear to decrease the hepatotoxic and CNS changes associated with exposure (Ergün et al., 2006; Dvorska et al., 2007). The best way to protect pet birds from exposure to mycotoxins is to feed only human-grade grain, corn, and peanut products; avoid spoiled foods; and store grain products in cool, dry places (Lightfoot and Yeager, 2008).

6. Preventative strategies

Such experiences have reaffirmed the need for manufacturers to devote extensive resources to documenting product quality. In many cases the processes already in place exceed the recognized standards within the industry. Nonetheless, most companies have increased the screening and sourcing control on ingredients used in pet foods. Regulatory standards are provided at several levels to ensure safety and adequacy of commercial products. In addition, the manufacture and regulation of pet foods is continually progressing forward, which should result in even more veterinary and consumer confidence in commercially manufactured foods (Anonymous: FDA, 2005; 2008, 2011).

A control program for mycotoxins from field to table should involve the criteria of an Hazard Analysis Critical Control Point (HACCP) approach which will require an understanding of the important aspects of the interactions of the toxigenic fungi with crop plants, the on-farm production and harvest methods for crops, the production of livestock using grains and processed feeds, including diagnostic capabilities for mycotoxicoses, and to the development of processed foods for consumption as well as understanding the marketing and trade channels including storage and delivery of foods to the consumer. A good testing protocol for mycotoxins is necessary to manage all of the control points for finally being able to ensure a food supply free of toxic levels of mycotoxins (Richard, 2007). This system could be applied to prevent the risks of mycotoxins in animals by the pet food industry.

Conventional detection methods for AFB_1 require trained personnel, a laboratory environment, expensive equipment and often several hours or days in analytical time. Several commercial rapid test kits for use in determining the aflatoxin concentration are present in market. These test kits are self contained and thus no additional equipment is required. The kit system provide all the necessary instructions to complete an analysis and it also enables visual evaluation of the results of grains samples on farm or at buying point. It is possible to detect AFB_1 in cereals, nuts, spices and their derived products. Food samples are prepared for analysis by simply shaking the sample by hand in the presence of an extraction solution. However, the biggest challenge is the detection of minimum level of aflatoxin on feed or ingredients. But, a representative sample is essential, because aflatoxins can be concentrated in a few kernels that contaminate an entire load. A multi-level probe sampling at several sites and depths will give the best results. AOAC approved methods generally agree that an initial sample weight of 10 pounds (5 kilograms) is desirable (Byrne, 2008; Phillips, 2007).

Pet food amelioration is often considered a practical solution for mycotoxin contamination. Food processing techniques such as sieving, washing, pearling, ozonation, and acid-based mold inhibition can reduce the mycotoxin content of cereal grains. Dietary supplementation with large neutral amino acids, antioxidants, and omega-3 polyunsaturated fatty acids as well as inclusion of mycotoxin-sequestering agents and detoxifying microbes may ameliorate the harmful effects of mycotoxins in contaminated pet food. Amelioration of pet food, however, should be used as an additional safety factor but not to replace the sound application of risk and safety determination (Leung et al., 2006).

Sorption methods for the detoxification of aflatoxins are being studied and applied for the enterosorption and inactivation of aflatoxins in the gastrointestinal tract. Hydrated sodium calcium aluminosilicate (HSCAS) is a phyllosilicate clay commonly used as an anticaking agent in animal feeds. HSCAS tightly and selectively adsorbs aflatoxin and it has been shown to prevent the adverse effects of aflatoxins in various animals when included in the diet. Studies have also confirmed that HSCAS can alter the bioavailability of aflatoxin in dogs. HSCAS does not interfere with the utilization of vitamins and micronutrients in the diet and protects dogs fed diets with even minimal aflatoxin contamination. However, it does not protect animals against other mycotoxins. Despite regular and careful ingredient screening for aflatoxin, low concentrations may reach the final product undetected. Therefore, HSCAS may provide the petfood industry further assurance of canine diet safety (Bingham et al., 2004).

Bingham et al. (2004) realized a crossover study, using six dogs randomly fed a commercial dog food (no-clay control) or coated with HSCAS (0.5% by weight) were subsequently administered a sub-clinical dose of aflatoxin B_1. Diets were switched and the process repeated. The HSCAS-coated diet significantly reduced urinary aflatoxin M_1 by 48.4% (+/-16.6 SD) versus the control diet. It was demonstrated that HSCAS protected dogs fed diets with even minimal aflatoxin contamination. Despite regular and careful ingredient screening for aflatoxin, low concentrations may reach the final product undetected. Therefore, HSCAS may provide the pet food industry further assurance of canine diet safety.

7. Conclusion

It is known that mycotoxin contamination in pet food poses a serious health threat to pets and recent problems with contamination, while affecting only a small percentage of commercial pet foods, impacted the entire pet food industry, affecting the confidence of veterinarians and owners. Long-lived, healthy consumers (pets) contribute to greater sales, so breakdowns in product quality can have catastrophic effect on profits or even company viability. More research is needed to better address the pet mycotoxin problem. Safety and efficacy of foods intended for animals are of prime interest to manufacturers because the health problems of pets are of a highly emotional concern, besides the pet food safety is the responsibility of the pet food industry. In the other hand, pet owners must care to store the animal's food at home with regard to avoid fungal contamination, putting the open bags in a clean and dry place, with aeration and protected against humidity from environment. The shelf-life of commercial products must be observed, even at home.

8. References

Association of American Feed Control Officials –AAFCO (2007). Industry Statistics & Trends. In: *American Pet Products Association, Inc.,* Accessed january 09, 2011, Available from:
<http://www.americanpetproducts.org/press_industrytrends.asp>.

Adams, C.L., Conlon, P.D. & Long, K.C. (2004). Professional and veterinary competencies: addressing human relations and the human–animal bond in veterinary medicine. *Journal of Veterinary Medical Education,* Vol. 31, pp. 66–71.

Food and Drug Administration- FDA (2005). Diamond Pet Food Recalled Due to Aflatoxin. In: *Recall – Firm Press Release.* Accessed december 19, 2010, Available from:
<www.fda.gov/Safety/Recalls/ArchiveRecalls/2005>.

Food and Drug Administration- FDA (2007). Pet Food Recall (Melamine)/Tainted Animal Feed Updated. In: *Recalls & Withdrawals,* Accessed december 19, 2010, Available from: <http://www.fda.gov/oc/opacom/hottopics/petfood.html>.

American Pet Products Association Inc. – APPA (2011). Industry Statistics & Trends. In: 2011 Pet Products Trend Report. Accessed january 16, 2011, Available from: <http://www.americanpetproducts.org/press_industrytrends.asp 2009/2010>.

American Pet Products Manufacturers Association – APPMA (2006). National Pet Owners Survey 2005/2006. In: *APPMA News,* Accessed january 21, 2010, Available from: < http://www.americanpetproducts.org/newsletter/october2006/index.html>.

Armbrecht B.H., Geleta J.N., Shalkop W.T. & Durbin C.G. (1971). A subacute exposure of beagle dogs to aflatoxin. *Toxicology and Applied Pharmacology,* Vol. 18, pp. 579–85.

Asao, T., Buchi, G., Abdel-Kader, M.M., Chang, S.B., Wick, E.L. & Wogan, G.N. (1963). Aflatoxins B and G. *Journal of the American Chemical Society,* Vol. 85, pp. 1706–1707.

Bailly J.D., Raymond I., Le Bars P., Leclerc J.L., Le Bars J. & Guerre P. (1997). Aflatoxine canine: cas clinique et revue bibliographique. *Revue de Médecine Vétérinaire.* Vol. 148, pp. 907–14.

Bárta, M. Adámková, T. Petr & J. Bártová. (1990). Dose and time dependence of chromosomal aberration yields of bone marrow cells in male Chinese hamsters after a single i.p. injection of aflatoxin B_1. *Mutation Research,* Vol. 244, pp. 189-195.

Bastianello, S.S., Nesbit, J.W., Williams, M.C. & Lange, A.L. (1987). Pathological findings in a natural outbreak of aflatoxicosis in dogs. *Onderstepoort Journal of Veterinary Research,* Vol. 54, pp. 635–40.

Bhat, R. (1999). Mycotoxin contamination of food and feeds. Mycotoxin contamination of foods and feeds. *Proceedings of Working document of the Third Joint FAO/WHO/UNEP.* Tunis, Tunisia, March 1999.

Bingham A.K., Huebner H.J., Phillips T.D. & Bauer J.E (2004). Identification and reduction of urinary aflatoxin metabolites in dogs. *Food and Chemical Toxicology*, Vol. 42, 11, pp. 1851-1858.

Boermans, H. J. & Leung, M. C.K. (2007). Mycotoxins and the pet food industry: Toxicological evidence and risk assessment. *International Journal of Food Microbiology*, Vol. 19, pp. 95-102.

Bradburn, R.D.C. & Blunden, G. (1994). The aetiology of turkey "x" disease. *Phytochemistry*, 35, 3, pp. 817.

Brera, C., Catano, C., de Santis, B., Debegnach, F., de Giacomo, M., Pannunzi, E. & Miraglia, M. (2006). Effect of industrial processing on the distribution of aflatoxins and zearalenone in corn-milling fractions. *Journal of Agricultural and Food Chemistry*, Vol. 54, pp. 5014–5019.

Busby, W.F., & G.N. Wogan (1984). Aflatoxins, In: *Chemical Carcinogens*, C.D. Searle, pp. 945-1136, ACS Monograph 182, Washington, DC.

Byrne, J. Aflatoxin detection kit validated by USDA (2008). In: *Quality and Safety*, Accessed december 23, 2010, Available from: <http://www.foodproductiondaily.com/Quality-Safety/Aflatoxin-detection-kit-validated-by-USDA>.

CAST - Council for Agricultural Science and Technology (2003). Mycotoxins: risks in plants animal and human systems. *Task Force Report 139*. Ames.

CAST - Council for Agricultural Science and Technology (1989). Mycotoxins: economic and health risks. *Task Force Report 116*. Ames, pp. 1–9.

Chaffee V.W., Edds G.T., Himes J.A. & Neal F.C. (1969). Aflatoxicosis in dogs. *American Journal of Veterinary Research,* Vol. 30, pp. 1737–49.

Covello, V.T. & Merkhofer, M.W. (1993). *Risk Assessment Methods Approaches for Assessing Health and Environmental Risks*. pp. 318, Plenum, New York.

Cowell, C. S., Stout, N. P., Brinkmann, M. F., Moser, E. A. & Crane, S. W. (2000). Making of commercial pet foods, In: *Small Animal Clinical Nutrition*, Hand, M. S., Thatcher, C. D., Remillard & R. L., Roudebush, pp. 127–146, Walsworth Publishing Company, Marceline, USA.

Cable News Network- CNN (2010). Kroger announces select pet food recall. In: *Pet Food*. Accessed january 14, 2011, Available from: <http://articles.cnn.com/2010-12-19/us/pet.food.recall_1_pet-food-aflatoxin-toxic-chemical-byproduct?_s=PM:US>.

Crane, S.W., Griffin, R.W. & Messent, P.R. (2000). Introduction to commercial pet foods, In: *Small Animal Clinical Nutrition*, Hand, M.S., Thatcher, C.D., Remillard, R.L. & Roudebush, P., pp. 111-126, Walsworth Publishing Company, Marceline, USA.

Croy, R.G., Essigman, J.M., Reinhold, V.N. & Wogan, G.N. (1978). Identification of the principal aflatoxin BrDNA adduct formed in vivo in rat liver, *Proceedings of the National Academy of Sciences,* Vol. 75.

Cullen J.M. & Newberne, P.M. (1994). Acute hepatotoxicity of aflatoxins, In: *The toxicology of aflatoxins: human health, veterinary and agricultural significance*. Eaton, D.L. & Groopman, J.D., pp. 3–26, Academic Press, San Diego, USA.

Degernes, L.A. (1995).Toxicities in waterfowl. *Seminars in Avian and Exotic Pet Medicine*, Vol. 4, (1), pp. 15–22.

Devegowda, G. & Castaldo, D. (2000). Mycotoxins: hidden killers in pet foods. Is there a solution? In: *Technical Symposium on Mycotoxins*. Alltech Inc, Nicholasville, Kentucky, USA.

Dog Food Advisor. In: Kroger Dog Food Recalled for Aflatoxin Accessed january 25, 2011, Available from: <http://www.dogfoodadvisor.com/dog-food-recall/kroger-dog-food-recall/>.

Dunn, K.L., Mehler, S.J. & Greenberg, H.S. (2005). Social work with a pet loss support group in a university veterinary hospital. *Social Work in Health Care*, Vol. 41, pp. 159–170.

Dvorska, J.E., Pappas, A.C., Karadas, F., Speake, B.K. & Surai, P.F. (2007). Protective effect of modified glucomannans and organic selenium against antioxidant depletion in the chicken liver due to T-2 toxincontaminated feed consumption. *Comparative Biochemistry and Physiology. Part C, Pharmacology, Toxicology and Endocrinology*, Vol. 145, (4), pp. 582–7.

Dzanis, D.A. (2008). Anatomy of a Recall. *Topical Review*, 23, 3, pp. 133 – 136.

Eaton DL, Ramsdell HS, Neal GE. (1994a). Biotransformation of aflatoxins. In: *The toxicology of aflatoxins: human health, veterinary and agricultural significance*, Eaton, D.L., Groopman, J.D., pp. 45–47, Academic Press, San Diego, USA.

Eaton, D.L. & Gallagher, E.P. (1994b). Mechanisms of aflatoxin carcinogenesis. *Annual Review of Pharmacology and Toxicology*, Vol. 34, pp. 135–72.

Eraslan, G., Essiz, D., Akdogan, M., E. Karaoz, M. Oncu & Z. Ozyildiz (2006). Efficacy of dietary sodium bentonite against subchronic exposure to dietary aflatoxin in broilers. *Bulletin of the Veterinary Institute in Pulawy*, Vol. 50, pp. 107-112.

Ergün, E., Ergün, L. & Esxsiz, D. (2006). Light and electron microscopic studies on liver histology in chicks fed aflatoxin. Dtsch Tierarztl Wochenschr, 113, 10, pp. 363–368.

FAO Food and Nutrition. (2004). Worldwide Regulations for Mycotoxins in Food and Feed in 2003, Paper. n° 81, pp. 180.

Faustman, E.M. & Omenn, G.S. (2001). Risk assessment. In: *Casarett and Doull's Toxicology: The Basic Science of Poisons*, Klaassen, C.D., pp. 83–104, McGraw-Hill, USA.

Food and Drug Administration – FDA (2005). Center for Food Safety and Applied Nutrition. In: *Diamond Pet Food Recalled Due to Aflatoxin*. Accessed january 14, 2011, Available from: <http://www.fda.gov/Safety/Recalls/ArchiveRecalls/2005/ucm111929.htm>.

Food and Drug Administration – FDA (2009). Center for Food Safety and Applied Nutrition. In: *Interim melamine and analogues safety/risk assessment*. Accessed january 14, 2011, Available from: <http://www.cfsan.fda.gov/_dms/melamra.html.>.

Federal Food, Drug and Cosmetic Act, as Amended. (2004). Superintendent of Documents. Washington, DC: U.S. Government Printing Office.

Feuell, A.J. (1966). Aflatoxin In groundnuts. Problems of detoxification. *Tropical Science*, Vol. 8, pp. 61-70.

Garland, T. & Reagor, J. (2001). Chronic canine aflatoxicosis and management of an epidemic, In: *Mycotoxins and Phycotoxins in Perspective at the Turn of the Millennium*, deKoe,W., Samson, R., van Egmond, H., Gilbert, J., Sabino, M., pp. 231–236, Ponsen & Looven, Wageningen, Netherlands.

Groopman, J.D., Cain, L.G. & Krusler, C.C. (1988). Aflatoxin exposure in human populations: measurements and relationship to cancer, *CRC Critical Reviews in Toxicology*, Vol. 19, pp. 113-145.

Haschek, W.M., Voss, K.A. & Beasley, V.R. (2002). In: *Selected Mycotoxins Affecting Animal and Human Health, Handbook of Toxicologic Pathology* (Second Edition), pp. 645-699, Academic Press, London.

Henke, S.E., Gallardo, V.C., Martinez, B. & Balley, R. (2001). Survey of aflatoxin concentrations in wild bird seed purchased in Texas. *J Wildl Dis*, 37, pp. 831–835.

Hooser, S.B. & Talcott, P.A. (2006). In: *Mycotoxins in Small Animal Toxicology* (Second Edition), Peterson, M.E. & Talcott, P.A., pp. 888-897, Elsevier Inc, USA.

Hunter, J.H. (1969). Growth and aflatoxin production in shelled corn by the *Aspergillus flavus* group as related to relative humidity and temperature. Ph.D. thesis. Purdue University, West Lafayette, IN.

Hussein, S.H. & Brasel, J.M. (2001). Review: Toxicity, metabolism, and impact of mycotoxins on humans and animals. *Toxicology*, Vol. 167, pp. 101–134.

Kabak, B., Dobson, A.D., Var, I. (2006). Strategies to prevent mycotoxin contamination of food and animal feed: a review. *Food Science and Nutrition*, Vol. 46, pp. 593–619.

Ketterer, P.J., Williams, E.S, Blaney, B.J. & Connolle, M.D. (1975). Canine aflatoxicosis. *Australian Veterinary Journal*, Vol. 51, pp. 355–357.

LaBonde, J. (1995). Toxicity in pet avian patients. *Seminars in Avian and Exotic Pet Medicine*, Vol. 4 (1), pp. 23–31.

Leung, M.C.K., Díaz-Llano, G. & Smith, T.K. (2006). Mycotoxins in Pet Food: A Review on Worldwide Prevalence and Preventative Strategies. *Journal of Agricultural and Food Chemistry*, Vol. 54, 26, pp. 9623–9635.

Lightfoot, T.L. & Yeager, J.M. (2008). Pet Bird Toxicity and Related Environmental Concerns., Vol. *Veterinary Clinics of North America: Exotic Animal Practice*, Vol. 11, pp. 245–246.

Miazzo, R., Peralta, M.F., Magnoli, C., Salvano, M., Ferrero, S. & Chiacchiera, S.M. (2005). Efficacy of sodium bentonite as a detoxifier of broiler feed contaminated with aflatoxin and fumonisin. *Poultry Science*, Vol. 84, pp. 1-8.

Milani, M.M. (1996). The importance of the human–animal bond. *Journal of the American Veterinary Medical Association*, Vol. 209, pp. 1064–1065.

Miller, D.M. & Wilson, D.M. (1994). Veterinary diseases related to aflatoxin. In: *The Toxicology of Aflatoxins: Human Health, Veterinary, and Agricultural Significance*. Eaton, D.L. & Groopman, J.D., pp. 347–364, Academic Press, San Diego, USA.

Miller, E.P. & Cullor, J.S. (2000). Food safety. In: *Small Animal Clinical Nutrition*, Hand, M.S., Thatcher, C.D., Remillard, R.L., pp. 183-198, Topeka, KS, Mark Morris Institute.

Moss, M.O. (1996). Mycotoxic fungi, In: *Microbial Food Poisoning*, Eley, A.R., pp. 75-93, Chapman & Hall, New York, USA.

MSNBS News Services (2006). Toxic pet food may have killed dozens of dogs. In: *Pet Health*. Accessed january 23, 2011, Available from:
    <http:// www.msnbc.msn.com/id/10771943/>.

Munkvold, G.P. (2003). Cultural and genetic approaches to managing mycotoxins in maize. *Annual Review of Phytopathology*,Vol. 41, pp. 99–11.

Neal, G.E., Judah, D.J. & Green, J.A. (1986). The in vitro metabolism of aflatoxin B_1 catalysed by hepatic microsomes isolated from control or 3-methylcholanthrene stimulated rats and quail, *Toxicology and Applied Pharmacology*, Vol. 82, pp. 454-460.

Newberne, P.M., Wogan, G.N. (1968). Sequential morphologic changes in aflatoxin B carcinogenesis in the rat. *Cancer Research*, Vol. 28, pp. 700–781.

Newberne, P.M., Russo, R. & Wogan, G.N. (1966). Acute toxicity of aflatoxin B_1 in the dog. *Veterinary Pathology*, Vol. 3, pp. 331–340.

National Research Council (NRC), 1993. Issues in Risk Assessment. National Academy Press, Washington, DC, USA.

Ominski, K.H., Marquardt, R.R., Sinah, R.N & Abramson, D. (1994). Ecological aspects of growth and mycotoxin production by storage fungi. In: *Mycotoxins in Grain-compounds other than Aflatoxin*, Miller, J.D. & Trenholm, H.L., pp. 287–312, Eagan Press, St. Paul.

Ortatali M. & Oguz H. (2001). Ameliorative effects of dietary clinoptilolite on pathological changes in broiler chickens during aflatoxicosis. *Research in Veterinary Science*, Vol. 71, pp. 59-66.

Paine, A.J., Marrs, T.C. (2000). The design of toxicological studies, In: *General and Applied Toxicology*, Ballantyne, B., Marrs, T., Syversen, T., pp. 321–334, MacMillan References Texts Ltd, London, United Kingdom.

Park, D.L., Lee, L.S. & Pohland, A.E. (1988). Review of the decontamination of afltoxins by ammoniation: current status and regulation. *Journal Association of Official Analytical Chemists*, Vol. 71, pp. 685–703.

Pereyra, M.L.G., Carvalho, E.C.Q., Tissera, J.L., Keller, K.M., Magnoli, C.E., Rosa, C.A.R., Dalcero, A.M. & Cavaglieri, L.R. (2008). An outbreak of acute aflatoxicosis on a chinchilla (*Chinchilla lanigera*) farm in Argentina. Case Reports. *Journal of Veterinary Diagnostic Investigation*, Vol. 20, 6, pp. 853-856.

Phillips, T. (2010). Aflatoxin insurance, In: *A clay called HSCAS can help petfood manufacturers avoid a crisis*. Accessed january 11, 2011, Available from:
  < http://www.petfoodindustry.com/4019.html>.

Pitt, J.I. & Miscamble, B.F. (1995). Water relations of *Aspergillus flavus* and closely related species. *Journal of Food Protection*, Vol. 58, pp. 86–90.

Plumlee, K.H. (2004). Chapter 23 – Mycotoxins. *Clinical Veterinary Toxicology*, pp. 231-281. Mosby Inc., St. Louis.

Price, W.D., Lovell, R.A. & McChesney, D.G. (1993). Naturally occurring toxins in feedstuffs: Center for Veterinary Medicine Perspective. *Journal of Animal Science*, Vol. 71, pp. 2556-2562.

Puschner, B. (2002). Mycotoxins. *Veterinary Clinics of North America: Small Animal Practice*, Vol. 32, pp. 409–419.

Rauber, R.H., Dilkin, P., Giacomini, L. Z., Araújo de Almeida, C. A. & Mallmann, C. A. (2007). Performance of turkey poults fed different doses of aflatoxins in the diet. *Poult Science*, Vol. 86, 8, pp. 1620–1624.

Razzai–Fazeli, E., 2005. Effects of mycotoxins on domestic pet species. In: *The Mycotoxin Blue Book*. Diaz, D., pp. , 77–92, Nottingham University Press, Nottingham, UK.

Richard, J.L. (2007). Some major mycotoxins and their mycotoxicoses—An overview. *International Journal of Food Microbiology*, Vol. 119, pp. 3–10.

Robens, J.F. & Richard, J.L. (1992). Aflatoxins in animal and human health. *Reviews of Environmental Contamination & Toxicology*, pp. 127.

Sahoo, P.K. & Mukherjee, S.C. (2001). Immunosuppressive effects of aflatoxin B_1 in Indian major carp (*Labeo rohita*). *Comparative Immunology, Microbiology & Infectious Diseases*, Vol. 24, (3), pp. 143-149.

Sakr, S.A., Lamfon, H.A. & El-Abd, S.F. (2006). Ameliorative effects of Lupinus seeds on histopathological and biochemical changes induced by aflatoxin-B_1 in rat liver. *Journal of Applied Science Research*, Vol. 2, pp. 290-295.

Schaafsma, A.W., Limay-Rios, V., Paul, D.E. & Miller, J.D. (2009). Mycotoxins in fuel ethanol co-products derived from maize: a mass balance for deoxynivalenol. *Journal of the Science of Food and Agriculture*, Vol. 89, pp. 1574–1580.

Schmale III, D.G. & Munkvold, G.P., (2009). Mycotoxins in Crops: A Threat to Human and Domestic Animal Health. In: *The Plant Health Instructor*, Accessed january 14, 2011, Available from:
  <http://www.apsnet.org/edcenter/intropp/topics/Mycotoxins/Pages/Reference s.aspx, The American Phytopathological Society>.

Schmidt, R.E. & Panciera, J.R. (1980). Effects of aflatoxin on pregnant hamsters and hamsters foetuses. *Journal of Comparative Pathology*, Vol. 90, pp. 339-346.

Schweitzer, S.H., Ouist, C.F., Grimes, G.L & Forest, D.L. (2001). Aflatoxin levels in corn available as wild turkey feed in Georgia. *Journal of Wildlife Diseases*, Vol. 37, 3, pp. 657–659.

Scudamore, K.A., Hetmanski, M.T., Chan, H.K. & Collins, S. (1997). Occurrence of mycotoxins in raw ingredientsused for animal feeding stuffs in the United Kingdom in 1992. *Food Additives & Contaminants*, Vol. 14, pp. 157-173.

Sharma, R. & Márquez, C. (2001). Determination of aflatoxins in domestic pet foods (dog and cas) using immunoaffinity column and HPLC. *Animal Feed Science and Technology*, Vol. 93, pp. 109-114.

Shi, X.D., Li, X.T., Fu, J.L., Sheng, J.J., Gui, Y.L., Hui-J.Y., Zhen, Y.D. & Yong, J.L. (2010). Toxic effects and residue of aflatoxin B_1 in tilapia (*Oreochromis niloticus* × *O. aureus*) during long-term dietary exposure. *Aquaculture*, Vol. 307, pp. 233-240.

Siame, B.A., Mpuchane, S.F., Gashe, B.A., Allotey, J. & Teffera, G. (1998). Occurrence of aflatoxins fumonisin B_1 and zearalenone in foods and feeds in Botswana. *Journal of Food Protection*, Vol. 61, pp. 1670-1673.

Stenske, K.A., Smith, J.R., Newman, S.J., Newman, L.B. & Kirk, C.A. (2006).Aflatoxicosis in dogs and dealing with suspected contaminated commercial foods. *Journal of the American Veterinary Medical Association*, Vol. 228, pp. 1686–1691.

Trucksees, M.W. & Wood, G.E. Recent method of analysis for aflatoxins in food and feeds. In: *The toxicology of aflatoxins: human health, veterinary and agricultural significance.* Eaton, D.L. & Groopman, J.D., pp. 73-88, Academic Press, San Diego, USA.

United States Environmental Protection Agency –USEPA (1995). The use of the benchmark dose approach. Health Risk Assessment. USEPA, Washington, DC, USA.

Van Der Zijden, A.S.M., Blanch Kolensmid, W.A.A. & Boldingh J. (1962). Isolation in crystalline form of a toxin responsible for turkey X disease. *Nature*, London, 195, pp. 1062.

Van Houweling, C.D., Bixler, W.B. & McDowell, J.R. (1977). Role of the Food and Drug Administration concerning chemical contaminants in animal feeds. *Journal of the American Veterinary Medical Association*, Vol. 171, pp. 1153-1156.

Wangikar, P.B., Dwivedi, P., Sinha, N., Sharma, A.K. & Telang, A.G. (2005). Effects of aflatoxin B_1 on embryo fetal development in rabbits, *Food and Chemical Toxicology*, Vol. 43, pp. 607–615.

Willard, M. D., Simpson, R. B. Delles, K. , Cohen, N. D., Fossum, T. W. Kolp, E. D. L. & Reinhart G. A. (1994). Preparation with Food Safety. Retrieved from < www.dogcathomeprepareddiet.com/food_safety_and_preparation.html>.

World Health Organization – WHO (1993). International Agency for Research on Cancer (IARC), Monographs on evaluation of carcinogenic risks to humans.

Yegani, M., Chowdhury, S.R., Oinas, N., MacDonald E. & Smith, T. (2006). Effects of feeding grains naturally contaminated with Fusarium mycotoxins on brain regional neurochemistry of laying hens, turkey poults, and broiler breeder hens. *Poultry Science*, Vol. 85 (12), pp. 2117–23.

Zain, M.E. Impact of mycotoxins on humans and animals. J. Saudi Chem. Soc. Article in Press, doi:10.1016/j.jscs.2010.06.006.

Zicker, S.C. (2008). Evaluating Pet Foods: How Confident Are You When You Recommend a Commercial Pet Food? *Topics in Companion Animal Medicine*, Vol. 23 (3), pp. 121-126.

Part 2

Measurement and Analysis

6

Aflatoxins: Their Measure and Analysis

Anna Chiara Manetta

Department of Food and Feed Science, University of Teramo
Italy

1. Introduction

Aflatoxins are natural secondary metabolites produced by some moulds (mainly *Aspergillus flavus* and *Aspergillus parasiticus*) and are contaminants of agricultural commodities in the field particularly in critical temperature and humidity conditions before or during harvest or because of inappropriate storage (Rustom, 1997; Sweeney & Dobson, 1998). Aflatoxins B1 (AFB1) and B2 (AFB2), produced by *A. flavus*, and aflatoxins G1 (AFG1) and G2 (AFG2), produced by *A. flavus* as well as *A. parasiticus*, can contaminate maize and other cereals such as wheat and rice, but also groundnuts, pistachios, cottonseed, copra and spices. Following the ingestion of contaminated feedstuffs by lactating dairy cows, AFB1 is biotransformed by hepatic microsomal cytochrome P450 into aflatoxin M1 (AFM1), which is then excreted into the milk (Frobish et al., 1986). Because of the binding of AFM1 to the milk protein fraction, in particular with casein (Brackett & Marth, 1982), it can be present also in dairy products manufactured with contaminated milk.

The WHO-International Agency for Research on Cancer (IARC) has classified AFB1, AFB2, AFG1, AFG2 and since 2002 also AFM1 as carcinogenic agents to humans (group 1) (IARC, 2002).

Considering their natural occurrence, it is impossible to fully eliminate their presence; so, coordinated inspection programmes aimed to check the presence and concentration of aflatoxins in feedingstuffs are recommended by the Commission of the European Communities.

National and international institutions and organizations such as the European Commission (EC), the US Food and Drug Administration (FDA), the World Health Organization (WHO) and the Food and Agriculture Organization (FAO) have recognized the potential health risks to animals and humans posed by consuming aflatoxin-contaminated food and feed.

To protect consumers and farm animals regulatory limits have been adopted. The current maximum residue levels (MRL) for aflatoxins set by the EC (Commission European Communities, 2006a) are 2 µg/kg for AFB1 and 4 µg/kg for total aflatoxins in groundnuts, nuts, dried fruits and cereals for direct human consumption. These have been extended to cover some species of spices with limits of 5 µg/kg and 10 µg/kg for AFB1 and total aflatoxins, respectively. These levels are about five times lower than those adopted in the USA. Limits of 0.1 µg/kg are established by the EC for AFB1 in baby foods and dietary foods. The current regulatory limit for AFM1 in raw milk is 0.05 µg/kg, while in baby foods and dietary foods has been set at 0.025 µg/kg. Taking into account the developments in Codex Alimentarius, recently EC has introduced the maximum accepted levels for aflatoxins in other foodstuffs, like oilseeds (2 µg/kg for AFB1 and 4 µg/kg for total aflatoxins), almonds, pistachios and apricot kernels (5 µg/kg for AFB1 and 10 µg/kg for total aflatoxins), hazelnuts and Brazil nuts (5 µg/kg for AFB1 and 10 µg/kg for total aflatoxins) (Commission European Communities, 2010).

About animal feeds, only AFB1 is regulated: EC has set a limit of 0.02 mg/kg in all feed materials and in most complete and complementary feedstuffs for cattle, sheep, goats, pigs and poultry, while it is 0.005 mg/kg in complete feedingstuffs for dairy animals and 0.01 mg/kg for complete feedingstuffs for calves and lambs (Commission European Communities, 2003).

Because of the toxicity of these molecules and considering the MRL set in food and in feedstuffs, analytical identification and quantification of such contaminants at these low levels has to be carried out with reliable methods: they must be able to provide accurate and reproducible results to allow an effective control of the possible contamination of food and feed commodities. For this reason, the EC has set also the performance criteria for the methods of analysis to be used for the official control of mycotoxins in general and aflatoxins in particular (Commission European Communities, 2006b).

Nowadays, many sensitive, specific, but also simple and rapid methods are available: in literature there is considerable attention to aflatoxin detection. As new analytical technologies have developed, they have been rapidly incorporated into mycotoxin testing strategies. Sometimes many works reflect advances in analytical science (the availability of mass spectrometry detectors is an example), but often modifications of existing methods are published to improve the analytical process. Several methods have been also validated for the determination of aflatoxins in various matrices, but the validation does not always comply with the more recent EC guidelines (Commission European Communities, 2006b; Commission European Communities, 2002; Commission European Communities, 2004).

Among these, Commission Decision 2002/657/EC has set the performance and the procedures for the validation of screening and confirmatory methods.

Numerous methods have been developed to meet analytical requirements from rapid tests for factories and grain silos to regulatory control in official laboratories. This review will focus upon different analytical methods used for aflatoxin determination. They include thin layer chromatography (TLC), high-performance liquid chromatography (HPLC) in combination with fluorescence detection with or without derivatisation, liquid chromatography tandem mass spectrometry (LC/MS) and immunochemicals methods, such as enzyme linked immunosorbent assay (ELISA), immunosensors, dipsticks, strip-test.

2. Chromatographic methods

Aflatoxins possess significant UV absorption and fluorescence properties, so techniques based on chromatographic methods with UV or fluorescence detection have always predominated.

Originally the chromatographic separation was performed by TLC: since when aflatoxins were first identified as chemical agents, it has been the most widely used separation technique in aflatoxin analysis in various matrices, like corn, raw peanuts (Park et al, 2002), cotton seed (Pons et al, 1980), eggs (Trucksess et al, 1977), milk (Van Egmond, 1978) and it has been considered the AOAC official method for a long period. This technique is simple and rapid and the identification of aflatoxins is based on the evaluation of fluorescence spots observed under a UV light. AFB1 and AFB2 show a blue fluorescence colour, while it is green for AFG1 and AFG2. TLC allows qualitative and semi-quantitative determinations by comparison of sample and standard analysed in the same conditions. Many TLC methods for aflatoxins were validated more than 20 years ago and also when there has been a more recent validation, the performance of the methods has often been established at contamination levels too high to be of relevance to current regulatory limits.

The combination of TLC methods with much-improved modern clean-up stage offers the possibility to be a simple, robust and relatively inexpensive technique (Vargas et al, 2001), that after validation can be used as viable screening method. Moreover, given the significant advantages of the low cost of operation, the potential to test many samples simultaneously and the advances in instrumentation that allow quantification by image analysis or densitometry, TLC can be used also in laboratories of developing countries in alternative to other chromatographic methods that are more expensive and require skilled and experienced staff to operate. Improvements in TLC techniques have led to the development of high-performance thin-layer chromatography (HPTLC), successfully applied to aflatoxin analysis (Nawaz et al, 1992).

Overpressured-layer chromatographic technique (OPLC), developed in the seventies, has been used for quantitative evaluation of aflatoxins in foods (Otta et al, 1998) and also in fish, corn, wheat samples that can occur in different feedstuffs (Otta et al, 2000).

Because of its higher separation power, higher sensitivity and accuracy, the possibility of automating the instrumental analysis, HPLC now is the most commonly used technique in analytical laboratories. HPLC using fluorescence detection has already become the most accepted chromatographic method for the determination of aflatoxins. For its specificity in the case of molecules that exhibit fluorescence, Commission Decision 2002/657/EC, concerning the performance of analytical methods, considers the HPLC technique coupled with fluorescence detector suitable confirmatory method for aflatoxins identification.

However, HPTLC and HPLC techniques complement each other: the HPTLC for preliminary work to optimize LC separation conditions during the development of a method or its use as screening for the analysis of a large number of samples to limit the HPLC analysis only to positive samples are not unusual.

Liquid chromatographic methods for aflatoxins determination include both normal and reverse-phase separations, although current methods for aflatoxin analysis tipically rely upon reverse-phase HPLC, with mixtures of methanol, water and acetonitrile for mobile phases.

Aflatoxins are naturally strongly fluorescent compounds, so the HPLC identification of these molecules is most often achieved by fluorescence detection. Reverse-phase eluents quench the fluorescence of AFB1 and AFG1 (Kok, 1994); for this reason, to enhance the response of these two analytes, chemical derivatisation is commonly required, using pre- or post-column derivatisation with suitable fluorophore, improving detectability.

The pre-column approach uses trifluoroacetic acid (TFA) with the formation of the corresponding hemiacetals (Stubblefield, 1987; Simonella et al, 1998; Akiyama et al, 2001) that are relatively unstable derivatives. The post-column derivatisation is based on the reaction of the 8,9-double bond with halogens. Initially, the post-column reaction used iodination (Shepherd & Gilbert, 1984), but it has several disadvantages, like peak broadening and the risk of crystallisation of iodine. An alternative method is represented by bromination by an electrochemical cell (Kobra Cell) with potassium bromide dissolved in an acidified mobile phase or by addition of bromide or pyridinium hydrobromide perbromide (PBPB) to mobile phase and using a short reaction coil at ambient temperature (Stroka et al, 2003; Manetta et al, 2005; Senyuva & Gilbert, 2005; Brera et al, 2007; Manetta et al, 2010). The bromination methods offer the advantage to be rapid, simple and easy to automate, improving reproducibility and ruggedness and reducing analysis time.

A post-column derivatisation method that seems analytically equivalent to iodination and bromination is the photochemical one: it is based on the formation of hemiacetals of AFB1 and AFG1 as the effect of the irradiation of the HPLC column eluate by a UV light (Joshua, 1993; Waltking & Wilson, 2006).

A method based on the formation of an inclusion complex between aflatoxins and cyclodextrins (CDs) has been recently developed (Chiavaro et al, 2001): specific CDs are added to mobile phase (water-methanol) including aflatoxins in their cyclic structure, enhancing AFB1 and AFG1 fluorescence (Aghamohammadi & Alizadeh, 2007).

The introduction of mass spectrometry and the subsequent coupling of liquid chromatography to this very efficient system of detection have resulted in the development of many LC-MS or LC-MS/MS methods for aflatoxin analysis. Because of the advantages of specificity and selectivity, chromatographic methods coupled to mass spectrometry continue to be developed: they improve detection limits and are able to identify molecules by means mass spectral fragmentation patterns. Some of them comprise a single liquid extraction and direct instrumental determination without clean-up step (Cappiello et al, 1995; Kokkonen et al, 2005; Júnior et al, 2008). This assumption relies on the ability of the mass analyser to filter out by mass any co-eluting impurities. However, many Authors assert that further sample preparation prior to LC-MS analysis would benefit analysis (Chen et al, 2005; Cavaliere et al, 2006; Lattanzio et al, 2007) because ionisation suppression can occur by matrix effects. A number of instrument types have been used: single quadrupole (Blesa et al, 2003), triple quadrupole (Chen et al, 2005), linear ion trap (Cavaliere et al, 2006; Lattanzio et al, 2007). Atmospheric pressure chemical ionisation (APCI) is the ionisation source that provides lower chemical noise and, subsequently, lower quantification limit than electrospray ionisation (ESI) also if this one, on the other hand, is more robust. The use of mass spectrometric methods can be expected to increase, particularly as they become easier to use and the costs of instrumentation continue to fall.

Despite the enormous progress in analytical technologies, methods based on HPLC with fluorescence detection are the most used today for aflatoxins instrumental analysis, because of the large diffusion of this configuration in routine laboratories.

The recent availability of analytical columns with reduced size of the packing material has improved chromatographic performance. Today, numerous manufacturers commercialize columns packed with sub-2 μm particles to use devices that are able to handle pressure higher than 400 bar, such as Ultra-Performance Liquid Chromatography® (UPLC). This strategy allows a significant decrease in analysis time: aflatoxins runs are completed in 3-4 min with a decrease of over 60% compared to traditional HPLC. In addition, solvent usage has been reduced by 85%, resulting in greater sample throughput and significant reduction of costs of analysis. UPLC system can be coupled to traditional detector or, using a mobile phase of water/methanol with 0.1% formic acid, to mass spectrometry detector.

For a short time capillary electrophoresis has been a technique of interest in aflatoxins separation, in particular its application as micellar electrokinetic capillary chromatography with laser-induced fluorescence detection (Maragos & Greer, 1997), but it has not found application in routine analysis.

In Table 1 some analytical methods for aflatoxin determination have been included with their performance characteristics.

Aflatoxin	Matrix	Method	Sample preparation	LOD (μg/kg)	LOQ (μg/kg)	R%	RSD$_R$ (%)	Reference
B1	Corn	HPLC/Fluor. Pre-column der. (TFA), Post-column (PBPB)	IAC	-	-	82-84	19-37	Brera et al, 2007
B1,B2, G1,G2	Corn, raw peanut, peanut butter	TLC/Densit.	SPE	-	-	95-139	26-84 (B1)	Park et al, 1994
B1,B2, G1,G2	Corn, raw peanut, peanut butter	HPLC/Fluor. post-column der. (iodine)	IAC	-		97-131	11-108	Trucksess et al, 1991
B1,B2 G1,G2 M1	Mould cheese	LC-MS/MS triple quadrupole (ESI source)	Only extraction	0.3(M1) 0.8(B-G)	0.6(M1) 1.6(B-G)	96-143	2-12	Kokkonen et al, 2005
B1,B2, G1,G2	Fish, corn, wheat	OPLC	Extraction and L-L partition	2	-	73-104	7-13 (RSDr)	Otta et al, 2000
B1	corn	Capillary electrophoresis / laser induced fluor.	SPE or IAC	0.5	-	85	-	Maragos & Greer, 1997
B1,B2 G1,G2	peanuts	HPLC/Fluor.	MSPD	-	0.125-2.5	78-86	4-7 (RSDr)	Blesa et al, 2003
M1	Milk	HPLC/Fluor. Pre column der. (TFA)	SPE or IAC	0.027-0.031	-	82-92	15-19 (RSDr)	Simonella et al, 1999
M1	Milk	colourimetric ELISA	none	0.006	-	100	11 (RSDr)	Simonella et al, 1999
M1	Milk, soft cheese	HPLC/Fluor. Post column der. (PBPB)	SPE	0.001-0.005	-	76-90	3-9 (RSDr)	Manetta et al, 2005
M1	Hard cheese	HPLC/Fluor. Post column der. (PBPB)	SPE	0.008	0.025	67	4-7 (RSDr)	Manetta et al, 2009
M1	Milk	HPLC/Fluor.	IAC	-	0.005	74	21-31	Dragacci et al, 2001
M1	Milk	HPLC/Fluor.	IAC	0.006	0.015	91	8-15	Muscarella et al, 2007
M1	Milk	Chemilumines cent ELISA	none	0.00025	0.001	96-122	2-8	Magliulo et al, 2005
M1	Milk	LC-MS/MS linear ion trap (ESI and APCI source)	carbograph-4 cartridge	-	0.006-0.012	92-96	3-8	Cavaliere et al, 2006

M1	Milk	Membrane-based flow through enzyme immunossay	IAC	0.05	-	97	-	Sibanda et al, 1999
M1	Milk	Electrochemical biosensor	none	0.01	-	-	-	Paniel et al, 2010
M1	Milk, milk powder	LC-MS/MS triple quadrupole (ESI source)	IAC	0.59-0.66	-	78-87	-	Chen et al, 2005
M1	Milk, milk powder	LC-MS/MS triple quadrupole (ESI source)	Multifunction column	9-14	-	7-16	-	Chen et al, 2005

Legend: Fluor.: fluorescence detection; Densit.: densitometry; der.: derivatisation.

Table 1. Performance characteristics of some analytical methods for aflatoxins

3. Sample preparation

Aflatoxins present in food and feed commodities must be extracted from the matrices by a suitable solvent or mixture of solvents and cleaned-up prior to analysis.

Sample preparation technology is one of the most relevant field of analytical science.

The pretreatment of sample (protein precipitation, defatting, extraction, filtration) is an important phase for removing many interferences and for having, in this way, extracts without impurities to allow accuracy and reproducibility in the subsequent instrumental step.

The first phase is the extraction of the toxins from the matrices: it generally involves chloroform, dichloromethane or aqueous mixtures of polar organic solvents as methanol, acetone, acetonitrile, the aqueous mixture being recently the most used ones because more compatible not only with environment but also with the antibodies involved in the subsequent step of clean-up with immunoaffinity columns that are increasingly utilised.

Supercritical fluid extraction (SFE) has some applications in food analysis because this system of extraction uses supercritical carbon dioxide and not organic solvents or involves them only in small amounts. However, in aflatoxins analysis this technique of extraction has not been successfully used because of the low recoveries of aflatoxins and the presence in the extracts of impurities such as lipids that are the main interferences with the purification step and with the chromatographic separation.

Clean-up is another very critical step. It is necessary for removing many of the co-extracted impurities and obtaining cleaner extracts for the subsequent instrumental determination, to have the most accurate and reproducible results. The traditional techniques, such as liquid-liquid partition or purification on conventional glass columns packed with silica, are time and solvent consuming. Nowadays, new sample preparation technologies, based on extraction by adsorbent materials, are available.

Solid-phase extraction (SPE) and Immunoaffinity Chromatography (IAC) represent very efficient systems that combine in one step filtration, extraction, adsorption and clean-up, allowing to obtain extracts without interferences, to reduce the analytical time and the volumes of solvents used, to improve the reproducibility and the accuracy, to be easily automatable. With these sample treatment techniques the analytes present in solutions can be concentrated, improving detection limit.

The SPE can be a powerful method for sample preparation: it represents a very significant improvement in the purification step. It is based on the separation mechanism of the modern chromatography: the sample extract is loaded on a cartridge packed with a selective adsorbent material, on which the analytes to be detected are adsorbed and then separated by elution with suitable solvent. In this process the molecules of interest that are in the sample are separated on the basis of its different partition between a liquid (solvent of extraction) and a solid (sorbent phase). The eluent and the adsorbent material compete in the affinity with the analytes: the components of the sample that have higher affinity for mobile phase are easy eluted, while the molecules with affinity for stationary phase are retained. In this technique one or more washing steps are necessary to remove the interferences co-adsorbed on a sorbent stationary phase.

Different types of adsorbent material are available, silica and octadecyl-bonded phase being the most used ones for aflatoxins B and G and for aflatoxin M, respectively.

Matrix solid-phase dispersion (MSPD) is the innovation of the SPE, although it has not yet found application in routine analysis. The MSPD has the advantage to combine extraction and clean-up in one step: the sample is homogenised in a specific sorbent phase in a mortar. Then, the mixture is transferred in a cartridge constricted between two frits and after the column has been washed with suitable solvents, the analytes are eluted for the subsequent instrumental detection. In literature some applications of MSPD to aflatoxins analysis are reported (Blesa et al, 2003; Hu et al, 2006) with high recoveries and satisfactory precision.

IAC is a very efficient technique of purification: it is based on the high specific interactions among biological molecules, so that such chromatography is able to complete the separation of complex mixture in one step. In a cartridge, like that used for SPE, the stationary phase is constituted by a ligand that is specific for the substance to be separated. The ligand is immobilized on a chromatographic bed material and it can be a policlonal or monoclonal antibody *vs* the analyte to be separate. When the sample is loaded into a cartridge, only the analytes of interest are retained, bound to their antibody, while the other components are eluted. The analyte is then eluted with suitable solvent that is generally methanol. The advantages of IAC is the effective and specific purification provided that allows to achieve cleaner eluates also starting from complex matrices. As a result, performances improve, especially in terms of detection and quantification limits; an added advantage is the limited use of organic solvents. So, IAC has become a major tool for mycotoxin analysis and, in particular, for aflatoxins determination. Another important advantage of this purification method is the fact that the extract of different matrices can be purified by essentially the same protocol. As a consequence, many methods developed to meet the requirements of the low EU maximum tolerated levels have relied on this purification technique and, perhaps for the same reason, many methods involved in collaborative studies and in validation protocols are based on the IAC purification step (Trucksess et al, 1991; Stroka et al, 2001; Dragacci et al, 2001; Stroka et al, 2003; Senyuva et al, 2005; Brera et al, 2007; Muscarella et al, 2007). IACs were thought to be more robust in terms of applicability to different matrices without the need for major adjustments to the method. Immunoaffinity columns offer the opportunity to concentrate large volumes of sample extract to achieve high sensitivity, which is for example the requirement for aflatoxins in baby foods. Moreover, immunoaffinity columns are less demanding in terms of the skills and the experience required.

Recently, IAC has been improved by the introduction of cartridges containing antibodies that are specific to more than one analyte, allowing the simultaneous clean-up of different classes of mycotoxins, like aflatoxins and ochratoxin A and zearalenone (Gobel & Lusky, 2004), and also aflatoxins, ochratoxin A, zearalenone, fumonisins, deoxynivalenol and T-2 toxin (Lattanzio et al, 2007).

In addition to the high cost of immunoaffinity columns, a critical factor in the IAC clean-up procedure is the fact that antibodies are sensitive to organic solvents; this is a problem because sample extracts generally contain high concentrations of acetonitrile, methanol or acetone, obligating to dilute them before application to the column. Acetonitrile, in particular, although it is a good extraction solvent used for SPE clean-up, is rarely used as an organic solvent for IAC because of the production of insoluble substances that can affect aflatoxins recovery (Patey et al, 1991). Very recently, some Authors have proposed a novel immunoaffinity column for aflatoxin analysis in roasted peanuts and some kinds of spices that shows satisfactory organic solvent tolerance, allowing acetonitrile extraction (Uchigashima et al, 2009).

In both SPE and IAC the final eluate can be concentrated evaporating the solvent, improving detection and quantification limits.

4. Immunological methods

High performance liquid chromatographic methods with fluorescent detection are mainly used in routine aflatoxins analysis. They are often laborious and time-consuming and require knowledge and experience of chromatographic techniques to solve separation and interference problems. The big demand in analytical chemistry to have sensitive, specific, but also simple and fast methods for an effective monitoring of aflatoxins in food and feed commodities, has produced analytical methods that combine simplicity with high detectability and analytical throughput. This can be realized by means of immunological methods in conjunction with a highly sensitive detection of the label.

As IAC methods, these assays involve antigen-antibody specific interactions at the surface of various supports. Previously conventional enzyme immunoassay for aflatoxin analysis use antibodies immobilized on well polystyrene microtiter plates: they are based on a competitive process involving antigen and antigen labelled with an enzyme (horseradish peroxidase, generally) and on colorimetric detection with chromogenic substrates (Thirumala-Devi et al, 2002). Enzyme-linked immunosorbent assay is the best established and the most available immunoassay in aflatoxin rapid detection, using the 96 well plate microtiter format. Many commercial companies have developed and commercialised ELISAs which applicability, analytical range and validation criteria are well defined. Despite the increasing use of LC-MS techniques, antibody-based methods for aflatoxins analysis continue to be investigated. The development of these immunochemical methods and their evolution from single to multiple analyte screening, including topics on ELISA, immunosensors, fluorescence polarization and rapid visual tests (lateral-flow, flow-through and dipstick) have been developed. In literature there are many applications to aflatoxins analysis by ELISA: AFB1 determination in deep-red pepper (Ardic et al, 2008), which requires a clean-up by IAC prior ELISA test; many commercial AFB1 screening test in feedstuffs often without purification; AFM1 in milk (Fremy & Chu, 1984; Thirumala-Devi et al, 2002), that needs only defatting step prior to analysis, resulting in a useful screening test for routine quality control of milk of different farms before mixing the different milk bulks, especially when the absence of AFM1 above the regulatory limit needs to be documented. Enzyme labels can be detected also by chemiluminescent substrates, such as the luminol/peroxide/enhancer system for horseradish peroxidase (HRP) or dioxetane-based substrates for alkaline phosphatase, resulting in a very sensitive detection system in immunoassay. Chemiluminescent detection allows the use of 384 well plates with an assay volume of 20 µl, which is at least five times lower than that used in the conventional 96 well microtiter format (Roda et al, 2000). A 5-fold reduction in antibody, labelled probe and chemiluminescent mixture volume reduce the costs of the assay, maintaining the same analytical performance. Thanks to the combination of the chemiluminescent detection of enzymatic activity with the use of a 384 well microtiter format, a highly sensitive, accurate, reproducible, simple and robust chemiluminescent enzyme immunoassay has been developed for AFM1 in milk samples (Magliulo et al, 2005) , with a reduction of costs and increased detectability compared with other immunological methods and commercial available kits for AFM1 in milk.

In the case of immunosensors for aflatoxins, antibodies are immobilized on the surface of a screen-printed electrode, magnetic beads held on the surface of a screen-printed electrode (Piermarini et al, 2009), on piezoelectric quartz crystal immunosensor with gold nanoparticles (Jin et al, 2009).

Typical competitive ELISA formats are surface-based; in fact, they require either a toxin-protein conjugate or an antibody to be immobilized onto a surface (membrane, well, electrode, sensor surface, *etc.*) to facilitate the separation of the 'bound' and "unbound" tracer: assays of this nature are termed "heterogeneous" and encompass the vast majority of mycotoxin immunoassays. The separation can be achieved in various ways, from washing (as in ELISAs), chromatographically (as in lateral flow test strips), or reagent flowing over a surface (as in certain biosensors).

Fluorescence polarization immunoassay (FPIA) is a homogeneous assay conducted in solution phase. It is based on the different rate of rotation of smaller and larger molecules. A molecule like a toxin can be covalently linked to a fluorophore to make a fluorescent tracer. The tracer competes with toxin (eventually present in the sample) for a limited amount of toxin-specific antibody. In the case the toxin is absent in the sample, the antibody binds only the tracer, reducing its motion and causing a high polarization. In presence of the toxin, less of the tracer is bound to the antibody and a greater tracer fraction exists unbound in solution, where it shows a lower polarization (Maragos, 2009). The significant advantage of fluorescence polarization over traditional ELISA techniques is that it is measured without the need for separating the free and bound tracer. In particular, it does not require additional manipulations, such as the washing steps of competitive ELISAs, making it simple, rapid, also field portable and, therefore, useful for screening purpose. A homogeneous assay for determining the aflatoxin content in agricultural products based on the technique of fluorescence polarization has been described (Nasir & Jolley, 2002). The disadvantage of this technique is that the aflatoxin contents are underestimated, probably because of the low cross-reactivity of the antibody with AFB2, AFG1 and AFG2.

The lateral flow device is one of the simplest and fastest immunoassay techniques have been developed. It is a screening test available in the format of strip or dipstick (Delmulle et al, 2005). Immunodipstick or lateral flow immunoassay has recently gained increasing attention because it requires simple and minimal manipulations and little or no instrumentations. Colloidal gold conjugated anti-aflatoxin antibodies are immobilised at the base of the stick. Aflatoxin present in the sample extract interacts with them; bound and unbound antibodies move along the membrane-based stick, pass a test line containing aflatoxin, which binds free antibodies, forming a visible line, indicating that the level of eventual aflatoxin contamination of the sample is below the test cut-off value.

Recently, an immunoassay-based lateral flow device for the quantitative determination of four major aflatoxins in maize, that can be completed in 10 min, has been developed (Anfossi et al, 2011). Even quantification is possible by acquiring images of the strip and correlating intensities of the coloured lines with analyte concentration by means of a calibration curve in matrix. Very simple sample preparation is required, making the method reliable, rapid for application outside the laboratory as a point-of-use test for screening purposes.

The immobilization of the antibodies on nanoparticles with a silver core and a gold shell enhances the sensitivity of the assay (Liao & Li, 2010).

Similarly, the membrane-based flow-through device is a qualitative test: the test line is generated by an enzyme-substrate colour reaction (Sibanda et al, 1999). Thanks to the simplicity of the material required, these methods are fit for using as portable rapid field assay for the early detection of aflatoxin-contaminated lots.

Immunological methods, based on antigen/antibody specific interaction, can give false positive results: although antibodies are specific for their antigens, they can react with other substances, similar to those in analysis, binding them as it happens in the antigen/antibody reaction. For this reason, in the case of a suspected non-compliant result, it shall be confirmed by confirmatory method (LC-fluorescence or LC-MS for aflatoxins), as it has been set by the Commission Decision 2002/657/EC.

The recent development of biosensors has stimulated their application also to aflatoxin analysis: in literature many examples are reported, like DNA biosensor (Tombelli et al, 2009), electrochemical immunosensor (Paniel et al, 2010), electrochemical sensor (Siontorou et al, 1998; Liu et al, 2006), fluorimetric biosensor (Carlson et al, 2000).

The advantages of biosensing techniques are: reduced extraction, clean-up analytical steps and global time of analysis (1 min or only few seconds); possibility of online automated analysis; low cost; skilled personnel not required. On the other side, sensitivity should be enhanced and their stability should be improved to allow long-term use.

Because of the ease of use of these devices, many commercial systems continue to be developed not only for aflatoxins, but also for all mycotoxins. For a long time many rapid assays were commercialized with no documentation on their performance characteristics. Since 2002, with Commission Decision 2002/657/EC, laboratories approved for official residue control can use for screening purposes only those techniques "for which it can be demonstrated in a documented traceable manner that they are validated". As a consequence, many screening test are now commercially available with documentation enclosed with validation parameters, like detection limit and cut-off, sensitivity, specificity, false negative and false positive rate.

5. Conclusions

For aflatoxins analysis several methods have been developed over the last 30 years. Because of the advances in technology, the better clean-up procedures and the combination of both, a higher sensitivity has been registered, HPLC with fluorescence detection becoming the most used analytical methodology in laboratory. Moreover, highly sophisticated methods based on liquid chromatography coupled with mass spectrometry have been developed, improving identification and accurate determination often without the need of sample preparation. Other advances have regarded the environment, as the replacing of chlorinated solvents, preferred in the past, by aqueous mixture of methanol or acetonitrile, that are also more compatible with antibodies, recently introduced in many applications . These reagents marked a turning point in the sample preparation step as well as in the identification phase, showing a high flexibility in many practical situations in which reliable, rapid and simple analyses are required to reduce costs. The choice of a method is made bearing in mind for what purpose aflatoxins analysis has to be performed. So, for example, if a yes/no or semi-quantitative response is considered satisfactory, the use of rapid test is suitable. On the other hand, official control laboratories, which are involved in the monitoring and risk-assessment studies and in official controls, have to apply methods that have been validated and adopted by AOAC International, CEN or ISO. As mycotoxins, not only aflatoxins, are a real problem for health, there will be always a big interest to them and, certainly, it is likely methods for their analysis will continue to improve.

Because of the potential co-occurrence of such contaminants, the challenge is to develop screening methods for their rapid simultaneous detection of multiple families of mycotoxins from the same sample. But the differences in their chemical and physical properties and of concentration range of interest have made simultaneous detection very difficult. In this regard HPLC technique coupled with mass spectrometry or multiple detectors has good prospects.

6. References

Akiyama H., Goda Y., Tanaka T. & Toyoda M. (2001). Determination of aflatoxins B1, B2, G1 and G2 in spices using a multifunctional column clean-up. *Journal of Chromatography A*, 932, pp. 153-157, ISSN: 0021-9673

Aghamohammadi M. & Alizadeh N. (2007). Fluorescence enhancement of the aflatoxin B1 forming inclusion complexes with some cyclodestrins and molecular modelling study. *Journal of Luminescence*, 127, pp. 575-582

Anfossi L., D'Arco G., Calderara M., Baggiani C., Giovannoli C. & Giraudi G. (2011). Development of a quantitative lateral flow immunoassay for the detection of aflatoxins in maize. *Food Additives and Contaminants*, 28, 2, pp. 226-234

Ardic M., Karakaya Y., Atasever M. & Durmaz H. (2008). Determination of aflatoxin B1 levels in deep-red pepper (isot) using immunoaffinity column combined with ELISA. *Food and Chemical Toxicology*, 46, pp. 1596-1599

Brackett R. E. & Marth E. H. (1982). Association of aflatoxin M1 with casein. *Zeitschrift fur Lebensmittel-Untersuchung and Forschung*, 174, pp. 439-441

Blesa J., Soriano J. M., Moltò J. C., Marìn M. & Mañes J. (2003). Determination of aflatoxins in peanuts by matrix solid-phase dispersion and liquid chromatography. *Journal of Chromatography A*, 1011, 1-2, pp. 49-54, ISSN: 0021-9673

Brera C., Debegnach F., Minardi F., Pannunzi E., De Santis B. & Miraglia M. (2007). Immunoaffinity column cleanup with liquid chromatography for determination of aflatoxin B1 in corn samples: interlaboratory study. *Journal of AOAC International*, 90, pp. 765-772

Cappiello A., Famiglini G. & Tirillini B. (1995). Determination of aflatoxins in peanut meal by LC/MS with a particle beam interface. Chromatographia, 40, Nn. 7-8, pp. 411-416

Carlson M. A., Bargeron C. B., Benson R. C., Fraser A. B., Phillips T. E., Velky J. T., Groopman J. D., Strickland P. T. & Ko H. W. (2000). An automated, handheld biosensor for aflatoxin. *Biosensor and Bioelectronics*, 14, 10-11, pp. 841-848

Cavaliere C., Foglia P., Pastorini E., Samperi R. & Laganà A. (2006). Liquid chromatography/tandem mass spectrometric confirmatory method for determining aflatoxin M1 in cow milk: comparison between electrospray and atmospheric pressure photoionization sources. *Journal of Chromatography A*, 1101, pp. 69-78, ISSN: 0021-9673

Chen C-Y., Li W-J. & Peng K-Y. (2005). Determination of aflatoxin M1 in milk and milk powder using high-flow solid-phase extraction and liquid chromatography-tandem mass spectrometry. *Journal of Agricultural and Food Chemistry*, 53, pp. 8474-8480

Chiavaro E., Dall'Asta C., Galaverna G., Biancardi A., Gambarelli E., Dossena A. & Marchelli R. (2001). New reversed-phase liquid chromatographic method to detect aflatoxins in food and feed with cyclodextrins as fluorescence enhancers added to the eluent. *Journal of Chromatography A*, 937, 31-40, ISSN: 0021-9673

Commission of European Communities (2003). Commission Directive 2003/100/EC. *Official Journal of the European Communities*, L 285, pp. 33-37

Commission of European Communities (2002). Commission Decision 2002/657/EC. *Official Journal of the European Communities*, L 221, pp. 8-36

Commission of European Communities (2004). Commission Regulation (EC) No. 882/2004. *Official Journal of the European Communities*, L 165, pp. 1-141

Commission of European Communities (2006a). Commission Regulation (EC) No. 1881/2006. *Official Journal of the European Communities*, L 364, pp. 5-24

Commission of European Communities (2006b). Commission Regulation (EC) No. 401/2006. *Official Journal of the European Communities*, L 70, pp. 12-34

Commission of European Communities (2010). Commission Regulation (EC) No. 165/2010. *Official Journal of the European Communities*, L 50, pp. 8-12

Delmulle B. S., De Saeger S. M. D. G., Sibanda L., Barna-Vetro I. & Van Peteghem C. H. (2005). Development of an immunoassay-based lateral flow dipstick for the rapid detection of aflatoxin B1 in pig feed. *Journal of Agricultural and Food Chemistry*, 53, 9, pp. 3364-3368

Dragacci S., Grosso F. & Gilbert J. (2001). Immunoaffinity column cleanup with liquid chromatography for determination of aflatoxin M1 in liquid milk: collaborative study. *Journal of AOAC International*, 84, 2, 437-443

Fremy J. M. & Chu F. S. (1984). Direct enzyme-linked immunosorbent assay for determining aflatoxin M1 at picogram levels in dairy products. *Journal of the Association of Official Analytical Chemists*, 67, pp. 1098-1101

Frobish R. A., Bradley B. D., Wagner D. D., Long-Bradley P. E. & Hairstone H. (1986). Aflatoxin residues in milk of dairy cows after ingestion of naturally contaminated grain. *Journal of Food Protection*, 49, pp. 781-785

Gobel R. & Lusky K. (2004). Simultaneous determination of aflatoxins, ochratoxin A, and zearalenone in grains by new immunoaffinity column/liquid chromatography. *Journal of AOAC International*, 87, 2, pp. 411-416, ISSN 1060-3271

Hu Y-Y., Zheng P., Zhang Z-H. & He Y-Z. (2006). Determination of Aflatoxins in High-Pigment Content Samples by Matrix Solid-Phase Dispersion and High-Performance

Liquid Chromatography. *Journal of Agricultural and Food Chemistry*, 54, pp. 4126-4130

International Agency for Research on Cancer (2002). *Some traditional herbal medicines, some mycotoxins, naphthalene and styrene. IARC monographs on the evaluation of carcinogenic risks to humans* (vol 82). Lyon, France: World Health Organization, pp. 1-556

Jin X., Jin X., Chen L., Jiang J., Shen G. & Yu R. (2009). Determination of aflatoxins in high-pigment content samples by matrix solid-phase dispersion and high-performance liquid chromatography. *Biosensors and Bioelectronics*, 24, 8, pp. 2580-2585

Joshua H. (1993). Determination of aflatoxins by reversed-phase high performance Liquid chromatography with post-column in-line photochemical derivatization and fluorescence detection. *Journal of Chromatography*, 654, pp. 247-254

Júnior J., Mendonça X. & Scussel V.M. (2008). Development of an LC-MS/MS method for the determination of aflatoxins B_1, B_2, G_1, and G_2 in Brazil nut. *International Journal of Environmental Analytical Chemistry*, 88, 6, pp. 425 - 433

Kok W. Th. (1994). Derivatization reactions for the determination of aflatoxins by liquid chromatography with fluorescence detection. *Journal of Chromatography B*, 659, 1-2, pp. 127-137

Kokkonen M., Jestoi M. & Rizzo A. (2005). Determination of selected mycotoxins in mould cheeses with liquid chromatography coupled to tandem with mass spectrometry. *Food Additives and Contaminants*, 22, 5, pp. 449-456

Lattanzio V. M. T., Solfrizzo M., Powers S. & Visconti A. (2007). Simultaneous determination of aflatoxins, ochratoxin A and Fusarium toxins in maize by liquid chromatography/tandem mass spectrometry after multitoxin immunoaffinity cleanup. *Rapid Communications in Mass Spectrometry*, 21, pp. 3253-3261

Liao J. Y. & Li H. (2010). Lateral flow immunodipstick for visual detection of aflatoxin B1 in food using immune-nanoparticles composed of a silver core and a gold shell. *Microchimica Acta*, 171, pp. 289-295

Liu Y., Qin Z. H., Wu X. F. & Hong J. (2006). Immune-biosensor for aflatoxin B1 based bio-electrocatalytic reaction on micro-comb electrode. *Biochemical Engineering Journal*, 32, pp. 211-217, ISSN: 1369-703X.

Magliulo M., Mirasoli M., Simoni P., Lelli R., Portanti O. & Roda A. (2005). Development and validation of an ultrasensitive chemiluminescent enzyme immunoassay for aflatoxin M1 in milk. *Journal of Agricultural and Food Chemistry*, 53, pp. 3300-3305

Manetta A. C., Di Giuseppe L., Giammarco M., Fusaro I., Simonella A., Gramenzi A., Formigoni A. (2005). High-performance liquid chromatography with post-column derivatisation and fluorescence detection for sensitive determination of aflatoxin M1 in milk and cheese. *Journal of Chromatography A*, 1083, pp. 219-222, ISSN: 0021-9673

Manetta A. C., Giammarco M., Di Giuseppe L., Fusaro I., Gramenzi A., Formigoni A., Vignola G. & Lambertini L. (2009). Distribution of aflatoxin M1 during Grana Padano cheese production from naturally contaminated milk. *Food Chemistry*, 113, 595-599, ISSN 0308-8146

Maragos C. M., Greer J. (1997). Analysis of aflatoxin B1 in corn using capillary electrophoresis with laser-induced fluorescence detection. *J. Agric. Food Chem.*, 45, pp. 4337-4341.

Maragos C. (2009). Fluorescence polarization immunoassay of mycotoxins: a Review. *Toxins*, 1, pp. 196-207, ISSN 2072-6651

Muscarella M., Lo Magro S., Palermo C. & Centonze D. (2007). Validation according to European Commission Decision 2002/657/EC of a confirmatory method for aflatoxin M1 in milk based on immunoaffinity columns and high performance liquid chromatography with fluorescence detection. *Analytica Chimica Acta*, 594, pp. 257-264, ISSN 0003-2670

Nawaz S., Coker R. D. & Haswell S. J. (1992). Development and evaluation of analytical methodology for the determination of aflatoxins in palm kernels. *Analyst*, 117, pp. 67-74

Nasir M. S. & Jolley M. E. (2002). Development of a fluorescence polarization assay for the determination of aflatoxins in grains. *Journal of Agricultural and Food Chemistry*, 50, pp. 3116-3121

Otta K. H., Papp E., Mincsovics E. & Záray Gy. (1998). Determination of aflatoxins in corn by use of the personal OPLC basic system. *Journal of Planar Chromatography*, 11, pp. 370-373

Otta K. H., Papp E. & Bagócsi B. (2000). Determination of aflatoxins in food by overpressured-layer chromatography. *Journal of Chromatography A*, 882, pp. 11-16, ISSN: 0021-9673

Paniel N., Radoi A. & Marty J. L. (2010). Development of an electrochemical biosensor for the detection of aflatoxin M1 in milk. *Sensors*, 10, pp. 9439-9448, ISSN 1424-8220

Park D. L., Trucksess M. W., Nesheim S., Stack M. E. &, Newell R. F. (1994). Solvent efficient thin-layer chromatographic method for the determination of aflatoxins B1, B2, G1 and G2 in corn and peanut products: collaborative study. *Journal of AOAC International*, 77, pp. 637-646

Patey A. L., Sharman M. & Gilbert J. (1991). Liquid chromatographic determination of aflatoxin levels in peanut butters using an immunoaffinity column cleanup method: International collaborative trial. *Journal of the Association of Official Analytical Chemists*, 74, pp. 76-81

Piermarini S., Volpe G., Micheli L., Moscone D. & Palleschi G. (2009). An ELIME-array for detection of aflatoxin B1 in corn samples. *Food Control*, 20, pp. 371-375

Pons W. A., Lee L. S. & Stoloff L. (1980). Revised method for aflatoxin in cottonseed products and comparison of thin layer chromatography and high pressure liquid chromatography determinative steps: collaborative study. *Journal of the Association of Official Analytical Chemists*, 63, pp. 899-906

Roda A., Manetta A. C., Piazza F., Simoni P. & Lelli R. (2000). A rapid and sensitive 384-microtiter wells format chemiluminescent enzyme immunoassay for clenbuterol. *Talanta*, 52, pp. 311-318

Rustom, I. Y. S. (1997). Aflatoxin in food and feed: Occurrence, legislation and inactivation by physical methods. *Food Chemistry*, 59, pp. 57-67

Senyuva H. Z. & Gilbert J. (2005). Immunoaffinity column cleanup with liquid chromatography using post-column bromination for determination of aflatoxins in hazelnut paste: interlaboratory study. *Journal of AOAC International*, 88, pp. 526-535

Shepherd M. J. & Gilbert J. (1984). An investigation of HPLC post-column iodination conditions for the enhancement of aflatoxin B1 fluorescence. *Food Additives and Contaminants*, 1, pp. 325-335

Sibanda L., De Saeger S. & Van Peteghem C. (1999). Development of a portable field immunoassay for the detection of aflatoxin M1 in milk. *International Journal of Food Microbiology*, 48, pp. 203-209

Simonella A., Scortichini G., Manetta A. C., Campana G., Di Giuseppe L., Annunziata L. & Migliorati G. (1998). Aflatossina M_1 nel latte vaccino: ottimizzazione di un protocollo analitico di determinazione quali-quantitativa basato su tecniche cromatografiche, di immunoaffinità e immunoenzimatiche. *Veterinaria Italiana*, Anno XXXIV, 27-28, pp. 25-39

Siontorou C. G., Nikolelis D. P., Miernik A. & Krull U. J. (1998). Rapid methods for detection of Aflatoxin M1 based on electrochemical transduction by self-assembled metal-supported bilayer lipid membranes (s-BLMs) and on interferences with transduction of DNA hybridization. *Electrochimica Acta*, 43, pp. 3611-3617.

Stroka J., Anklam E., Joerissen U. & Gilbert J. (2001). Determination of aflatoxin B1 in baby food (infant formula) by immunoaffinity column cleanup liquid chromatography with postcolumn bromination: collaborative study. *Journal of AOAC International*, 84, pp. 1116-1123, ISSN 1060-3271

Stroka J., von Host C., Anklam E. & Reutter M. (2003). Immunoaffinity column cleanup with liquid chromatography using post-column bromination for determination of aflatoxin B_1 in cattle feed: collaborative study. *Journal of AOAC International*, 86, 6, pp. 1179-1186, ISSN 1060-3271

Stubblefield R. D. (1987). Optimum conditions for formation of aflatoxin M1-trifluoroacetic acid derivative. *Journal of the Association of Official Analytical Chemists*, 70, pp. 1047-1049

Sweeney, M. J. & Dobson A. D. W. (1998). Mycotoxin production by Aspergillus, Fusarium and Penicillium species. *Inernational Journal of Food Microbiology*, 43, pp. 141-158

Thirumala-Devi K., Mayo M. A., Hall A. J., Craufurd P. Q., Wheeler T. R., Waliyar F., Subrahmanyam A. & Reddy D. V. (2002). Development and application of an indirect competitive enzyme-linked immunoassay for aflatoxin M(1) in milk and milk-based confectionery. *Journal of Agricultural and Food Chemistry*, 50, pp. 933-937

Tombelli S., Mascini M., Scherm B., Battacone G. & Migheli Q. (2009). DNA biosensors for the detection of aflatoxin producing *Aspergillus flavus* and *A. parasiticus*. *Monatshefte für Chemie / Chemical Monthly*, 140, pp. 901-907.

Trucksess M. W., Stoloff L., Pons W. A., Cucullu A. F., Lee L. S. & Franz A. O. (1977). Thin layer chromatographic determination of aflatoxin B1 in eggs. *Journal of the Association of Official Analytical Chemists*, 60, 4, pp. 795-798

Trucksess M. W., Stack M. E., Nesheim S., Page S. W., Albert R., Hansen T. J. & Donahue K. F. (1991). Immunoaffinity column coupled with solution fluorometry or liquid chromatography postcolumn derivatization for determination of aflatoxins in corn, peanuts and peanut butter: collaborative study. *Journal of the Association of Official Analytical Chemists*, 74, 1, pp. 81-88

Uchigashima M., Saigusa M., Yamashita H., Miyake S., Fujita K., Nakajima M. & Nishijima M. (2009). Development of a novel immunoaffinity column for aflatoxin analysis using an organic solvent-tolerant monoclonal antibody. *Journal of Agricultural and Food Chemistry*, 57, pp. 8728-8734

Van Egmond H. P., Paulsch W. E. & Schuller P. L. (1978). Confirmatory test for aflatoxin M1 on thin layer plate. *Journal of the Association of Official Analytical Chemists*, 61, pp. 809-812

Vargas E. A., Preis R. A., Castro L. & Silva C. M. G. (2001). Co-occurrence of aflatoxins B1, B2, G1, G2, zearalenone and fumonisin B 1 in Brazilian corn. *Food Additives and Contaminants*, 18, 11, pp. 981-986

Waltking A. E. & Wilson D. (2006). Liquid chromatographic analysis of aflatoxin using post-column photochemical derivatization: collaborative study. *Journal of AOAC International*, 89, pp. 678-692, ISSN 1060-3271

7

Aflatoxin Measurement and Analysis

Imtiaz Hussain

Govt. Post Graduate College Samundri, Faisalabad,
Pakistan

1. Introduction

Much research work has been devoted over the last 40 years for developing methods for detection and determination of aflatoxins in foods and agriculture commodities (Chu, 1991; Holcomb, et al., 1992). This effort is continuing and keeping pace with the progress in analytical chemistry. Methods for aflatoxins are required to meet the legislation, monitoring and survey work, and for research. Different highly efficient and sophisticated techniques have been developed in the recent years for the determination of aflatoxins in different commodities. Presently the most commonly used methods for detection of aflatoxins are: high-performance liquid chromatography (HPLC), thin-layer chromatography (TLC) and enzyme-linked immunosorbent assay (ELISA) (Lee et al., 2009) and fluorometeric method (Hansen, 1990). All analytical procedures include the steps: sampling, extraction, clean- up (purification) and determination (identification and quantification). The analytical detail in this chapter has been discussed in three sub-groups: sample preparation techniques, detection techniques and typical complete procedures.

2. Sample preparation techniques

Sampling and sample preparation is of utmost importance in the analytical identification of aflatoxins. It certainly affects the final conclusion. For the determination of aflatoxins at the parts-per-billion level, the systematic approaches to sampling, sample preparation and analysis are absolutely necessary. European Union has formed specific plans for certain commodities e.g. corn and peanuts. The performance of sampling plans for aflatoxin in granular feed products, such as shelled maize (Johansson, et al., 2000) and cotton seed (Whitaker et al., 1976) has been evaluated, while there has been little evaluation of sampling plans to detect aflatoxin in milk.

In case of sampling of solid commodities the entire primary sample must be ground and mixed so that the analytical test portion has the same concentration of toxin as the original sample. In case of sampling of liquid commodities like milk, due to homogeneous distribution of aflatoxins in liquid milk, there is less uncertainty in aflatoxin measurement in milk. After proper sampling, there are the steps of extraction and clean-up. Sometimes extraction and clean-up is the same step and sometimes extraction is different step and clean-up is a different step. Extraction of samples, together with effective clean-up step, is an essential step in the analysis of aflatoxins. The analyte migrates into the extraction solvent. The interfering compounds are removed by clean-up step. Common extraction solvents for aflatoxins are acetonitrile-water and methanol-water.

In addition to conventional technique of liquid-liquid extraction, there was need to develop new techniques due to its time consuming and tedious to apply nature. The new approaches have been developed to lessen the problems. A number of clean-up columns, using different principles such as solid phase extraction and immunoaffinity techniques, have been developed. The new techniques are easy to use and easily available. The immunoaffinity columns enhance selectivity, as only the analyte is retained in the column which can be eluted easily. On the other hand, in Mycosep columns the analyte is passed and all the other interfering contaminants are retained. A view of extraction techniques is forthcoming.

2.1 Liquid-liquid separation
The liquid-liquid separation is a conventional process and it is based on the partition of organic compounds between aqueous phase and immiscible organic solvent which may be non-polar or slightly polar. Extraction, in most cases, involves conventional procedures using acetone, chloroform and methanol *etc*. Small amounts of water give better extraction efficiencies. Hexane and cyclohexane are frequently used for compounds with aliphatic properties, whereas dichloromethane and chloroform are used for medium polar contaminants.

In a typical case of liquid-liquid separation, methanol and water were used as the extraction solvents in the first effective method for the determination of aflatoxin in fluid milk (Jacobson, et al., 1971). This method was modified by McKinney (1972) and others. Stubblefield and Shannon (1974) accomplished extraction with acetone and water, precipitation with lead acetate solution to de-proteinize the milk, and a de-fating step with hexane. TLC with fluorescence detection was applied for ultimate separation, detection, and quantification. The collaborative study proved the method to be successful and the method became an official AOAC method for aflatoxin M_1 (AOAC Official Method 974.17, 1990). AOAC is abbreviation for Association of Official Analytical Chemists.

In another case of liquid-liquid separation, extraction of aflatoxin from liquid milk was made with chloroform in a separating funnel and then extract was cleaned-up over a small silica gel column. Finally the separation was made by TLC and detection was made with fluorescence (Stubblefield, 1979). After modifications, this method was applied for determination of aflatoxin in cheese, in which two-dimensional TLC was applied to improve separation of the aflatoxin spots from the background. An AOAC/ IUPAC collaborative study evaluated the method (Stubblefield, et al., 1980) and it became an official AOAC method for aflatoxin M_1 in milk and cheese (AOAC Official Method 980.21, 2000).

The various mixtures of methanol – water (Masri, et al., 1969), acetone - water and acetone - chloroform – water (Purchase & Steyn, 1967) were used for the extraction of aflatoxins.

Liquid-liquid separation is a simple procedure and involves inexpensive equipment. Its disadvantages include contamination and loss of sample by adsorption to the glassware, as there are several steps. Large volumes of solvents are used and have to be disposed and these create pollution problems. In trace analysis, solvents with high purity have to be used which are highly costly.

2.2 Solid phase extraction (SPE)
The most significant recent improvement in the purification step, in aflatoxin analysis, is the use of solid-phase extraction (SPE). The use of solid-phase extraction with C-18 cartridges is now well established in aflatoxin determination. Solid phase extraction is suitable for the

analysis of aqueous samples. It can be performed on-line as well as off-line. Solid phase extraction process starts with conditioning of the column by activating it with the solvent. The sample is then applied and the analyte is trapped in the column. The interferences are removed by rinsing step. Finally, the analyte is eluted and then pre-concentration step is employed by evaporating excess solvent with nitrogen. A number of samples can be prepared simultaneously with the use of vacuum manifold. A vacuum manifold is shown in Photograph 1.

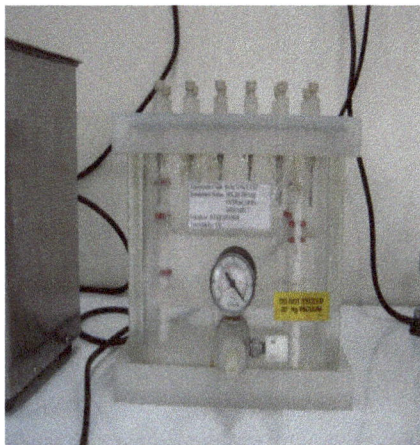

Photograph 1. A Vacuum manifold

Most frequently C-8 and C-18 bonded silica columns are used and these are very pressure resistant and give reproducible results. There is no significant drawback in case of SPE as compared to liquid-liquid separation. Its advantages include the consumption of less solvent, less time, and the possibility of automation. Photograph 2 shows some SPE C-18 columns.

A typical case example of C-18 cartridge use in aflatoxin analysis is that of the study of Bijl et al. (1987). They proposed a simple and sensitive method for the determination of aflatoxin M_1 in cheese. The ground cheese sample is extracted with acetone-water mixture (3+1). Acetone is evaporated under vacuum, and the aqueous phase is passed through a C-18 disposable cartridge. After cartridge is washed with acetonitrile-water mixture (1+9), the toxin is eluted with acetonitrile. The extract is then cleaned up on a silica cartridge. Final analysis is performed by two dimensional thin layer chromatography combined with fluoro-densitometry or by liquid chromatography on a reverse phase C-18 column with fluorescence detection. Recovery is greater than 90%, the coefficient of variation is 6% or less. The detection limit is in the range 10 ng/kg.

Application of C-8 (SPE) clean-up was shown by Manetta et al. (2005). They developed a new HPLC method with fluorescence detection using pyridinium hydrobromide perbromide as a post-column derivatizing agent to determine aflatoxin in milk and cheese. The detection limits for milk and cheese were 1 ng/ kg and 5ng/ kg respectively. The calibration curve was linear from 0.001 to 0.1 ng injected toxin. The method includes a preliminary C-8 (SPE) clean-up. The average recoveries of aflatoxin M_1 from milk and cheese, spiked at levels of 25-75 ng/ kg and 100-300 ng/ kg, respectively, were 90 and 76%.

The precision (RSD$_r$) ranged from 1.7 to 2.6% for milk and from 3.5 to 6.5% for cheese. The method is rapid and easily automatable and therefore is useful for accurate and precise screening of aflatoxin in milk and cheese.

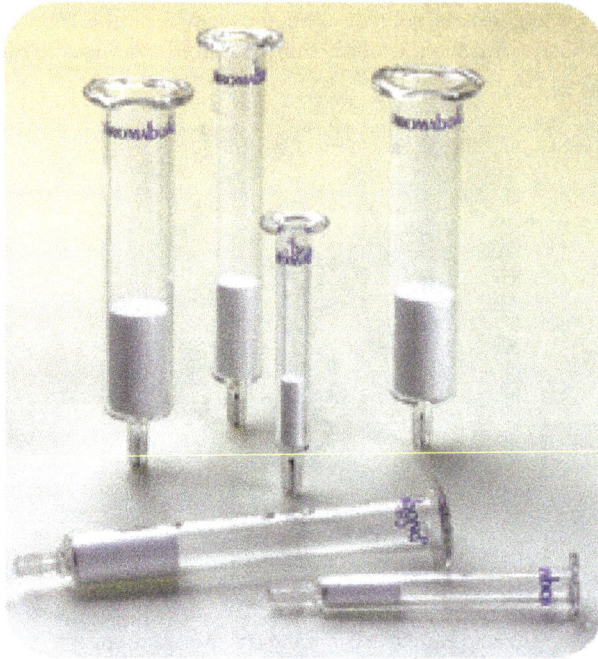

Photograph 2. Solid phase extraction C-18 columns

2.3 Immunoaffinity columns (IACS)

The immunoaffinity clean-up procedure was expanded in order to encompass successfully the determination of aflatoxins. Now, immunoaffinity columns have become increasingly popular in recent years for clean-up purposes, because these offer high selectivity and are easy to use. These can be applied for purification of samples that are contaminated with different aflatoxins. Aflatoxins are low weight molecules and they are only immunogenic if they are bound to a protein carrier. Antibodies are produced for aflatoxins. These antibodies are bound to an agarose, sepharose, or dextran carrier and packed in a column. The analyte molecules (aflatoxins) are bound selectively to the antibodies in the column. The matrix components do not interact with the antibodies and a rinsing (washing) step removes most of the possible interferences. The toxin can be eluted with a solvent causing antibody denaturation. Immunoaffinity columns have higher recovery than liquid-liquid partitioning. Single analyte columns are available and multifunctional columns for simultaneous determination of a number of mycotoxins are also available. Major disadvantages include the high costs and the fact that a column can be used once due to the denaturation of antibodies during elution step. Immunoaffinity columns are available commercially. Immunoaffinity column (AflaTest-Vicam, USA) is shown in Photograph 3.

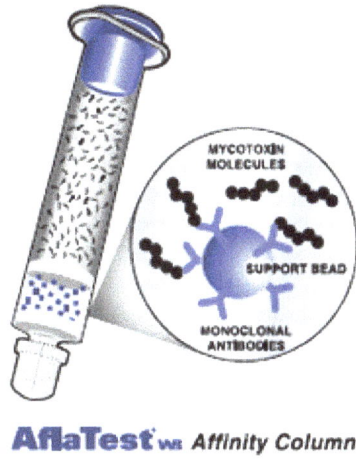

AflaTest we *Affinity Column*

Photograph 3. Immunoaffinity column (AflaTest-Vicam, USA)

2.4 Mycosep™ columns

Mycocep™ columns, which remove matrix components with efficiency and can produce a purified extract within a short time, are also available. The Mycosep™ multifunctional clean-up columns (Romer Labs Inc., Union, MT, USA) consist of a number of adsorbents (charcoal, celite, ion exchange resins and others) which are packed in a plastic tube. On the

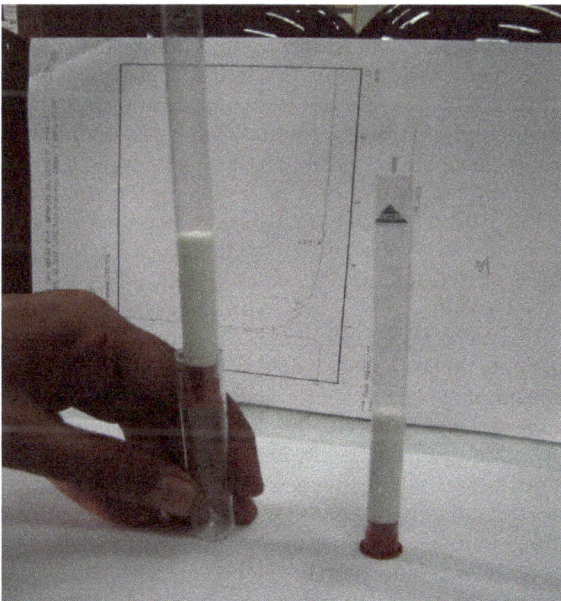

Photograph 4. Mycosep™ Columns of Romer Labs Inc., USA

lower end of the Mycosep™ column, there is a rubber flange, a porous frit and one-way valve which allow the extract to force through the packing material, when the column is inserted into the culture tube (glass tube). The purified extract appears on the top of the plastic tube with in seconds. Almost all interfering substances are retained on the column, whereas the analyte does not show significant affinity to the packing material. No additional washing steps are required as in solid phase extraction. Columns are available for a range of mycotoxins and are usually suitable for one analyte. Photograph 4 shows Mycosep™ column of Romer Labs Inc., USA.

3. Detection techniques

After the extraction of the analyte (aflatoxin) from the sample and applying a clean-up procedure to remove interferences, then comes identification and quantification in the last in the analytical methodology. For the detection of aflatoxins, three main types of assays have been developed. These include biological, analytical and immunological methods. The biological methods were used when analytical and immunological methods were not available for routine analysis. Biological assays are non-specific and time consuming and are qualitative in nature.

3.1 Analytical methods
Many analytical methods have been developed and are available for estimation of aflatoxins in agricultural commodities. These include: thin-layer chromatography, high performance thin-layer chromatography, and high-performance liquid chromatography.

3.1.1 Thin-layer chromatography (TLC)
Thin layer chromatography is also known as flat-bed chromatography or planar chromatography and is one of the most widely used techniques in aflatoxin analysis. TLC is a chromatographic technique which is used for the separation, purity assessment and identification of aflatoxins. TLC can identify and quantify aflatoxins at levels as low as 1ng/g. Thin-layer chromatography consists of a stationary phase immobilized on a glass or plastic plate and a solvent acting as a mobile phase. The sample, either liquid or dissolved in a volatile solvent, is applied in the form of a spot on the stationary phase. Then the chromatographic plate is placed vertically in a solvent reservoir and the solvent moves up the plate by capillary action. When the solvent front reaches a certain limit of the stationary phase, the plate is removed from the solvent reservoir. The separated spots are then visualized with ultraviolet light or by spraying with a suitable reagent. The contents of a sample can be identified by running standards simultaneously with the unknown spots. The different components in a mixture move up the plate at different rates due to differences in their partitioning behavior between the mobile liquid phase and the stationary phase. The R_f value for each spot is calculated. It is the ratio of the distance (cm) from start to centre of sample spot and distance (cm) from start to solvent front. R_f stands for "ratio of fronts" or "retardation factor". It is characteristic for a given compound on the same stationary phase using the same mobile phase under same conditions of development of the plate. For identification purposes, R_f values of standards are compared to those of unknown samples. A number of methods have been developed for the determination of aflatoxins by TLC. Silica plates are mostly used with a number of solvent mixtures. Mostly the solvent systems are based on chloroform and small amounts of methanol or acetone. Now-a-days,

less toxic and environmental friendly solvent mixtures (e.g. toluene/ethyl-acetate or acetone/ iso-propanol) are also employed. Aflatoxins are strongly fluorescent (exitation λ= 365 nm, detection or emission λ= 430 nm) by themselves and can easily be detected by fluorodensitometry.

Thin layer chromatography is the standard AOAC method for aflatoxin analysis since 1971, AOAC Official Method 971.24, First Action 1971 and Final Action 1988 (AOAC Official Method 971.24, 2000). TLC separation of aflatoxins provided basis for sensitive analytical techniques. TLC quantification method gives a reasonable level of selectivity and sensitivity to separate aflatoxins from other interfering compounds. TLC is the method of choice for rapid screening of aflatoxins and for situations where advanced techniques equipments are not available.

A typical case application of TLC in aflatoxin analysis is that of Van Egmond et al. (1978). They confirmed the identification of aflatoxin M_1 on thin layer plate by reacting aflatoxin M_1 with trifluoroacetic acid (TFA). In the method the plate was developed with chloroform-methanol-acetic acid-water (92+8+2+0.8) mixture. The R_f value of the blue fluorescent derivative was compared with that of the aflatoxin M_1 standard which was also spotted on the TLC plates.

3.1.2 High performance thin-layer chromatography (HP-TLC)

There is lack of precision associated with TLC procedures due to the introduction of possible errors during the sample application, plate development, and plate interpretation steps. High performance thin-layer chromatography methods improve the precision by automating the sample application and plate interpretation steps. This technique is less commonly used as compared to HPLC, which is more sophisticated as compared to this.

3.1.3 High performance liquid chromatography (HPLC)

Analytical laboratories moved away from TLC to HPLC determination with advances in HPLC methods in 1980s. High performance liquid chromatography is a very precise and highly automated quantification technique for aflatoxins analysis with high selectivity and sensitivity. Now-a-days, HPLC methods are widely used because of their superior performance and reliability as compared with TLC. HPLC methods have been developed for all major mycotoxins in cereals and other agricultural commodities. In the field of analysis of aflatoxins, HPLC is mainly used for final separation and detection of the analyte of the interest and extraction and clean-up techniques have to be applied prior to detection with HPLC.

In HPLC, a liquid mobile phase or solvent is used to move the sample through the column. An immobilized liquid stationary phase is packed in the column. The analyte is then partitioned between the two phases as it passes through the column and thus leading to the separation of compounds due to the difference in partitioning coefficients. Two types of HPLC methods are commonly used i.e., normal phase chromatography and reversed phase chromatography. In normal phase chromatography, a polar stationary phase e.g. silica gel and a non-polar solvent e.g. hexane are used. Whereas reversed-phase chromatography (RP-HPLC) employs non-polar stationary phase e.g., C-8 or C-18 hydrocarbons and polar mobile phase e.g. water, methanol or acetonitrile. In HPLC, detection is mainly accomplished by using ultra violet (UV) detector, diode array detector (DAD) or a fluorescence detector (FLD). Fluorescence detection utilizes the emission of light (435 nm) from molecules that have been excited to higher energy levels by absorption of electromagnetic radiation (365 nm) for aflatoxins. Fluorescence detection has superior sensitivity than other detection systems and sometimes derivatization of the analyte has to be performed which enhances the sensitivity. Fluorescence detection is possible in the range of microgram/kg. Choice of detector usually depends on the nature of the sample.

RP-HPLC is commonly performed for determination of aflatoxins in foods. Stationary phase for aflatoxins include C-18 material. Pre- or post-column derivatization is necessary for low-level detection. For aflatoxins, derivatization is performed with strong acids or oxidants e.g., Br_2, I_2 or trifluoro-acetic acid. This results in increase of fluorescence by a factor 20. Sometimes, a pre-column is employed to avoid heavy contamination or subsequent blocking of main separation column.

The HPLC systems of Shimadzu (Japan) and Agilent (USA) are very commonly used and these are highly sophisticated. All the HPLC systems are comprised of many components. The main components are: liquid pump, column oven, system controller, detectors (fluorescent detector, ultra violet (UV) detector, diode array detector i.e., DAD), communication bus module i.e., CBS and data acquisition software. Photograph 5 shows HPLC System.

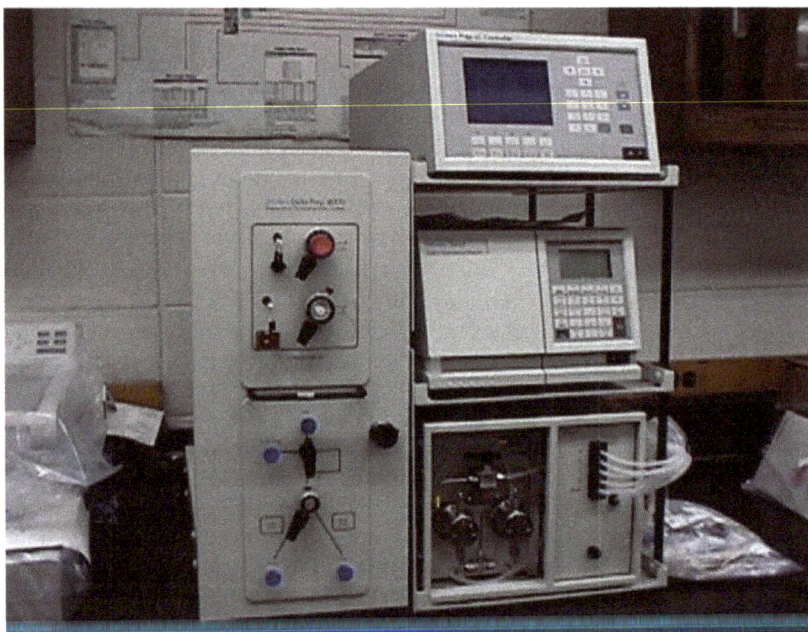

Photograph 5. High-performance liquid chromatography system

HPLC system gives result in the form of chromatogram. A chromatogram gives two types of analytical informations: one is qualitative and the other one is quantitative . In the HPLC chromatogram, retention time is given on x-axis, while on y-axis height of the peak is given. Retention time is used for identification purposes and area of the peak is used for quantitative purposes. A sample HPLC chromatogram is shown in the Fig. 1. The Fig. 1 shows two graphs, i.e., "A" and "B". Graph "A" is for standard and graph "B" is for sample. By comparing the two graphs, identification of the unknown compound is made. After identification, quantitation is done from the area of the peaks.

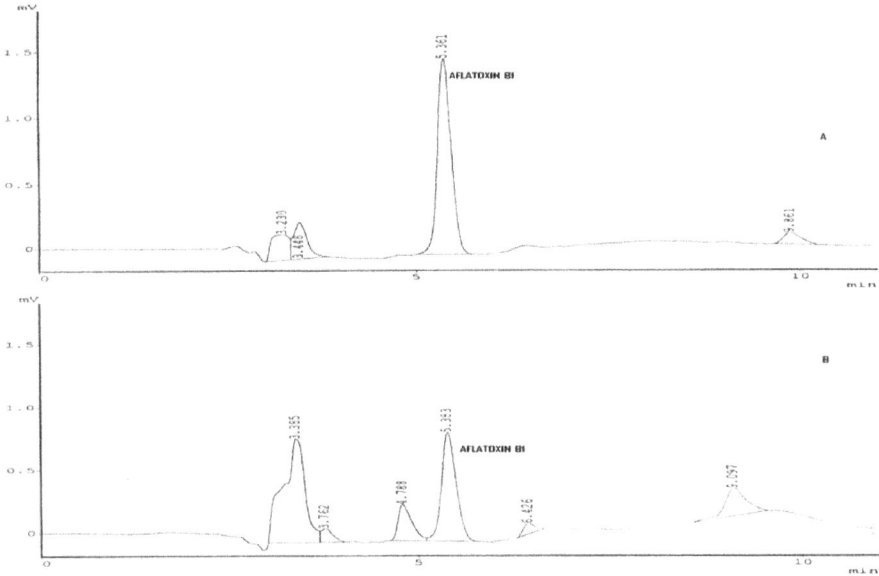

Fig. 1. A sample HPLC chromatogram, (A) for standard and (B) for sample

3.1.4 Liquid chromatography with mass spectrometric detection

Liquid chromatography with mass spectrometric detection (LC-MS) is fairly a recent development in aflatoxins detection and it is one of the most advanced techniques. It is time-consuming and requires expert knowledge. In mass spectrometric detection, extraction and clean-up techniques have to be applied before detection. In LC-MS, the HPLC effluent enters an ionization chamber via a nebulizer. There are several techniques for ionization, namely electrospray, thermo-spray, chemical and fast atom bombardment. Fragmentation takes place in a collision chamber. The fragments then enter the high vacuum region of the mass spectrometer, where detection takes place. Several set-ups are available for optimal identification and quantification. Ion trap instruments are more suitable for identification than triple quadruple instruments (higher MSn power), whereas triple quadruple instruments provide better information for quantification with faster scanning and higher sensitivity. There are also available hybrid instruments that provide a linear ion trap in a triple quadruple instrument to get the best results out of both set-ups. LC-MS methods have their applications in determination of aflatoxins in corn, milk and samples of other commodities. In Selection-Ion-Monitoring (SIM) mode, detection can be made at pico-grams levels.

3.2 Immunological methods

Immunological methods are based on the affinities of the monoclonal or polyclonal antibodies for aflatoxins. Due to the advancement in biotechnology, highly specific antibody-based tests are now commercially available for measuring aflatoxins in foods in less than ten minutes. There are two major requirements for immunological methods. First requirement is high quality antibodies and second is methodology to use the antibodies for the estimation of aflatoxins. Being low molecular weight molecules, aflatoxins cannot stimulate the immune system for the production of antibodies. Such molecules of low molecular weight, which cannot evoke the immune system, are called haptens. Therefore, before immunization, aflatoxins must be conjugated to a carrier molecule which is a larger molecule like proteins. Bovine serum albumin (BSA) is most commonly used as a carrier protein and hapten is conjugated with it. The three types of immunochemical methods are: immunuaffinity column assay (ICA), radioimmunoassay (RIA), and enzyme-linked immunosorbent assay (ELISA). Immunoaffinity columns are mainly used for clean-up purposes and RIA has limited use in aflatoxins analysis. ELISA is most commonly used for the estimation of aflatoxins.

Many rapid tests, using specific antibodies for isolation and detection of mycotoxins in food have been discussed and applied by various workers (Newsome, 1987; Groopman & Donahue, 1988). Use of immunoaffinity cartridges is a more recent advance in quantitative extraction of aflatoxin. Monoclonal antibodies specific for aflatoxin are immobilized on Sepharose® and packed into small cartridges. The work of Mortimer et al. (1987) is very important as it is the first published method for aflatoxin M_1 with immunoaffinity columns. For the aflatoxin determination, a milk sample is loaded onto the affinity column. The antigen i.e., aflatoxin is selectively complexed by the specific antibodies on the solid support to form antigen-antibody complex. Then, the column is washed with water to remove all other matrix components of the sample. A small volume of pure acetonitrile is used to elute the aflatoxin and the eluate is concentrated and analyzed by HPLC coupled with fluorescence detection.

Many collaborative studies were done to develop the immunological methods; especially for aflatoxin M_1. Immunoaffinity-based methods for aflatoxin M_1 were modified and subsequently published and studied collaboratively under the auspices of the International Dairy Federation and AOAC international by groups of mainly European laboratories that could determine aflatoxin M_1 in milk at concentrations equal to 0.05 µg/ L. The collaborative study of Tuinstra et al. (1993) led to International Dairy Federation Standard 171. Another collaborative study (Dragacci et al. 2001) was conducted to evaluate the effectiveness of an immunoaffinity column clean-up liquid chromatographic for determination of aflatoxin M_1 in milk at proposed European regulatory limits. The procedure included centrifugation, filtration, and application of the test portion to an immunoaffinity column. Then the column was washed with water and aflatoxin was eluted with pure acetonitrile. Aflatoxin was separated by reversed-phase liquid chromatography and detection was made with fluorescence detector. Liquid milk samples (frozen), both naturally contaminated with aflatoxin M_1 and blank samples for spiking, were sent to 12 collaborators in 12 different European countries. Test portions of milk samples were spiked at 0.05ng aflatoxin M_1 per mL. After the removal of two non-compliant sets of results, the mean recovery of aflatoxin M_1 was 74%. The relative standard deviation for repeatability (RSDr) ranged from 8 to 18%, based on results of spiked samples (blind pairs at 1 level) and naturally contaminated samples (blind pairs at 3 levels). The relative standard deviation for reproducibility (RSDR) ranged from 21 to 31%. As evidenced by HORRAT values at the low level of aflatoxin M_1 contamination, the method showed acceptable within and between laboratory precision data for liquid milk. The collaborative study resulted in approval of AOAC Official Method 2000.08 (AOAC Official Method 2000.08, 2005).

3.2.1 Enzyme-linked immunosorbent assay (ELISA)

The enzyme-linked immunosorbent assay is most widely used test to detect aflatoxins, due to its simplicity, sensitivity and adaptability. There are two types of enzyme-linked immunosorbent assay, which are direct competitive enzyme-linked immunosorbent assay and indirect competitive enzyme-linked immunosorbent assay. In direct competitive enzyme-linked immunosorbent assay method, specific antibody is coated to a solid phase such as a microtiter plate, whereas in indirect competitive enzyme-linked immunosorbent assay method, toxin-protein conjugate is coated onto the microtiter plate. In aflatoxin analysis, direct competitive enzyme-linked immunosorbent assay is used. The enzyme-linked immunosorbent assay is detection and quantification of an antigen (aflatoxin) in a sample by using an enzyme labeled toxin and antibodies specific to aflatoxin. The enzyme-linked immunosorbent assay is based on antigen-antibody reaction (Aycicek et al., 2005). Antigen is that substance which can elicit production of antibodies when introduced into warm blooded animals. Whereas antibodies are glycoproteins which are produced as a result of an immune response, after introduction of antigens, leading to the production of a specific antigen-antibody complex. In the direct competitive enzyme-linked immunosorbent assay, specific antibodies for aflatoxin are coated on to the wells in the microtiter strip. The test samples or aflatoxin standards are added to the wells. After incubation and washing, enzyme conjugate (a conjugate of aflatoxin and bovine serum albumin is attached with an enzyme molecule, such as, horseradish peroxidase or penicillinase or alkaline phosphatase) is added to the wells. Free aflatoxin and aflatoxin enzyme conjugate compete for the aflatoxin antibody sites in the wells. Washing step removes any unbound enzyme conjugate. Then substrate/chromogen is added to the wells and incubated. The bound enzyme conjugate converts the colorless chromogen into a blue product. The stop solution is added which leads to color change from blue to yellow. Then measurement is made photometrically at 450 nm in an ELISA reader. The absorbance is inversely proportional to the aflatoxin concentration in the sample i.e., the lower the absorbance, the higher the aflatoxin concentration.

The main instrument used in enzyme-linked immunosorbent assay is the ELISA reader. It is basically a photometric instrument which gives the absorbance of the solution at the end of the process. The whole process has been described with complete details in the past paragraph. An ELISA reader is shown in Photograph 6.

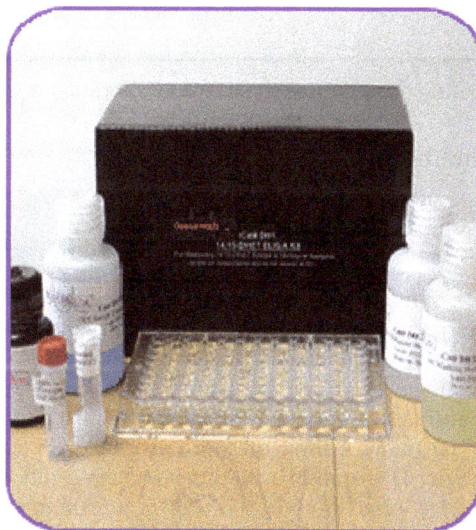

Photograph 6. An ELISA reader

A sample enzyme-linked immunosorbent assay calibration curve is shown in the Fig. 2. The ELISA reader gives absorbance readings from which % absorbance is calculated. For standard solutions, the % absorbance is plotted against aflatoxin concentration to get the calibration curve. The aflatoxin concentration is on x-axis and % absorbance is on y-axis. From the calibration curve, aflatoxin concentration is calculated for samples.

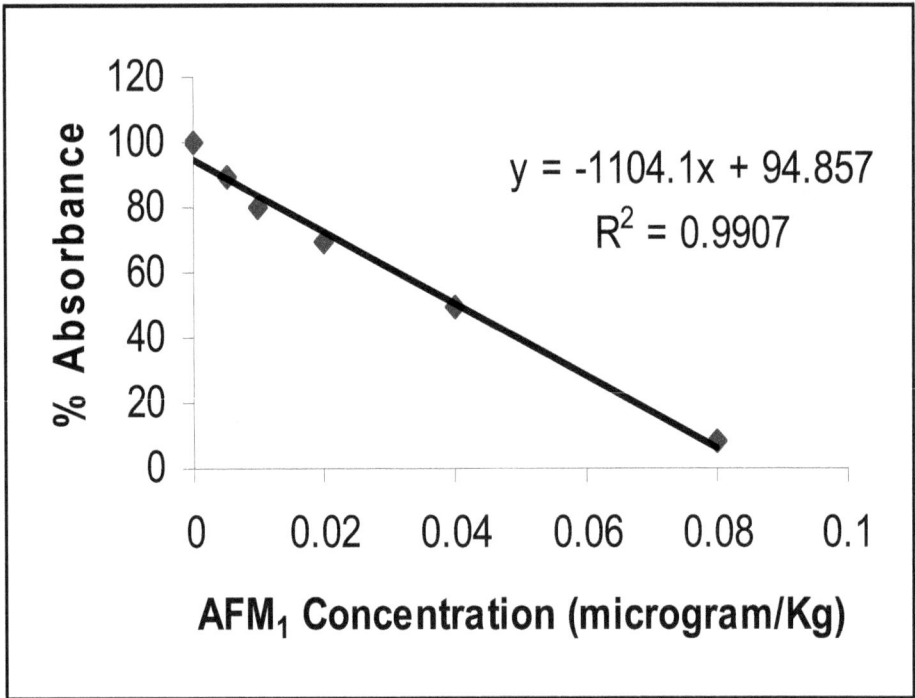

Fig. 2. A sample enzyme-linked immunosorbent assay calibration curve

4. Typical complete methods

Some typical methods are given completely to make the understanding of the process of aflatoxin analysis.

4.1 Determination of aflatoxin M_1 with fluorometer

This is very simple and efficient method. It is also a specific method. The analysis is carried out with Fluorometer along with the use of affinity chromatography columns for clean-up step according to the method described by Hansen (1990). Before analysis, sample is brought to room temperature. To remove cream from the milk sample, it is centrifuged at $2000 \times g$ for 10 minutes. The 10 mL sample of skim milk is passed through AflaTest affinity column of the Vicam, USA. These affinity columns contain antibodies to aflatoxin. The column is then washed twice with 10 mL portions of 10% methanol and the aflatoxin M_1 is eluted from the affinity column by passing 1.0 mL of 80% methanol. All the sample eluate (1.0 mL) is collected in a glass cuvette.

The concentration of aflatoxin M_1 is measured in a fluorometer (Vicam, USA) with the option of 360 nm excitation filter and 440 nm emission filter. The results are recorded using digital Fluorometer readout with automatic printing device.

4.2 Determination of aflatoxin M_1 by HPLC

A very competent method used for determination of aflatoxin M_1 is that of the AOAC Official Method 2000.08 (AOAC Official Method 2000.08, 2005). Details of the method are given in coming lines.

4.2.1 Extraction procedure

After warming at about 37°C in water bath, liquid milk is centrifuged at 2000×g to separate fat layer and then filtered. The prepared test portion of 50 mL is transferred into syringe barrel attached with immonoaffinity column (IAC) and passed at slow steady flow rate of 2-3 mL/ min. The washing of column is done with 20 mL water and then it is blown to dryness and afterwards aflatoxin M_1 is eluted with 4 mL pure acetonitrile by allowing it to be in contact with the column at least 60 seconds. The eluate is evaporated to dryness using gentle stream of nitrogen and at the time of LC (liquid chromatography) determination it is diluted with the mobile phase.

4.2.2 LC Determination with fluorescence detection

The HPLC system of Agilent 1100 series (Agilent, USA), equipped with an auto sampler LAS G1313A and a fluorescence detector FLD G1321A with excitation and emission wavelength of 365nm and 435nm respectively, may be used for aflatoxin M_1 determination. Any other suitable system may be used instead of the above mentioned system. The ZORBAX Eclipse XDB-C18 (Octadecyl silane chemically bonded to porous silica) column (Agilent, USA), 4.6×150 mm with particle size 5 µm in diameter, may be used. Acetonitrile in ratio of 25% with 75% water is used as mobile phase. The flow rate is 0.8 mL/min. Calibration curve is determined using a series of calibration solutions of aflatoxin M_1 in acetonitrile. The concentrations of calibration curves may be in the range of 0.05, 0.1, 0.5, 1.0, 5.0, and 10.0 µg/ L. The retention time for aflatoxin M_1 may be in the range of 6.1-6.5 min.

4.2.3 Calculations

Calculations are made according to the following equation:

$$W_m = W_a \times (V_f/V_i) \times (1/V_s)$$

Where W_m = amount of aflatoxin M_1 in the test sample in µg/L; W_a = amount of aflatoxin M_1 corresponding to area of aflatoxin M_1 peak of the test extract (ng); V_f = the final volume of re-dissolved eluate (µL); V_i = volume of injected eluate (µL); V_s = volume of test portion (milk) passing through the column (mL).

4.3 Determination of aflatoxin B_1 by HPLC

An important method for the determination of aflatoxin B_1 is that of the AOAC Official Method 994.08 which has been described here with small modifications (AOAC Official Method 994.08, 2000).

4.3.1 Extraction and clean-up procedure
A test portion of 50.0g and 100mL extraction solvent (850mL acetonitrile with 150mL deionized water) is taken in 250mL Erlenmeyer flask and placed in a shaker for 1 hour at high speed. After filtration, 8mL extract is taken with pipette in 10mL glass tube. MycoSep® column (rubber flange end) is pushed slowly into the tube. As column is pushed into the tube, extract is forced through frit, through 1-way valve, and through packing material and is collected in column reservoir. The purified extract (2mL) is transferred quantitatively from top of column to screw cap vial (derivatization vial) and is evaporated under nitrogen.

4.3.2 Aflatoxin derivatization
After adding n-hexane (200µL) in the derivatization vial to re-dissolve aflatoxin, 50µL of trifluoroacetic acid is added and it is mixed on vortex mixer for 30 seconds. After five minutes, 1.95mL of deionized water: acetonitrile (9:1) mixture is added and again mixed on vortex mixer for 30 seconds. Layers are allowed to separate and aqueous layer (lower layer) containing aflatoxins is removed, filtered through 0.45µm syringe filter and then injected onto LC column.

4.3.3 LC Determination with fluorescence detection
The high-performance liquid chromatography equipment (LC-10, Shimadzu, Japan), comprising liquid pump LC-10AS, column oven CTO-10A, system controller SCL-10A, fluorescence detector RF-530, communication bus module CBM-101, and data acquisition software class LC-10A may be used for aflatoxin B1 determination. Any other suitable system may also be used instead of the above said one. The excitation wavelength of 365nm and emission wavelength of 435nm is set during analysis. The stainless steel column Discovery® C-18 of Supelco (Bellifonte, PA, USA) with dimensions of 25cm×4.6mm (id) and with particle size of 5 µm diameter may be used. The mobile phase (acetonitrile: methanol: deionized water in the ratio of 20:20:60) is degassed with sonicator before use. The flow rate is 1.0 mL/ min. Calibration curve is determined using a series of calibration solutions of aflatoxin B_1 in acetonitrile. The concentrations of calibration solutions may be 0.5, 1.5, 2.5, 5.0, 10.0, and 15.0 µg/ L. The retention time for aflatoxin B_1 is near to 5.36 minutes or may be slightly different by changing conditions or instrument.

4.3.4 Calculations
Aflatoxin B_1 peak is identified in derivatized extract chromatogram by comparing its retention time with corresponding peak in the standard chromatogram. The quantity of the aflatoxin B_1, 'C' is determined in the derivatized extract (injected) from the respective standard curves. The concentration of aflatoxin B_1 is calculated in test sample as follows:

$$\text{Aflatoxins } B_1 \text{ ng/g} = C/W$$

$$W = 50g \times (2mL/\ 200mL) \times (0.02mL/\ 2mL) = 0.005g$$

Where W = equivalent weight of test portion (in 20µL) injected into LC; C = ng aflatoxin (in 20µL) injected into LC.

4.4 Determination of aflatoxin M_1 in cheese by ELISA

An ELISA method for the determination of aflatoxin M_1 in cheese is given here as described by the protocol provided with RIDASCREEN® ELISA kit (RIDASCREEN® Aflatoxin M_1 30/15, 2007).

4.4.1 Sample preparation

Cheese (2.0g) samples are first of all triturated.   Extraction is completed with 40 mL dichloromethane by shaking for 15 minutes. The suspension is filtered and 10 mL of the extract is evaporated at 60°C under weak N_2-stream. The oily residue is re-dissolved in 0.5 mL methanol, 0.5 mL PBS buffer and 1 mL n-heptane. After mixing thoroughly, it is centrifuged for 15 minutes at 2700 × g. The upper heptane layer is removed completely. From the lower methanolic-aqueous phase, 100µL is taken and diluted with 400 mL buffer 1 and 100µL of it is used per well in the test.

4.4.2 Test procedure

The standard solutions (100µL) and prepared samples (100µL) are added into the microtiter well placed in the microwell holder. Gentle mixing is accomplished by shaking the plate manually and incubated for 30 minutes at room temperature (20-25°C) in the dark. The liquid is poured out of the wells and microwell holder is tapped vigorously upside down against adsorbent paper to ensure complete removal of liquid from the wells. The wells are washed by adding 250 µL washing buffer in each well and poured out the liquid again. Washing step is repeated for two times. Then 100 µL of the diluted enzyme conjugate is added and mixed gently by shaking the plate manually and incubated for 15 minutes at room temperature in the dark. After incubation the wells are washed again. The 100 µl of substrate/chromogen is added and mixed gently by shaking the plate manually and incubated for 15 minutes at room temperature in the dark. Now stop solution (100µL) is added in each well. Mixing is done by shaking the plate manually. The absorbance is measured photometrically at 450 nm against an air blank with in 15 minutes after the addition of stop solution.

4.4.3 Calculations

The following formula is used to measure the % absorbance.
(Absorbance of standard or sample / absorbance of zero standard) × 100 = % absorbance
The zero standard is made equal to 100 % and absorbance values are taken in percentages. A calibration curve is obtained by plotting %absorbance values for the standards against the aflatoxin M_1 concentration (µg/L).   The concentration of aflatoxin M_1 in samples is calculated from the calibration curve.

5. Conclusion

The methods of measurement and analysis of aflatoxins have been discussed in this chapter. Some photographs were taken by the author himself, while others were downloaded from internet. Some analytical studies in the aflatoxin analysis have been included to have the insight of methods' application and their development. Typical complete methods have been included as exemplary methods, so the understanding of the process of aflatoxin analysis may become clear.

6. Acknowledgement

I am profoundly grateful to Almighty Allah and bow my head to Him for giving me light to complete this manuscript.

7. References

AOAC Official Method 974.17 (1990). Aflatoxin M_1 in dairy products, thin-layer chromatographic method. Natural Poisons-chapter 49 (pp. 1199-1200). *Official Methods of Analysis of the AOAC,* 15th edition, AOAC Inc. Arlington, Virginia 22201, USA

AOAC Official Method 971.24 (2000). AOAC Official Method 971.24: Aflatoxins in coconut, copra, and copra meal. Natural Toxins-chapter 49 (pp. 14-15). *Official Methods of Analysis of AOAC International,* 17th edition, volume I, AOAC International, Gaithersburg, Maryland 20877-2417, USA

AOAC Official Method 980.21 (2000). Aflatoxin M_1 in milk and cheese, thin-layer chromatographic method. Natural Toxins-chapter 49 (pp. 37-38). *Official Methods of Analysis of AOAC International,* 17th edition, volume II, AOAC International. Gaithersburg, Maryland 20877-2417, USA

AOAC Official Method 994.08 (2000). Aflatoxins in corn, almonds, Brazil nuts, peanuts, and pistachio nuts, multifunctional column (Mycosep) method. Natural toxins-chapter 49 (pp. 26-27). *Official Methods of Analysis of AOAC International,* 17th edition, volume II, AOAC International, Gaithersburg, Maryland, USA

AOAC Official Method 2000.08 (2005). Aflatoxin M_1 in liquid milk, immunoaffinity column by liquid chromatography. Natural Toxins-chapter 49 (pp. 45-47). *Official Methods of Analysis of AOAC International,* 18th edition, AOAC International. Gaithersburg, Maryland 20877-2417, USA

Aycicek, H., Aksoy, A. & Saygi, S. (2005). Determination of aflatoxin levels in some dairy and food products which consumed in Ankara, Turkey. *Food Control,* Vol.16, No.3, pp. 263-266

Bijl, J. P., Van Peteghem, C. H., & Dekeyser, D. A. (1987). Fluorimetric determination of aflatoxin M_1 in cheese. *Journal of AOAC,* Vol.70, No.3, pp. 472-475

Chu, F. S. (1991). Detection and determination of mycotoxins. In: *Mycotoxins and Phytoalexins,* R. P. Sharma and D. K. Salunkhe, Boca Raton, Florida(Eds.), pp. 33-79, CRC Presss

Dragacci, S., Grosso, F., & Gilbert, J. (2001). Immunoaffinity column clean-up with liquid chromatography for determination of aflatoxin M_1 in liquid milk: Collaborative study. *Journal of AOAC International,* Vol.84 No.2, pp. 437-443

Groopman, J. D. & Donahue, K. F. (1988). Aflatoxin, a human carcinogen: determination in foods and biological samples by mono-colonial antibody affinity chromatography. *Journal of Association of Official Analytical Chemists,* Vol.71, pp. 861-867

Hansen, T.J. (1990). Affinity column cleanup and direct fluorescence measurement of aflatoxin M_1 in raw milk. *Journal of Food Protection,* Vol.53, No.1, pp. 75-77

Holcomb, M., Wilson, D. M., Trucksess, M. W. & Thompson, H. C., Jr, (1992). Determination of aflatoxins in food products by chromatography. *Journal of Chromatography,* Vol.624, pp. 341-352

Jacobson, W. C., Harmeyer, W. C., & Wiseman, H.G. (1971). Determination of aflatoxins B_1 and M_1 in milk. *Journal of Dairy Science,* Vol.54, pp.21-24

Johansson, A. S., Whitaker, T. B., Giesbrecht, F. G., Hagler, W. M., Jr, & Young, J. H. (2000). Testing Shelled corn for aflatoxin, Part III: Evaluating the performance of aflatoxin sampling plans. *Journal of AOAC International,* Vol.83, pp. 1279-1284

Lee, J. E., Kwak, B. M., Ahn, J. H., & Jeon, T. H. (2009). Occurrence of aflatoxin M_1 in raw milk in South Korea using an immunoaffinity column and liquid chromatography. *Food Control,* Vol.20, No.2, pp. 136-138

Manetta, A. C., Giuseppe, L. D., Giammarco, M., Fusaro, I., Simonella, A., Gramenzi, A., & Formigoni, A. (2005). High-performance liquid chromatography with post-column derivatisation and fluorescence detection for sensitive determination of aflatoxin M_1 in milk and cheese. *Journal of Chromatography A,* Vol.1083, No.1, pp. 219-222

Masri, M. S., Page, J. R., & Gracia, V. C. (1969). Modification of method for aflatoxins in milk. *Journal of Association of Official Analytical Chemists,* Vol.52, pp. 641-643

McKinney,J. D. (1972). Determination of aflatoxin M_1 in raw milk: A modified Jacobson, Harmeyer, and Wiseman method. *Journal of American and Oil Chemical Society,* Vol.49, pp. 444-445

Mortimer, D. N., Gilbert, J., & Shepherd, M. J. (1987). Rapid and highly sensitive analysis of aflatoxin M_1 in liquid and powdered milks using affinity column clean-up. *Journal of Chromatography,* Vol.407, pp. 393-398

Newsome, W. H. (1987). Potential and advantages of immunochemical methods for analysis of foods. *Journal of Association of Analytical Chemists,* Vol.69, pp. 919-923

Purchase, I. F. H., & Steyn, M. (1967). Estimation of aflatoxin M_1 in milk. *Journal of Association of Official Analytical Chemists,* Vol.50, pp. 363-364

RIDASCREEN® Aflatoxin M_1 30/15 (2007). Enzyme immunoassay for the quantitative analysis of aflatoxins M_1. *RIDASCREEN® Aflatoxin M_1 30/15,* Instruction booklet, pp. 1-18, R-Biopharm AG, Darmstadt, Germany

Stubblefield, R. D. (1979) The rapid determination of aflatoxin M_1 in dairy products. *Journal of American Oil and Chemical Society,* Vol.56, pp. 800-802

Stubblefield, R. D., & Shannon, G. M. (1974). Collaborative study of methods for the determination of aflatoxin M_1 in dairy products. *Journal of Association of Official Analytical Chemists,* Vol.57, pp. 852-857

Stubblefield, R. D., van Egmond, H. P., Paulsch, W. E., & Schuller, P. L. (1980). Determination and confirmation of identity of aflatoxin M_1 in dairy products: collaborative study. *Journal of Association of Official Analytical Chemists,* Vol.63, pp. 907-921

Tuinstra, L. G. M. T., Roos, A. H., & Van Trijp, J. M. P. (1993). Liquid chromatographic determination of aflatoxin M_1 in milk powder using immunoaffinity columns for clean-up: Interlaboratory study. *Journal of AOAC International,* Vol.76, pp. 1248-1254

Van Egmond, H. P., Paulsch, W. E., & Schuller, P. L. (1978). Confirmatory test for aflatoxin
    M₁ on thin layer plate. *Journal of AOAC,* Vol.61, No.4, pp. 809-812

Whitaker, T. B., Dickens, J. W., & Manroe, R. J. (1976). Variability associated with testing
    cottonseed for aflatoxin. *Journal of American Oil and Chemical Society,* Vol.53, pp.
    502-505

8

Biosensors for Aflatoxins Detection

Lucia Mosiello and Ilaria Lamberti
*ENEA, Italian National Agency for New
Technologies, Energy and the Environment, Rome,
Italy*

1. Introduction

The availability of rapid and reliable methods for rapid determination of small molecules, as contaminants in food samples, including Aflatoxins, is an increasing need also for human health. In order to monitoring food contaminants, as Mycotoxins (MTXs) Gas Chromatographic (GC) and High Pressure Liquid Chromatography (HPLC) methods are generally utilized, due to their high detection sensibility and selectivity. However, GC and HPLC analyses are time consuming and needs sample pre-treatment or pre-concentration procedures. Immunoassays and biosensors are becoming a recognized alternative or complementary to conventional analytical techniques for the detection of mycotoxins, as Aflatoxins.

Recently, biosensors based on the use of monoclonal or polyclonal antibodies have seen a great development in the field of small molecules analytical determination and specifically in the mycotoxins analyses. Among biosensors for mycotoxins monitoring, optical or electrochemical devices for Aflatoxins detection were described by different authors. The present Chapter describes the different biosensors for Aflatoxins developed and utilized in food analysis. The absence of cross-reactivity obtained with most of these biosensor, the possibility of on-line measurement, the absence of sample pre-treatment, can really put it in competition with other conventional systems such as HPLC and ELISA.

Chapter describes also main biosensors features and vantages for these innovative devices and various examples of biosensors and reviews some biosensors for Aflatoxins and other mycotoxins detection methods, as microarray.

In particular, we will focus our attention on biosensors developed for mycotoxins detection that utilize immunoglobulins or aptamer showing affinity for a correspondent analyte, associated to various transduction elements. Various biosensing platforms will be introduced, including, but not limited to, surface plasmon resonance and quartz microbalance crystals. Examples of biosensors array, as microarray, detecting Aflatoxin and Fumonisin will be also presented. Some of these biosensing devices were developed in our laboratories and the sensing performance of each device will be evaluated and compared in terms of sensitivity and detection limit.

Analytical methods used for mycotoxins determination are mainly based on TLC, HPLC or ELISA. Actually biosensor and microsystem technologies are used for different applications including studies of human and veterinary diseases, drug discovery, genetic screening, clinical and food diagnostics. According to these approaches the aim of many authors was to transfer the methods of the immunological assay from microtiter plates into a biosensor format in order to develop a rapid, sensitive and inexpensive method for the detection of mycotoxins for food safety applications. Microarray and biosensor technology enables the fast and parallel analysis of a multitude of biologically relevant parameters. Not only nucleic acid-based tests, but also peptide, enzyme and antibody assays using different formats of biosensor evolved within the last decade. Antibody-based microarrays are a powerful assay technology that can be used to generate rapid detection of analytes in complex samples which, in our opinion, is also potentially useful for the generation of rapid immunological assay of food contaminants.

2. Aflatoxins

Mycotoxins are secondary metabolites that moulds produce naturally. Due to their ubiquitous presence in foodstuffs and their potential risk for human health, prompt detection is important. It is estimated that approximately 25% of the world's crops are contaminated to some extent with mycotoxins. Some mycotoxins (e.g.,aflatoxins) have been designated biowarfare agents due to their potential carcinogenicity. (Prieto-Simón et al., 2007) .

Aflatoxins are highly toxic and carcinogenic secondary metabolites produced mainly by three anamorphic species of the genus *Aspergillus*: *A. flavus*, *A. parasiticus* and *A. nomius* (Ehrlich et al., 2003). They are the most potent, naturally-occurring carcinogens known and have been linked to liver cancer and several other maladies in animals and humans (Turner *et al.*, 2003; Valdivia et al., 2001; Otim et al., 2005).

When aflatoxin B1 (AFB1), the most toxic aflatoxin, is ingested by cows through contaminated feed, it is transformed into aflatoxin M1 (AFM1) through enzymatic hydroxylation of AFB1 at the 9a-position (see below) and has an approximate overall conversion rate equal to 0.3 to 6.2%.

AFM1 is secreted in milk by the mammary gland of dairy cows. Even though it is less toxic than its parent compound, AFM1 has hepatotoxic and carcinogenic effects. This toxin, initially classified as a Group 2B agent, has now been reclassified as Group 1 by the International Agency for the Research on Cancer (IARC).

Another important class of MTX are those produced by *Fusarium moniliforme,* a prevalent fungus that infects corn and other cereal grains.

Fumonisin B_1 (FB$_1$) is the most common mycotoxin produced by *F. moniliforme*, suggesting it has toxicologic significance. Ingestion of moldy corn infected by *F. moniliforme* or closely related fungi is linked to a higher incidence of primary liver cancer (Ueno et al., 1997) and esophageal cancer in regions of South Africa and China.

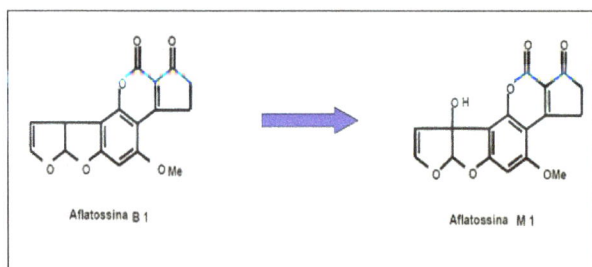

Aspergillus flavus

Aflatoxin molecules

Fig. 1. Aflatoxins B1 and M1 and *Aspergillus* fungus

3. Biosensors

A biosensor is an analytical device for the detection of an analyte that combines a biological component with a physicochemical detector component. The most commune biosensor scheme is reported in Fig.2 and it is consists of 3 parts:

- the *sensitive biological element,* or biological material (e.g. tissue, microorganisms, organelles, cell receptors, enzymes, antibodies, nucleic acids, etc.), a biologically derived material. The sensitive elements can be created by biological engineering.
- the *transducer* or the *detector element* (works in a physicochemical way; optical, piezoelectric, electrochemical, etc.) that transforms the signal resulting from the interaction of the analyte with the biological element into another signal (i.e., transducers) that can be more easily measured and quantified;
- associated electronics or signal processors that are primarily responsible for the display of the results in a user-friendly way.

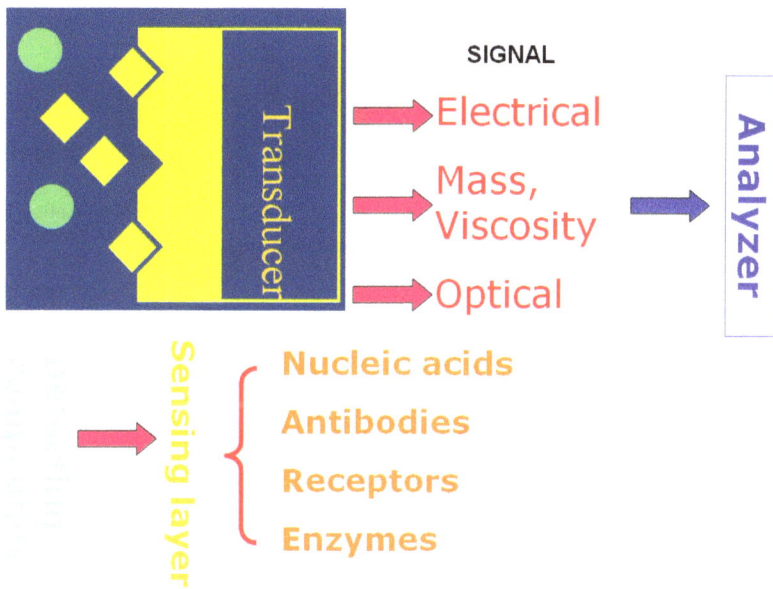

Fig. 2. Scheme of a Biosensor

Main advantages of biosensors technology in comparison with traditional analytical methods are fast detection (minutes) and response (seconds), high sensitivity (typically nM, improved sensitivity with nanoparticles pM and better), their high selectivity, easy preparation and operation assay method. In addition most of these devices are reusable and show low cost assay (Typically less then 10 EUR/sensor).

The methodology of surface chemistry is the basic knowhow for obtaining reproducible results with biosensors and various strategies can be used (Gagliardi et al, 2007).

The key points to consider when selecting an appropriate surface and coating procedure are a low degree of unspecific binding sites and uniform distribution of functional groups on the substrate surface.

For this reason during biosensor development and testing particular attention would be focused on

- Surface (on wich sensing layer will be coated) characterisation
- Biological reagent (immunoglobulin, nucleic acid, ecc.) characterisation
- Uniformity of biological element
- Standard solution preparation
- Calibration and Standard Curve construction

Among biosensors piezo-electric devices are sensors that integrate a biological element with a physiochemical transducer to produce an electronic signal proportional to an analyte which is then conveyed to a detector.

Mass sensitive piezoelectric transducers are usually based on AT-cut quartz crystal covered by gold electrodes. The external alternating voltage induces oscillation of the quartz. The frequency of this oscillation depends on the transducer thickness (Fig.3).

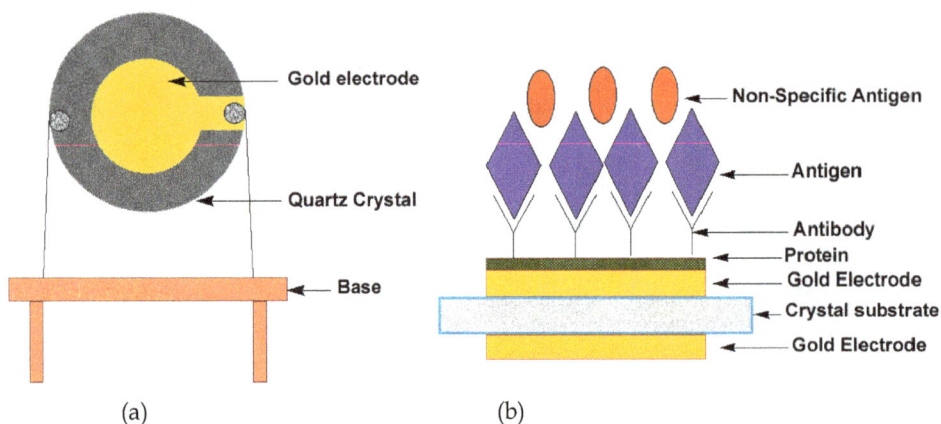

Fig. 3. (a) Mass piezoelectric trasducer; (b) A biosensor antibody-based

In these biosensors the frequency value of the oscillation of the quartz is proportional to the mass of the crystal following the Sauerbrey law and decreases with increasing of the mass (Equation 1, Sauerbrey equation).

$$\Delta f = -2.26 \times 10^{-6} f_0^2 (\Delta m / A) \tag{1}$$

Responding to the need to achieve high sensitivity and move to the use of disposable probes, several electrochemical immunosensors have recently been reported in literature for the detection of AFB1 (Aflatoxin B1) in corn and barley and AFM1 (Aflatoxin M1) in milk.

In particular, for AFB1 determination, an indirect competitive electrochemical immunoassay has been developed using disposable screen-printed carbon electrodes.

In an another work was presented a biosensing method for detection of aflatoxin B_1 and type-A trichothecenes, based on the use of indirect competitive ELISA format coupled with a 96-well screen-printed microplate.

Electrochemical immunoassays for AFB_1, T-2, and HT-2 were performed and the activity of the alkaline phosphatase or horseradish peroxidase labeled enzymes were measured using intermittent pulse amperometry (IPA) as electrochemical technique. Using standard solutions of the target analyte the LOD of the assays were 0.3 and 0.2 ng ml⁻¹ for T-2 and AFB₁ respectively, while the sensitivity was 1.2 ng ml⁻¹ for both. For Aflatoxin B₁, a stability study of electrochemical plate was also performed. Moreover, the matrix effect was evaluated using two different extraction treatments from corn.

The specificity of the assay was assessed by studying the cross-reactivity of the MAb (Monoclonal Antibody) towards other aflatoxins. The results indicated that the MAb could readily distinguish AFB1 from other toxins, with the exception of AFG1 (Piermarini, et al., 2007).

In the field of enzymatic/amperometric biosensor application an electrochemical immunosensor for the detection of ultratrace amounts of aflatoxin M1 (AFM1) in food products was developed.

This sensor was based on a competitive immunoassay using horseradish peroxidase (HRP) as a tag. Magnetic nanoparticles coated with antibody (anti-AFM1) were used to separate the bound and unbound fractions. The samples containing AFM1 were incubated with a fixed amount of antibody and tracer [AFM1 linked to HRP (conjugate)] until the system reached equilibrium. Competition occurs between the antigen (AFM1) and the conjugate for the antibody. Then, the mixture was deposited on the surface of screen-printed carbon electrodes, and the mediator [5-methylphenazinium methyl sulphate (MPMS)] was added.

The enzymatic response was measured amperometrically. A standard range (0, 0.005, 0.01, 0.025, 0.05, 0.1, 0.25, 0.3, 0.4 and 0.5 ppb) of AFM1-contaminated milk from the ELISA kit was used to obtain a standard curve for AFM1. To test the detection sensitivity of our sensor, samples of commercial milk were supplemented at 0.01, 0.025, 0.05 or 0.1 ppb with AFM1.

Immunosensor for Afla M1 described has a low detection limit (0.01 ppb), which is under the recommended level of AFM1 [0.05 μg L-1 (ppb)], and has good reproducibility.

Recently an innovative amperometric biosensor for AflatoxinB1 was described. This biosensor was developed using the enzyme conjugate aflatoxin-oxidase (AFO), embedded in sol-gel, linked to multiwalled carbon nanotubes (MWCNTs)-modified Pt electrode and was reported for the first time.

The covalent linkage between AFO and MWCNTs retained enzyme activity and responsed to the oxidation of afltoxin B₁ (AFB₁). Its apparent Michaelis-Menten constant for AFB₁ was 7.03 μmol·L⁻¹, showing a good affinity. The sensor exhibited a linear range from 3.2 nmol·L⁻¹ to 721 nmol·L⁻¹ (1 ng/ml to 225 ng/ml) with limits of detection of 1.6 nmol·L⁻¹ (signal-to-noise ratio = 3), an average response time of 44 s (less than 30 s when AFB₁ Conc. is bigger than 45 ng/ml), and a high sensitivity of 0.33×10^2 A mol⁻¹·L cm⁻². The active energy was 18.8 kJ mol⁻¹, demonstrating the significant catalyzation of AFO for oxidation of AFB₁ in this biosensor.

These results demonstrated that AFO act at the unsaturated carbon bond of bisfuran ring in AFB₁, to primarily form an unstable compound: oxygen additive product and hydrogen peroxide. This makes a clear choice to use AFO as a recognition receptor for biosensors to detect this mycotoxin (Li *et al.*, 2011).

4. SPR biosensor for aflatoxins

A promising technology for rapid Afaltoxins detection is the surface plasmon resonance biosensor. The principle of surface plasmon resonance is based on the detection of a change of the refractive index of the medium when an analyte binds to an immobilised partner molecule (antibody).

Optical sensors based on excitation of surface plasmons, commonly referred to as surface plasmon resonance (SPR) sensors, belong to the group of refractometric sensing devices. Development of SPR sensors for detection of chemical and biological species has gained considerable momentum, and the number of publications reporting applications of SPR biosensors for detection of analytes related to medical diagnostics, environmental monitoring, and food safety and security has been rapidly growing.

SPR affinity biosensors are sensing devices which consist of a biorecognition element that recognizes and is able to interact with a selected analyte and an SPR transducer, which translates the binding event into an output signal. The biorecognition elements are immobilized in the proximity of the surface of a metal film supporting a surface plasmon.

Analyte molecules in a liquid sample in contact with the SPR sensor bind to the biorecognition elements, producing an increase in the refractive index at the sensor surface, which is optically measured.

The change in the refractive index produced by the capture of biomolecules depends on the concentration of analyte molecules at the sensor surface and the properties of the molecules. If the binding occurs within a thin layer at the sensor surface of thickness, the sensor response is proportional to the binding-induced refractive index change. (Homola, 2008). The SPR principle is reported in Fig.4.

These biosensors show several advantages such as small sample volumes (µL volumes) and reusable metal chips.

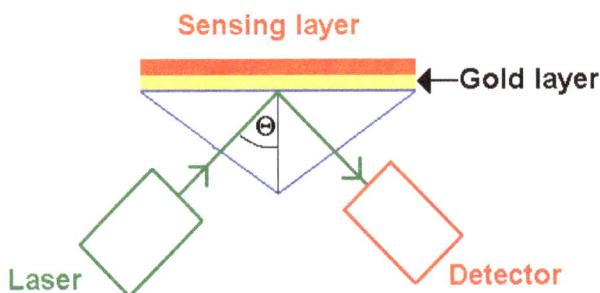

Fig. 4. SPR biosensor principle, surface plasmons are excited by polarised laser beam at certain angle Θ and the intensity of reflected light is measured.

Authors published data obtained using a SPR biosensor for Aflatoxin detection in maize extracts (Cuccioloni et al.).

In this paper different dilutions of Aflatoxin-containing and Aflatoxin-free fractions were added to the elastase-functionalized surface, and each response kinetic was routinely followed and analyzed as described above. The regeneration of the elastase monolayer was carried out as previously described. Detection procedures were replicated on different days on both the same and different elastase-functionalized surfaces. Additionally, the assessment of the number of regeneration cycles that a sensor surface can withstand without a significant loss of the sensitivity and accuracy of the assay and the stability of the sensing surface throughout multiple measurements were evaluated.

Limits of Detection and Quantitation: In compliance with the IUPAC rules, the limit of detection (LOD) was calculated as three times the standard deviation of the blank measurements. The limit of quantification (LOQ) is calculated as 10 times the standard deviation of the blank measurements. Aflatoxin-free certified T400A maize sample was used as a blank matrix.

Calibration Curve: Biosensor-based assay was applied for the determination of AFB1 using spiked maize samples (Figure 3). The analysis of the binding of AFB1 to elastase over the concentration range 1–50 µg/kg reported that the response for the optimized assay was linear in the range between 1.67 and 17.8 µg/kg. The calibration procedure was replicated on three different days. The experimentally measured lower limit of the linear range was 1.67 µg/kg of AFB1, whereas the KD was 0.91 µM (≈250 µg/kg) AFB1. The detection limits reached allow us to use this assay for detection of AFB1 in maize within the regulatory limits.

Recently some authors presented during a Nanotechology Conference a SPR biosensor for Aflatoxin B1 developed using fusion proteins as a linker.

Because one of the main goal in the development of SPR immunosensors is efficient immobilization of antibodies. Conventional methods, such as self-assembled monolayers (SAMs) of alkanethiols cause antibodies to be random oriented. To improve antibody linker and orientation in their work, the authors constructed a novel fusion protein by genetically fusing gold binding polypeptides (GBP) to protein A (ProA).

The resulting GBP-ProA protein was directly self-immobilized onto SPR gold chip surfaces via the GBP portion, followed by the oriented binding of anti-AFB_1 antibodies onto the ProA domain and AFB_1 in series. Consequently, a low detection limit (10 ng/mL) has been achieved for mycotoxin SPR immunosensor by using GBP-ProA fusion proteins as a crosslinker. (Ko et al., 2010).

5. QMC biosensor for others mycotoxins

A Quarz Crystal Microbalance (QCM) consists of a thin quartz disk with a electrodes plated. The application of an external electrical potential to a piezoelectric material produces internal mechanical stress. As the QCM is piezoelectric, an oscillating electric field applied across the device induces an acoustic wave that propagates through the crystal and meets minimum impedance when the thickness of the device is a multiple of a half wavelength of the acoustic wave. Deposition of a thin film on the crystal surface decreases the frequency in a portion to the mass of the *film*.

As described, the mycotoxins, such are Aflatoxins are toxical fungal metabolites that can occur in primary food products. In order to new biosensor development we focused our attention also on Ochratoxin A (OTA), which was discovered as a metabolite of *Aspergillus Ochraceus* (Van der Merwe et al., 1965). This mycotoxin generally appears during storage of cereals, coffee, cocoa, dried fruit, pork etc. and occasionally in the field of grapes. It may also be present in blood and kidneys of animals that have been fed on contaminated feeds. Animal studies indicated that this toxin is carcinogenic (Turner et al., 2009). Therefore, the European Commission has fixed maximum concentration of OTA in foodstuffs: 3 µg/kg (7.4 nM) for cereal products and 5 µg/kg (12.4 nM) for roasted coffee, respectively (Commission Regulation No. 1881/2006, 19 December 2006).

The establishment of efficient method of this analyte detection is therefore of high importance. In addition to traditional, but expensive and time-consuming methods such as liquid chromatography, new trends consist in development portable and easy to use biosensors (Siontorou et al. ,1998).

Most of the biosensors for this analyte detection developed so far were based on electrochemical detections such as oxidation of OTA at glassy carbon electrode (limit of detection (LOD) 0.26 μM) (Oliveira et al., 2007) or reduction of horseradish peroxidase (LOD 0.25 nM) (Alonso-Lomillo et al., 2010).

Immunosensor based on quartz crystal microbalance (QCM) was recently reported as well (Tsai *et al.*,2007). In this sensor anti-OTA antibodies were immobilised on a surface of 16-mercaptohexadecanoic acid. The detection based on the competitive binding between free OTA and that conjugated with BSA provided LOD 40 nM.

Recently a DNA aptamer sensitive to OTA has been developed (Cruz-Aguado et al., 2008). This aptamer was able to recognize OTA with sensitivity in a ppb level and with high selectivity. The electrochemiluminiscence biosensor using aptamers as receptors was recently developed (LOD 17 pM).

Thus most of the biosensors for mycotoxin OTA reported were based on indirect detection methods. Would be, however, rather useful to develop biosensor based on direct method that do not require additional modification of receptor or complicated multi stage assay. In a recent work Prof.T.Hianik (Comenius University, Bratislava, Slovakia) made therefore attempt to develop biosensor for OTA based on thickness shear mode acoustic method (TSM) using biotinylated DNA aptamers immobilised on a surface of quartz crystal transducer covered by neutravidin (Lamberti et al. 2011) .

TSM is certain analogy of QCM, however, in addition to mass, the TSM determines also the viscosity contribution arising from the friction between biolayer and the surrounding buffer (Fig.5).

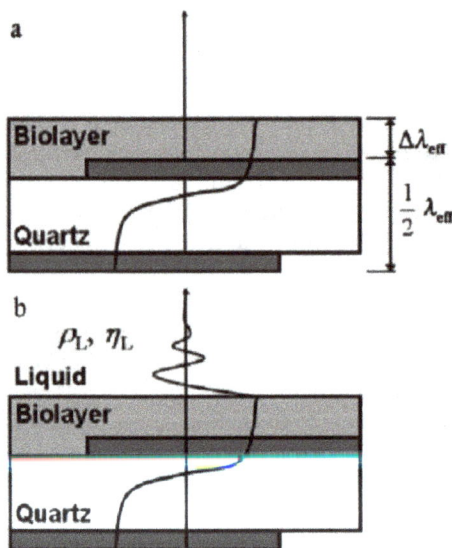

a) **Biolayer in an air**
No energy dissipation into the surrounding environment. However, due to the viscoelasticity of biolayer, certain part of energy dissipates.

b) Biolayer in liquid
Substantial dissipation takes place due to viscosity effect.

Fig. 5. Propagation of acoustic wave from the sensor surface

This is important for detection of small molecules, such are mycotoxins for which the QCM detection is difficult due to small molecular weight of the analyte. We showed that TSM allowing detecting this mycotoxin with LOD 30 nM and with good selectivity. He also studied the stability of DNA aptamers depending on concentration of calcium ions, that are important for binding OTA to DNA aptamer. In our opinion the described biosensor would be applied also for Aflatoxins detection.

6. Microarray for aflatoxin B1 detection

Microarrays provide a powerful analytical tool for the simultaneous detection of multiple analytes in a single experiment and consist of a biosensor *micro* o *nano* arrays.

Research on microarrays as multianalyte biosystems has generated increased interest in the last decade. The main feature of the microarray technology is the ability to simultaneously detect multiple analytes in one sample by an affinity-binding event at a surface interface. In some cases immunoanalytical microarrays have the potential to replace conventional chromatographic techniques. They are applied if the number of samples is high or analysis by current methods is difficult and/or expensive. Therefore, microarray platforms have a great potential as monitoring systems for the rapid assessment of water or food samples. Antibody-based microarrays are a powerful tool for analytical purposes, also for Aflatoxins detection application. Immunoanalytical microarrays are a quantitative analytical technique using antibodies as highly specific biological recognition elements. They can be designed for a variety of analytical applications producing rapid results with low limits of detection (LOD).

For these reasons in association to some biosensors for Aflatoxins examples, we reported in this Chapter also a feasibility study, made in our laboratories, on application of antibodies microarrays for simultaneous analysis of two different mycotoxins (Aflatoxin-B1 and Fumonisin B1). In this work we developed a competitive immunoassay in a microarray format and with the described method observed different microarray patterns in samples containing Aflatoxin-B1 or Fumonisine or either analytes at a ppb concentration range (Lamberti et al., 2009). The quality of the microarray data is comparable to data generated by a microplate-based immunoassay (ELISA), but further investigations are needed in order to better characterize these methods when applied for food contaminants determination. In any case we hope that our results can confirm the feasibility to develop hapten microarrays as for the immunochemical analysis of mycotoxins, as above described for others small organic molecules (e.g. bacterial toxins or biological warfare agents).

Enzyme linked immunosorbent assay (ELISA) and fluorescence immunoassay (FIA) are excellent survey tools for many analitycal purpose because of their high-throughput, user friendliness, and field portability. These important characteristics make immunoassays attractive tools for food testing by regulatory agencies to ensure food safety. Immunoassay is traditionally performed as individual test, however in many cases it is necessary to perform a panel of tests on each sample (detection of drug residues). To address this requirement, microarray-based immunoassay technologies have been developing utilizing microarray platform (multianalyte analysis) and classic immunoassay (multi-samples analysis).

In recent years, the antibody microarray technology has made significant progress, going from proof -of-concept designs to established high-performing technology platforms capable of targeting non-fractionated complex samples, as proteoma (Blohm & Guiseppi-Elie., 2001).

Microarrays consist of immobilized biomolecules spatially addressed on planar surfaces, microchannels or microwells, or an array of beads immobilized with different biomolecules.

Biomolecules commonly immobilized on microarrays include oligonucleotides, polymerase chain reaction (PCR) products, proteins, lipids, peptides and carbohydrates. Ideally, the immobilized biomolecules must retain activity, remain stable, and not desorb during reaction and washing steps. The immobilization procedure must ensure that the biomolecules are immobilized at optimal density to the microarray surface for efficient binding (Venkatasubbarao, S. et al., 2004).

Some microarray applications are focused on current trends in the movement of this technology from being a purely research method to becoming an analytical instrument applicable in the clinic and as well as in human health (Koppal, T. , 2004).

According to this trend we have tried to transfer the immunoassay method from microtiter plates into a microarray format in order to develop a multiparametric, rapid, sensitive and inexpensive method for the the detection of mycotoxins for food safety application.

To perform our test and check the feasibility of this format, we focused our studies on the most popular mycotoxins Aflatoxin B1 (AFB1) and Fumonisin B1 (FB1) and developed a competitive immunoassay in a microarray format, using the Dr.Chip platform provided by Life Line Lab Co. (Pomezia, Italy) and used also for other applications (Lamberti, 2010).

Microarray platform is equipment to create microarrays and to read the final results, via densitometric detection, based on the enzymatic and colorimetric assay. In Fig.6 are reported a detail of the plastic probe tray for protein spotting and pins. In the same picture is also shown the scheme of the glass treated with functional protein linker.

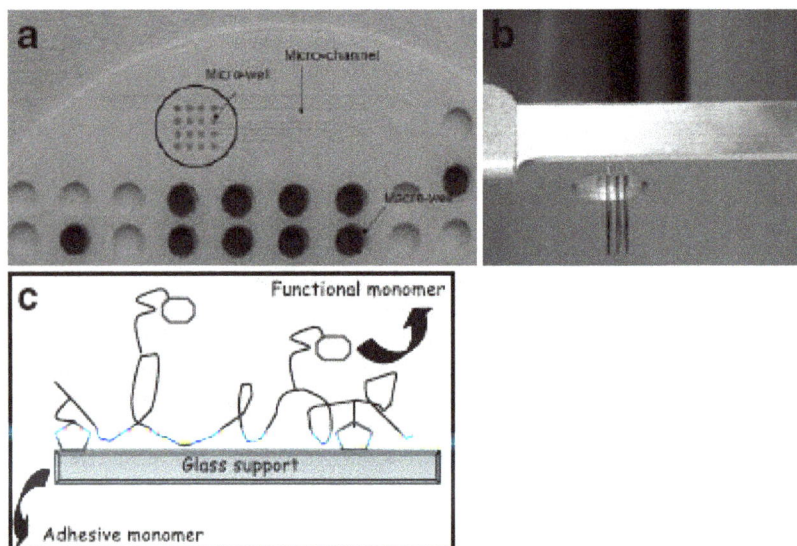

Fig. 6. Microarray spotting platform used for Aflatoxin B1 detection

As in other conventional competitive immunoassay the color intensity and corrispondent grey values obtained from antigen microarrays BSA-Afla B1, prepared as described in this paper and used in our immunological tests, are in inverse proportion to antigen concentration in standard solutions. Assay method for Aflatoxin is described in Fig.7.

SUMMARY OF ANTIBODY MICROARRAY OF MYCOTOXIN	
STEP 1: PROTEIN SPOTTING AND BLOCKING	MYCOTOXIN-BSA DILUTED IN PRINTING BUFFER ⇓ SPOT MYCOTOXIN-BSA ON CHIP ⇓ INCUBATE CHIP AT FOR 2 HRS 37°C ⇓ BLOCKING BUFFER WITH BSA 0.9% ⇓ DRY CHIP FOR 30' AT 37°C ⇓ THE CHIP IS READY TO USE
STEP 2: ASSAY PROCEDURE	PRE-INCUBATION: ANTI- MYCOTOXIN + MYCOTOXIN FREE ⇓ BINDING TO CHIP PROTEIN ⇓ WASH STEP ⇓ ADDING ANTI IgG BIOTIN CONJUGATE ⇓ WASH STEP ⇓ STREP AP BINDING ⇓ WASH STEP ⇓ ADDING AP SUBSTRATE ⇓ COLORIMETRIC DEVELOPMENT AND RESULT ANALYSIS

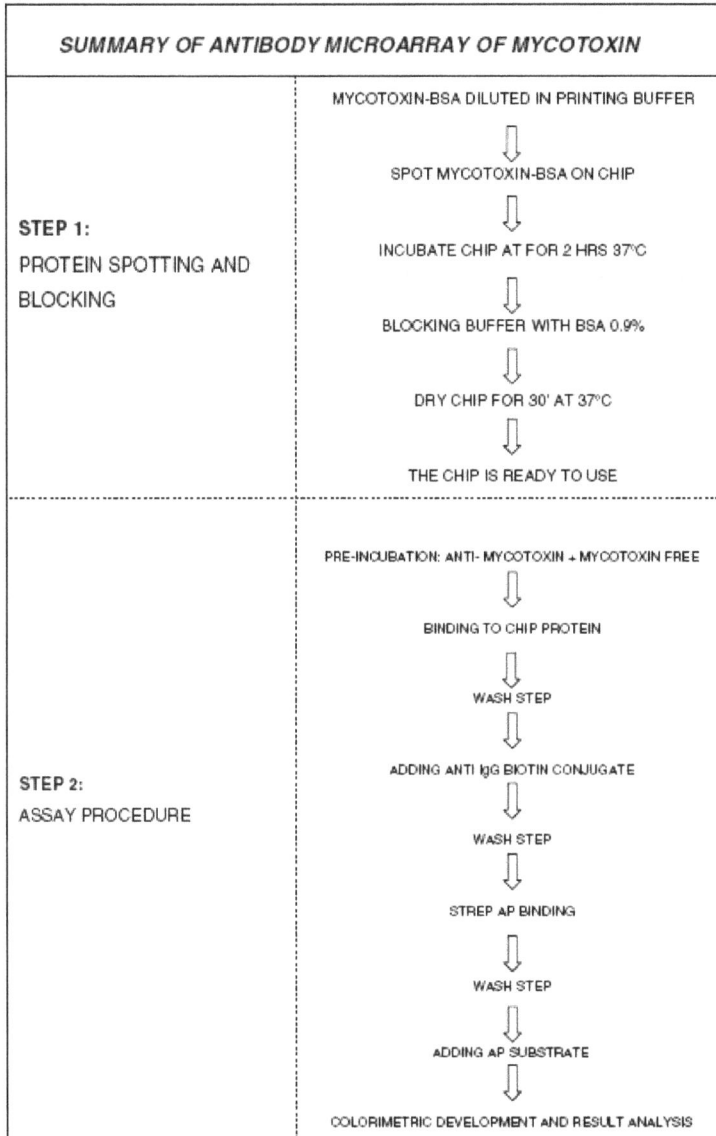

Fig. 7. Scheme of sensing assay for Aflatoxin B1 (Lamberti et al., 2009)

7. Conclusion

Mycotoxin analysis in food and feed is generally a multi-step process comprised of sampling, sample preparation, toxin extraction from the matrix (usually with mixtures of water and polar organic solvents), extract clean-up and finally detection and quantitative determination, for these reasons.

The availability of rapid and reliable methods for "in field" determination of fungine contamination, as Aflatoxins and other MTXs (mycotoxins) identification in foods is an increasing need for human health and food safety purposes. Gas chromatographic (GC) methods for Aflatoxins detection are generally used, due to their low detection limits and high selectivity. Laboratories generally have to analyze a large number of samples requiring adequate storage conditions and time-consuming sample pre-treatment and pre-concentration procedures. The establishment of efficient method for Aflatoxins detection is therefore of high importance and new trends consist in development portable and easy to use biosensors.

Recently, biosensors have seen a great development in the field of small molecules analytical determination and these methods are upon constant improvement also for MTXs and for Aflatoxins, but actually generally restricted to feasibility studies.

At this developmental stage, biosensors for Aflatoxins and MTX (Mycotoxin) detection could be very useful as a qualitative/semiquantitative "field test" for identifying "positive" samples, reducing the number of samples to be reanalyzed in the laboratory, according to analytical standard methods (GC).

Biosensor assays are rapid, easy to perform, and inexpensive and could be advantageous in comparison with ELISA or GC/MS for analysis on food, but in our opinion, further improvements of analytical parameters such as precision, accuracy, and detection limits (especially for Aflatoxins biosensor applications) are required.

8. Acknowledgment

The authors are grateful to Prof.Tibor Hianik of Nuclear Physics and Biophysics, Comenius University, Bratislava, Slovakia and to Prof.Caterina Tanzarella, University of RomaTre Department of Biology, Rome, Italy for their advice and scientific support.

9. References

Alonso-Lomillo, M.A., Domínguez-Renedo, O., Ferreira-Gonçalves, L., & Arcos-Martínez, M.J. (2010). Sensitive enzyme-biosensor based on screen-printed electrodes for Ochratoxin A. *Biosensors and Bioelectronics*, Vol. 25, No 6, pp. 1333-1337, 09565663,

Blohm, D.H., & Guiseppi-Elie, A. (2001). New developments in microarray technology. *Current Opinion in Biotechnology*, Vol. 12, No 1, pp. 41-47, 0958-1669.

Cruz-Aguado, J.A., & Penner, G. (2008). Determination of ochratoxin A with a DNA aptamer. *Journal of Agricultural and Food Chemistry*, Vol. 56, No 22, pp. 10456-10461, 0021-8561,,

Cuccioloni, M., Mozzicafreddo, M., Barocci, S., Ciuti, F., Pecorelli, I., Eleuteri, A.M., Spina, M., Fioretti, E., & Angeletti, M. (2008). Biosensor-based screening method for the detection of aflatoxins B 1-G1. *Analytical Chemistry*, Vol. 80, No 23, pp. 9250-9256, 0003-2700

Ehrlich, K.C., Montalbano, B.G., & Cotty, P.J. (2003). Sequence comparison of aflR from different Aspergillus species provides evidence for variability in regulation of Aflatoxin production. *Fungal Genetics and Biology* Vol. 38, pp. 63-74, 1087-1845.

Gagliardi, S., Rapone, B., Mosiello, L., Luciani, D., Gerardino, A., & Morales, P. (2007). Laser-assisted fabrication of biomolecular sensing microarrays, *IEEE Transactions on Nanobioscience*, Vol. 6, No 3, pp. 242-248, 1536-1241.

Homola J., 2008, Surface Plasmon Resonance Sensors for Detection of Chemical and Biological Species, *Chemical Review*, Vol. 108, 462-493, 0009-2665.

Ko, S., Kim, C.J., & Kwon, D.Y. (2010). Detection of aflatoxin B1 by SPR biosensor using fusion proteins as a linker. *Proceedings of the 2010 NSTI Nanotechnology Conference and Expo*, 978-1439834183, Anaheim, 978-1-4398-3418-3, CA, June 2010.

Koppal, T. (2004) Microarrays: migrating from discovery to diagnostics. *Drug Discovery Development*, Vol. 7, pp. 30–34.

Lamberti I., Tanzarella C., Solinas I., Padula C. & Mosiello L. (2009). An antibody-based microarray assay for the simultaneous detection of aflatoxin B1 and fumonisin B1. *Mycotoxin Research*, Vol. 25 No 4, pp. 193-200, 0178-7888.

Lamberti, I., Mosiello, L., Cenciarelli, C., Antoccia, & A., Tanzarella, C. (2010). A novel based protein microarray for the simultaneous analysis of activated caspases. *Lecture Notes in Electrical Engineering*, Vol. 54, pp. 323-326, 1876-1100.

Lamberti, I., Mosiello L., & Hianik, T. (2011). Development of thickness shear mode biosensor based on DNA aptamers for detection ochratoxin. *Chemical Sensors*, Vol 1, pp. 11-15, 2231-6035.

Li, S.C., Chen, J.H., Cao, H., Yao, D.S., & Liu, D.L. (2011). Amperometric biosensor for aflatoxin B1 based on aflatoxin-oxidase immobilized on multiwalled carbon nanotubes. *Food Control*, Vol. 22, No 1, pp. 43-49, 0956-7135.

Oliveira, S.C.B., Diculescu, V.C., Palleschi, G., Compagnone, D., & Oliveira-Brett, A.M. (2007). Electrochemical oxidation of ochratoxin A at a glassy carbon electrode and in situ evaluation of the interaction with deoxyribonucleic acid using an electrochemical deoxyribonucleic acid-biosensor. *Analytica Chimica Acta*, Vol. 588, No 2, pp. 283-291, 0003-2670.

Otim, M.O., Mukiibi-Muka, G., Christensen, H. & Bisgaard, M. (2005). Aflatoxicosis, infectious bursal disease and immune response to Newcastle disease vaccination in rural chickens. *Avian Pathology* Vol. 34: pp. 319-323, 0307-9457.

Piermarini, S., Volpe, G., Ricci, F., Micheli, L., Moscone, D., Palleschi, G., Führer, M., Krska, R., & Baumgartner, S. (2007). Rapid screening electrochemical methods for aflatoxin B1 and type-A trichothecenes: A preliminary study. *Analytical Letters*, Vol. 40 No 7, pp. 1333-1346, 0003-2719.

Prieto-Simón, B., Noguer, T., & Campàs, M. , 2007. Emerging biotools for assessment of mycotoxins in the past decade. TrAC. *Trends in Analytical Chemistry*, Vol. 26 No 7, pp. 689-702, 0165-9936.

Siontorou, C.G., Nikolelis, D.P., Miernik, A., & Krull, U.J. (1998). Rapid methods for detection of Aflatoxin M1 based on electrochemical transduction by self-assembled metal-supported bilayer lipid membranes (s-BLMs) and on interferences with transduction of DNA hybridization. *Electrochimica Acta*, Vol 43 No23, pp. 3611-3617, 0013-4686.

Tsai, W.-C., & Hsieh, C.-K. (2007). QCM-based immunosensor for the determination of ochratoxin A. *Analytical Letters*, Vol. 40 No 10, pp. 1979-1991.

Turner, N.W., Subrahmanyam, S., & Piletsky, S.A. (2009). Analytical methods for determination of mycotoxins: A review. *Analytica Chimica Acta*, Vol. 632 No 2, pp. 168-180, 0003-2670.

Turner, P.C., Moore, S.E., Hall, A.J., Prentice, A.M. & Wild, C.P. (2003). Modification of immune function through exposure to dietary aflatoxin in Gambian children. *Environmental Health Perspectives*, Vol. 111, pp. 217-220.

Ueno Y, Iijima K, Wang S-D, Sugiura Y, Sekijima M, Tanaka T, Chen C, & Yu S-Z. (1997). Fumonisin as a possible contributory risk factor for primary liver cancer: a 3-year study of corn harvested in Haimen, China by HPLC and ELISA. *Food and Chemical Toxicology* Vol. 35, pp. 1143-1150, 0278-6915.

Valdivia, A.G., Martinez, A., Damian, F.J., Quezada, T., Ortiz, R., Maryinez, C., Llamas, J., Rodriguez, M.L., Yamamoto, L., Jaramillo, F., Loarca-Pina, M.G. & Reyes, J.L. (2001). Efficacy of N-acetylcysteine to reduce the effects of aflatoxin B1 intoxication in broiler chickens. *Poultry Science,* Vol 80, pp. 727-734, 0032-5791.

Van Der Merwe, K.J., Steyn, P.S., Fourie, L., Scott, D.B., Theron, J.J. (1965). Ochratoxin A, a toxic metabolite produced by Aspergillus ochraceus Wilh Nature, Vol. 205, No 4976, pp. 1112-1113.

Venkatasubbarao, S. (2004). Microarrays – status and prospects. *Trends in Biotechnology,* Vol. 22 No12, pp. 630-637, 0167-7799.

Methods for Detection and Quantification of Aflatoxins

Alejandro Espinosa-Calderón, Luis Miguel Contreras-Medina,
Rafael Francisco Muñoz-Huerta, Jesús Roberto Millán-Almaraz,
Ramón Gerardo Guevara González and Irineo Torres-Pacheco
C.A. Ingeniería de Biosistemas, División de Estudios de Posgrado, Facultad de Ingeniería,
Universidad Autónoma de Querétaro, Querétaro, Qro.,
México

1. Introduction

Mycotoxins are fungal toxic metabolites which naturally contaminate food and feed. aflatoxins (AFs), a kind of mycotoxins, are the main toxic secondary metabolites of some *Aspergillus* moulds such as *Aspergillus flavus*, *Aspergillus parasiticus* and the rare *Aspergillus nomius* (Ali et al., 2005, Alcaide-Molina et al., 2009). Such toxins can be separated into aflatoxins B1, B2, G1, B2a and G2a. Its order of toxicity is B1 > G1 > B2 > G2. Letters 'B' and 'G' refer to its blue and green fluorescence colors produced by these compounds under UV light. Numbers 1 and 2 indicate major and minor compounds, respectively (Weidenbörner, 2001; Hussein & Brasel, 2001). *A. flavus* only produces B aflatoxins, while *A. parasiticus* and *A. nomius* also produce G aflatoxins (Alcaide-Molina et al., 2009).

Aflatoxins are produced on various grains and nuts, e.g., corn, sorghum, cottonseed, peanuts, pistachio nuts, copra, cereals, fruits, oilseeds, dried fruits, cocoa, spices and beer in the field and during storage. AFs occur mainly in hot and humid regions where high temperature and humidity are optimal for moulds growth and toxins production (Ventura et al., 2004; Zollner & Mayer-Helm, 2006). Its presence is enhanced by factors as stress or damage to the crop due to drought before harvest, insect activity, soil type and inadequate storage conditions (Alcaide-Molina et al., 2009).

Aflatoxins, when ingested, inhaled or adsorbed through the skin, have carcinogenic, hepatotoxic, teratogenic and mutagenic effects in human and animals (rats, ferrets, ducks, trout, dogs, turkeys, cattle and pigs) (Anwar-Ul_Haq & Iqbal, 2004) even at very small concentrations. When aflatoxins B1 is ingested by cows, it is transformed into its hydroxylated product, AFs M1 and M2. Such aflatoxins is secreted in the milk and is relatively stable during milk pasteurization, storage, and preparation of various dairy products (Stroka & Anklam, 2002).

Among the more than 300 known mycotoxins, aflatoxins represent the main threat worldwide. After 1975 there has been an increased concern about the possibility of the presence of carcinogenic mold metabolites, particularly aflatoxins in food and animal feed products. Although aflatoxins are regulated in more than 80 countries, their legislation is not yet completely harmonized at the international level (Cucci et al., 2007). Several institutions around the world have classified and regulated aflatoxins in food. The European Union (EU) has the most rigorous regulations concerning mycotoxins in food. The limits of AFB1 and total AF in foods are 5 and 10 µg/kg, respectively, in more than 75 countries around the world whilst they are 2 and 4 µg/kg in the European Union (EU) (Herzallah, 2009). The maximum residue levels for total AFs and also for the most toxic of them (AFB1) according to the EU Commission Regulations are 2 and 4 g/kg, respectively. The maximum legal limit for AFM1 in milk is set at 0.05µg/kg (50 ppt) for all EU Member States, and 25 ppt for baby food (Cucci et al., 2007). The European Committee Regulations (ECR) has established the maximum acceptable level of AFB1 in cereals, peanuts and dried fruits for direct human consumption in 4ng/g for total aflatoxins (AFB1, AFG1, AFB2, AFG2) and 2ng/g for AFB1 alone (Ricci et al., 2007). The International Agency for Research on Cancer (IARC) classified aflatoxins as Group 1 of human carcinogens (Alcaide-Molina et al., 2009). In USA, the U.S. Department of Agriculture and the U.S. Food and Drug Administration (FDA) have established an "actionable" level of 15-20 ppb of AFs in animal feed products.

Because of such facts, several methodologies for detection and quantification of AFs have been developed. The principal immunochemical based assay is the widespread enzyme linked immunosorbent assay (ELISA). Other methodologies base their performance upon electrochemical and optical principles such as: chromatography, UV-absorption, spectrometry, fluorescence and immunochemical assay tests. The aforementioned methods require well equipped laboratories, trained personnel, harmful solvents and several hours to complete an assay. Novel methods for detection of aflatoxins try to avoid these disadvantages. Among such novel methods, it can be found: biosensors, electrokinetics, electrochemical transduction, amperometric detection, and adsorptive stripping voltammetry. Each of the aforementioned methodologies has its own advantages and limitations according to sensitivity, easiness of use and cost-effectiveness. The objective of this chapter is to provide a general overview of the different methodologies to detect and quantify aflatoxins in the food analysis field.

2. Electrochemicals techniques

Aflatoxins can be measured by the use of electricity and electrochemical immunosensors. These immunosensors consist of a pair of electrodes (measuring and reference), implemented by using the screen-printing technique. The measuring electrode is coated with specific antibodies which will retain interest aflatoxins in the sample, whereas the other electrode (reference) is commonly made of a combination of Ag / AgCl.

The measurement procedure is similar to that carried out by the ELISA test (Enzyme Linked immunoabsorbent Assay). ELISA process is done by taking a sample of the substance to be measured and mixed with a known portion of conjugated aflatoxins with a special enzyme in a microtiter plate hole, and then it is inserted the measuring electrode. In this way, free aflatoxins in the sample compete for fill the places available (antibodies) in the measuring electrode. After some stabilization time, the measuring electrode is removed from the sample, washed with a buffer solution that removes all traces of the sample and leaves intact the electrode coating with aflatoxins that were captured but are not conjugate. After cleaning procedure, the electrode is introduced in a substrate solution that reacts with enzymes in aflatoxins conjugate, changing the electrical conductivity of the substrate depending on the amount of labeled aflatoxins antibodies attached to the electrode. Thus, the greater will see the effect of aflatoxins marked, the lower the concentration of free aflatoxins in the sample.

However, the electrodes developed by Tan et al. (2009), were coated with conjugate aflatoxins instead of being coated with specific antibody, whereas the sample was mixed with the antibody. In this manner, some antibodies will be captured by free aflatoxins in the sample and some others by those attached to the electrode. Following that, the electrode is washed and it is placed into a solution with antibodies conjugated with alkaline phosphatase enzyme that binds to the antibodies that are bound to conjugate aflatoxins onto the electrode. After that, the electrode is immersed in the substrate solution in order that antibodies conjugate react and cause a change in electrical conductivity.

Some methods have been reported the use of simple electrodes (Rameil et al., 2010; Tan et al., 2009), while others have made use of multiple electrodes (Neagu et al., 2009; Piermarini et al., 2007), where the latter has shown to have advantages over the first in that: it is more user friendly; it is possible to carry out many experiments in parallel with different samples; and it reduces the time required for new procedures (Piermarini et al., 2007).

In order to measure the electrical conductivity in the electrodes there are different techniques, such as intermittent pulse amperometric (IPA), potentiometry, or linear sweep voltage (LSV).

The intermittent pulse amperometric technique involves the application of a periodic pulse of some duration fixed voltage across the electrodes coated and reference measurement, while the measured current varies depending on the conductivity of the substance. Moreover, in the technique of potentiometry, the measuring electrode coated is immersed in substrate solution without contaminating aflatoxins until a stable electrical potential is obtained, called the potential base. This potential varies depending on the amount of aflatoxins contained in the sample. In the linear sweep voltage technique, the sample is fed with a voltage which changes linearly, with a fixed slope.

The ability of these techniques to detect aflatoxins depends on many factors, including the type of substrate solution that is used, as is the case reported by Rameil et al. (2010), where it was shown that the use of 3 - (4-hydroxyphenyl) propionic acid (p-HPPA), being a little toxic substance and does not require the use of organic solvents, can increase the conductivity of the substrate in potentiometry to measure aflatoxins M1 in milk. Another factor is the concentration of antibodies in the lining of the electrode, since the higher concentration of these, it can be reached higher peak current than in IPA technique, although the relationship between antibody concentration and electric current conducted is linear in a certain range, such as Tan et al. (2009) work suggests, where the linear range extends from a dilution of 1:30000 to 1:10000 of antibody against aflatoxins B1 found in rice, being the latter dilution which gave the best results.

Another point to consider is the detection limit, defined as the maximum decrease in signal equal to three times the standard deviation measured in the absence of aflatoxins to be determined. Detection limits down to 1 pg/ml have been obtained in the measurement of aflatoxins M1 in milk (Neagu et al., 2009); meanwhile, detection of aflatoxins B1 in rice has reached the limit 0.06 ng/ml (Tan et al., 2009).

There are also other measurement devices, as the case of piezoelectric immunosensors. Piezoelectricity is the property possessed by certain materials in which either generates a potential difference from applied mechanical deformation or vice versa (Webster 1999), so that materials that have this feature can resonate at certain frequencies. One of the most common piezoelectric materials is quartz crystal, used by Jin et al. (2009) as a sensor for measuring aflatoxins B1 in milk. In this case, the crystal was treated to bind aflatoxins AFB1-BSA conjugate to the material for later subjecting to a similar procedure as mentioned by Tan et al. (2009), differing from this one in that the antibodies attached to conjugate aflatoxins attached to crystal, were marked with gold nanoparticles coated with antibody detector first. The concentration of aflatoxins will be reflected in this case as a change in the resonant frequency of the crystal, as reported by Jin et al. (2009) for the case of aflatoxins B1, where there is a linear relationship between the frequency of resonance and the logarithm of the concentration of aflatoxins.

3. Chromatography

Chromatography is one of the most popular methods to analyze mycotoxins such as aflatoxins. The most common techniques of chromatography are Gas chromatography (GC), liquid chromatography (LC), High performance liquid chromatography (HPLC) and Thin-layer chromatography (TLC). From these methods, LC and HPLC are the most used. In many cases, they are followed by fluorescence detections stage (Cavaliere et al., 2006). LC, TLC and HPLC are the most used quantitative methods in research and routine analysis of aflatoxins (Vosough et al., 2010); these techniques offer excellent sensitivities but they frequently require skilled operators, extensive sample pretreatment and expensive equipment (Sapsford et al., 2006).

3.1 Liquid chromatography

At the beginning the only separative method was GC, nevertheless, it is restricted to a small set of biological molecules for instance. Those should not be volatiles or should be derivatizated (Roux et al., 2011). LC is other separative method which offers good sensitivity, high dynamic range, versatility and soft ionization conditions that permit access to the molecular mass of intact biological molecules. LC is usually coupled to fluorescence detection stage (FLD), UV absorption and amperometric detection (Elizalde-González, 1998) with pre-column derivatization or post-column derivatization. Extraction and clean up procedures for aflatoxins analysis typically rely on solid phase extraction (SPE) with different absorbent materials. A particular case of SPE is immunoaffinity columns. Improvements have been done, creating techniques based on LC, such as: TLC and Reversed-phase high performance liquid chromatography (RP-HPLC) (Elizalde-González, 1998). LC coupled with fluorescence stage use the aflatoxins fluorescence properties to quantify them. So that, by improve this property it can be obtained better sensibility for aflatoxin detection. The most common techniques to improve fluorescence properties are the use of pre-column derivatization with trifluoretic acid and post-column derivatization with iodine or bromine (Elizalde-González, 1998). Other studies have been done in order to obtain enhancement of the fluorescence emissions of aflatoxins. Franco et al. (1998) collected emission data for AFQ1, AFM1, AFP1 in solvents usually used for their chromatography separation in absence and in presence of different cyclodextrins. Such experiment was made in order to be applied principally in liquid chromatography.

3.2 Thin-Layer Chromatography (TLC)

Thin-layer chromatography is widely used in laboratories throughout the world for food analysis and quality control. Applications of TLC have been reported in areas of food composition, intentional additives, adulterants, contaminants, etc. TLC has been used to analyze agricultural products and plants. It has advantages as: simplicity of operation; availability of many sensitive and selective reagents for detection and confirmation without interference of the mobile phase; ability to repeat detection and quantification; and cost effectiveness analysis, because many samples can be analyzed on a single plate with low solvent usage, and the time that TLC employs to analyze the sample is less that LC method (Sherma, 2000; Fuch et al., 2010). The most important differences between TLC and HPTLC are: the different particular size of stationary phase; the care used to apply the samples; and the way to process the obtained data (Fuch et al., 2010).

Diprossimo et al. (1996) present a work where show that TLC was superior to the methods of BF (Best food) CB-RCS-Mod (modified CB method-Rapid Modification of the Cottonseed Method) in terms of less fluorescence interferences, better solvent efficiency, and lower detection levels. Results obtained using TLC method compared to HPLC and enzyme-linked immunosorbent assay (ELISA) was found to agree among method but TLC was least expensive (Schaafsma et al., 1998).

Papers that use TLC methods to detect and quantify aflatoxins use sample clean-up based on immunoaffinity columns. Therefore, they avoid interfering compounds and allow visual quantification of aflatoxins at concentrations of less than 1 ng/g (Stroka et al., 2000). Immnunoaffinity procedures provide very clean extracts because the sample is cleaned of interference substances. It also permits an easy aflatoxins determination, since they are applicable for automated sample clean-up (Stroka et al., 2002). Because of the advantages of this method, researches have been focused on them to develop new techniques to improve the methodologies for quantification of aflatoxins.

3.3 High Performance Liquid Chromatography (HPLC)

As aforementioned, HPLC is one of the most common methods to detect and quantify aflatoxins in food. It has been used jointly with techniques such as UV absorption, fluorescence, mass spectrometry and amperometric detectors. Elizalde-González et al. (1998) analyzed aflatoxins B1,B2, G1 and G2 based on HPLC and amperometric detection, and report that it is possible to detect 5 ng of all four aflatoxins. This proposed method is recommended for detection and quantification of the less toxic aflatoxin B2, which is presented in grains. Quinto et al. (2009) proposed a new method for determine aflatoxins B1, B2, G1, and G2 in cereal foods. This method is based on solid phase microextraction coupled with HPLC and a post-column photochemical derivatization to improve the fluorescence of analytes and fluorescence detection. Such method is fast compared with the complete analytical process that uses Immunoaffinity column. However, its sensibility is below the legal limits. Vosugh et al. (2009) present a work that uses HPLC in conjunction with diode array detector (DAD) and a second order iterative algorithm called parallel factor analysis (PARAFAC). Such method is used for quantifing aflatoxins B1, B2, G1, and G2 in pistachio nuts, this work also use a solid phase extraction stage as a clean-up procedure. Manneta et al. (2005) presents a new method with fluorescence detection using pyridinium hydrobromide perbromide as a post-column derivatization agent to determine aflatoxin M1 in milk and cheese. The detection limits obtained were of 1 ng/kg for milk and 5 ng/kg for cheese that are 50-fold lower than the maximum residue level (MRL) for AFM1 in milk and 40-fold than MRL for AFM1 in cheese set by various European countries.

An interesting application of HPLC is the combination of immobilized enzyme reactor (IMER) in on-line high performance liquid chromatography. This combination allows the selectivity, rapidity and non-destructive, reproducibility of this chromatographic system to be combined with the specification and sensitivity for an enzymatic reaction (Girelli & Mattei, 2005). Derivatization with a fluorophore enhances the natural fluorescence of aflatoxins and improves detectability. The pre-column approach uses the formation of the corresponding hemiacetals using trifluoroacetic acid (TFA), while the post-column one utilizes either bromination by an electrochemical cellor in addition of bromide, or pyridinium hydrobromide perbromide, for the mobile phase and the formation of an iodine derivative.

Even though the optical devices have dominated the traditional methods for HPLC, the present trend is to use mass detectors in the different HPLC types and configurations. This is because of the universal, selective and sensitive detection they provide (Alcaide-Molina, 2009).

There are several techniques that use chromatography for aflatoxin analysis in food (principally in milk, cheese, corn, peanuts, nuts). Commonly the quantification of the aflatoxins is made by a fluorescence detector that takes advantage of fluorescence properties of aflatoxins under determined wavelength. As a result, researchers have been focused on improving these fluorescence properties to develop more sensitive methods than the commonly used so far. Currently techniques such as pre-column derivatization and post-column derivatization are commonly used to improve aflatoxins fluorescence properties. They also have a clean-up stage to obtain a more pure sample, permiting a better quantification. Some of the common methods used in the clean-up stage are: immunoaffinity column and solid phase extraction.

3.4 Electrokinetics

HPLC is a method for detection of aflatoxins which often is enhanced by other techniques, resulting on alternative chromatographic methods. Accomplishing techniques related to electrokinetics are: Micellar electrokinetic chromatography (MEKC), reversed flow micellar electrokinetic chromatography (RFMEKC), and capillary electrokinetic chromatography (CEKC) with multiphoton excited fluorescence (MPE) detection, among others (Gilbert & Vargas, 2003).

Electrokinetics consists on an interfacial double layer of charges effect in heterogeneous fluids (Rathore and Guttman, 2003). Such effect generates the motion of the fluid due to an external force. This external force may be of different natures, but it is called electrophoresis when the force is an electric field; and capillary osmosis when the force is a chemical potential gradient and the motion of liquid happens in a porous body.

Capillary electrophoresis is a technique that although not been widely available as an alternative in many laboratories which routinely conduct HPLC, it has the advantage that it avoids the use of organic solvents. aflatoxin B1 can be determined by capillary electrophoresis (CEKC) with laser-induced fluorescence (LIF) detection (Maragos & Greer, 1997) after a clean-up process comparable to that required for HPLC, and with a very similar sensitivity to it. Besides, Electrophoresis does not require derivatization of aflatoxins, being that an advantage over HPLC. Sensitivity on CEKC can be further improved by using multiphoton excitation. Detection at levels 104 better than previously achieved by capillary separation in less than 90 seconds can be reached, which demonstrates the potential of this technique (Wei et al., 2000).

Micellar electrokinetic chromatography (MEKC) is conducted in polyacrylamide-coated capillaries under almost complete suppression of electroosmotic flow (Janini et al., 1996).

When small amounts of organic solvents are used in the buffer system good separation of aflatoxins are achieved. Nonetheless, it has been probed only with standard buffers (Gilbert & Vargas, 2003).

4. Fluorescence

All the aflatoxins have a maximum absorption around 360 nm (Akbas and Ozdemir, 2006). Letters 'B' and 'G' of the aflatoxins refer to its blue (425nm) and green-blue (450nm) fluorescence colours produced by these compounds under Ultra Violet (UV) light. AFB1 is the most common aflatoxin; it is followed by the AFB2. AFG is fairly rare. The fluorescence emission of the G toxin is more than 10 times greater than that for the B toxin (Alcaide-Molina et al., 2009).

Different techniques for detection of AFs related to fluorescence are exposed bellow.

4.1 Black light test

The black light test is a method which correctly identifies negative AFs samples with minimum expenditure of time and money. It consists on the illumination of the sample with a UV lamp. Tests should be made in a darkened area for best contrast. Fluorescence may be bright or dim, depending on the amount of fluorescing agent present. Polished metal surfaces reflect blue light, thus, users must be careful distinguishing fluorescence from such reflection. It is highly recommended to use safety goggles when working with the black light test. These goggles eliminate blue haze resulting from eye fluorescence caused by reflected longwave UV radiation.

However, fluorescence does not happen exclusively when aflatoxins are present. There are other substances in food that fluoresce under long wave UV radiation. Fungi as *Aspergillus niger*, various *Penicillium* species, *Aspergillus repens* and other species do not produce aflatoxins, but may produce fluorescent harmless metabolites. Then, it can be said that fluorescence is not a specific indication of the presence of aflatoxins, although it may indicate that conditions have been favourable for growth of toxic molds (B-100 Series Ultraviolet Lamps, UVP).

Furthermore, fluorescence is not stable. It disappears in 4 to 6 weeks of continuous exposure to visible or UV radiation although the toxin remains. Therefore, fresh samples must be taken. Hence, the reliability of the method depends on the size of the sample taken for analysis and how it is taken. A sample must be large enough to be representative of the entire lot and must be taken from all parts of the lot (B-100 Series Ultraviolet Lamps, UVP).

The black light test is commonly applied on animal feed. However, it is only a preliminary confirmatory test; it does not give a quantitative indication. Thus confirmatory and quantitative measurements are needed to be applied to those samples that reacted positively to the black light test. Non-fluorescing samples need not be subjected to this. A quantitative screening test which commonly follows the black light test is small chromatographic column (mini-column) (B-100 Series Ultraviolet Lamps, UVP). After the quantitative test a judgment can be made as to whether or not accept a lot.

4.2 Laser-Induced Fluorescence (LIF) screening method

LIF detection technique was pioneered by Yeung (Novotny & Ishii, 1985). This screening method consists on a mobile phase which contains an eluted sample of aflatoxins. Such mobile phase passes through a detection window in the LIF detector. Thus, the whole fluorescence induced by the laser is collected by the detector (Alcaide-Molina et al., 2009). In LIF detection, the number of molecules that are photo-degraded is inversely proportional to the velocity of the fluorophore in front of the laser beam (Simeon et al., 2001). The scheme of a la LIF sensor is shown on Fig. 1.

It has been said that AFB1 is the most toxic and one of the less fluorescent of the aflatoxins. However, the poorest sensitivity of the method may correspond to some other AF. Sensitivity tests should be applied for different AFs to select the one with the lowest sensitivity. The system should be calibrated with the curve of such aflatoxin; thereby, a signal provided by other AF is going to be translated into a higher concentration of this AF, leading to a confirmatory analysis on the screening method. This strategy, then, eliminates false negatives (Alcaide-Molina et al., 2009).

Thus, LIF detection shows as an appropriate detection technique with applications on very low concentrations of sample with native fluorescence or that fluoresce after derivatization (Simeon et al., 2001). However, LIF detection is a technique restricted to a limited number of laboratories because the high cost of the lasers, and because most of the analyte molecules have to be labelled with dyes that match the laser wavelength. Moreover, when the labelling reactions are not well understood, they can lead to contradictory results (Lalljie & Sandra, 1995).

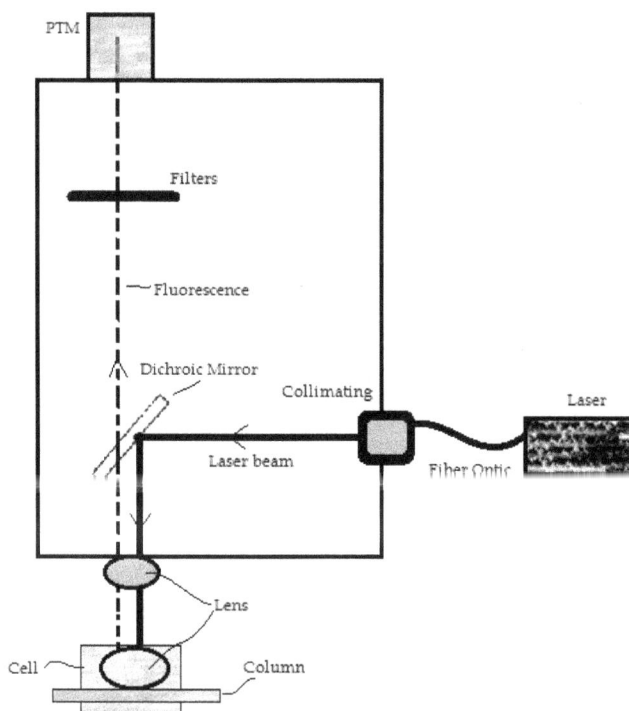

Fig. 1. Scheme of a LIF detector (adapted from Simeon et al., 2001)

4.3 High-Performance Liquid Chromatography (HPLC) with LIF

Fluorescence detection and electrochemical detection are the two sensitive detection means most commonly used for quantitative studies in HPLC. This happens because the sensitivity levels of those hybrid techniques are much better than the ones observed with conventional fluorescence. It has been demonstrated the usefulness of LIF for sensitive detection in HPLC and micro high-performance liquid chromatography (µHPLC) in sensing very low concentrations of substances that can be excited in the near-UV range (325 nm) after labelling at nanomolar concentrations (Folestad et al., 1985; Diebold et al., 1979). Thus, LIF-HPLC method has become very popular and an essential detection technique in capillary electrophoresis (CE). Its sensitivity has been increased by the use of photoactivation devices (Reif & Metzger, 1995). Its popularity is due to its capability to detect substances at lower ranges than the micromolar (Bayle et al., 2004). For more information about HPLC refer to section 3.

It has been said that in LIF detection, the number of molecules that are photo-degraded is inversely proportional to the velocity of the fluorophore in front of the laser beam. On the other side, the sensitivity of detection in HPLC depends on the inner diameter of the capillary connected to the output of the column. Therefore, at a constant flow-rate, the sensitivity depends on the velocity of the fluorophore in front of the laser beam of the LIF, and the solid angle of fluorescence collection by the optical arrangement (Simeon et al., 1999). As a result, the union of LIF and HPLC offers a good compromise between sensitivity and dead volumes (Simeon et al., 2001).

In flow injection experiments with LIF-HPLC systems, at a given diameter, the detector signal will increase when increasing flow-rates if photochemical degradation is a limiting factor (Simeon et al., 2001). Conversely, if the flow-rate is fixed, an increase in diameter is expected to lead to a quadratic increase in the detector volume, generating also a quadratic increase in the number of detectable molecules. Then can be said that if a larger volume is irradiated at a larger capillary diameter, the efficiency of fluorescence collection is less important than in the case of smaller capillaries (Simeon et al., 2001).

4.4 Photomultipliers (PTM)

Since Fluorescence systems have a wide sensitivity, they are a useful tool to measure AFM1 in milk, which legal limit is very low (about 50 ppt). These systems are suitable for preliminary screening at the earlier stages of the industrial process, and make it possible to discard contaminated milk stocks before their inclusion in the production chain (Cucci et al., 2007). PMTs are highly sensitive photomultipliers based flow through detection system suited for ultra low fluorescence, chemiluminescence or bioluminescence measurements (PMT-FL, FIAlab Instruments). Their photon counting photon counting sensor has a blue-green (280–630 nm) spectral response with a peak of quantum efficiency at 400 nm and ultra-low dark counts. The high sensitivity of these devices reaches parts per trillion, permitting measurements of extremely low fluorescence signals. These devices may work with an internal excitation lamp, a LED source or an SMA terminated fiber optic cable for use with an external lamp. They also count with removable emission and excitation filters, allowing placing the most suitable emission filter for selecting the spectral region of interest. The output of the PTMs is expressed in photo-counts, and corresponds to the entire signal integrated in the transmission spectral band of the emission filter. Therefore, the signal acquired from a sample can also include a background contribution due to the solvent. In principle, the latter can contribute to the actual fluorescence of the substance under analysis with a spurious signal of intrinsic fluorescence or Raman, depending on the excitation wavelength (Cucci et al., 2007).

The use of cyclodextrin (CD) as fluorescence enhancer for aflatoxins detection is widely reported in the literature (Zhilong, G. & Zhujun, 1997; Dall'asta et al., 2003), nevertheless, an increased error bar affects measurements due to the CD scattering effects.

The signal-to-noise ratio of these fluorescence measurements strongly depends on the type of cuvette used for containing the liquid sample. The cell geometry and its constituting material give rise to different effects, such as multiple reflections and stray-light. Small sample volumes and darkened walls are mandatory to achieve a better signal-to-noise ratio. Plastic cuvettes without the use of an additional fluorescence enhancer are not useful for the implementation of an early-warning system. Conversely, quartz cuvettes perform very well (Cucci et al., 2007).

Then, PTMs are compact and easy-to-handle sensors for the rapid detection of low concentrations of AFM1 in liquid solutions without the need for pre-concentration of the sample. They can be used as quick "threshold indications" and as an "early warning system", so as to rapidly single out risk/alarm situations (Cucci et al., 2007).

5. Ultra violet absorption

It has been said that all the aflatoxins have a maximum absorption around 360 nm with a molar absorptivity of about 20,000 cm^2/mol (Akbas & Ozdemir, 2006). But, even though aflatoxins could be detected by UV absorption, the sensitivity of such systems is not sufficient to detect these compounds at the parts per billion (ppb) levels required for food analyses (Alcaide-Molina et al., 2009). The detection limit of UV sensors reaches micromolar ranges (Couderc et al., 1998). This is why fluorescence (FL) techniques have become more popular for AFs detection.

For overcoming the named limitation, UV absorption technique is usually combined with HPLC systems. Experimental results indicate that the detection limit of aflatoxins is enhanced by the proper method of extraction and clean-up process (Göbel & Lusky, 2004; Ali et al., 2005). For example, the selected clean-up and extraction procedures should minimize the interfering substances and matrix effect on the elution and separation of aflatoxins (Akiyama et al., 2001). Such important factors, correctly applied, may be of great importance to help the less sophisticated laboratories with HPLC instruments equipped with UV detector to detect aflatoxins with a precision that complies with the international guidelines and regulations.

Then, even though, HPLC-UV systems still are less sensitive than HPLC-FL systems, especially at low AF levels (Herzallah, 2009), HPLC-UV systems indicate to be accurate, precise, and consequently, reliable enough for determination of aflatoxins in food, with low duration and running cost.

6. Spectrometry

6.1 Ion mobility spectrometry

The Ion mobility spectrometry is a technique that is used in the characterization of chemicals on the basis of speed acquired by the gas-phase ions in an electric field. This technique has been used to determine the concentration of aflatoxins, as evidenced by the work of Sheibani et al. (2008) in which are detected and quantified the concentration of aflatoxins B1 and B2 in pistachio. It has certain advantages in common with the FT-NIR, and low detection limit, fast response, simplicity, portability, low cost.

To detect aflatoxins in a sample, this is evaporated and mixed with a carrier gas. Then it is entered into the Ion Mobility Spectrometer (IMS) where the mixture is ionized and passed through an electric field gradient, where ions of different substances will travel at different speeds. The study by Sheibani et al. (2008) shows that using this technique is impossible to quantify as low as 0.25 ng.

6.2 Fourier Transform Near Infrared (FT-NIR) spectrometry

This technique has been underutilized for the detection of aflatoxins due to calibration requirements required against standard reference chemical processes (Tripathi & Mishra, 2009). Despite of the aforementioned limitations, this technique has some advantages, such as: fast and easy equipment operation, good accuracy, precision, performing nondestructive analyzes, prediction of chemical and physical sample from a single spectrum parameters from a single spectrum enabling several components to be determined simultaneously based on the use of multivariate calibrations.

It basically consists of measuring the absorbance of the sample to light whose wavelength varies in the range known as the Near Infrared (NIR). In the work of Tripathi & Mishra (2009) it is found that for the correct quantification of aflatoxins B1 in chili powder network readings were taken in the range of 6900.3 - 4998.8 cm^{-1} and also in the range of 4902.3 - 3999.8 cm^{-1}, excluding the water absorption bands (5155 and 7000 cm-1). Good results were obtained with respect to chemical techniques such as High Performance Liquid Chromatography (HPLC) and Thin Layer Chromatography (TLC), although its detection range is between 15 to 500 mg / kg which is slightly high compared to these techniques.

7. Biosensors

The term biosensors refers generally to a small, portable and analytical device based on the combination of recognition biomolecules with an appropriate transducer, and able of detecting chemical or biological materials selectively and with a high sensitivity (Paddle, 1996). Its principle of detection is the specific binding of the analyte of interest to the complementary biorecongnition element immobilized on a suitable support medium. When the analyte binds the element, there happens a specific interaction which results in a change of one or more physico-chemical properties. Such properties may be: pH, electron transfer, mass, or heat transfer that are detected and can be measured by a transducer. Depending of the method of signal transduction, biosensors can be divided into different groups: electrochemical, optical, thermometric, piezo-electric or magnetic. In the case of aflatoxin detection, electrochemical and optical are the most commonly used (Velasco-Garcia & Mottram, 2002). Until 1996 only few biosensors for toxins were recorded and most of them were based on ELISA. The goal of the more recent studies is to simplify and expedite the method of detection while maintenance and improvement of sensitivity is attempted (Sapsford et al., 2006).

A method that has gained popularity is the use of antibodies to clean-up samples prior to measurement by LC of HPLC. Carlson et al. (2000) present an immune-affinity fluorometric biosensor where the sample was filtered through a column containing sepharosa beads to which the polyclonal aflatoxin-specific antibodies were joined. The beads with attached aflatoxins were subsequently rinsed to remove any impurities and interference. Posterior, an eluant solution was passed through the beads causing antibodies to release the bound aflatoxins. The analyte was collected and placed in a fluorometer. This system consists essentially of two subsystems a fluidics subsystem in charge of mechanical-handling and processing and an electro-optical system that add a miniature fluorometer.

Sapsfor et al. (2006) present a system to detect and quantify foodborne contaminants using an array biosensor. It is capable of measuring large pathogens such as the bacteria Campylobacter jejuni and small toxins (mycotoxins ochratoxin A, fumonisin B, aflatoxin B_1 and deoxynivalenol). The system is capable of multiple detections of aflatoxins in a short time.

Aflatoxins have inhibitory effect on acetylcholinesterase (AchE) and their detection is coupled with the decrease in the activity of AchE which is measured using a choline oxidase amperometric biosensor (Nayak et al., 2009). Amperometric methods allow the detection of low aflatoxin concentration that cannot be detected by classical spectrophotometry because of the omission of the dilution step used in classical method.

Wang et al. (2009) present an implementation of Long range surface Plasmon – enhanced fluorescence spectroscopy (SPFS) in an immunoassay based biosensor for the highly sensitive detection of AFM1 in milk samples (LRSP). Here fluoropore-labeled molecules captured on the sensor are exited with surface plasmons (SPs) and the emitted fluorescence light is measured. The system takes the advantage of the electromagnetic intensity improvement occurring upon the resonance excitation of SPs that increase the intensity of fluorescence signal. This technique is based on surface Plasmon resonance which is becoming popular for the detection of chemical and biological species.

Others tendencies are the use of nanotechnology to detect aflatoxins such as the paper presented by Xiulan et al. (2004) where colloidal gold particles and antibodies were combined and used to develop an immunochromatographic (IC) method for aflatoxin B1 analysis. The result of this was that the analysis could be completed in less than 10 minutes and the lower test limit was 2.5ng/ml for aflatoxin B1. Such limit was increased in two times of ELISA.

When aflatoxins are consumed by cattle, they are transformed into their hydroxylated product, AFM1 that is known for its hepatotoxic and carcinogenic effects. To date, aflatoxins are regulated in many countries because of the milk intake in infants is high and when they are young the vulnerability to toxins is higher. Because of this, it is necessary to monitor AFM1 in milk at ultra low level, so that, analytical methods with high detectability and analytical throughput are required. Kanungo et al. (2011) present a novel approach where a highly sensitive microplate sandwich ELISA was developed and integrated with Magnetic nanoparticles (MNPs) which could detect ultra trace amount of AFM1 in milk. Sandwich-type immunoassay is an effective bioassay due to the high specificity and sensitivity. MNPs were used as an affinity capture column wherein immobilized antibodies on their surface could capture AFM1 from milk sample.

According to the aforementioned, the new trends could be the use of nanoparticles in combinations of the commonly used techniques such as LC, HPLC, TLC and immunoassay techniques. These combinations are to improve the detection at ultra low level of compounds in order to diminish the risk that this kind of mycotoxins causes to humankind. For doing this it is necessary to use methods that combine simplicity with high detectability.

8. Adsorptive stripping voltammetry

Adsorptive Stripping Voltammetry is a method based on accumulation and reduction of AFB1 and AFB2 species on the surface of hanging mercury drop electrode (HMDE). Such electrode offers both sensitivity and selectivity. The pioneers on this method applied to detection of aflatoxins are Hajian and Ensafi (2009), for more information refer to their article.

Voltammetry is an electro-analytical method. It obtains information about the sample by measuring a current while the potential is varied (Komorsky et al., 1992). The voltammetry used in the experiment of Hajian and Ensafi had three-electrodes containing hanging mercury drop electrode as a working electrode, a carbon rod as an auxiliary electrode and an Ag/AgCl (3.0 M KCl) reference electrode. This method was proved only on AFs B1 and B2, where both aflatoxins were found to adsorb and undergo irreversible reduction reaction at the working mercury electrode (Rodriguez et al., 2005).

Adsorptive Stripping Voltammetry is an electrochemical method which has no or very low dependence on pH. This dependence displayed only for B1 in the pH range of 5.0 to 6.0 (Sun et al., 2005).

As it is expected in adsorption processes, by increasing accumulation time, the peak currents for both of the aflatoxins are increased and then leveled off because of the saturation of electrode surface (Hajian & Ensafi, 2009). Therefore, an accumulation time of 60 seconds is recommended for improving sensitivity. It is also recommended to use the extraction and clean-up method for aflatoxins that was used by Garden and Strachan (2001). Such extraction and clean-up method try to obtain the highest yield of aflatoxins with the minimum matrix effect.

This method uses single standard addition method by spiking 10 ng / ml of standard aflatoxin followed with general procedure for voltammetric analysis. The total determination of aflatoxins is based on the next formula reported by Hajian and Ensafi (2009):

$$Aflatoxin\left(\frac{ng}{ml}\right) = \frac{P'}{P} * C * 20 \qquad (1)$$

Where: P' is peak current of sample (nA), P is peak current of standard aflatoxin (after subtract from P') (nA), C is the concentration of aflatoxin spiked in the cell (ng/ ml) and 20 is a factor value after the sample weight, volume of methanol/water used in the extraction and preparation of injection sample have been considered (Hajian & Ensafi, 2009).

Adsorptive Stripping Voltammetry is a suitable method for determination of total aflatoxins (B1 and B2) in food. This method has some advantages such as high sensitivity, extended linear dynamic range, simplicity and speed (Hajian & Ensafi, 2009). The reliability of this method for determination of total aflatoxins is comparable to HPLC.

9. Miscellaneous methods

Different techniques have tried to offer new options for screening aflatoxins. Screening consists on rapid and/or *in situ* detection. There are two main difficulties for an effective screening method: the necessity of a very high sensitivity, which in fact is a necessity of any technique; and the demand of preliminary sample preparations. Some of these techniques, which are commented ahead, present a lack of applications because of their practical inconveniences or because they have not been proved yet with real samples (Gilbert & Vargas, 2003).

Optical-fiber: Modular separation based on a fiber-optic sensor (Dickens & Sepaniak, 2000) has been tested in buffers, showing enough sensitivity (0.005 ng/ml for detection of aflatoxin B1). Unfortunately, it is limited to handling only liquid matrices.

Electrochemical transduction: The interaction of the aflatoxin M1 with bilayer lipid membranes can be sensed electrochemically (Andreou & Nikolelis, 1997; Andreou et al., 1997) reaching a good specificity and speed of response. But, its principal negative factor is its detection limit 750 ng/ml, which is very unpractical.

Flow injection monitoring: Stabilized systems of filter-supported membranes are capable of achieving significantly improved sensitivity (Andreou & Nikolelis, 1998). These membranes have been proposed for use in detecting aflatoxin M1 in cheese (Siontorou et al., 2000). Single strand DNA oligomers have been incorporated into the membranes to control surface electrostatic properties. This incorporation led to achievement a sensitivity much closer to regulatory limits, and with the ability to analyze four cheese samples per minute. Even though this technique appears to be a good option for in situ testing, it does not have yet many applications (Gilbert & Vargas, 2003).

10. Conclusion

Different methods for detection and quantification of aflatoxins have been discussed along this document. Through the researching made for this document, it has been found that the most popular methods are: ELISA, electrochemical immunosensors, chromatography and fluorescence. Even though ELISA is the most common and widespread technique, it has the disadvantage of requiring well equipped laboratories, trained personnel, harmful solvents and several hours to complete an assay. The detection and quantification of aflatoxins by using electrochemical immunosensor has proven to be efficient, easy to use and able to detect very low levels of these substances. Chromatography is a method which needs immunoaffinity columns and phase solid extraction need to be used to clean-up the sample, and also pre-column and post-column derivatization to enhance the aflatoxins fluorescence properties. So that, by improving these characteristics, it is possible to obtain a better quantification and sensibility. Fluorescence detection is a very good alternative to the conventional techniques used today. It has a very high sensitivity, especially when is combined with other techniques as HPLC. Some fluorescence techniques are used even in portable sensors, resulting on in situ measurements. Techniques such as FT-NIR spectrometer and IMS have proven to be quick, inexpensive and user-friendly, however, the FT-NIR technique shows lack of sensitivity when detecting low concentrations of aflatoxins. New techniques in this field are being developed in order to give a rapid and/or in situ detection of these toxins. Some examples of these new techniques are: optical-fiber, electrochemical transduction, low injection monitoring and biosensors. All of these, except for the biosensors, still present a lack of applications because of their practical inconveniences. The biosensors have been designed to overcome the drawbacks that the common tools employed to detect and quantify aflatoxins presents. They use the inherent fluorescence property that aflatoxins have to improve the detection, that in combination with optical and immunochemical techniques used to clean-up the samples achieve a better quantification.

Due to the risk that the aflatoxins represent to humans, the researchers all over the word are looking for methods to detect and quantify them. Apparently, the measurement of aflatoxins in the future tends to be the combination of optical, immunchemical, and fluorescence techniques.

11. Acknowledgements

Authors give thanks to Consejo Nacional de Ciencia y Tecnología (CONACyT), in Mexico, for its financial support through the scholarships with Registration Numbers: 239421 (AEC), 201401 (LMCM), 209021 (RFMH) and 207684 (JRMA). We also express our gratitude to CONACYT for funding the project CB-2008-01.000000000106133.

12. References

B-100 Series Ultra Violet Lamps, Ultra Violet Products (UVP). Nuffield Road, Cambridge, UK.

PMT-FL fluorometer, FIAlab Instruments. Bellevue WA, USA.

Akbas, M. and Ozdemir, M. (2006). Effect of different ozone treatments on aflatoxin degradation and hysicochemical properties of pistachios. Journal of the Science of Food and Agriculture, 86(13):2099–2104.

Akiyama, H., Goda, Y., Tanaka, T., and Toyoda, M. (2001). Determination of aflatoxins B1, B2, G1 and G2 in spices using a multifunctional column clean-up. Journal of Chromatography A, 932(1-2):153–157.

Alcaide-Molina, M., Ruiz-Jiménez, J., Mata-Granados, J., and Luque de Castro, M. (2009). High through-put aflatoxin determination in plant material by automated solid-phase extraction on-line coupled to laser-induced fluorescence screening and determination by liquid chromatography-triple quadrupole mass spectrometry. Journal of Chromatography A, 1216(7):1115–1125.

Ali, N., Hashim, N., Saad, B., Safan, K., Nakajima, M., and Yoshizawa, T. (2005). Evaluation of a method to determine the natural occurrence of aflatoxins in commercial traditional herbal medicines from Malaysia and Indonesia. Food and chemical toxicology, 43(12):1763–1772.

Andreou, V. and Nikolelis, D. (1997). Electro-chemical transduction of interactions of aflatoxin M1 with bilayer lipid membranes (BLMs) for the construction of one-shot sensors. Sensors and Actuators B: Chemical, 41(1-3):213–216.

Andreou, V. and Nikolelis, D. (1998). Flow injection monitoring of aflatoxin M1 in milk and milk preparations using filter-supported bilayer lipid membranes. Analytical chemistry, 70(11):2366–2371.

Bayle, C., Causs´e, E., and Couderc, F. (2004). Determination of aminothiols in body fluids, cells, and tissues by capillary electrophoresis. Electrophoresis, 25(10-11):1457–1472.

Carlson, M., Bargeron, C., Benson, R., Fraser, A., Phillips, T., Velky, J., Groopman, J., Strickland, P., and Ko, H. (2000). An automated, handheld biosensor for aflatoxin. Biosensors and Bioelectronics, 14(10-11):841–848.

Cavaliere, C., Foglia, P., Pastorini, E., Samperi, R., and Lagana, A. (2006). Liquid chromatography/tandem mass spectrometric confirmatory method for determining aflatoxin M1 in cow milk:: Comparison between electrospray and atmospheric pressure photoionization sources. Journalof Chromatography A, 1101(1-2):69–78.

Couderc, F., Caussé, E., and Bayle, C. (1998). Drug analysis by capillary electrophoresis and laser-induced fluorescence. Electrophoresis, 19(16-17):2777–2790.

Cucci, C., Mignani, A., Dall'Asta, C., Pela, R., and Dossena, A. (2007). A portable fluorometer for the rapid screening of M1 aflatoxin. Sensors and Actuators B: Chemical, 126(2):467–472.

Dallasta, C., Ingletto, G., Corradini, R., Galaverna, G., and Marchelli, R. (2003). Fluorescence enhancement of aflatoxins using native and substituted cyclodextrins. Journal of Inclusion Phenomena and Macrocyclic Chemistry, 45(3):257–263.

Dickens, J. and Sepaniak, M. (2000). Modular separation-based fiber-optic sensors for remote in situ monitoring. Journal of Environmental Monitoring, 2(1):11–16.

Diebold, G., Karny, N., and Zare, R. (1979). Determination of zearalenone in corn by laser fluorimetry. Analytical Chemistry, 51(1):67–69.

DiProssimo, V. and Malek, E. (1996). Comparison of three methods for determining aflatoxins in melon seeds. Journal of AOAC International, 79(6):1330–1335.

Eiceman, G. and Karpas, Z. (2005). Ion mobility spectrometry. Number v. 1. CRC Press.

Elizalde-Gonzalez, M., Mattusch, J., and Wennrich, R. (1998). Stability and determination of aflatoxins by high-performance liquid chromatography with amperometric detection. Journal of Chromatography A, 828(1-2):439–444.

Franco, C., Fente, C., Vazquez, B., Cepeda, A., Mahuzier, G., and Prognon, P. (1998). Interaction between cyclodextrins and aflatoxins Q1, M1 and P1:: Fluorescence and chromatographic studies. Journal of Chromatography A, 815(1):21–29.

Folestad, S., Galle, B., and Josefsson, B. (1985). Small-bore LC/laser fluorescence. Journal of chromatographic science, 23(6):273– 278.

Fuchs, B. et al. (2010). Lipid Analysis by Thin-Layer Chromatography-A Review of the current State. Journal of Chromatography A.

Gilbert, J. and Vargas, E. (2003). Advances in sampling and analysis for aflatoxins in food and animal feed. Toxin Reviews, 22(2-3):381–422.

Girelli, A. and Mattei, E. (2005). Application of immobilized enzyme reactor in on-line high performance liquid chromatography: a review. Journal of Chromatography B, 819(1):3–16.

Göbel, R. and Lusky, K. (2004). Simultaneous determination of aflatoxins, ochratoxin A, and zearalenone in grains by new immunoaffinity column/liquid chromatography. Journal of AOAC International, 87(2):411–416.

Hajian, R. and Ensafi, A. (2009). Determination of aflatoxins B1 and B2 by adsorptive cathodic stripping voltammetry in ground-nut. Food Chemistry, 115(3):1034–1037.

Herzallah, S. (2009). Determination of aflatoxins in eggs, milk, meat and meat products using HPLC fluorescent and UV detectors. Food Chemistry, 114(3):1141–1146.

Hussein, H. and Brasel, J. (2001). Toxicity, metabolism, and impact of mycotoxins on humans and animals. Toxicology, 167(2):101–134.

Kanungo et al., 2010] Kanungo, L., Pal, S., and Bhand, S. (2010). Miniaturised hybrid immunoassay for high sensitivity analysis of aflatoxin M1 in milk. Biosensors and Bioelectronics.

Kok, W. (1994). Derivatization reactions for the determination of aflatoxins by liquid chromatography with fluorescence detection. Journal of Chromatography B: Biomedical Sciences and Applications, 659(1-2):127–137.

Janini, G., Muschik, G., and Issaq, H. (1996). Micellar electrokinetic chromatography in zero-electroosmotic flow environment. Journal of Chromatography B: Biomedical Sciences and Applications, 683(1):29–35.

Jin, X., Jin, X., Chen, L., Jiang, J., Shen, G., and Yu, R. (2009). Piezoelectric immunosensor with gold nanoparticles enhanced competitive immunoreaction technique for quantification of aflatoxin B1. Biosensors and Bioelectronics, 24(8):2580–2585.

Komorsky-Lovric, S., L. M. and Branica, M. (1992). Peak current–frequency relationship in adsorptive stripping square-wave voltammetry. Journal of Electroanalytical Chemistry, 335(1-2):297–308.

Lalljie, S. and Sandra, P. (1995). Practical and quantitative aspects in the analysis of FITC and DTAF amino acid derivatives by capillary electrophoresis and LIF detection. Chromatographia, 40(9):519–526.

Manetta, A., Di Giuseppe, L., Giammarco, M., Fusaro, I., Simonella, A., Gramenzi, A., and Formigoni, A. (2005). High-performance liquid chromatography with post-column derivatisation and fluorescence detection for sensitive determination of aflatoxin M1 in milk and cheese. Journal of Chromatography A, 1083(1-2):219–222.

Nayak, M., Kotian, A., Marathe, S., and Chakravortty, D. (2009). Detection of microorganisms using biosensors–A smarter way towards detection techniques. Biosensors and Bioelectronics, 25(4):661–667.

Neagu, D., Perrino, S., Micheli, L., Palleschi, G., and Moscone, D. (2009). aflatoxin M1 determination and stability study in milk samples using a screen-printed 96-well electrochemical microplate. International Dairy Journal, 19(12):753–758.

Novotny, M. and Ishii, D. (1985). Microcolumn separations: columns, instrumentation, and ancillary techniques. Elsevier Science Ltd.

Paddle, B. (1996). Biosensors for chemical and biological agents of defence interest. Biosensors and Bioelectronics, 11(11):1079–1113.

Piermarini, S., Micheli, L., Ammida, N., Palleschi, G., and Moscone, D. (2007). Electrochemical immunosensor array using a 96-well screen-printed microplate for aflatoxin B1 detection. Biosensors and Bioelectronics, 22(7):1434–1440.

Quinto, M., Spadaccino, G., Palermo, C., and Centonze, D. (2009). Determination of aflatoxins in cereal flours by solid-phase microextraction coupled with liquid chromatography and post-column photochemical derivatization-fluorescence detection. Journal of Chromatography A, 1216(49):8636–8641.

Rameil, S., Schubert, P., Grundmann, P., Dietrich, R., and Märtlbauer, E. (2010). Use of 3-(4-hydroxyphenyl) propionic acid as electron donating compound in a potentiometric aflatoxin M1-immunosensor. Analytica chimica acta, 661(1):122–127.

Rastogi, S., Das, M., and Khanna, S. (2001). Quantitative determination of aflatoxin B1-oxime by column liquid chromatography with ultraviolet detection. Journal of Chromatography A, 933(1-2):91–97.

Rathore, A. and Guttman, A. (2003). Electrokinetic phenomena: principles and applications in analytical chemistry and microchip technology. CRC Press.

Reif, K. and Metzger, W. (1995). Determination of aflatoxins in medicinal herbs and plant extracts. Journal of Chromatography A, 692(1-2):131–136.

Ricci, F., Volpe, G., Micheli, L., and Palleschi, G. (2007). A review on novel developments and applications of immunosensors in food analysis. Analytica chimica acta, 605(2):111–129.

Rodriguez, J., Berzas, J., Castaneda, G., and Rodriguez, N. (2005). Voltammetric determination of Imatinib (Gleevec) and its main metabolite using square-wave and adsorptive stripping square-wave techniques in urine samples. Talanta, 66(1):202–209.

Roux, A., Lison, D., Junot, C., and Heilier, J. (2010). Applications of liquid chromatography coupled to mass spectrometry-based metabolomics in clinical chemistry and toxicology: A review. Clinical Biochemistry.

Sapsford, K., Ngundi, M., Moore, M., Lassman, M., Shriver-Lake, L., Taitt, C., and Ligler, F. (2006a). Rapid detection of food-borne contaminants using an array biosensor. Sensors and Actuators B: Chemical, 113(2):599–607.

Sapsford, K., Taitt, C., Fertig, S., Moore, M., Lassman, M., Maragos, C., and Shriver-Lake, L. (2006b). Indirect competitive immunoassay for detection of aflatoxin B1 in corn and nut products using the array biosensor. Biosensors and Bioelectronics, 21(12):2298–2305.

Schaafsma, A., Nicol, R., Savard, M., Sinha, R., Reid, L., and Rottinghaus, G. (1998). Analysis of Fusarium toxins in maize and wheat using thin layer chromatography. Mycopathologia, 142:107–113.

Sheibani, A., Tabrizchi, M., and Ghaziaskar, H. (2008). Determination of aflatoxins B1 and B2 using ion mobility spectrometry. Talanta, 75(1):233–238.

Sherma, J. (2000). Thin-layer chromatography in food and agricultural analysis. Journal of Chromatography A, 880(1-2):129–147.

Simeon, N., Chatelut, E., Canal, P., Nertz, M., and Couderc, F. (1999). Anthracycline analysis by capillary electrophoresis: Application to the analysis of daunorubicine in Kaposi sarcoma tumor. Journal of Chromatography A, 853(1-2):449–454.

Simeon, N., Myers, R., Bayle, C., Nertz, M., Stewart, J., and Couderc, F. (2001). Some applications of near-ultraviolet laser-induced fluorescence detection in nanomolar-and subnanomolar-range high-performance liquid chromatography or micro-high-performance liquid chromatography. Journal of Chromatography A, 913(1-2):253–259.

Siontorou, C., Andreou, V., Nikolelis, D., and Krull, U. (2000). Flow Injection Monitoring of aflatoxin M1 in Cheese Using Filter-Supported Bilayer Lipid Membranes with Incorporated DNA. Electroanalysis, 12(10):747–751.

Stroka, J., Otterdijk, R., and Anklam, E. (2000). Immunoaffinity column clean-up prior to thin-layer chromatography for the determination of aflatoxins in various food matrices. Journal of Chromatography A, 904(2):251–256.

Stroka, J. and Anklam, E. (2002). New strategies for the screening and determination of aflatoxins and the detection of aflatoxin-producing moulds in food and feed. TrAC Trends in Analytical Chemistry, 21(2):90–95.

Sun, N., Mo, W., Shen, Z., and Hu, B. (2005). Adsorptive stripping voltammetric technique for the rapid determination of tobramycin on the hanging mercury electrode. Journal of pharmaceutical and biomedical analysis, 38(2):256–262.

Tan, Y., Chu, X., Shen, G., and Yu, R. (2009). A signal-amplified electrochemical immunosensor for aflatoxin B1 determination in rice. Analytical biochemistry, 387(1):82–86.

Tripathi, S. and Mishra, H. (2009). A rapid FT-NIR method for estimation of aflatoxin B1 in red chili powder. Food Control, 20(9):840–846.

Velasco-Garcia, M. and Mottram, T. (2003). Biosensor technology addressing agricultural problems. Biosystems engineering, 84(1):1–12.

Ventura, M., Gomez, A., Anaya, I., Diaz, J., Broto, F., Agut, M., and Comellas, L. (2004). Determination of aflatoxins B1, G1, B2 and G2 in medicinal herbs by liquid chromatography-tandem mass spectrometry. Journal of Chromatography A, 1048(1):25–29.

Vosough, M., Bayat, M., and Salemi, A. (2010). Matrix-free analysis of aflatoxins in pistachio nuts using parallel factor modeling of liquid chromatography diode-array detection data. Analytica chimica acta, 663(1):11–18.

Wang, Y., Dost'alek, J., and Knoll, W. (2009). Long range surface plasmon-enhanced fluorescence spectroscopy for the detection of aflatoxin M1 in milk. Biosensors and Bioelectronics, 24(7):2264–2267.

Webster, J. (1999). The measurement, instrumentation, and sensors handbook. The electrical engineering handbook series. CRC Press.

Wei, J., Okerberg, E., Dunlap, J., Ly, C., and Jason, B. (2000). Determination of biological toxins using capillary electrokinetic chromatography with multiphoton-excited fluorescence. Analytical chemistry, 72(6):1360–1363.

Weidenbörner, M. (2001). Encyclopedia of food mycotoxins. Springer Verlag.

Xiulan, S., Xiaolian, Z., Jian, T., Zhou, J., and Chu, F. (2005). Preparation of gold-labeled antibody probe and its use in immunochromatography assay for detection of aflatoxin B1. International journal of food microbiology, 99(2):185–194.

Yaacob, M., Yusoff, M., Rahim, A., and Ahmad, R. (2008). Square wave cathodic stripping voltammetric technique for determination of aflatoxin B1 in ground nut sample. The Malaysian Journal of Analytical Sciences, 12(1):132–141.

Zhilong, G. and Zhujun, Z. (1997). Cyclodextrin-based optosensor for determination of tryptophan. Microchimica Acta, 126(3):325–328.

Zollner, P. and Mayer-Helm, B. (2006). Trace mycotoxin analysis in complex biological and food matrices by liquid chromatography-atmospheric pressure ionisation mass spectrometry. Journal of Chromatography A, 1136(2):123–169.

10

Enzymatic Sensor for Sterigmatocystin Detection and Feasibility Investigation of Predicting Aflatoxin B₁ Contamination by Indicator

Da-Ling Liu[1,3] et al.[*]
[1]Institute of Microbial Biotechnology, Jinan University, Guangzhou,
[2]National Engineering Research Center of Genetic Medicine,Guangzhou,
[3]Guangdong Provincial Key Laboratory of Bioengineering Medicine,Guangzhou,
[4]Institutes of Biomedicine and Health, Chinese Academy of Sciences, Guangzhou,
[5]IACM Laboratory, Macau,
P.R.China

1. Introduction

1.1 Enzymatic sensory detection of sterigmatocystin

The development of fast and sensitive sensor for mycotoxins' detection has drawn a great attention in resent years (Prieto-Simom, B. et al., 2007). However, to construct anti-interference biosensor for the practical samples is still challenge.

Sterigmatocystin, a biogenic precursor of aflatoxin B₁, has been classified as group 2B by the International Agency for Research on Cancer (IARC). Its chemical structure consists of a xanthone nucleus attached to bisfuran and it bears a close structural similar to aflatoxin B₁ (Fig. 1) (Versilovskis et al., 2008). The toxicity of sterigmatocystin is primarily confined to the liver and kidney and closely correlated to the occurrence of hepatocellular carcinoma, gastric carcinoma and esophagus carcinoma (Purchase & van der Watt, 1970). Contamination of cereals with *Aspergillus* fungi refers to harmfulness, due to the potential of sterigmatocystin production by these fungi. Sterigmatocystin is similar to aflatoxin B₁ both in the carcinogenicity and fluorescence excitability. While the fluorescence of sterigmatocystin is not so strong as aflatoxin B₁ and the sterigmatocystin-antibody not commercially available, the detection of sterigmatocystin is harder or/and cost more. Several methods for the detection of sterigmatocystin have been established, including thin-layer chromatography (TLC), high-performance liquid chromatography (HPLC), liquid

---
[*] Hui-Yong Tan[1], Jun-Hua Chen[1,4], Ada Hang-Heng Wong[1], Meng-Ieng Fong[5], Chun-Fang Xie[1,2], Shi-Chuan Li[1], Hong Cao[1] and Dong-Sheng Yao[1,2]

1 Institute of Microbial Biotechnology, Jinan University, Guangzhou, P.R.China,
2 National Engineering Research Center of Genetic Medicine,Guangzhou, P.R.China,
3 Guangdong Provincial Key Laboratory of Bioengineering Medicine,Guangzhou, P.R.China,
4 Institutes of Biomedicine and Health, Chinese Academy of Sciences, Guangzhou, P.R.China,
5 IACM Laboratory, Macau, P.R.China.

chromatography with mass spectrometry (LC-MS), gas chromatography with mass spectrometry (GC-MS), high-performance liquid chromatography–tandem mass spectrometry (LC-MS/MS) (Versilovskis et al., 2007; Turner et al., 2009;). Although accurate and sensitive, most of the chromatographic methods are often considered laborious and time intensive, requiring expensive equipments and extended cleanup steps. Therefore, developing a rapid and sensitive method for sterigmatocystin detection is urgently needed. Due to the advantages of enzymatic recognition which offer the response signal with diplex recognitions: the selective binding coupled with the catalytic action of the enzyme toward its substrate, the false results might occur less compared with immuno-sensor or ELISA (enzyme-linked immunosorbent assay) methods which has been concerned the false results (Lim et al., 2007; Massart et al., 2008; DeForge et al., 2010).

Aflatoxin-oxidase (AFO), confirmed to possess oxidation activity toward sterigmatocystin, was utilized as bio-recognition element to constructing the enzymatic biosensor for sterigmatocystin detection. Our previously reported AFO biosensors for fast detection of sterigmatocystin have indicated their potential practicability (Yao et al., 2006; Chen et al., 2010). However, to develop anti-interference enzymatic biosensor for the practical food samples is an arduous target. Recently, we have developed a Prussian blue-base AFO biosensor which revealed effective anti-interferent quality (detailed investigations are going to be published else where). Prussian blue, a prototype of mixed-valence transition metal hexacyanoferrates, has been extensively used as an electrontransfer mediator in amperometric biosensors due to its excellent electrocatalysis toward the reduction of hydrogen peroxide (Karyakin et al., 1994; Ricci & Palleschi, 2005; Zhao et al., 2005; Ricci et al., 2007; Liu et al., 2009). Because of its selective catalysis of hydrogen peroxide in the presence of oxygen and other interferents, Prussian blue is regarded as "artificial peroxidase" (Itaya et al., 1984; Karyakin et al., 1998, 1999, 2000; Karyakin & Karyakina, 1999). The extremely low applied potential of 0.0 V and effectively perselective barrier effect of the Prussian blue - chitosan composite were supposed to be a major attribution towards the interferents from real samples. Here reports the procedure of the biosensor (chitosan – AFO - Single wall carbon nanotubes / Prussian blue – chitosan / L-Cysteine / Au) construction and the results for the sensor's practical use.

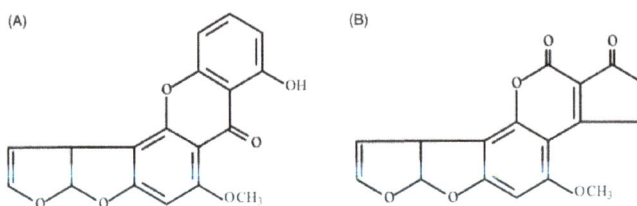

Fig. 1. Chemical structures of sterigmatocystin (A) and aflatoxin B_1 (B).

1.2 Predictive detection of aflatoxins

The prompt and fast method is valuable for food safety and feed. However, the early awareness may be more informative for both consumers and producers. Versicolorin A is the first compound having the toxic bisfuran structure in biosynthesis of aflatoxin B_1. The possibility and feasibility to predict the contamination of aflatoxin B_1 using versicolorin A as the indicator have been reported in the present chapter, also.

Aflatoxins are secondary metabolites produced by filamentous fungi *Aspergillus*, particularly *flavus* and *parasiticus*, which are ubiquitous and can grow extensively in crops and their products. The carcinogenic and immuno-suppresant toxicity of aflatoxins is a serious health risk both to human beings and animals. Among the aflatoxins variants, aflatoxin B_1 is the most toxic and is strictly controlled under food and feed safety regulations in many countries. As is known, mycotoxins may occur at any stage of crops' growth, harvest, storage, transport and marketing. The "fast detection" is still not fast enough to assure life safety and diminish the economic loss since the detection is "after-event" (detectable after the contamination occurred). Development of pre-alert or early-awareness methods has aroused general interests, especially in a time of constant climate changes and food and feed shortage. There is an extensive demand to develop methods for the early identification of emerging hazards to food safety (Concina et al., 2009; Kleter & Marvin, 2009; Marvin & Kleter, 2009).

Biosynthesis of aflatoxins is a complex process (Fig. 2) (Shier et al., 2005), with more than 20 genes involved. Yu (Woloshuk et al., 1994; Yu et al., 1995) revealed that most of these genes were located on the aflR gene (aflatoxin biosynthetic pathway regulatory gene), and that their physical order and distance is highly correlated to the aflatoxin biosynthetic pathway. This gene cluster has been further investigated and expanded (Yu et al., 2004).

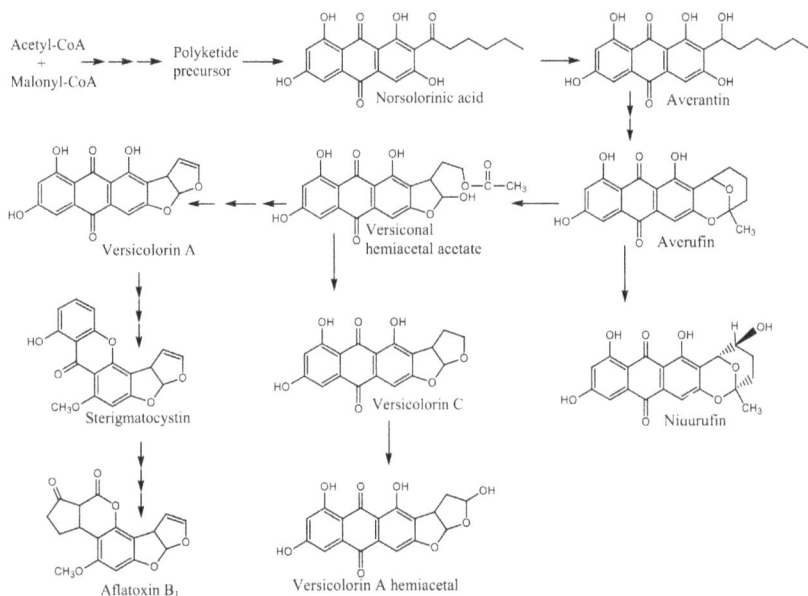

Fig. 2. Presumed biosynthetic pathway of aflatoxin by Shier et al., 2005

It has been proposed that aflatoxicosis is caused by the oxidation of the bisfuran group on aflatoxin B_1 and its variants to yield the ultimate carcinogen aflatoxin B_1-exo-8,9-epoxide in the liver (Jones & Stone, 1998; Smela et al., 2002). Versicolorin A, a precursor of aflatoxin B_1 in the biosynthetic pathway of aflatoxins (Ehrlich et al., 2003; Woloshuk et al., 1994; Yu et al., 1995, 2004), is a member of this toxic group of bisfurans along with its succeeding metabolites sterigmatocystin and aflatoxin B_1. Versicolorin A, the metabolic precursor of aflatoxin B_1, was first separated by Lee (Lee et al., 1975) in a mutant strain of *Aspergillus parasiticus* named *Aspergillus versicolor*. Its molecular formula is $C_{18}H_{10}O_7$ with a molecular weight of 338. Its physical and chemical properties have been fully characterized (Lee et al., 1975; Shier et al., 2005). Some paper have reported the positive mutagenicity of Versicolorin A, and Versicolorin A has shown less mutagenic than aflatoxin B_1 (about 1.5% or 5% toxic of aflatoxin B_1) in Ames test. However, incomprehensive reports of the toxicity of Versicoloring A had been published (Wong et al., 1977; Dunn et al., 1982; Mori et al., 1985), thus we looked for the minimum dose of Versicoloring A mutagenicity using Ames tests with four tester strains and a human peripheral lymphocytes test.

As mentioned above, in the aflatoxin B_1 biosynthesis procedures versicolorin A is a key precursor and far away from the end product of aflatoxin B_1 with a lower toxicity. Versicolorin A might be a candidate indicator for pre-alert of aflatoxin B_1 pollution. This study expands report of versicolorin A and aflatoxin B_1 levels in pure cultures of *A. flavus* and *A. parasiticus* on different culture media, *A. parasiticus* inoculated white rice, and local (Guangdong province, China) commercial feed samples. To evaluate whether versicolorin A is feasible to pre-alert aflatoxin B_1 pollution, 34 feed samples (corn dregs) previously considered safe (aflatoxin $B_1 \leq 25$ μg/kg, China regulation [GB13078-2001]) but with a high level of versicolorin A (≥ 50 μg/kg) were chosen. The storage tests were performed. The final aflatoxin B_1 was determined and the relationships between original versicolorin A and the final aflatoxin B_1 have been analyzed.

2. Materials and methods

2.1 Enzymatic sensory detection of sterigmatocystin
2.1.1 Chemicals
Sterigmatocystin and L-Cysteine were obtained from Sigma-Aldrich Co. (St. Louis. USA). Single wall carbon nanotubes (SWCNTs) (95% purity) were purchased from Shenzhen Nanotech Port Co. (Shenzhen, China). Chitosan (CS) (95% deacetylation,) and other chemicals were of analytical grade without further purification. Phosphate buffer solution (PBS, 0.05 M) consisting of K_2HPO_4, KH_2PO_4 and 0.1 M KCl was employed as supporting electrolyte. The double-distilled water was used throughout. The preparation of aflatoxin-oxidase (AFO) followed a similar procedure according to the literature (Liu et al., 2001), and with corresponding specific enzyme activity of 320 U/mg (1 U was equal to the amount of enzyme that can decrease 1 nmol of sterigmatocystin per minute). Measurements were performed using CHI660C electrochemical workstation (CH Instrument, USA). The electrochemical system consists of gold working electrode, a platinum wire as the auxiliary electrode, and an Ag/AgCl (saturated with KCl) electrode as the reference electrode. All experiments were conducted at room temperature in a 10 ml electrochemical cell with respect to Ag/AgCl. The amplitude of the applied sine wave potential was 5 mV, with a formal potential 0.24 V. The current-time curves were recorded at 0.0 V under stirring.

2.1.2 Preparation of sterigmatocystin biosensor
Gold electrodes (2 mm in diameter, CH Instruments Inc.) were cleaned following the reported protocol (Zhang et al., 2007) and then rinsed with water. After flowing dry with nitrogen, electrodes were immediately immersed into 0.02 M L-Cysteine solutions for 6 h at 4 °C to form self-assembly monolayer modified electrode. Extensively washed with water to

remove the unbound L-Cysteine (Cys), the self-assembly monolayer modified electrodes were denoted as Cys/Au. A 0.2 wt.% CS solution was prepared by dissolving chitosan (CS) powder in 1% (V/V) acetic acid solution with magnetic stirring for about 2 h followed with filter removal of the undissolved particles and adjusting the pH to 5.5 with condense NaOH. The prepared and CS solution was then stored in 4 °C. The Prussian blue-chitosan (PB-CS) hybrid film was deposited onto the Cys/Au modified electrode according to the following four steps:

1. Preparation of the film: A PB-CS solution consisting of 2.5 mM K$_3$[Fe(CN)$_6$], 2.5 mM FeCl$_3$, 0.1 M KCl, 0.1 M HCl, and 0.01% CS was deoxygenated by purging high-purity nitrogen for 10 min. PB-CS was then electrodeposited onto Cys/Au by applying a constant potential of 400 mV (*vs.* Ag/AgCl) for 40 s.
2. Activation of the film: The PB-CS layer was then further activated in an electrolyte solution containing 0.1 M KCl and 0.1 M HCl, which was used for film growth by successive cyclic scanning from -50 mV to 350 mV for 30 cycles at 50 mV/s.
3. Drying of the film: After carefully rinsed with doubly distilled water, the modified electrode was then baked at 100 °C for 1 h since it was reported in the literature (Ricci et al., 2003) that a more stable and active layer of Prussian blue (PB) could be obtained with 1 h baking at 100°C.
4. Conditioning of the film: A potential of -50 mV was applied for 600 s in 0.05 M PBS consisting of K$_2$HPO$_4$, KH$_2$PO$_4$ and 0.1 M KCl (pH 6.5). And then a 20 cycles of scan from -50 mV to 350 mV at 50 mV/s was followed.

After the four steps procedure, the electrode, constructed with PB-CS electrically depositing onto Cys/Au modified electrode, was referred to as PB-CS/Cys/Au electrode. For the enzyme biosensor, the modification was carried out by dropping 10 µl of an aqueous suspension containing 0.5 mg/ml carboxylated single wall carbon nanotubes (SWCNTs), 2.5 mg/ml aflatoxin-oxidase (AFO), and 0.2 wt.% CS on the PB-CS/Cys/Au electrode. Before used, SWCNTs were carboxylated in a 3:1 (V/V) mixture of concentrated H$_2$SO$_4$/HNO$_3$ with sonication at 60 °C according to the literature (Zhang et al., 2008). The AFO-modified electrode (referred to as CS-AFO-SWCNTs/PB-CS/Cys/Au) was then dried at 4 °C in a refrigerator for 24 h. The enzyme electrodes must be washed thoroughly with PBS before experiments and store at 4 °C when not in use.

2.1.3 Rice samples preparation

Non-infected rice sample (purchased from the local market) was first grounded in a household blender. Aliquots (0.5 g) of the rice powder were then spiked with sterigmatocystin at different concentrations and mixed in a vortex mixer. After adding 4 ml of extraction solvent (80% methanol), the sample was fully mixed by shaking for 30 min, and then, centrifuged at 6000 g for 10 min at 4 °C. The supernatant was carefully removed and diluted with PBS (1:10, V/V) for further analysis.

2.1.4 Safety conditions

Sterigmatocystin is a very potent carcinogen, so great care should be taken to avoid personal exposure. It is necessary to wear lab dresses, gloves, and mask when doing experiments. All laboratory glassware and consumables contaminated with sterigmatocystin were soaked overnight in a 5% sodium hypochlorite solution containing 5% acetone. The decontamination solution was allowed a minimum of 30 min before disposal.

2.2 Feasibility investigation on predictive detection of aflatoxin B_1
2.2.1 Preparation of pure versicolorin A

Aspergillus versicolor ATCC 36537 (Lee et al., 1975) (purchased from ATCC) was regenerated on slants under 24 °C in darkness according to the product manual. After 5d activation in liquid growth medium (malt extract 20g, glucose 20g, peptone 1g, distilled water 1L) twice, it was cultured in YES medium (sucrose 200g, yeast extract 20g, distilled water 1L) at 24 °C in darkness without agitation for 7-10d for versicolorin A production. The mycelial mass was extracted with acetone until colorless and the combined extracts were filtered, dried with anhydrous sodium sulfate and evaporated to dryness at 50 °C in a rotary evaporator. For each 1L culture, 10ml petroleum ether and 250ml 30% acetone-water was added to re-dissolve the residue and transferred to a separatory funnel, followed by partitioning with 100ml hexane thrice. Finally, the hexane partition was pooled and evaporated at 50 °C until dryness and the residue was dissolved in 20ml methanol and stored at 4 °C in darkness.

Crude versicolorin A was further purified by preparative HPLC (Billington & Hsieh, 1989) using 95: 5 methanol: water at a constant flow of 10ml/min on a 50×250mm 10μm Agilent Prep-C_{18} column mounted on Agilent 1100 series installed with a DAD detector. Versicolorin A was eluted at 18.432min detected by absorbance at 214nm and 290nm. Pure versicolorin A was lyophilized and stored at -20 °C in darkness. Pure versicolorin A powder was re-dissolved in methanol and verified by LC-MS when in use.

2.2.2 Mutagenicity tests

Ames tests with *Salmonella typhimurium* TA97, TA98, TA100 and TA102 tester strains and the human peripheral lymphocytes test were carried out with pure Versicoloring A in the Guangzhou Disease Prevention and Control Center, Guangzhou, China. In the Ames tests, we used the positive controls of 50μg/plate Dexon in the TA97 and TA98 tests, 1.5μg/plate NaN3 in the TA100 test, and 0.5μg/plate Mitomycin C in the TA102 test in the absence of S9 mix; for S9+ tests, 10μg/plate 2-aminofluorene served as positive control in the TA97, TA98 and TA100 tests, and 60μg/plate Chrysazin in the TA102 test. The experiment group consisted of Versicoloring A at variable concentrations of, 20.0, 10.0, 5.0 and 2.5 μg/plate. A blank control and a negative control of DMSO were also included. Experiments were repeated twice in triplicate. The TA98 test was repeated twice in triplicate with Versicoloring A concentrations of 0.8, 0.6 and 0.4 μg/plate.

In the human peripheral lymphocytes test, Versicoloring A concentrations of 1.6, 0.8, 0.4 0.2 and 0.1 μg/mL were used with peripheral lymphocytes of 8 healthy patients in parallel. A blank control, a negative control of DMSO and a positive control of 40μg/mL Mitomycin C were also included.

2.2.3 Detection of versicolorin A and aflatoxin B_1 production time course in pure medium cultures

Pure cultures of *A. flavus* and *A. parasiticus* on different culture media were studied. The three media used were: CAO (sucrose 30g, $MgSO_4$ 0.5g, $FeSO_4$ 0.01g, K_2HPO_4 1g, $NaNO_3$ 3g, KCl 0.5g, distilled water 1L), YES (as described above) and PG (peptone 100g, glucose 10g, distilled water 1L). 1ml 1.0 x 10^6 CFU/ml *A. flavus* or *A. parasiticus* spore suspension fluid was inoculated in 100ml liquid medium and cultured at 28 °C without agitation in darkness. Toxins were extracted according to a protocol previously described by Bennett (Lee et al., 1975, 1976; Bennett et al., 1976). TLC developed by toluene: ethyl acetate: glacial acetic acid at a ratio of 50:30:4 (V/V/V) on a 12×12cm silica plate was used for detecting metabolites in crude extract with reference to reported R_f values (Shier et al., 2005).

Enzymatic Sensor for Sterigmatocystin Detection and Feasibility Investigation of Predicting
Aflatoxin B₁ Contamination by Indicator

167

2.2.4 Detection of versicolorin A and aflatoxin B₁ production time course in contaminated white rice

Versicolorin A and aflatoxin B$_1$ was detected in pure cultures of *A. parasiticus* on white rice by thin layer chromatography (TLC) or high performance liquid chromatography (HPLC). Commercial bulk white rice was purchased in a supermarket and exposed to UV prior to fungal contamination. 1ml of 1.0×10^6 CFU/ml *A. parasiticus* spore suspension fluid was inoculated on 20g rice. Culture conditions were indicated in the context or under the diagrams. Toxins were extracted and detected by TLC as described above. Otherwise, quantifications of crude samples were made with HPLC on 4.6×150mm 5μm Shimadzu ODS-C$_{18}$ column mounted on Shimadzu 6AD series installed with a DAD and fluorescence detector. 10μl sample was loaded and eluted with solvent A (10mM ammonium acetate, 20μM sodium acetate in water) and solvent B (10mM ammonium acetate, 20μM sodium acetate in methanol) by a two-step gradient of 85%B for 10min and 100%B for 10min respectively at a constant flow of 0.3mL/min. Versicolorin A was eluted at 23.163min detected by absorbance at 222nm and 288nm; aflatoxin B$_1$ was eluted at 11.973min detected by fluorescence at an excitation wavelength of 365nm and an emission wavelength of 435nm.

2.2.5 Detection of versicolorin A and aflatoxin B₁ on commercial feed samples

A set of 100 animal feeds samples (corn dregs) were analyzed. Feed samples of 20g were crushed with blender. Aflatoxin B$_1$ and versicolorin A were extracted and determined by HPLC procedures as 2.2.4 described. Data analyzed by using the statistic soft ware of SPSS13.0.

2.2.6 Detection of the original versicolorin A and the after-storage aflatoxin B₁ for the samples which concern safe originally

Aflatoxin B$_1$ in 200 feeds samples were determined by ELISA (Aflatoxin Tube Kit, Beacon, USA) according to instructions in the product manual. Those which aflatoxin B$_1$ were not more than 25 μg/kg were screened. And followed by the determination of versicolorin A by HPLC method described in 2.2.4. Thirty-four samples with high levels of versicolorin A (≥ 50 μg/kg) of them were chose for the following storage tests.

The 34 chosen samples have divided into two groups. Seventeen samples of them were stored under darkness at 22 ± 2 °C with relative humidity $70 \pm 2\%$ for 10 days, and the rest were stored under darkness at 28 °C with 80% relative humidity for 4 days. After determinations of the final aflatoxin B$_1$ and versicolorin A content by HPLC methods, data of versicolorin A and aflatoxin B$_1$ before and after storage have been analyzed, and statistical soft ware of SPSS13.0 were used.

3. Results and discussions

3.1 Enzymatic sensory detection of sterigmatocystin

3.1.1 Analytical performance of the enzyme electrode for sterigmatocystin detection

Fig. 3 (A) shows the cyclic voltammograms of sterigmatocystin detected by CS–AFO–SWCNTs/PB–CS/Cys/Au electrode in 0.05 M PBS (pH 6.5) at a scan rate of 50 mV/s. With the addition of certain amount of sterigmatocystin, the cyclic voltammograms changed obviously with an increase in the cathodic peak current and a concomitant decrease in the anodic peak current. The possible interferent usually appeared in drink and food samples were selected for interference studies to investigate the selectivity of the as-prepared biosensor. As shown in Fig. 3 (B), the biosensor shows no observable change of the response to 4 g/ml glucose, methanol, oleic acid, phenol, L-tryptophan, and ascorbic acid; in contrast, the biosensor exhibits very strong response to the successive addition of 20 ng/ml sterigmatocystin in the presence of the interfering substances.

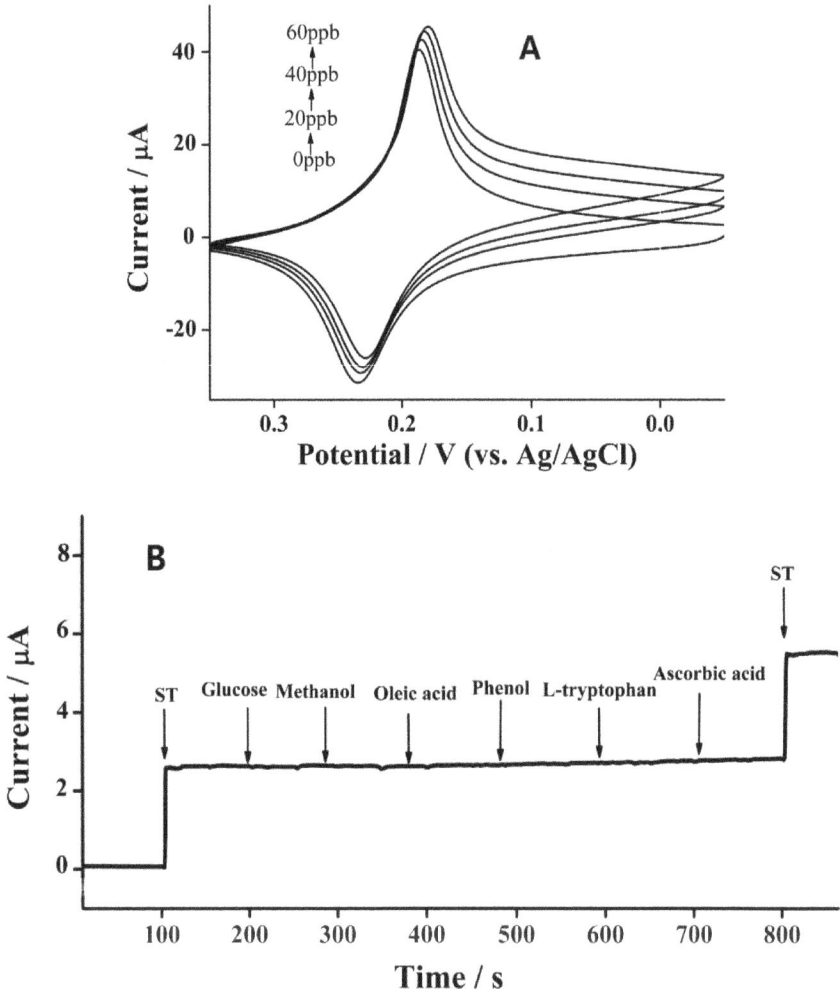

Fig. 3. (A) The cyclic voltammograms of CS-AFO-SWCNTs/PB-CS/Cys/Au electrode in 0.05 M PBS (pH 6.5) in the presence of different concentration of sterigmatocystin (ST). Scan rate: 50 mV/s. (B) Amperometric current-time curve illustrating the interferences free sensing of ST at the proposed biosensor in 0.05 M pH 6.5 PBS. ST (20 ng/ml) and the potential interfering substances (4 g/ml) were added at regular intervals as indicated by the arrows. Applied potential: 0.0 V.

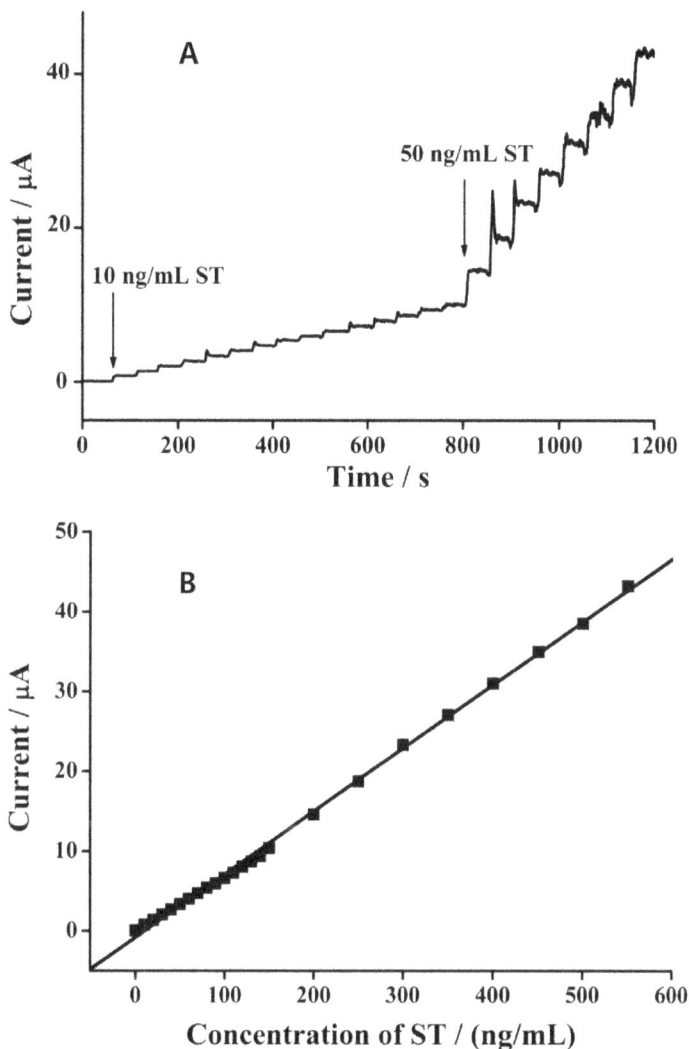

Fig. 4. (A)Typical amperometric current-time curve of CS-AFO-SWCNTs/PB-CS/Cys/Au electrode to successive addition different concentration of sterigmatocystin (ST) in 0.05 mol/L pH 6.5 PBS at 0.0V. (B) The corresponding calibration curve of the electrode.

Fig. 4 (A) shows the amperometric current-time responses of the biosensor on successive step changes of sterigmatocystin concentration in a continuous stirring electrolytic cell at 0.0 V. As Fig. 4 (B) shown, the response current increased linearly with the sterigmatocystin concentration in the range of 10 to 950 ng/ml (correlation coefficient of 0.9985) with a sensitivity of 2.64 $A \cdot g^{-1} \cdot ml \cdot cm^{-2}$ and a detection limit of 2 ng/ml (S/N=3). The 95% of the steady-state current can be obtained within 8 s by using the CS-AFO-SWCNTs/PB-CS/Cys/Au electrode, indicating a fast response to sterigmatocystin change.

3.1.2 Rice samples analysis with enzymatic sensor

The bioelectrode has been used to determine the recoveries of 15 various concentrations of sterigmatocystin by standard addition in real corn samples. As Table 1 shown, satisfactory values between 82.0 and 115.0 % for sterigmatocystin were obtained for the recovery. This biosensor electrode is convenient in use with quick response and trustworthy results. Besides this merit, the uncomplicated procedure of the sample preparation may also appeal to users.

Sample number	Added (ng/mL)	Detected (ng/mL)	R.S.D (%)	Recovery (%)
1	10	11.5	4.9	115.0
2	15	13.2	9.2	88.0
3	20	17.1	4.3	85.5
4	25	21.4	6.4	85.6
5	30	27.8	3.8	92.7
6	35	28.7	4.5	82.0
7	40	34.6	4.6	86.5
8	45	47.9	8.3	106.4
9	50	55.1	4.2	110.2
10	60	62.2	5.8	103.7
11	70	64.2	8.5	91.7
12	80	73.9	10.8	92.4
13	90	98.9	4.2	109.9
14	100	97.2	10.6	97.2
15	150	161.7	8.6	107.8

Table 1. The detection of sterigmatocystin in rice sample using CS-AFO-SWCNTs/PB-CS/Cys/Au electrode. The data reported in the table represents the average of four measurements.

3.2 Feasibility investigation on predictive detection of aflatoxin B_1
3.2.1 Versicolorin A and aflatoxin B_1 content time course for the pure culture of A. flavus and A. parasiticus

Pure cultures of *A. flavus* and *A. parasiticus* on different culture media revealed that versicolorin A can be detected in significant amounts after 7d while aflatoxin B_1 might not, depending on the culture conditions (Table 2). Similarly, versicolorin A and aflatoxin B_1 production in pure cultures of *A. parasiticus* on white rice demonstrated that versicolorin A but not aflatoxin B_1 was detected in early fungal contamination using TLC (Fig.5-1 and 5-2). However, analysis by HPLC revealed the existence of both metabolites on Day 3. Additionally, the amount of aflatoxin B_1 was significantly lower than that of versicolorin A in all samples (Fig. 6). Furthermore, HPLC analysis of versicolorin A and aflatoxin B_1 in commercial animal feeds demonstrated the same phenomena (Fig.7).

Fungus	Medium	Versicolorin A		Aflatoxin B$_1$	
		Day 7	Day 10	Day 7	Day 10
A. flavus	CAO	+	+	-	-
	YES	+	+	-	-
	PG	+	+	-	-
A. parasiticus	CAO	+	+	-	-
	YES	+	+	+	+
	PG	+	+	-	+

"+" denotes positive and "-" denotes negative; detection limit for aflatoxin B$_1$ is 5ng.

Table 2. Results of versicolorin A and aflatoxin B$_1$ production in pure cultures of *A. flavus* and *A. parasiticus* on different culture media incubated under 28 °C and ambient humidity without agitation in darkness and detected by TLC.

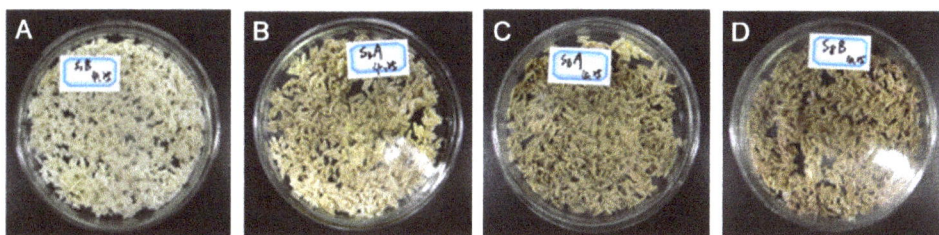

Fig. 5.1. Observation of versicolorin A and aflatoxin B$_1$ production in pure cultures of *A. parasiticus* on white rice under 35 °C and ambient humidity in darkness over 14 days by TLC. Photographs of rice samples taken on Day 2 (A), Day 5 (B), Day 7 (C) and Day 14 (D) after fungus inoculation.

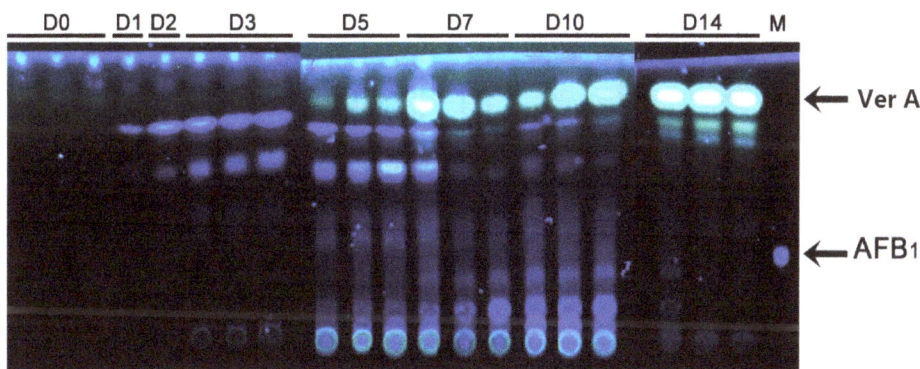

Fig. 5.2. Observation of versicolorin A and aflatoxin B$_1$ production in pure cultures of *A. parasiticus* on white rice under 35 °C and ambient humidity in darkness over 14 days by TLC. TLC detection of versicolorin A and aflatoxin B$_1$ in rice samples on respective days indicated above after fungus inoculation. Experiments were performed in triplicate.

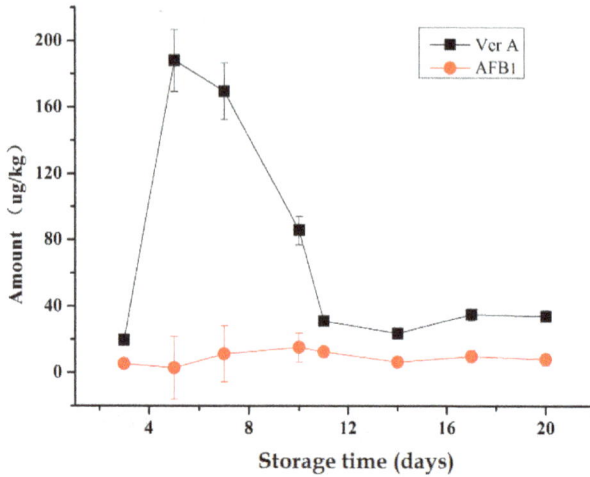

Fig. 6. Observation of versicolorin A and aflatoxin B₁ production in pure cultures of *A. parasiticus* on white rice at 28 °C and 80% relative humidity in darkness over 20d by HPLC. All experiments were performed in triplicate.

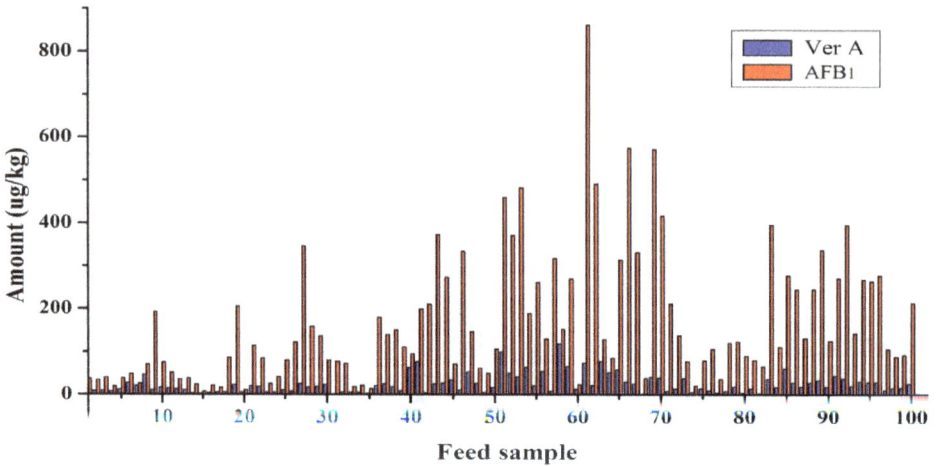

Fig. 7. Detection of versicolorin A and aflatoxin B₁ on commercial animal feeds by HPLC.

3.2.2 Statistical analysis of versicolorin A and aflatoxin B₁

From the 100 feed samples data, it's indicated that they are significantly logarithmic relative as Fig. 8 shown.

$$y = 0.658x + 1.240 \quad (y = \lg \text{Conc. AFB}_1, \ x = \lg \text{Conc.Ver A}) \qquad \text{(Equation 1)}$$
$$R = 0.637 , \ Rsq = 0.405 , \ P < 0.001 \ (\text{by SPSS13.0 soft ware})$$

Analyses of versicolorin A and aflatoxin B$_1$ in white rice contaminated with *A. parasiticus* and in commercial animal feeds purchased from the market revealed that the two metabolites were co-existent. We deduced that the observed phenomenon was caused by the immediacy in their biosynthesis and the heterogeneity of the fungal contamination. However, we could not rule out the possibility that aflatoxin B$_1$ production lags behind versicolorin A in other circumstances because of the complex pathway of aflatoxin biosynthesis. In addition, our investigations on different culture conditions of *A. flavus* and *A. parasiticus* demonstrated that toxin production differs under different nutritional compositions and culture temperatures. It is apparent that the time relationship between sequential product of aflatoxin B$_1$ metabolites depends on the choice of sample of interest and culture conditions.

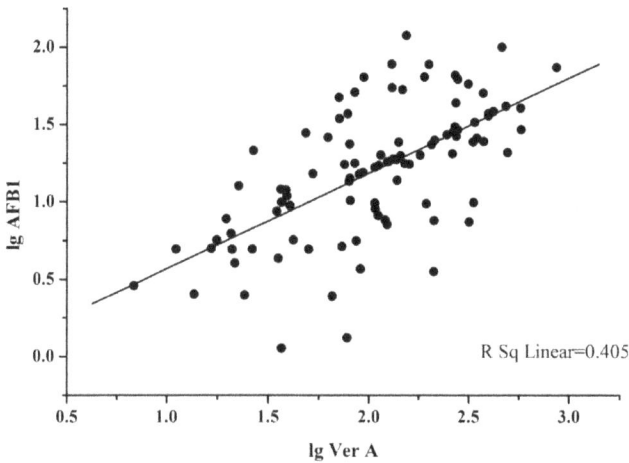

Fig. 8. Statistical analysis for content of versicolorin A and aflatoxin B$_1$ (sample pool:100)

In this study, pure cultures of *A. flavus* and *A. parasiticus* on different culture media revealed that versicolorin A was detected in significant amounts by TLC, but aflatoxin B$_1$ might not be detected under the same culture conditions. HPLC analysis of *A. parasiticus*- contaminated white rice on different days after fungal inoculation showed that versicolorin A was detected in amounts 2 to 28 times higher than that of aflatoxin B$_1$. Analysis of commercial 100 feed samples also showed that versicolorin A quantities were 1.2~59 times higher than that of aflatoxin B$_1$. Therefore, it could be concluded that versicolorin A existed concurrently and in significantly higher amounts as compared to aflatoxin B$_1$ in aflatoxin B$_1$-positive samples. The content of versicolorin A has shown significant relative to the content of aflatoxin B$_1$.

Assays for determination of aflatoxins are diverse. Aflatoxin B$_1$ is the major biomarker for aflatoxin contamination in food and feed. Aflatoxin B$_1$ determination methods include TLC, HPLC, ELISA, etc (Turner et al., 2009). However, each of these methods has their pros and cons (Jiang et al., 2005). For instance, TLC is fast and convenient but the detection limit is high. HPLC is more suitable for quantification but chemical derivatization and fluorescence detectors are required for high sensitivity (Kok, 1994). Additionally, cleanup with affinity columns is essential for a majority of food and feed samples (Jiang et al., 2005). On the other hand, versicolorin A can be detected by simple HPLC coupled with fixed wavelength UV detector (222nm or 288nm, or both of them if DAD detector is available). Moreover, it was found to exist concurrently and in significantly larger quantities than aflatoxin B_1 in our studies. Thus, it offers the alternative to a sensitive and cost efficient indicator of aflatoxin contamination.

3.2.3 The content changed for storage of versicolorin A and aflatoxin B_1

The seventeen chosen samples with aflatoxin B_1 lower than 25ug/kg while versicolorin A more than 50ug/kg were stored under darkness with 22 ± 2 °C and relative humidity $70 \pm 2\%$ for 10 days. The content changed as shown by Fig. 9, 10. The trends of the decrease of versicolorin A with the increase of aflatoxin B_1 after storage are clearly presented.

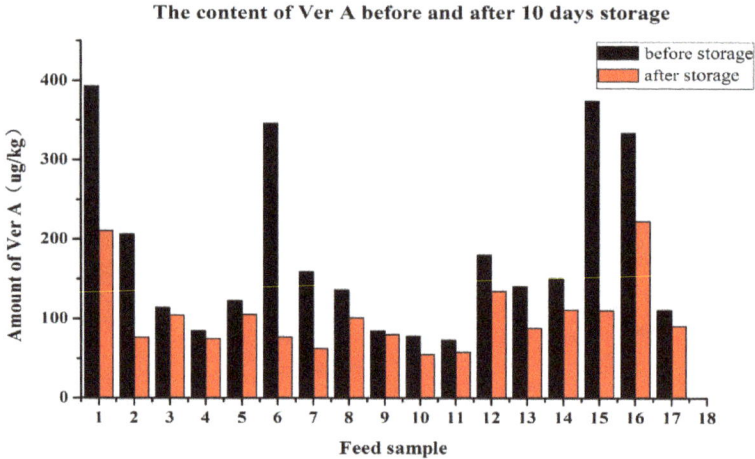

Fig. 9. The content of versicolorin A before and after 10d storage (darkness with 22 ± 2 °C and $70 \pm 2\%$ relative humidity)

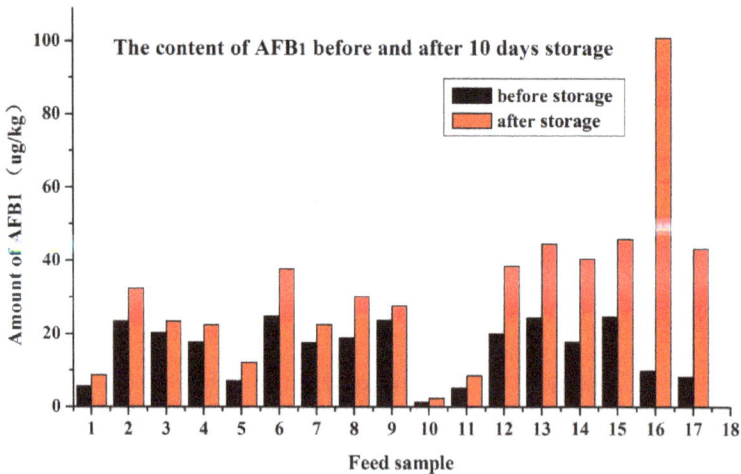

Fig. 10. The content of aflatoxin B_1 before and after 10d storage (darkness with 22 ± 2 °C and $70 \pm 2\%$ relative humidity)

3.2.4 Statistical analysis of the versicolorin A before storage and aflatoxin B₁ after storage

To reveal whether the versicolorin A content is meaningful of subsequent contamination of aflatoxin B_1, statistical analysis of the original versicolorin A against with aflatoxin B_1 after-10d-storage ($22 \pm 2\,^{\circ}C$ and $70 \pm 2\%$ relative humidity) has been performed. Results indicated that they are significantly relative in a negative reciprocal relationship shown as Fig. 12 and equation 2 display.

$$\text{Conc. AFB}_1^{\text{subs.}} = -2890.631 \, (1 \, / \, \text{Conc.Ver A}^{\text{ori.}}) + 50.919 \qquad \text{(Equation 2)}$$
$$R = 0.791 \, \text{,} \; Rsq = 0.626 \quad \text{(by SPSS13.0 soft ware)}$$
$$(\text{10D storage with } 22 \pm 2\,^{\circ}C \text{ and relative humidity } 70 \pm 2\%)$$

Fig. 11 shows a threshold for the original versicolorin A about 67 μg/kg. From the equation 2, it can be calculated that if the original versicolorin A level were about 67 μg/kg or 132 μg/kg, after 10d storage (darkness with $22 \pm 2\,^{\circ}C$ and $70 \pm 2\%$ relative humidity) the aflatoxin B_1 content were approximately 10 μg/kg or 30 μg/kg, respectively.

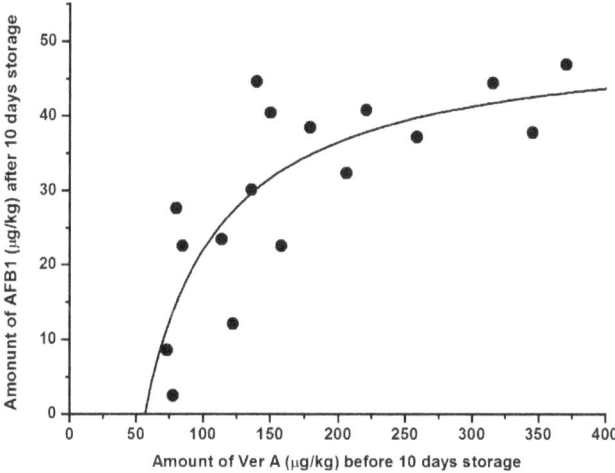

Fig. 11. Statistical analysis of original versicolorin A and subsequent aflatoxin B_1 after 10d storage (darkness with $22 \pm 2\,^{\circ}C$ and $70 \pm 2\%$ relative humidity)

Another group of the same chosen samples have been investigated under the 4d storage at 28 °C with relative humidity 80% for. Results were showed in Fig. 12, 13 and equation 3. Under the fungi growth optimum condition (28 °C with relative humidity 80%), the subsequent aflatoxin B_1 showed a linear relationship with the original versicolorin A content.

These storage investigation results suggested the contamination progress rate may be various depending on the storage conditions, and to investigate the content of original versicolorin A and subsequent aflatoxin B_1 after-storage may reveal the various contamination pattern for a certain storage condition.

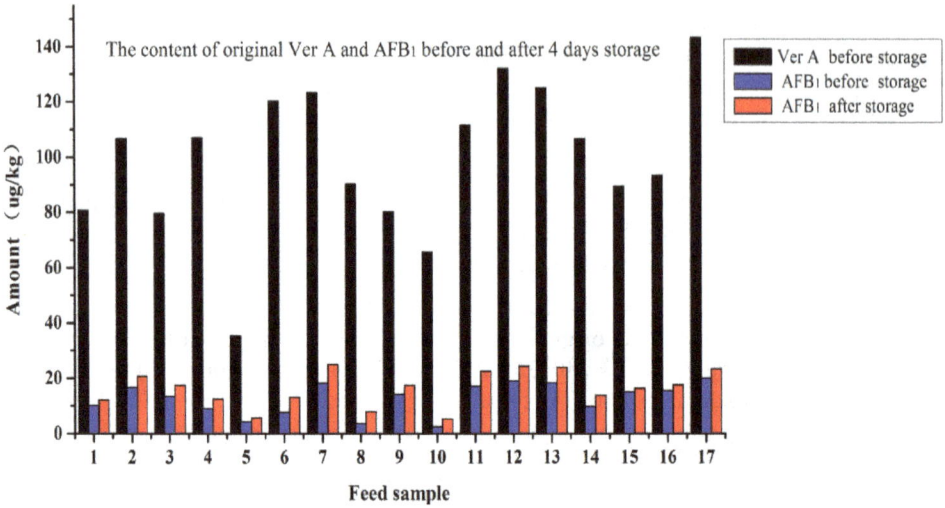

Fig. 12. The content of original versicolorin A and aflatoxin B_1 before and after 4 days storage (darkness with 28 °C and relative humidity 80%)

$$\text{Con. AFB}_1^{\ subs.} = 0.216 \text{ Con.Ver A}^{ori.} - 4.731 \qquad \text{(Equation 3)}$$

R=0.885, Rsq=0.784, P<0.001 (statistics significant)

(For 4D storage with 28°C and relative humidity 80%)

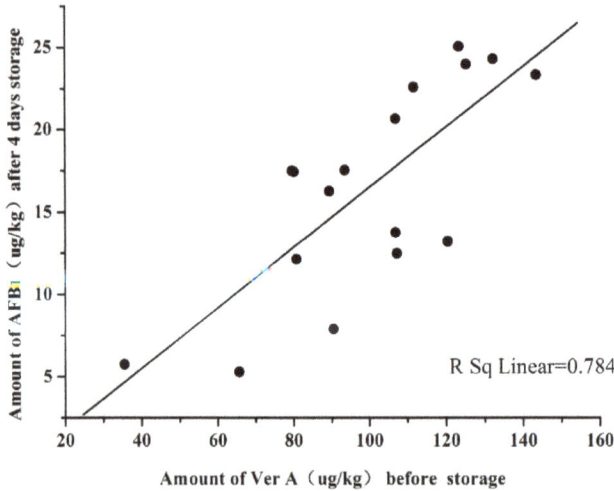

Fig. 13. Statistical analysis for original content of versicolorin A and aflatoxin B_1 before and after 4d storage (darkness with 28 °C and relative humidity 80%)

3.2.5 Mutagenicity tests

Results of the Ames tests with with *Salmonella typhimurium* TA97, TA98, TA100 and TA102 tester strains demonstrated that VerA exhibited mutagenicity on the TA98 tester strains at the concentration of 0.6μg/plate and above. (Figure 14).

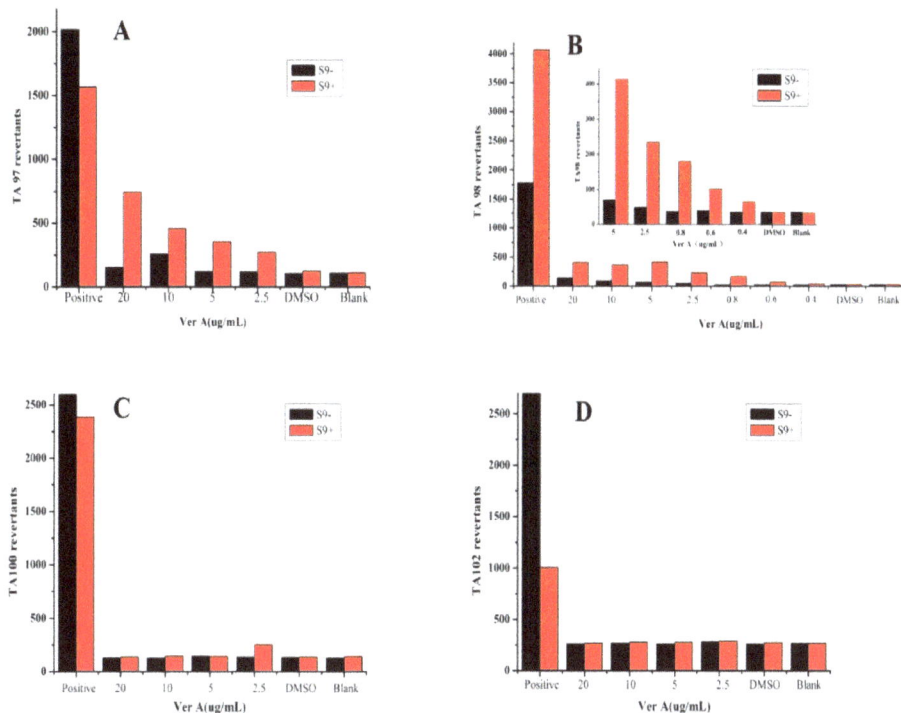

Fig. 14. Ames tests results of VerA with *Salmonella typhimurium* (A) TA97, (B) TA98 and (inset) at VerA concentrations between 5 and 0.4 μg/plate, (C) TA100, and (D) TA102 tester strains. All experiments were repeated twice in triplicate.

On the other hand, the human peripheral lymphocytes test indicated genotoxicity for VerA at the concentration of 1.6μg/mL, which is 25 times of Mitomycin C ($P<0.01$) (shown as Fig. 15). Hence, VerA may be confirmed to be a mutagen towards humanbeings.

4. Conclusions

4.1 Enzymatic sensory detection of sterigmatocystin

Due to the low detection potential (0.0 V) and the role of selective recognition by the enzyme, the biosensor exhibited sensitive and creditable response in corn samples analysis with resistant to glucose, methanol, oleic acid, phenol, L-tryptophan and ascorbic acid. The sensor has given values of recovery in the range of 82.0% - 115.0% and RSD of 4.2% - 10.8% with a simple two-step sample-preparation of 80% methanol extraction followed by centrifugation.

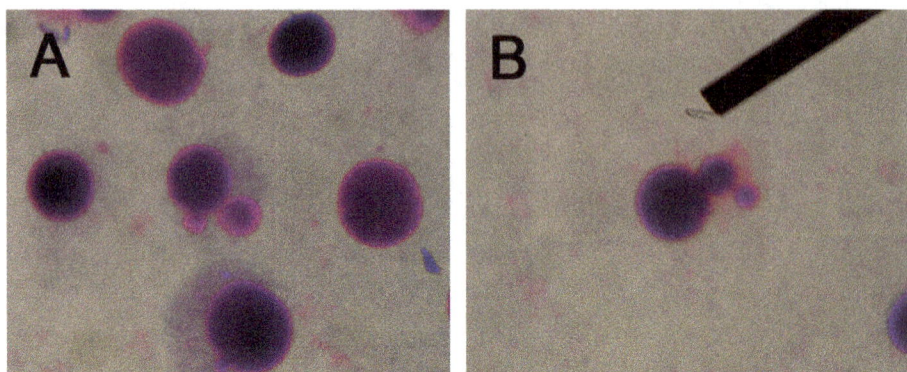

Fig. 15. Photographs of the human peripheral lymphocytes test. (A) Human peripheral lymphocytes from 8 donors were incubated with different concentrations of pure VerA for 24h and stained with 3% Giemsa. (B) A micronucleus-containing lymphocyte is indicated by the pointer. Nucleus was shown in red and cytosol in blue.

4.2 Feasibility investigation on predictive detection of aflatoxin B₁

Based upon the results of this investigation, we conclude that versicolorin A may exist prior to or concurrent with aflatoxin B_1. Although in other cases, in various cereals at diverse conditions, it would be rational to suggest that they are closely relative. In case of versicolorin A detected (even if aflatoxin B_1 not found or at very low level) in some samples, to stop the storage is highly recommended and timely treatment is required.

The mutagenicity test results manifested that Versicoloring A exhibited mutagenicity with the minimum VerA concentration causing mutagenicity in the study was 0.6μg/plate at an induction factor of 3.4 as compared to the negative control. This value is lower than the minimum dose of 0.8μg/plate reported previously (Wong et al., 1977). Nevertheless, Versicoloring A exhibited lower mutagenic effect as compared to 25ng/plate for AFB₁ (Green et al., 1982). On the other hand, Versicoloring A induced significant micro-nuclei at the concentration of 1.6μg/mL in the human peripheral lymphocytes test, which is 25 times that of positive control Mitomycin C (P<0.01). Notwithstanding, it manifested mutagenicity in absence of S9 mix in concentration of 5.0 μg/plate in the TA98 test, which implied Versicoloring A, when it is at a high concentration, may toxic without oxidative active by animal liver.

Besides, with the known of mutagenetic toxicity of versicolorin A (Dunn et al., 1982; Mori et al., 1985), requisite detection of versicolorin A is recommended in food and feed safety regulatory guidelines. Versicolorin A should be considered in food and feed safety guidelines and could also be monitored as a prediction indicator of aflatoxin B_1 contamination.

5. Acknowledgement

The authors thank Key Project of National Development (863) Program (2007AA100605), and Guangdong Provincial Technologies R&D Program (2005B20401004) for financial support of this research.

6. References

Bennett, J.W.; Lee, L.S. & Gaar, G.G. (1976). Effect of acetone on production of aflatoxins and versicolorin pigments by resting cell cultures of Aspergillus parasiticus, *Mycopathologia*, Vol.58, pp. 9-12,

Billington J.A. & Hsieh D.P.H. (1989) .Purification and analysis of versiconal hemiacetal acetate and versicolorin A using low-pressure liquid chromatography and high-performance liquid chromatography, *J. Agric. Food Chem.*, Vol.37, pp. 676–679,

Concina, I.; Falasconi, M.; Gobbi, E.; Bianchi, F., Musci, M., Mattarozzi, M., Pardo, M., Mangia, A., Careri, M. & Sberveglieri, G. (2009). Early detection of microbial contamination in processed tomatoes by electronic nose, *Food Control*, Vol.20, pp. 873–880,

Chen, J.H.; Liu, D.L.; Li, S.C. & Yao, D.S. (2010). Development of an amperometric enzyme electrode biosensor for sterigmatocystin detection, *Enzyme and Microbial Technology*, Vol.,47, pp.119–126,

Dalby, T.; Seier-Petersen, M.; Kristiansen, M.P.; Harboe, Z.B. & Krogfelt, K.A. (2009). Problem solved: a modified enzyme-linked immunosorbent assay for detection of human antibodies to pertussis toxin eliminates false-positive results occurring at analysis of heat-treated sera, *Diagnostic Microbiology and Infectious Disease*, Vol. 63, pp. 354–360,

DeForge, L.E.; Loyet, K.M.; Delarosa, D.; Chinn, J.; Zamanian, F.; Chuntharapai, A.; Lee, J.; Hass, P.; Wei, N.; Townsend, M.J.; Wang, J.Y. & Wong, W.L.T. (2010). Evaluation of heterophilic antibody blocking agents in reducing false positive interference in immunoassays for IL-17AA, IL-17FF, and IL-17AF. *Journal of Immunological Methods*, Vol. 362, pp. 70–81,

Dunn, J.J.; Lee, L.S. & Ciegler, A. (1982). Mutagenicity and toxicity of aflatoxin precursors, *Enviromental mutagenesis*, Vol.4, pp.19-26,

Ehrlich, K.C.; Montalbano, B.G. & Cotty, P.J. (2003). Sequence comparison of aflR from different Aspergillus species provides evidence for variability in regulation of aflatoxin production, *Fungal Genet Biol.*, Vol.38, pp.63-74,

Green, C.E.; Rice, D.W.; Hsieh, D.P.H. & Byard, J.L. (1982). The comparative metabolism and toxic potency of aflatoxin B1 and aflatoxin M1 in primary cultures of adult-rat hepatocytes, *Food and Chemical Toxicology*, Vol. 20, No. 1, pp. 53-60,

Itaya, K.; Shoji, N. & Uchida, I. (1984). Catalysis of the reduction of molecular oxygen to water at Prussian blue modified electrodes, *J. Am. Chem. Soc.*, Vol.106, No.12, pp. 3423-3429,

Jiang, H.; Xiong.Y.H. & Xu Y. (2005). Research progress in analysis methods of Aflatoxins, *Wei Sheng Yan Jiu*, Vol.34, N0.2, (March 2005), pp. 252-255

Jones W.R. & Stone M.P. (1998) Site-specific targeting of aflatoxin adduction directed by triple helix formation in the major groove of oligodeoxyribonucleotides, *Nucleic Acids Res.*, Vol.26, pp.1070-1075,

Karyakin, A.A.; Gitelmacher, O.V. & Karyakina, E.E. (1994). A high sensitive glucose amperometric biosensor based on Prussian Blue modified electrodes, *Analytical Letters*, Vol.27, No.15, (December 1994), pp. 2861-2869,

Karyakin, A.A.; Karyakina, E.E. & Gorton, L. (1998). The electrocatalytic activity of Prussian blue in hydrogen peroxide reduction studied using a wall-jet electrode with continuous flow, *Journal of Electroanalytical Chemistry*, Vol.456, No.1-2, pp. 97-104,

Karyakin, A.A. & Karyakina, E.E. (1999). Prussian Blue-based 'artificial peroxidase' as a transducer for hydrogen peroxide detection, Application to biosensors. *Sensors and Actuators B: Chemical*, Vol.57, No.1-3, (7 September 1999), pp. 268-273,

Karyakin, A.A.; Karyakina, E.E. & Gorton, L. (1999). On the mechanism of H2O2 reduction at Prussian Blue modified electrodes, *Electrochemistry Communications*, Vol.1, No.2, (1 February 1999), pp. 78-82,

Karyakin, A.A.; Karyakina, E.E. & Gorton, L. (2000). Amperometric Biosensor for Glutamate Using Prussian Blue-Based "Artificial Peroxidase" as a Transducer for Hydrogen Peroxide, *Anal. Chem.*, Vol.72, No.7, pp. 1720-1723,

Kleter, G.A. & Marvin, H.J.P. (2009). Indicators of emerging hazards and risks to food safety: A review, *Food and Chemical Toxicology*, Vol.47, pp.1022-1039,

Kok, W. (1994). Derivatization reactions for the determination of aflatoxins by liquid chromatography with fluorescence detection, *J Chromatogr B Biomed Appl*, Vol.659, pp.127-137,

Lee, L.S.; Bennett, J.W.; Cucullu, A.F. & Stanley, J.B. (1975). Synthesis of versicolorin A by a mutant strain of Aspergillus parasiticus deficient in aflatoxin production, *J. Agric. Food Chem.*, Vol.23, pp.1132-1134,

Lee, L.S.; Bennett, J.W.; Cucullu, A.F. & Ory, R.L. (1976). Biosynthesis of aflatoxin B1. Conversion of versicolorin A to aflatoxin B1 by Aspergillus parasiticus, *J. Agric. Food Chem.*, Vol.24, pp.1167-1170,

Lim, Z.Y.; Ho, A.Y.L.; Devereux, S.; Mufti, G.J. & Pagliuca, A. (2007). False positive results of galactomannan ELISA assay in haemato-oncology patients: A single centre experience, *Journal of Infection*, Vol. 55, pp. 201-204,

Liu, Y.; Chu, Z.; Zhang, Y. & Jin, W. (2009). Amperometric glucose biosensor with high sensitivity based on self-assembled Prussian Blue modified electrode, *Electrochimica Acta*, Vol.54, No.28, (1 December 2009), pp. 7490-7494,

Marvin, H.J.P. & Kleter, G.A. (2009). Early awareness of emerging risks associated with food and feed production: Synopsis of pertinent work carried out within the SAFE FOODS project, *Food and Chemical Toxicology*, Vol.47, pp. 911-914,

Massart, C.; Corcuff, J.B. & Bordenave, L. (2008). False-positive results corrected by the use of heterophilic antibody-blocking reagent in thyroglobulin immunoassays, *Clinica Chimica Acta*, Vol. 388, pp. 211-213,

Mori, H.; Kitamura, I.; Sugie, S.; Kawai, K. & Hamasaki, T. (1985). Genotoxicity of fungal metabolites related to aflatoxin B1 biosynthesis, *Mutat Res.*, Vol.43, pp.121-125,

Prieto-Simón, B.; Noguer, T. & Campàs, M. (2007). Emerging biotools for assessment of mycotoxins in the past decade. *TrAC Trends in Analytical Chemistry*, 26, 689-702.

Purchase, I.F.H. & van der Watt, J.J. (1970). Carcinogenicity of sterigmatocystin, *Food Cosmet. Toxicol.*, Vol. 8, pp. 289-295,

Ricci, F.; Amine, A.; Moscone, D. & Palleschi, G. (2007). A probe for NADH and H2O2 amperometric detection at low applied potential for oxidase and dehydrogenase based biosensor applications Biosens, *Biosensors and Bioelectronics*, Vol.22, No.6, (15 January 2007), pp. 854-862,

Ricci, F. & Palleschi, G. (2005). Sensor and biosensor preparation, optimisation and applications of Prussian Blue modified electrodes, *Biosensors and Bioelectronics*, Vol.21, No.3, (15 September 2005), pp. 389-407,

Shier, T.W.; Lao, Y.; Steele, T.W.J. & Abbas, H.K. (2005). Yellow pigments used in rapid identification of aflatoxin-producing *Aspergillus* strains are anthraquinones associated with the aflatoxin biosyntheticpathway, *Bioorganic chemistry*, Vol.33, pp. 426-438,

Smela, M.E.; Hamm, M.L.; Henderson, P.T.; Harris, C.M.; Harris, T.M. & Essigmann, J.M. (2002) The aflatoxin B(1) formamidopyrimidine adduct plays a major role in causing the types of mutations observed in human hepatocellular carcinoma, *Proc Natl Acad Sci U S A.*, Vol.99, pp. 6655-60,

Turner, N.W.; Subrahmanyam, S. & Piletsky, S.A. (2009). Analytical methods for determination of mycotoxins: a review, *Anal. Chim. Acta.*, Vol. 632, pp.168–180,

Versilovskis, A.; Bartkevics, V. & Mikelsone, V. (2008). Sterigmatocystin presence in typical Latvian grains, *Food Chem.*, Vol.109, pp. 243–8,

Versilovskis, A.; Bartkevics, V. & Mikelsone, V. (2007). Analytical method for the determination of sterigmatocystin in grains using high-performance liquid chromatography–tandem mass spectrometry with electrospray positive ionization, *J. Chromatogr. A*, Vol.1157, pp.467–471,

Wheeler, L.; Hamm, M.L. & Demeo, M. (1981). A comparison of aflatoxin B1-induced cytotoxicity, mutagenicity and prophage induction in Salmonella typhimurium mutagen tester strains TA1535, TA1538, TA98 and TA100, *Mutat Res.*, Vol.81, pp.39-48,

Woloshuk, C.P.; Foutz K.R.; Brewer, J.F.; Bhatnagar, D.; Cleveland, T.E. & Payne, G.A. (1994). Molecular characterization of aflR, a regulatory locus for aflatoxin biosynthesis, *Appl Environ Microbiol.*, Vol.60, pp. 2408-2414,

Wong J.J.; Sing R. & Hsieh D.P.H. (1977). Mutagenicity of fungal metabolites related to aflatoxin biosynthesis, *Mutat Res.*, Vol.44, pp. 447-50,

Yao, D.S.; Cao, H.; Wen, S.M.; Liu, D.L.; Bai, Y. & Zheng, W.J. (2006). A novel biosensor for sterigmatocystin constructed by multi-walled carbon nanotubes (MWNT) modified with aflatoxin–detoxifizyme (ADTZ), *Bioelectrochemistry*, Vol.68, pp.126 – 133,

Yu J.; Chang P.K.; Cary, J.W.; Wright, M.; Bhatnagar, D.; Cleveland, T.E.; Payne, G.A. & Linz, J.E. (1995). Comparative mapping of aflatoxin pathway gene clusters in *Aspergillus parasiticus* and *Aspergillus flavus*, *Appl Environ Microbiol.*, Vol.61, pp. 2365–2371,

Yu J.; Chang P.K.; Ehrlich, K.C. ; Cary, J.W.; Bhatnagar, D.; Cleveland, T.E.; Payne, G.A.; Linz, J.E.; Woloshuk, C.P. & Bennett, J.W. (2004). Clustered Pathway Genes in Aflatoxin Biosynthesis, *Applied and Environmental Microbiology*, Vol.70, pp.1253-1262,

Zhang, G.; Sun, S.; Yang, D.; Dodelet, J.P. & Sacher, E. (2008). The surface analytical characterization of carbon fibers functionalized by H2SO4/HNO3 treatment, Carbon, Vol.46, No.2, (February 2008), pp.196-205,

Zhang, J.; Song, S.; Wang, L.; Pan, D.& Fan, C. (2007). A gold nanoparticle-based chronocoulometric DNA sensor for amplified detection of DNA, *Nat. Protoc.*, Vol.2, pp.2888-2893,

Zhao, G.; Feng, J.J.; Zhang, Q.L.; Li, S.P. & Chen, H.Y. (2005).Synthesis and Characterization of Prussian Blue Modified Magnetite Nanoparticles and Its Application to the Electrocatalytic Reduction of H2O2, *Chem. Mater.*, Vol.17, No.12, pp. 3154–3159

11

Aflatoxin Measurement and Analysis

Peiwu Li and Qi Zhang et al.*
Key Laboratory of Biotoxin Analysis of Ministry of Agriculture,
Key Laboratory of Oil Crops Biology of the Ministry of Agriculture,
Oil Crops Research Institute, Chinese Academy of Agricultural Sciences, Wuhan,
China

1. Introduction

Aflatoxin is a group of secondary metabolites produced by fungi *Aspergillus* species, such as *A. flavus* and *A. parasiticus;* in particular, *A. flavus* is common in agriculture. *A. bombycis*, *A. ochraceoroseus*, *A. nomius*, and *A. pseudotamari* are also aflatoxin-producing species, but they are encountered much less frequently (Bennett and Klich, 2003).

Aflatoxin contamination can be occurred very widely. They can be found in over a hundred kinds of agro-products and foods,such as peanut, corn, rice, soy sauce, vinegar, plant oil, pistachio, tea, Chinese medicinal herb, egg, milk, feed etc,. Also some of them in animal organism can be detected. Besides these, aflatoxin can spread and accumulated in environment, for example, river and agricultural field.

Aflatoxins are highly toxic, mutagenic, teratogenic, and carcinogenic compounds, a group of difuranocoumarin derivatives, consisted of a coumarin and a double-furan-ring of molecule usually. Aflatoxin B1, for example, its toxicity is ten times of potassium cyanide, 68 times of arsenic and 416 times of melamine. Furthermore, their carcinogenicity is over 70 times than that of dimethylnitrosamine and 10000 times of Benzene Hexachloride (BHC). And International Agency for Research on Cancer (IARC) of the World Health Organization (WHO) accepted that aflatoxin should be classified as a Group 1 carcinogen in 1987, and then AFB1 is classified as Group 1 (carcinogenic to humans) by the WHO–IARC in 1993 (Li, Zhang & Zhang, 2009). According to the nearest researches by University of Pittsburgh, aflatoxin may play a causative role in 4.6–28.2% of all global HCC cases (Liu and Wu, 2010).

To protect agricultural environment, estimate quality of commercials of agro-products and food, and safeguard safety of consumers' health and lives, over seventy countries setup maximum limits in agro-products, and analytical methods for determination of aflatoxin, play a great role for monitoring and estimation of the contaminants.

There are a variety of well established methodologies reported for analysing aflatoxins in many different foodstuffs, such as thin layer chromatography, high-performance liquid chromatography, ultra-pressured layer chromatography, immunoaffinity chromatography-high-performance liquid chromatography, near infrared spectroscopy and immunoassay

* Daohong Zhang, Di Guan, Xiaoxia, Ding Xuefen Liu, Sufang Fang, Xiupin Wang and Wen Zhang
Key Laboratory of Biotoxin Analysis of Ministry of Agriculture, Key Laboratory of Oil Crops Biology of the Ministry of Agriculture, Oil Crops Research Institute, Chinese Academy of Agricultural Sciences, Wuhan, China.

methods. We here will not only demonstrate current such analytical methods for aflatoxins, but also illuminate tomorrow's trends on analysis of aflatoxins. To help readers understand them well, some basic information of these methods were also presented, including principle of developing, choosing and using these methods.

2. Pretreatment of sample

2.1 Immunoaffinity or multipurification column

The immunoaffinity column (IAC) occupies a special place among the immune analytical approaches, being used many years as a method of sample purification and concentration in the aflatoxin analysis (Scott & Trucksess, 1997). The principle of the IAC is that an antibody (polyclonal or monoclonal) recognized the analyte is immobilized onto a solid support such as agarose or silica in phosphate buffer, all of which is contained in a small column.

The clean-up procedures are completed in four steps (Figure 1):

Condition. The column is initially conditioned with phosphate buffered saline (PBS) and reaches room temperature.

Loading of the sample. The crude sample extract is applied to the IAC containing specific antibodies to aflatoxin at slow steady flow rate of 2-3 mL/min. Gravity or vacuum system can be used to control flow rate. The aflatoxin binds to the antibody and is retained in the IAC. The crude sample extract must be in aqueous solution because organic solvents can damage the antibody and can interfere with the antibody-aflatoxin interaction. The binding strength of the antibody-aflatoxin will influence recovery of the IAC. The specificity of antibody is important to remove the structurally closely compounds which can cause interferences in the quantitation of aflatoxin. The capacity of the IAC (the total number of antibody sites available for binding aflatoxin) is also important as overloading the column will lead to poor recovery (Senyuva & Gilbert, 2010).

Washing. The column is washed with washing solution (water or phosphate buffered saline) to remove impurities. After washing completely, the IAC is blown to dryness by N2 stream.

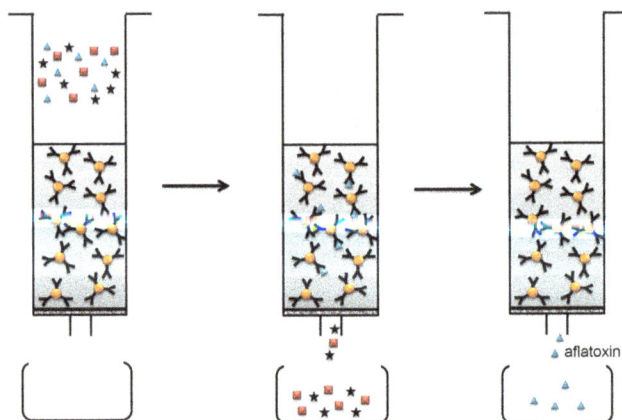

Fig. 1. Scheme of aflatoxin immunoaffinity column for sample pretreatment (clean-up and enrichment).

Elution. By passing a solvent such as acetonitrile through the IAC, breaking the antibody-aflatoxin bond, the captured aflatoxin is removed from the antibody and thus eluted from the column. The big volume of sample loading and the small volume of solvent eluting make the analyte concentrate. The eluate containing aflatoxin is then further developed by addition of fluorescence enhancer or directly measured by HPLC method.

The principle of solid phase extraction (SPE) columns is a variation of chromatographic techniques that uses a solid phase and a liquid phase to isolate one, or one type, of analyte from a solution. The columns contain different packing materials, ranging from silica gel, C-18 (octadecylsilane), florisil, phenyl, aminopropyl, ion exchange materials, both anionic and cationic, and molecular imprinted polymers (Giraudi et al, 2007; Jornet et al, 2000; Mateo et al, 2002; Vatinno et al, 2008; Yu & Lai, 2010; Zambonin et al, 2001). The generally procedure is to load the sample into column, retain the analyte, wash away impurities, and then elute the analyte. A MycoSep multifunctional cleanup column has been developed for one step clean-up of aflatoxin (Figure 2). The MycoSep clean-up column is pushed into a test tube (containing the sample), forcing the sample to filter upwards through the packing material of the column. The interferences adhere to the chemical packing in the column and the purified extract, containing the aflatoxin of interest, passes through a membrane (frit) to the surface of the column. The method is rapid, simple and economical due to the fact that the clean-up of aflatoxin from the column is a single pass procedure using the extract solvent as the eluting solvent. The column has a long shelf-life because it contains no biological reagents, and can be stored at room temperature. However, unlike immunoaffinity columns, the MycoSep clean-up column cannot concentrate the analyte during the clean-up procedure, and also the recovery may vary depending upon the complexity of the food samples (Zheng et al, 2006).

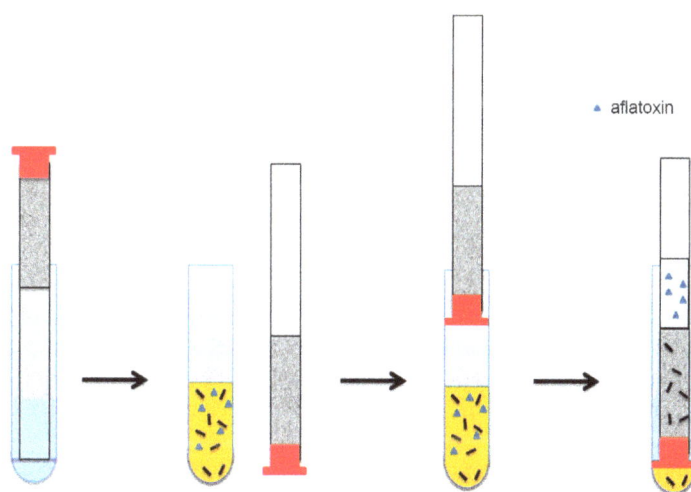

Fig. 2. Scheme of aflatoxin multifunctional cleanup column for sample pretreatment (clean-up).

2.2 How to simplify current protocol

The selection of pretreatment methods for samples depends mainly on two aspects: one is the analytical methods adopted, another is samples to be analyzed. The former is more important with great differences according to the kinds of analytical methods. Complexity, time consuming and cost are the main factors contributed the popular degree by operators and practicability in on-site use. Among these factors complexity degree is most concerned for the exposure hazards of aflatoxins.

Sample pretreatment for instrumental analysis (e.g., HPLC, GC, LC/MS and GC/MS) is very tedious, expensive and time consuming, and needs well equipped laboratories to accomplish it, e.g., frequently involving in large-scale equipment, large sample volumes, extensive extraction or derivatization steps (Tang et al., 2008), complicate clean-up and concentration, and multiple centrifugation, etc. While for immunoassay (for instance, enzyme-linked immunosorbent assay, ELISA) it is usually easier, cheap and rapid generally without derivation but still need clean-up and concentration. How to simplify current pretreatment protocol is a question to extend the methods for aflatoxins detecting outside the laboratory. As an alternative, lateral-flow immunochromatographic assay combines chromatography with immunoassay with less interference due to chromatographic separation, offers the advantages of most simple, cheap and time-saving, requiring only a simple extraction step (Tanaka et al., 2006) or even no need for extraction (e.g., detection of aflatoxin M1 in milk). Therefore, the pretreatment protocol of sample can be simplified by adopting suitable analytical methods, e.g., immunochromatographic assay.

3. Sample analysis

3.1 High fidelity methods
3.1.1 HPLC (UPLC) with fluorescence detector

Since the late 1960's, High Performance Liquid Chromatography (HPLC) had developed, HPLC is by far the most reported chromatographic method using a variety of detection strategies. It was developed rapidly in recent years, about 80% of the world organic compounds (health food efficacy composition, nutritional fortifiers, vitamins, protein etc.) use HPLC for separation and determination. The assessment of the quality of foods using this method provides an acceptable, accurate, and alternative method to establish guidelines and to evaluate the status of aflatoxins in contaminated foods.

HPLC analysis of aflatoxins

HPLC have high efficiency, high sensitivity (HPLC-FLD with as low as 0.1 pg (ng kg^{-1}) detecting limit (Herzallah, 2009) and high resolution. And the chromatographic column can be used repeatedly. So modern analysis of components relies heavily on HPLC employing various adsorbents depending on the physical and chemical structure of different components.

The most commonly found detectors for HPLC are fluorescence detectors (FLD), which rely on the presence of a chromophore in the molecules. A number of toxins already have natural fluorescence (e.g. aflatoxins) and can be detected directly by HPLC–FLD. Determination for aflatoxins by HPLC with fluorescence detections is often the method of the choice. The use of the HPLC in determination of aflatoxins and their metabolites showed higher levels of accuracy and lower detection limits when using CN activate Solid Phase extraction (SPE-CN) or immunoaffinity column (IAC) combined with application of FLD(Brera et al, 2007; Edinboro, & Karnes, 2005; Jaimez & Fente, 2000)

Chromatography columns were the most important part of the HPLC, normal and reversed-phase columns were used for separation and purification of toxins depending on their polarity. Reversed-phase C18 columns with methanol–water or acetonitrile–water mobile phases, is most commonly used for aflatoxins in most laboratories.

Modern analysis of mycotoxins relies heavily on HPLC employing various adsorbents depending on the physical and chemical structure of the mycotoxins. The use of the HPLC in determination of aflatoxins and their metabolites showed higher levels of accuracy and lower detection limits when using SPE-CN or IAC regardless of the HPLC detectors used. Zhao used UPLC for determinations of Aflatoxins B_1, B_2, G_1 and G_2 (AFB1, AFB2, AFG1 and AFG2), and the detection limits (S/N = 3) for B1, B2, G1 and G2 were 0.32, 0.19, 0.32 and 0.19µg kg-1, the corresponding quantification limits (S/N = 10) were 1.07, 0.63, 1.07 and 0.63µg kg-1, respectively (Fu et al, 2008).

Fluorescence enhancement methods of aflatoxins

Derivative with a suitable fluorophore can enhance the natural fluorescence of aflatoxins,which can improve the fluorescence detection sensitivity. The present needs for HPLC fluorescence detection of aflatoxins determination in food and feedstuffs are an emphasis on the improvement of the sampling and extraction steps to lead to more accurate determinations, and further investigations of non-destructive pre-column or post-column derivative methods appears to be a large unexplored field. Some aflatoxins like aflatoxin B1, aflatoxin G1, because of its low signal or its easy quenching signals, several derivation reagents were used during the detection procedure.

There are mainly three kinds of derivatizations: TFA, halogen, and its derivatives, metal ions (Hg2+), cyclodextrine and its derivatizations. The enhancement mechanisms varies with different kinds of derivatizations.

AFB1 derivative method is mainly based on hydrolysis of the second furan ring in acidic solution, and AFB1 is transformed into B2a ,which makes a fluorescent greatly enhanced.This mechanism is commonly used by TFA,halogen,and its derivatives (PBPB) etc.(Francis et al., 1988; Joshua, 1993; Braga et al, 2005)

Dr. Ma (2007) had studied on the metal ions (Hg2+) enhancement for aflatoxins and proposed the probably mechanism was that AFB1 can be chelated with Hg2+, the propose of the complexes fluorescence can be enhanced, the speculate metal complexes electronic transition occurred ligand AFB1 to employed by Hg (II), the charge transfer transition metal ions, namely ligand-to-metal charge transition (LMCT) transition. LMCT transition with high energy, and its absorb is in the UV area, LMCT transition is occurred against bonding σ orbital, electronic horizontally inspire with ligand AFB1 oxidation and reduction of metal, occurred by electron reaction. Metal ions are two ligand simultaneously electronic warp reduction. Speculation that ligand AFB1 is probably in the form of ·L base separation formed 2·L or formed new molecular L - L or L - M2+ -L, reactant system rigid structure to strengthen or conjugated system increased, fluorescent intensity was greatly enhanced (Ma, 2007).

The main reaction procedure may be described by the next response equations:

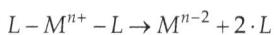

$$L - M^{n+} - L \rightarrow M^{n-2} + 2 \cdot L$$

$$L - M^{n+} - L \longrightarrow M^{n-1} - L + \cdot L$$
$$M^{n+1} - L \longrightarrow M^{n-2} + \cdot L \quad (A)$$
$$2M^{n-1} - L \longrightarrow L - M^{n+} - L + \cdot M^{n-2} \quad (B)$$

$$L - M^{n+} - L \longrightarrow M^{n-2} + 2 \cdot L \quad (C)$$
$$L - M^{n+} - L \longrightarrow M^{n-2} + L - L \quad (D)$$

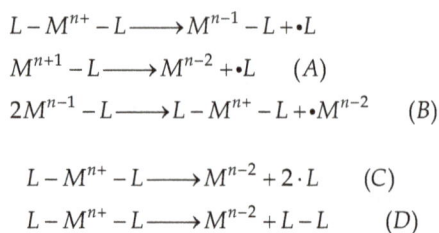

The proposed mechanism of inclusion allows explaining data previously reported on fluorescence emission enhancement for AFB1 in presence of β-cyclodextrines (β-CDs), the region of AFB1 exhibiting the most hydrophobic character is constituted by the methoxy group and by the portion of the coumarinic and cyclopentanone ring opposite to the carbonyl groups. However, the methoxy group alone is probably too small to produce a good fitting, displacing all water molecules placed within the β-CD cavity. The hydrophobic portion of coumarinic and cyclopentanone rings cannot be included into β-CD for steric reasons. β-CDs and AFB1 main composed a Host and guest system in this way β-CD can protect AFB1 from come into contact with some reagents which can lead to fluorescence signals quenching, and in this way it is consistent with the observed enhancement of AFB1 fluorescence emission in presence of β-CDs, and this system may explained by Hydropathic analysis. The inclusion of the bifuranic system of AFB1 into the β-CD cavity allows for fluorescence enhancement due to the protection of the fluorophore from the quenching and also in this case a variation in the circular dichroism spectrum. The affinity of AFTs to β-CD is rather low, being the calculated binding constants for the AFT: CD complexes around 10-3 M. Although the enhancement of AFTs native fluorescence, due to inclusion into CDs, has already been successfully employed in HPLC analysis for increasing the sensitivity, the low affinity of the formed complex cannot lead to a specific chemosensor for mycotoxin detection in acomplex matrix such as food (Manetta et al, 2005).

Derivatisation can also be performed by employing either pre- or post-column. Bromine (Br2), TFA (trifluoroacetic acid) are common used for pre-column derivative; Post-column reaction with iodide or bromide by an electrochemical cell (Kobra Cell) or addition of bromide or pyridinium hydrobromide perbromide (PBPB) (Akiyama et al, 2001; Stroka et al, 2003) to the mobile phase coupled with fluorescence detection has yielded sensitive determinations of aflatoxins: these reactions and others have been extensively reviewed, like β-cyclodextrine, is also used for post-column derivatisations. Aghamohammadi showed the methods which are based on the enhanced fluorescence of AFB1 by β-CD in 10% (v/v) methanol–water solution, For concentrations ranging from 0 to 15 μg kg-1 of AFB1 in pistachio samples as prediction set, the values of root mean square difference (RMSD) and relative error of prediction (REP) using multiple linear regressions (MLR) were 0.328 and 4.453%, respectively were observed (Aghamohammadi & Hashemi, 2007).

AFB and AFG were commonly derived in most experiment because of its low and easy quenching signals. A. Cepeda et al., (1996) was also studied using of cyclodextrin (CD) inclusion compounds showed an analytical method based on the incorporation post-column of a CD solution that promotes the greatest enhancement of AFB and AFG fluorescence (Figure 4).

From the figure 4 the different chromatograms we can see that with the addition of CD and its derivatives AFB1, AFB2 and AFG1, AFG2 were obtained greatly fluorescence enhancement.

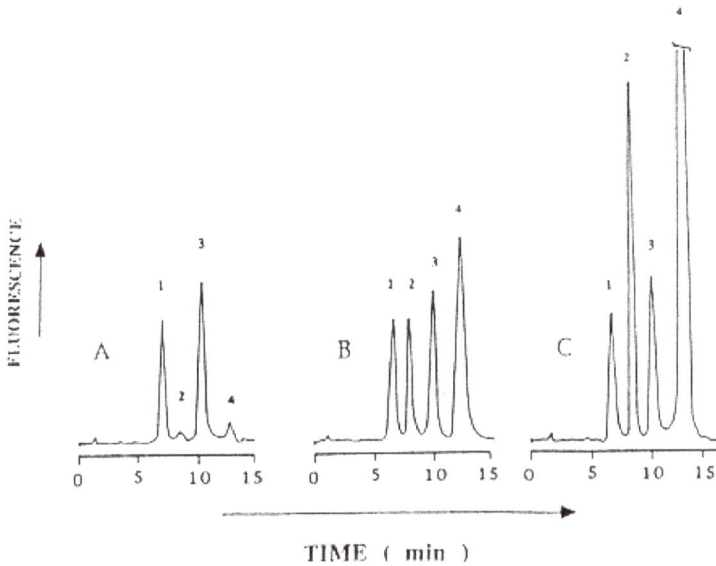

Fig. 3. Comparison of the different chromatograms: (A) without CD; (B) with addition of 10 - 2 M CD; (C) with addition of 10 -2 M DM-CD. Peaks: 1 =AfG2; 2=AfG1; 3=AFB2; 4=AFB1.

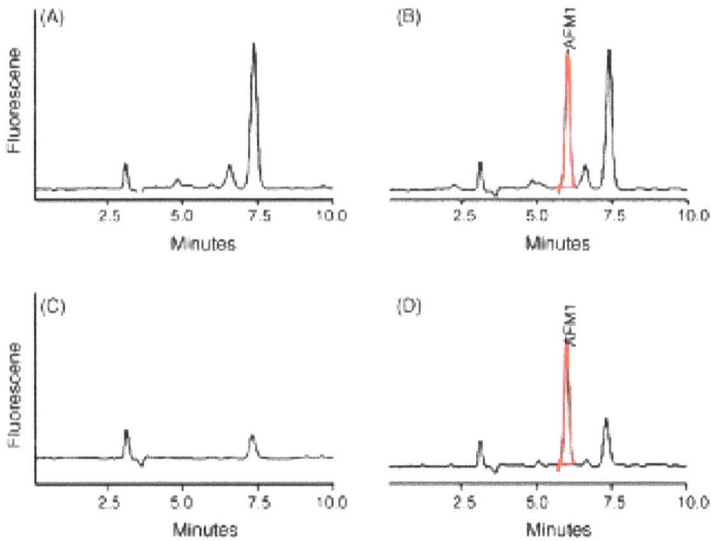

Fig. 4. Chromatograms of AFM1-free milk (A); milk spiked with AFM1 at 200 ng kg-1 (D); mobile phase, acid/acetonitrile/2-propand deicerized water (2: 10: 10.78), flow rate was 1.2 ml min-1

Besides AFB and AFG, fluorescence enhancement for sensitive detection could also be used for AFM1 analysis. Anna Chiara Manetta (2005) reported HPLC method with fluorescence detection by using pyridinium hydrobromide perbromide as a post-column derivatising agent had been developed to determine aflatoxin M1 in milk and cheese. The detection limits were 1 ng kg-1 for milk and 5 ng kg-1 for cheese. The calibration curve was linear from 0.001 to 0.1 ng injected. The method included a preliminary C18-SPE clean-up and the average recoveries of Aflatoxin M1 from milk and cheese, spiked at levels of 25–75 ng kg-1 and 100–300 ng kg-1, respectively, were 90 and 76%; the precision (RSD) ranged from 1.7 to 2.6% for milk and from 3.5 to 6.5% for cheese.

Chromatograms (Figure 5) and the data result showed that use of CD for detect AFM1 can significantly improve the detection sensitivity.

3.1.2 HPLC-MS-MS

High performance liquid chromatography (HPLC) combined with fluorescence detection is proved to be very accurate and has been extensively studied in different materials. However, in order to improve detection limits of AFB_1 and AFG_1, a tedious pre- or post-column derivatization must be done in conventional HPLC methods (Huang et al, 2009; Tassaneeyakul et al, 2004). These problems have been successfully solved in the present study by introducing HPLC-MS method.

As shown in Figure 6, a HPLC-MS system was equipped with an autosampler, the HPLC system, the ionization source (which interfaces the LC to the MS) and the mass spectrometer. There are various types of ionization sources that can be used as the interface between the HPLC and the mass spectrometer. Both electrospray ionization (ESI) and atmospheric pressure chemical ionization (APCI) are the two most common ionization sources. For both ESI and APCI, the ionization occurs at atmospheric pressure, so these sources are often referred to as atmospheric ionization (API). As shown in Figure 7, there are several types of mass spectrometers available for interfacing with HPLC. Single quadrupole mass spectrometer (Figure 7a) is a common system used for the HPLC-MS, this system can provide a mass spectrum for each chromatographic peak that elutes from the LC column and is analyzed by the MS system. Time-of-flight (TOF) mass spectrometer (Figure 7b), which has the added capability of providing a higher mass resolution spectrum from each component that is assayed. The triple quadrupole MS-MS system (Figure 7c) and ion-trap mass spectrometer (Figure 7d) are important tools in quantitative analysis and qualitative analysis. HPLC-ESI-MS/MS has become the most emerging analytical tool for the determination of aflatoxins and their metabolites (Cavaliere et al, 2007; Sulyok et al, 2010; Huang et al, 2010). Single quadrupole mass spectrometer (Nonaka et al, 2009) and ion-trap mass spectrometer (Cavaliere et al, 2006) were also used in the determination aflatoxins. LC-MS provides decisive advantages in performing identification as well as determination of analytes at trace levels.

Matrix effects, however, limit the potential of LC-MS. Molecules originating from the sample matrix that coelute with the compounds of interest can interfere with the ionization process in the mass spectrometer, causing ionization suppression or enhancement, which is the so-called matrix effect (Fan et al, 2011). Ion suppression (or enhancement) might be encountered due to matrix components that co-elute with the analyte of interest. If available, internal standards can often successfully amend these effects. Other possible strategies including the use of matrix matched standards or very careful validation of certain toxin/matrix combinations to exactly sample can determine the matrix effect.

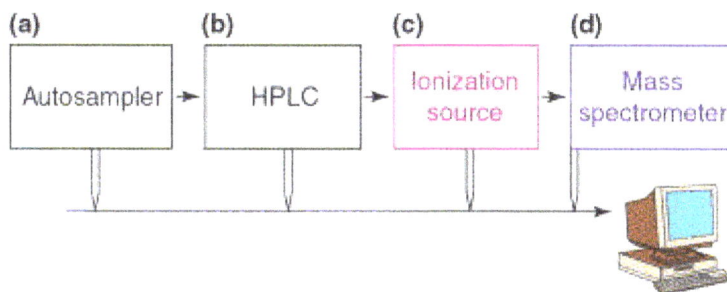

Fig. 5. The elements of an LC-MS system. (a) Autosampler; (b) HPLC; (c) ionization source; (d) Mass spectrometer.

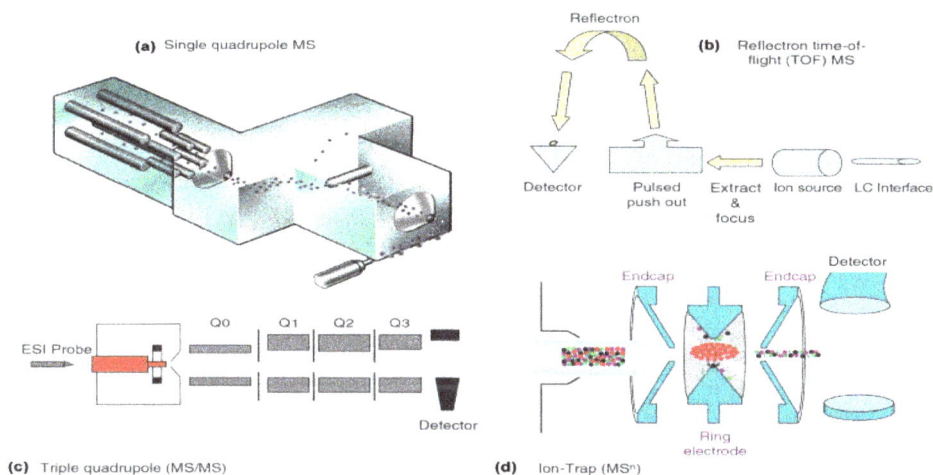

Fig. 6. Types of mass spectrometers which can be used in LC-MS. (a) Single quadruple MS; (b) Time-of-flight MS; (c) Triple quadrupole MS; (d) Ion-Trap MS

In general, all aflatoxins exhibit good ESI ionisation efficiency in the positive ion mode with abundant protonated molecules $[M+H]^+$ and sodium adduct ions, but practically no fragmentation in the full scan spectra (Blesa et al, 2003; Ventura et al., 2004). The formation of sodium adduct ions can easily be suppressed by the addition of ammonium ions to the mobile phase leading to a better MS sensitivity (Cavaliere et al, 2006). Reports about the utility of APCI interfaces are inconsistent and ionisation efficiencies in this mode seem to be highly dependent on the aflatoxin subgroup and the APCI interface geometry (Abbas et al, 2002). In this respect, only the structurally related sterigmatocystin offers strikingly better sensitivity with an APCI interface in the positive ion mode than with ESI (Scudamore et al, 1996), and consequently only Abbas et al. applied APCI for the detection of AFBs in the low ppb range(Abbas et al., 2002). According to recent investigations, autospheric pressure photoionization (APPI) seems to be a more reliable alternative to ESI. Since this interface offers strikingly lower levels of chemical noise and ion suppression than ESI it was found to be two to three times more sensitive(Cavaliere et al., 2006). The product ion spectra of the protonated aflatoxin species contain a number of abundant product ions reflecting bond cleavages and rearrangement reactions of the polycyclic ring system along with loss of water, carbon monoxide and carbon dioxide(Cavaliere et al., 2006). Despite this favorable fragmentation behaviour, only the quantitative single stage of LC-MS can not meet the EU criteria concerning unambiguous compound identification in residue analysis(Zllner & Mayer-Helm, 2006). In this respect, Cavaliere et al. demonstrated that the QTrap technology opens a new dimension of MS analyte confirmation and quantification. Its operation in the quadrupole linear ion trap configuration (enhanced product ion scans) produces complete product ions mass spectra even close to the LOQ which guarantees accurate analyte quantification simultaneously to unambiguous analyte confirmation(Cavaliere et al., 2006).

Cavaliere et al. compared the calibration curves set up in standard solution and in sample matrix and found close similarity of both slopes, proving that the influence of matrix components on the analyte signal was negligible and matrix effects could be excluded. Alternatively, Edinboro and Karnes infused post-column the aflatoxin analyte into a blank sample injection. As they did not find any dips in the baseline they concluded that ion suppression was absent in the analyte elution zone(Edinboro & Karnes, 2005).

Direct comparison of LC/MS and LC-FL revealed in most cases good correlation of quantitative results(Blesa, et al., 2003) though LC/MS method robustness and sensitivity seem to be inferior to LC-FL. In this context, Vahl and Jorgensen reported large variations of the recovery rates in different spices. They attributed this observation to severe matrix effects that are not compensated by the applied internal standard AFM1 and by a calibration curve set up in standard solution(Vahl & Jrgensen, 1998). Besides, Blesa et al. demonstrated in peanut samples that LC/MS is less sensitive than LC-FL(Blesa et al., 2003) though this can be partly explained by the use of single quadrupole instrumentation in the SIM mode that is inferior to a tandem MS and SRM recording(Cavaliere et al., 2006; Vahl et al., 1998).

3.2 Rapid assay methods
3.2.1 Portable tester

Due to high toxicity and extensive pollution of aflatoxins, some special portable tester and corresponding assay techniques were developed for rapid, sensitive, quantitative and convenient on-site determination of aflatoxins. The rapid tester device is based on chromatography and fluorescence spectrometric technologies, including clean-up and concentration with an immunoaffinity column, derivatization for fluorescence enhance and fluorescence excited at 360 nm. Ma et al. (2007) developed a rapid method for detecting aflatoxin B1 with an immunoaffinity column and portable rapid tester (Li et al., 2005; Li et al, 2006; Ma et al, 2007), which was obtained from Beijing Chinainvent Instrument Tech. Co. Ltd. (Beijing, China). Using the assay method developed, the results of showed the linear range of the method was 0.3–25 lg/kg, the average recovery was above 90% with CV being under 5%, the LOD for AFB1 from peanut and its related products was 0.3 lg/kg, the time for whole test process was about 45 min and the cost of detection was lower than other instruments and methods. Chiavaro et al. (Chiavaro et al, 2005) detected AFB1 and AFM1 in pig liver with portable tester obtained from VICAM (Watertown, MA, USA). The detection limit was 1.0 mg/kg for AFB1 and AFM1. Mean recoveries were 80.7 ± 9.0% for AFB1 spiked at 1.0–9.7 mg/kg levels and of 76.7 ± 6.6% for AFM1 spiked at 1.0–5.5 mg/kg levels. Considering its low price, portability and reliable quantification, the rapid tester dedicated to aflatoxins is suitable to use in the field, particularly in Third World countries.

Nowadays, the light sources of rapid tester are mainly LED, Xenon light for fluorescence assay. Due to the lack of fluorescence intensity, the aflatoxin has to be derived to enhance fluorescence using toxic and environmentally unfriendly solvents such as bromine. To address this issue, a laser is applied as excitation resource of portable tester. This light resource can provide steady light and can induce aflatoxins at ppt level without enhancer derivatization. Although the price of laser is higher than LED and Xenon light, the advantages of laser resource will make it have more widely applicable and a bright future.

3.2.2 Biosensor

Immunosensors are designed to improve sensitivity and to simplify determination. There are at least four classification of immunoassay at present: optical, electrochemical, piezoelectric (PZ) and micromechanical (Raman Suri et al., 2009), all of which depend on Abs and sensitive components. Two kinds of immunosensor have been developed for determination of aflatoxin (i.e. electrochemical and optical).

Competitive and non-competitive assays have both been used to develop electrochemical immunosensors for determination of aflatoxins. One type of electrochemical immunosensor is based on competitive ELISA. In this assay system, specific Ab or Ag (hapten-protein conjugate) is immobilized on the electrode, and enzyme conjugate is free. After competitive reaction, a different density of enzyme due to different concentration of analyte will bind to the electrode. Finally, the binding enzyme density can be shown by current produced from the catalytic oxidation reaction of the enzyme with substrates. Many such immunoassays have been described for aflatoxins (Ammida et al, 2004; Micheli et al, 2005; Parker & Tothill, 2009; Tan et al, 2009; Vig, et al, 2009) and they all had high sensitivities (LOD 0.01–0.4 ng/mL). With a non-competitive immunoassay, the formation of the Ab–Ag complex by a simple one-step immunoreaction between the immobilized enzyme-Ab conjugate and analytes in sample solution introduced a barrier of direct electrical communication between the immobilized enzyme and the electrode surface, so local current variations could be detected by the enzyme bioelectrocatalytic oxidation reaction with substrates. Sun et al. (Sun et al, 2008) and Liu et al. (Liu et al, 2006) developed such immunoassays for aflatoxin B1, whose linear ranges of detection were 0.1–12 ng/mL and 0.5–10 ng/mL, respectively. Using no enzyme and substrate, Owino et al. (2007) developed a non-competitive immunoassay with an LOD of 100 mg/L for aflatoxin B1 through a variation of electrochemical-impedance spectroscopy.

Optical immunosensors developed for determination of aflatoxins include mainly surface plasmon resonance (SPR) and some array devices. SPR, which is a well-known physical phenomenon, is surface electromagnetic waves that propagate in a direction parallel to the metal/dielectric (or metal/vacuum) interface. Since the wave is on the boundary of the metal and the external medium (air or water for example), these oscillations are very sensitive to any change of this boundary, such as the adsorption of molecules to the metal surface (El-Sherif, 2010). For biomolecular-interaction analysis, SPR sensors are valued for their ability to monitor molecular binding without labels and in real-time (Amarie et al, 2010). In a SPR of antibody-antigen interaction system, specific antibodies are immobilized on a sensitive optical component (i.e. layer of Au on a glass surface). When the antibodies capture analytes specifically, SPR occurs through the sensitive component. The angle of SPR is increased in line with the increase in the amount of analyte binding to the Au. Based on SPR method, immunoassays for aflatoxin B1 have been described by Daly et al. (2000) and Wang et al. (Wang & Gan, 2009), and their linear ranges were 3.0–98.0 ng/mL and 0.3–7.0 ng/mL, respectively. An outstanding characteristic of these immunoassays depends on a one-step reaction of Ab and analyte with a non-competitive format. To increase the sensitivity of detection, Wang et al. (Wang et al, 2009) developed a novel biosensor using long-range surface-plasmon-enhanced fluorescence spectroscopy. In this system, the binding of fluorophore-labeled molecules to the sensor surface is probed with surface plasmons and the emitted fluorescence light is detected. This approach takes advantage of the enhanced intensity of electromagnetic field occurring upon the resonant excitation of surface plasmons, which directly increases the fluorescence signal. Using this novel sensor, they obtained the lowest reported LOD for aflatoxin M1 (0.6 pg/mL). Solid-array sensors often depend on a competitive assay format. Specific Abs or Ags are immobilized on a solid surface (e.g., waveguide surface) and fluorescence-labeled conjugates are presented in the competitive system. Using an indirect competitive procedure, Sapsford et al. (2006) developed such an immunoassay for aflatoxin B1 with LODs for AFB1 0.3 ng/mL in buffer, 1.5 ng/g and 5.1 ng/g in corn, and 0.6 ng/g and 1.4 ng/g in nut products. Array sensor is a good tool for multiple compounds. For determination of aflatoxin B1 and ochratoxin A in the same operation, Adányi et al. (Adányi et al, 2007) devised a solid-array sensor with a sensitive detection range of 0.5–10 ng mL-1 using a competitive detection method.

3.2.3 Microplate reader

Microtiter plate and reader-based immunoassays mainly use competitive assays. Microtiter plates should have the features of binding proteins uniformly (e.g., Ags or Abs against aflatoxins or secondary Abs). 96-well polystyrene is used most commonly (Table 1). Microtiter readers can report optical absorbance or intensity of chemiluminescence or fluorescence, and they often contain data processing software that can build assay standard curves and equations and report amounts of analytes. In the past, most immunoassays developed were microtiterplate and reader based (Zhang, Li, Zhang, et al, 2009; Li, Zhang, Zhang, et al., 2009; Guan, Li, Zhang, et al, 2011). Some new materials (e.g., magnetic nanoparticles) have been used in aflatoxin-ELISA (Radoi et al, 2008). ELISA is the rapid test method most used today. ELISA kits have been commercial and used widely for aflatoxins in foods and agricultural products. Chemiluminescence immunoassay (CLIA) developed based on ELISA. Generally, chemiluminescence immunoassay can reach higher sensitivity than ELISA. With 384-well black polystyrene microtiter plates, a secondary Ab labeled with HRP and a luminol-based substrate, Magliulo et al. (2005) reported a chemiluminescence immunoassay for aflatoxin M1 in milk, that the limit of quantification was 1 ppt, so they thought that the developed method was suitable for accurate, sensitive, high-throughput screening of aflatoxin M1 in milk samples with a reduction of costs and increased detectability, as compared with previously developed immunoassays.

Fluorescence labels were also developed in ELISA format for analysis of aflatoxins, which is called Time-Resolved Fluoroimmunoassay (TRFIA). The labels used in this assay are lanthanide chelates such as Eu, Tb, and Sm. Lanthanide chelate labels offer the potentially significant advantage of a strong fluorescence with long decay time. As the measurement time is extended, the background noise is substantially reduced when the short-lived, non-specific background interference has disappeared. Moreover, the labels have a large Stock shift between the excitation and emission wavelength. The advantages of lanthanide chelate labels greatly increase the sensitivity of TRFIA. Huang et al. (2009) developed a TRFIA method for aflatoxin B1 using Eu3+ chelates as label. The sensitivity of the method was 0.02 µg/L and dynamic range of 0.02–100 µg/L. The intra- and inter-batch coefficient of variation was 3.2 and 7.3%, respectively, and the average recovery rate was 88.1%.

The advantage of microtiter plate-based immunoassays may be that they can be used to detect a large number of samples with a 96-well or 384-well plate at one time. These methods are used as quantitative or semi-quantitative assays for high through-put screening of aflatoxin samples.

Type	Label	Plate	Microplate reader
ELISA	HRP	Polystyrene, 96 well, clear	Absorbance, 450 nm
CLIA	HRP	Polystyrene, 96/384 well, black	Chemiluminescence, CCD
TRFIA	Lanthanide chelate	Polystyrene, 96 well, black	Fluorescence, 613 nm

Table 1. The parameters of immunoassay based on microplate

3.2.4 Lateral flow strip

Lateral flow strip assay is a new immunochromatographic technology combining chromatography with immunoassay and has attracted great interest in recent years. Nanoparticles are usually selected as the detector reagent, e.g., nanogold (Au) is most applied. A lateral flow strip comprises three membrane pads: absorbent pad, conjugate-release pad, sample pad and a nitrocellulose (NC) membrane, as shown in Figure 8. With capillary action, test buffer containing analytes is introduced to the absorbent pad from the bottom of the strip. After reaching the Au conjugate-release pad, the Au-labeled Ab can bind analytes specifically. The complex is then transferred by the flow to the nitrocellulose membrane and reacted with the immobilized Ag for the generation of signals. If the test buffer contains analytes, the complex migrates along the membrane and binds to the secondary Abs on the control line and no red signal can be observed on the test line. If the analyte is absent, some the Au-labeled Abs bind to the immobilized Ag (aflatoxin-protein conjugate) on the test line and the rest of the Au-labeled Abs flow to and bind control Abs (Li, Zhang, & Zhang, 2009).

Lateral flow strip assay has many advantages, such as:

1. requiring only a sample extraction step before use;
2. simplicity of procedure with single step, e.g., only adding test solution to the sample pad on the strip;
3. rapid on-site detection within a few minutes (5-15 min);
4. concentration levels of target analytes can be observed directly with the naked eyes;
5. user-friendly format no need for skill personnel;
6. less interference due to chromatographic separation; and
7. low cost

Because of these advantages, lateral flow strip assay has become one of the commercial and widely-used immunoassays for rapid determination of mycotoxins, such as ochratoxin A (Lai et al, 2009; Liu, Tsao, Wang, & Yu, 2008; Wang, Liu, Xu, Zhang, & Wang, 2007; Cho et al., 2005), deoxynivalenol (Kolosova, De Saeger, Sibanda, Verheijen, & Van Peteghem, 2007; Xu et al., 2010; Kolosova et al., 2008), T-2 Toxin (Molinelli et al., 2008), zearalenone (Kolosova, De Saeger, Sibanda, Verheijen, & Van Peteghem, 2007), fumonisin B1 (Wang, Quan, Lee, & Kennedy, 2006), aflatoxins (Sun, Zhao, Tang, Zhou, & Chu, 2005; Sheibani, Tabrizchi, & Ghaziaskar, 2008) and so on.

The visual detection limit (VDL), defined as the minimum concentration producing the color on the test line significantly different or weaker to that on the test line of negative control strip without aflatoxin (Li, Wei, Yang, Li, & Deng, 2009; Zhou et al., 2009), was used to express the sensitivity of the lateral flow strip assay. The visual detection limit of published conventional lateral flow strip assay for aflatoxins are summaried in Table 2.

References	Aflatoxins	VDL [a] (ng/g)
(Delmulle, De Saeger, Sibanda, Barna-Vetro, & Van Peteghem, 2005)	AFB1	2.0
(Sun, Zhao, Tang, Zhou, & Chu, 2005)	AFB1	0.5
(Shim et al., 2007)	AFB1	0.1
(Zhang, Li, Zhang, Zhang, 2011)	AFB1	0.03
	AFB2	0.06
	AFG1	0.12
	AFG2	0.25

[a] The VDLs here were selected out from the original as defined above.

Table 2. VDLs of published conventional lateral flow strip assay for aflatoxins.

Challenges in test strip production include adjusting the flow properties of the test strip and, as already mentioned, reducing matrix background interference by optimization of multiple parameters including (Krska & Molinelli, 2009):
1. type and pore size of analytical membrane;
2. type and concentration of blocking agent for blocking membrane binding sites after spraying of reagents;
3. type of buffer, pH range and ionic strength; and
4. use of surfactants and modifiers for pre or post treatment of test strip materials

Similar to ELISA, optimization with a selection of reagents (concentrations), materials and assay conditions is necessary.

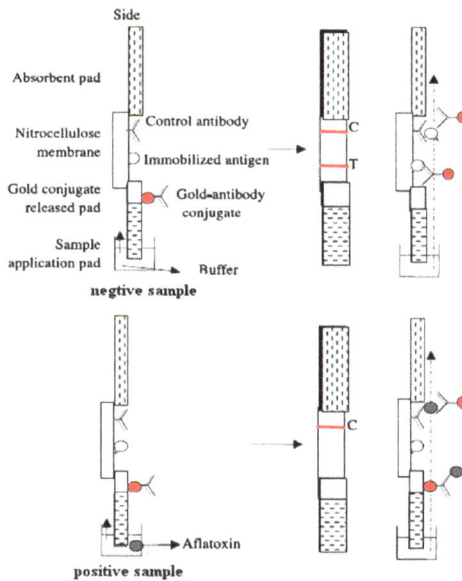

Fig. 7. Construction of lateral flow strip, which comprises three pads (from top to bottom): absorbent pad, gold-conjugate release pad and sample application pad, and a nitrocellulose (NC) membrane. The sample is introduced by capillary action from the bottom of the strip. On reaching the gold-labeled antibody pad, the antigen-Ab reaction takes place. The binding complex is then transferred by the flow to the NC membrane and then reacted with the immobilized antigen to generate signals. Signals generated from the sample without aflatoxin (negative sample) and with aflatoxin (positive sample) are shown in panels (Sun, Zhao, Tang, Zhou, & Chu, 2005).

3.3 Other methods
Besides the above, both of layer chromatography (TLC) and generic fluorospectrophotometry are two traditional methods for determination of aflatoxin content. And there are several standard methods published previously (http://www.aoac.org/omarev1/2005_08.pdf; Van Egmond and Jonker, 2004). Recently they were used by fewer and fewer laboratories with occurring of so many modern equipments and protocols. Maybe, lack of automatism and high possibility to be harmful to operators and environment are the main reason.They are not described with more details here.

4. New trends

4.1 Quantitative strip assay
As description above, lateral flow assays are currently widely used in a wide range. However, most of the strip tests developed are qualitative tests (Molinelli, Grossalber, & Krska, 2009) with a simple yes/no response to the levels of the target analytes. Although the conventional quanlitative analysis may be suitable for verifying certain analyte (e.g., for a preganancy test), it is not adequate when the level of an analyte is important (Liu et al.,

2007), e.g., most clinical decision for illness progression require known concentrations of pathogens; the countermeasures for contaminated foods and feeds need be taken according to the contamination level. A trend can be seen towards (semi-) quantitative strip tests driven by a strong demand from industry (Molinelli, Grossalber, & Krska, 2009). To meet the requirement, two kinds of approaches have appeared depending on the need of detector or not. With advanced nanotechnologies, a few methods have integrated chromatographic separation and electrochemical (Wang, Quan, Lee, & Kennedy, 2006), fluorescence (Sun, Zhao, Tang, Zhou, & Chu, 2005) or optical detectors (Sheibani, Tabrizchi, & Ghaziaskar, 2008) for rapidly quantitative detection. Compared with conventional strips which just based on visual judgment, these approaches offer a greater sensitivity and dynamic range as well as a better quantitative capability (Kim, Oh, Jeong, Pyo, & Choi, 2003). However, these approaches can lead to environmental pollution from heavy metal (e.g., mercury, Hg), or may suffer from optical interference (e.g., photobleaching), the rising costs due to the use of detector, and the complex software for imaging and analysis (Liu et al., 2007); all of these potential problems limit their well application on spot. As detector-free approaches, a one-step competitive ICA for semiquantitative determination of lipoprotein (a) in plasma is developed (Lou, Patel, Ching, & Gordon, 1993), the dose ranges can be simply encoded to different numbers of a colored ladder bar that had fully developed color on the assay strip, and a pH sensitive dye is used as the end-of-assay indicator. A potential problem could arise that the time of end-of-assay with a pH sensitive indicator may vary from people to people and cause a disparity in result determination. Subsequently, a dipstick test determined microalbuminnuria in patients with hypertension (semi-) quantitively by comparing the colored singal with a standard color chart (Gerber, Johnston, & Alderman, 1998) such as with pH paper. However, the color indication of the assay is not self-confirmative, and may also show an error in matching intensity (Cho & Paek, 2001).

According to the description above, although problems exist in two kinds of approaches, the detector-free methods seem to have more potential on-site application value considering convenience, low-cost and no interferences from instrument itself. To overcome the disadvantages of published detector-free methods, a novel strategy for detector-free (semi-) quantitive strip (DFQ-strip) assay is proposed just like a novel "ruler" for content measurement of target analyte. The illustration design of the DFQ-strip was shown in Figure 9. The DFQ-strip consisted of five parts similar as the traditional ones with three pads (sample, conjugate release, and absorbent pads), a NC membrane and a plastic backing plate. On NC membrane, three scale lines defined as SL- I , II and III constituted the measuring bar which played a role as a ruler. After reaction different number of scale line appeared indicating the concentration (range) of analyte, in other words, every scale line's disappearing represented a concentration (expressed as threshold level) playing a role as scale on the ruler, while the visual detection limit played a role as an unlined out scale. As a detector-free approach, the strategy spurned the traditional method with just one test line for one analyte or multi-test line for multianalyte, three scale lines were designed to offer multiple dynamic ranges for one analyte. Therefore, compared with the traditional qualitative tests, the DFQ-strip assay not only expresses yes/no response but also offer the content (range) of target analytes. For a negative sample, three color bands (scale lines) are formed in the test zone of DFQ-strip (figure 10a) and the color intensity is graded with the weakest color in SL- I and deepest color in SL-III. For positive samples, with migration, the free probe became less and less, which is more and more favorable for the competition of analyte. The intensity of the color is inversely proportional to the analyte concentration in sample. Thus, during the competitive reaction, SL- I will disappear fistly, and then SL- II and SL-III at last. Consequently, a positive sample, in accordance with the amount of analyte in sample, will result in three, two, only one weaker red band or no color mark in test zone compared with those of negative control (figure 10b). But, similar as the traditional strip assay, in any case, if no red line appears at the control zone, the test result is considered invalid (figure 10c).

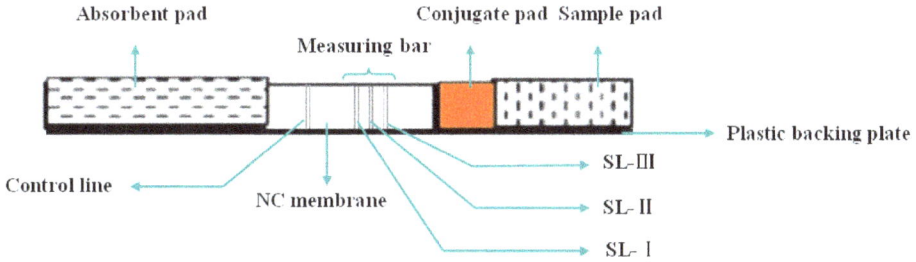

Fig. 8. Illustration of the DFQ-strip design. The DFQ-strip consisted of five parts similar as the traditional ones with three pads (sample, conjugate release, and absorbent pads), a NC membrane and a plastic backing plate and the differences lay in lines on NC membrane. There were four lines, one control line and three scale lines on NC membrane. The measuring bar which played the role as a ruler was comprised by SL- I , SL- II and SL-III.

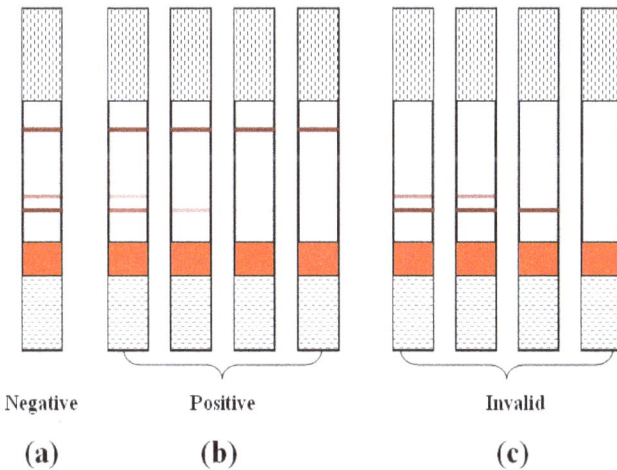

Fig. 9. Illustrations of DFQ-strip assay results for negative, positive and invalid.

A comprehensive model system of DFQ-strip is constructed taking aflatoxin B1 as target analyte. The visual detection limit (VDL, unlined out scale) of the DFQ-strip assay was 0.06 ng/mL, and the threshold levels (scales) for SL-I, II and III were 0.125, 0.5 and 2.0 ng/mL, respectively (the data will be published recently). Moreover, all results supported the feasibility of the idea with high sensitivity, precision and accuracy, multiple dynamic ranges, as well as good (semi-) quantitative capability, stability. Besides, this DFQ-strip assay had good practicability, great application value for toxic or harmful substances (e.g., mycotoxins) in-situ monitoring but still posessed the advantages of conventional strips such as procedure simplicity, rapid operation, immediate results, low cost, and no requirement for skilled technicians or expensive equipment, etc. The strategy is proposed as an alternative idea for sensitive, rapid, convenient and (semi-) quantitative detection of analyte on site.

4.2 Green immunoassay

Aflatoxin standards and their derivate have been considered as high poison. So there is high possibility that these compounds using in analytical processes induce second contamination of environment. How to reduce or eliminate the use of hazardous substances? For this, green immunoassay strategies will be introduced as the below.

On the one hand, nontoxic surrogates of aflatoxin can be designed in ELISA system. As is known, aflatoxin calibration curves must be used for every plate to reduce differences in plate-to-plate variability and improve accuracy. Furthermore, the pure toxin, used as the calibrator, is hazardous to operators and the environment. According to the reaction principle of antibody and hapten, we can design some mater, such as second antibody, which can also bind the active area of the specific antibody against aflatoxin. Such compounds will act as calibrator and be named as surrogate.

There are usually four steps for development of a green immunoassay with nontoxic surrogates:

- to prepare specific antibodies (the first antibody) against targets;
- to produce F(ab')2 fragments of the target antibody
- to produce anti-idiotype antibodies (the second antibody) to the target;
- to establish an calibrator curve for detection.

As an example developed in our laboratory, a green enzyme-linked immunosorbent assay (ELISA) to measure aflatoxin M1 (AFM1) in milk was developed and validated with a surrogate calibrator curve. Polyclonal anti-idiotype (anti-Id) antibody, used as an AFM1 surrogate, was generated by immunizing rabbits with F(ab')2 fragments from the anti-AFM1 monoclonal antibody (mAb). The rabbits exhibited high specificity to the anti-AFM1 mAb, and no cross-reactivity to either of the other anti-aflatoxin mAbs or the isotype matched mAb was observed. After optimizing the physicochemical factors (pH and ionic strength) that influence assay performance, a quantitative conversion formula was developed between AFM1 and the anti-Id antibody ($y = 31.91x - 8.47$, $r = 0.9997$). The assay was applied to analyze AFM1 in spiked milk samples. The IC50 value of the surrogate calibrator curve was 2.4 µg mL-1, and the inter-assay and intra-assay variation was less than 10.8%; recovery ranged from 85.2 to 110.9%. A reference high-performance liquid chromatography method was used to validate the developed method, and a good correlation was obtained ($y = 0.81x + 9.82$, $r = 0.9922$).

On the other hand, how can we develop some immunoassay using no target standards? In our previous review (Li et al, 2009), noncompetitive immunoassay format was described. And this immunoassay's signal can be found stronger and stronger with increasing of target content, which means "no signal, no target". This kind of noncompetitive immunoassay is especially suggested for fast screening of samples without any use of the toxin standards, although, when developing this assay, toxin standards will have to be used for evaluation of sensitivity. Recently, some noncompetitive immunoassays, such as SPR assay and sandwich assay for other small chemicals were developed.However, there are still no reports for analysis of aflatoxins in agro-products or in environment.

5. Outlook

Facing so many kinds of current analytical methods for aflatoxin, how can we choose them for our sample analysis?

Actually, each method has its own features. In our opinion, there are three classes of analytical methods, (1) High fidelity method, (2) qualitative rapid method and (3) quantitative rapid method.

1. High fidelity method means they have been authorized with high sensitivity, accuracy and precision, and especially means traditional chromatographic technology with high extent of efficiency and intelligentization. Considering its mature and vive methodologies, there has been many standard method set by governments or international organizers. Usually, such methods have been using to make impartiality data for inspection reports. For analysis of aflatoxin, HPLC with detector of mass or fluorescence belongs to high fidelity method. Disadvantageously, these methods depend on very expensive instruments which can only be sited on some special room. The room also needs to meet some special requirement of environment. Usually, their process need to spend so many organic solvent and total cost of sample measurement is relatively high.

2. Qualitative rapid method means it needs few time to finish a test process and it can only tell operator "positive" or "negative" data. A typical such method is nanogold particle-based immunochromatographic assay. Usually, these methods especially fit for screening of a great number of samples or on-site analysis. The main advantage is simple, rapid, convenient, detector-free and low-cost for sample analysis. Its main disadvantage is lack of content details and it is generally not considered to make data for inspection report on agro-products.

3. Quantitative rapid method means it can be used to get content details with high sensitivity; however it has lower accuracy and/or precision than that of high fidelity method some time. Here, it especially means quantitative immunoassays including ELISA, portable tester-based immunoassay, immunosensor and so on (Li, zhang & zhang, 2009). These methods have been considering as important valuable complement for high fidelity method (HPLC-MS/FLD). They have attractive features including high sensitivity and specificity, simple operating, short time consuming, the possibility of analysis of difficult matrices without extensive pre-treatment, and low costs. According to the previous discussion, these technologies facing the following challenges,

    1. preparation of more specific antibodies against aflatoxins via inducing of novel structural hapten, screening of mAb or rAb, or mending antibody of engineering,
    2. exploring of non-animal antibody preparation techniques, such as development of rAb or some simulative antibodies mentioned above,
    3. researches on use of novel labels, such as sensitive nanoparticles (quantum dots, gold particle, magnetic beads, etc),
    4. development of noncompetitive immunoassays with one reaction step for faster, simpler and more sensitive assay, and
    5. legalization of immunoassay methods. Comparing the amount of immunoassay kinds, there are only few methods have been constituted as test standard. In China,

for example, there are just some ELISA standards for determination of aflatoxins. So, we think legalization of immunoassay methods may become one of important tast in the future.

The second and the third above maybe become the main research trends. And, rationally, we predict immunoassay devices such as portable fast tester special for aflatoxins will be used in wide fields.

For analytical works, our aims need to be clear firstly, which means "why the samples need to be determination of aflatoxins?" Generally, there are three kinds of aims: (1) For justifying only with or without target contaminants; (2) For getting qualitative extent of contamination with low or high content of aflatoxin; (3) For quantitative evaluation on contaminant in samples. And then, to reach the aim, an appropriate method need to be chosen with the principle of saving (time and/or cost) and speed of measurement.

With developing of analytical technologies, sensitivities of methods will be enhanced. To meet requirement of on-site assay, many novel analytical devices, representing automatization, minization and high throughput, will be developed and improved. It means that tomorrow analytical methods will be of simplification, intelligentization and portability. Also, future assay protocols will use fewer and fewer poison chemicals including toxin standards and organic solvent. These methods will make great importance on analysis of aflatoxin, to protect agricultural environment, to estimate quality of commercials of agro-products and food, and to safeguard safety of consumers' health and lives.

6. Acknowledgements

The studies were supported by the Key Project of Ministry of Agri- culture (2010-G1, 2011-G5), the earmarked fund for Modern Agro-industry (Rapeseed) Technology Research System (nycytx-005), the Project in the National Science & Technology Pillar Plan (2010BAD01B07, 2011BAD02D02) and Special Foundation of President of the Chinese Agricultural Academy of Sciences

7. References

Abbas, H., Williams, W., Windham, G., Pringle III, H., Xie, W., & Shier, W. (2002). Aflatoxin and fumonisin contamination of commercial corn (Zea mays) hybrids in Mississippi. *J. Agric. Food Chem*, 50(18), 5246-5254.

Adányi, N., Levkovets, I. A., Rodriguez-Gil, S., Ronald, A., Váradi, M., & Szendro, I. (2007). Development of immunosensor based on OWLS technique for determining Aflatoxin B1 and Ochratoxin A. *Biosensors and Bioelectronics*, 22(6), 797-802.

Aghamohammadi, M. &, Hashemi, J. (2007). Enhanced synchronous spectro- fluorimetric determination of aflatoxin B_1 in pistachio samples using multivariate analysis. *Anal. Chim. Acta*, 582, 288–294.

Akiyama H., Goda Y., Tanaka T., & Toyoda M.(2001). Determination of aflatoxins B_1, B_2, G_1 and G_2 in spices using a multifunctional column clean-up. *Journal of Chromatography A*, 932 ,153.

Amarie, D., Alileche, A., Dragnea, B., & Glazier, J. A. (2010). Microfluidic Devices Integrating Microcavity Surface-Plasmon-Resonance Sensors: Glucose Oxidase Binding-Activity Detection. *Analytical chemistry*, 82(1), 343-352.

Ammida, N. H. S., Micheli, L., & Palleschi, G. (2004). Electrochemical immunosensor for determination of aflatoxin B1 in barley. *Analytica chimica acta, 520*(1-2), 159-164.

Bennett, J.W. & Klich, M. (2003). Mycotoxins. *Clin. Microbiol. Rev.*, 16: 497 – 516

Blesa, J., Soriano, J., Molto, J., Marin, R., & Ma es, J. (2003). Determination of aflatoxins in peanuts by matrix solid-phase dispersion and liquid chromatography* 1. *Journal of Chromatography A, 1011*(1-2), 49-54.

Braga S.M., de Medeiros F.D., de Oliveira E.J., &Macedo R.O.(2005). Development and validation of a method for the quantitative determination of aflatoxin contaminants in Maytenus ilicifolia by HPLC with fluorescence detection. *Phytochemical Analysis,* 16(4), 267-271.

Brera, C., Debegnach, F., Minardi, V., Pannunzi, E., De Santis, B., & Miraglia, M. (2007). Immunoaffinity column cleanup with liquid chromatography for determination of aflatoxin B$_1$ in corn samples: interlaboratory study. *Journal of AOAC International, 90*(3), 765-772.

Cavaliere, C., Foglia, P., Pastorini, E., Samperi, R., & Lagan , A. (2006). Liquid chromatography/tandem mass spectrometric confirmatory method for determining aflatoxin M1 in cow milk: Comparison between electrospray and atmospheric pressure photoionization sources. *Journal of Chromatography A, 1101*(1-2), 69-78.

Cavaliere, C., Foglic, P., Guarino, C., et al. (2007). Determination of aflatoxins in olive oil by liquid chromatography-tandem mass spectrometry. *Analytica Chimica Acta, 596*, 141-148.

Cepeda A.1, Franco C.M., Fente C.A., Vazquez B.I., Rodriguez J.L., Prognon P., & Mahuzier G. (1996). Postcolumn excitation of aflatoxins using cyclodextrins in liquid chromatography for food analysis. *J. Chrornatogr. A., 721*, 69-74.

Chiavaro, E., Cacchioli, C., Berni, E., & Spotti, E. (2005). Immunoaffinity clean-up and direct fluorescence measurement of aflatoxins B1 and M1 in pig liver: Comparison with high-performance liquid chromatography determination. *Food Additives & Contaminants: Part A, 22*(11), 1154-1161.

Cho, J. H., & Paek, S. H. (2001). Semiquantitative, bar code version of immunochromatographic assay system for human serum albumin as model analyte. *Biotechnology And Bioengineering,* 75, 725-732.

Cho, Y. J., Lee, D. H., Kim, D. O., Min, W. K., Bong, K. T., Lee, G. G., et al. (2005). Production of a monoclonal antibody against ochratoxin A and its application to immunochromatographic assay. *Journal of Agricultural and Food Chemistry,* 53, 8447-8451.

Daly, S. J., Keating, G. J., Dillon, P. P., Manning, B. M., O'Kennedy, R., Lee, H. A., & Morgan, M. R. A. (2000). Development of surface plasmon resonance-based immunoassay for aflatoxin B1. *Journal of Agricultural and Food Chemistry, 48*(11), 5097-5104.

Delmulle, B.S., De Saeger, S. M. D. G., Sibanda, L., Barna-Vetro, I., & Van Peteghem, C. H. (2005). Development of an immunoassay-based lateral flow dipstick for the rapid detection of aflatoxin B1 in pig feed. *Journal of Agricultural and Food Chemistry, 53*, 3364-3368.

Edinboro, L. E., & Karnes, H. T. (2005). Determination of aflatoxin B1 in sidestream cigarette smoke by immunoaffinity column extraction coupled with liquid chromatography/mass spectrometry. *J. Chromatogr. A, 1083*(1-2), 127-132.

El-Sherif, M. (2010). Fiber-Optic Chemical and Biosensors. *Optical Guided-wave Chemical and Biosensors II, 8,* 109-149.

Fan, S.F., Wang, X.P., Li, P.W., Zhang, W., Zhang, Q. (2011). Simultaneous determination of 13 phytohormones in oilseed rape tissues by liquid chromatography-electrospray tandem mass spectrometry and the evaluation of the matrix effect. *Journal of Separation Science, 34,* 640-650.

Fu, Z.H.,Huang ، X.X.. &Min,S.G. (2008). Rapid determination of aflatoxins in corn and peanuts. *Journal of Chromatography A,* 1209, 271–274.

Francis O.J., Kirschenheuter G.P., Ware G.M., et al. (1988). Beta-cyclodextrin post-column fluorescence enhancement of aflatoxins for reverse-phase liquid chromatographic determination in corn. *J. Assoc. Off. Anal. Chem.* 71:725-728.

Gerber, L. M., Johnston, K., & Alderman, M. H. (1998). Assessment of a new dipstick test in screening for microalbuminnuria in patients with hypertension. *American Journal Of Hypertension, 11,* 1321-1327.

Giraudi, G., Anfossi, L., Baggiani, C., Giovannoli, C., & Tozzi, C. (2007). Solid-phase extraction of ochratoxin A from wine based on a binding hexapeptide prepared by combinatorial synthesis. *Journal of Chromatography A, 1175*(2), 174-180.

Guan, D., Li, P., Zhang, Q., Zhang, W., Zhang, D., Jiang, J. (2011). An ultra-sensitive monoclonal antibody-based competitive enzyme immunoassay for aflatoxin M1 in milk and infant milk products. *Food Chem.,* 125: 1359-2364.

Herzallah,S.M. (2009). Determination of aflatoxins in eggs, milk, meat and meat products using HPLC fluorescent and UV detectors. Food Chemistry, 114, 1141–1146.

http://www.aoac.org/omarev1/2005_08.pdf

Huang, B., Xiao, H., Zhang, J., Zhang, L., Yang, H., Zhang, Y., & Jin, J. (2009). Dual-label time-resolved fluoroimmunoassay for simultaneous detection of aflatoxin B1 and ochratoxin A. *Archives of toxicology, 83*(6), 619-624.

Huang, B.F., Han, Z., Cai, Z.X., Wu, Y.J., Ren, Y.P. (2010). Simultaneous determination of aflatoxins B1, B2, G1, G2, M1 and M2 in peanuts and their derivative products by ultra-high-performance liquid chromatography-tandem mass spectrometry. *Analytica Chimica Acta, 662,* 62-68.

Jaimez J., Fente C.A. (2000). Application of the assay of aflatoxins by liquid chromatographywith fluorescence detection in food analysis. *J. Chromatogr. A,* 882, 1–10.

Jornet, D., Busto, O., & Guasch, J. (2000). Solid-phase extraction applied to the determination of ochratoxin A in wines by reversed-phase high-performance liquid chromatography. *Journal of Chromatography A, 882*(1-2), 29-35.

Joshua, H. (1993). Determination of aflatoxins by reversed-phase high- performance liquid chromatography with post-column on-line photochemical derivatization and fluorescence detection. *J Chromatogr A,.* 654, 247.

Kim, Y. M., Oh, S. W., Jeong, S. Y., Pyo, D. J., & Choi, E. Y. (2003). Development of an ultrarapid one-step fluorescence immunochromatographic assay system for the quantification of microcystins. *Environmental Science and Technology, 37,* 1899-1904.

Kolosova, A. Y., De Saeger, S., Sibanda, L., Verheijen, R., & Van Peteghem, C. (2007). Development of a colloidal gold-based lateral-flow immunoassay for the rapid simultaneous detection of zearalenone and deoxynivalenol. *Analytical and bioanalytical chemistry, 389,* 2103-2107.

Kolosova, A. Y., Sibanda, L., Dumoulin, F., Lewis, J., Duveiller, E., Van Peteghem, C., et al. (2008). Lateral-flow colloidal gold-based immunoassay for the rapid detection of deoxynivalenol with two indicator ranges. *Analytica Chimica Acta, 616*, 235-244.

Krska, R., & Molinelli, A. (2009). Rapid test strips for analysis of mycotoxins in food and feed. *Analytical and bioanalytical chemistry, 393*, 67–71.

Lai, W. H., Fung, D. Y. C., Xu, Y., Liu, R. R., & Xiong, Y. H. (2009). Development of a colloidal gold strip for rapid detection of ochratoxin A with mimotope peptide. *Food Control, 20*, 791-795.

Li, D. W., Wei, S., Yang, H., Li, Y., & Deng, A. P. (2009). A sensitive immunochromatographic assay using colloidal gold-antibody probe for rapid detection of pharmaceutical indomethacin in water samples. *Biosensors and Bioelectronics, 24*, 2277-2280.

Li, P. W., Zhang, Q., & Zhang, W. (2009). Immunoassays for aflatoxins. *Trac-Trends in Analytical Chemistry, 28*, 1115-1126.

Li, P., Zhang, Q., Zhang, W., Zhang, J., Chen, X., Jiang, J., Xie, L., Zhang, D. (2009). Development of a class-specific monoclonal antibody-based ELISA for aflatoxins in peanut. *Food Chem.*, 115: 313-317

Li, P., Zhang, W., Ding, X., Yang, J., Xie, L., Yang, M., & Li, G. (2005). A fast detection method for aflatoxin B1. *P.R. China Patent*, CN200510018555.X.

Li, P., Zhang, X., Zhang, W., Xie, L., & Ding, X. (2006). A fast test device for aflatoxin B1, *P.R. China Patent*, ZL2006 2 20097322.3.

Liu, B. H., Tsao, Z. J., Wang, J. J., & Yu, F. Y. (2008). Development of a monoclonal antibody against ochratoxin A and its application in enzyme-linked immunosorbent assay and gold nanoparticle immunochromatographic strip. *Analytical Chemistry, 80*, 7029-7035.

Liu, G., Lin, Y. Y., Wang, J., Wu, H., Wai, C. M., & Lin, Y. (2007). Disposable electrochemical immunosensor diagnosis device based on nanoparticle probe and immunochromatographic strip. *Analytical Chemistry, 79*, 7644-7653.

Liu, Y., Qin, Z., Wu, X., & Jiang, H. (2006). Immune-biosensor for aflatoxin B1 based bio-electrocatalytic reaction on micro-comb electrode. *Biochemical Engineering Journal, 32*(3), 211-217.

Liu, Y. & Wu, F. (2010). Global Burden of Aflatoxin-Induced Hepatocellular Carcinoma: A Risk Assessment. *Environ Health Perspect*, 118 (6): 818–824.

Lou, S. C., Patel, C., Ching, S., & Gordon, J. (1993). One-step competitive immunochromatographic assay for semiquantitative determination of lipoprotein(a) in plasma. *Clinical Chemistry, 39*, 619-624.

Ma, L. (2007). Study on Detection Technology for Determination of Aflatoxin B$_1$ with High Sensitivity. Doctoral Dissertation, China.

Ma, L., Li, P., Zhang, W., Xie, L., Ding, X., & Chen, X. (2007). Study on the fluorescence fast detection of aflatoxin B1 in peanut and related products purified by immuno-affinity column. *Chinese Journal of Oil Crop Sciences, 29* (2), 93-97.

Magliulo, M., Mirasoli, M., Simoni, P., Lelli, R., Portanti, O., & Roda, A. (2005). Development and validation of an ultrasensitive chemiluminescent enzyme immunoassay for aflatoxin M1 in milk. *Journal of Agricultural and Food Chemistry, 53*(9), 3300-3305.

Manetta, A.C., Giuseppe, L.D.,Giammarco, M., &Fusaro, I. (2005). Explaining cyclodextrin–mycotoxin interactions using a 'natural' force field. *J. Chrornatogr. A.*, 1083, 219-222.

Mateo, J. J., Mateo, R., Hinojo, M. J., Llorens, A., & Jimenez, M. (2002). Liquid chromatographic determination of toxigenic secondary metabolites produced by Fusarium strains. *Journal of Chromatography A, 955*(2), 245-256.

Micheli, L., Grecco, R., Badea, M., Moscone, D., & Palleschi, G. (2005). An electrochemical immunosensor for aflatoxin M1 determination in milk using screen-printed electrodes. *Biosensors and Bioelectronics, 21*(4), 588-596.

Molinelli, A., Grossalber, K., Führer, M., Baumgartner, S., Sulyok, M., & Krska, R. (2008). Development of qualitative and semiquantitative immunoassay-based rapid strip tests for the detection of T-2 toxin in wheat and oat. *Journal of Agricultural and Food Chemistry, 56*, 2589-2594.

Molinelli, A., Grossalber, K., & Krska, R. (2009). A rapid lateral flow test for the determination of total type B fumonisins in maize. *Analytical and bioanalytical chemistry, 395*, 1309-1316.

Nonaka, Y., Saito, K., Hanioka, N., Narimatsu, S., Kataoka, H. (2009). Determination of aflatoxins in food sanples by automated on-line in-tube solid-phase microextraction coupled with liquid chromatography-mass spectrometry. *Journal of Chromatography A*, 1216, 4416-4422.

Owino, J. H. O., Ignaszak, A., Al-Ahmed, A., Baker, P. G. L., Alemu, H., Ngila, J. C., & Iwuoha, E. I. (2007). Modelling of the impedimetric responses of an aflatoxin B1 immunosensor prepared on an electrosynthetic polyaniline platform. *Analytical and bioanalytical chemistry, 388*(5), 1069-1074.

Parker, C. O., & Tothill, I. E. (2009). Development of an electrochemical immunosensor for aflatoxin M1 in milk with focus on matrix interference. *Biosensors and Bioelectronics, 24*(8), 2452-2457.

Radoi, A., Targa, M., Prieto-Simon, B., & Marty, J. L. (2008). Enzyme-Linked Immunosorbent Assay (ELISA) based on superparamagnetic nanoparticles for aflatoxin M1 detection. *Talanta, 77*(1), 138-143.

Raman Suri, C., Boro, R., Nangia, Y., Gandhi, S., Sharma, P., Wangoo, N., Rajesh, K., & Shekhawat, G. S. (2009). Immunoanalytical techniques for analyzing pesticides in the environment. *TrAC Trends in Analytical Chemistry, 28*(1), 29-39.

Sapsford, K. E., Taitt, C. R., Fertig, S., Moore, M. H., Lassman, M. E., Maragos, C. M., & Shriver-Lake, L. C. (2006). Indirect competitive immunoassay for detection of aflatoxin B1 in corn and nut products using the array biosensor. *Biosensors and Bioelectronics, 21*(12), 2298-2305.

Saqer M. Herzallah.J. (2009).Determination of aflatoxins in eggs, milk, meat and meat products using HPLC fluorescent and UV detectors. *Food Chemistry,* 114 , 1141-1146.

Scott, P. M., & Trucksess, M. W. (1997). Application of immunoaffinity columns to mycotoxin analysis. *Journal of AOAC International, 80*(5), 941-949.

Scudamore, K., Hetmanski, M., Clarke, P., Barnes, K., & Startin, J. (1996). Analytical methods for the determination of sterigmatocystin in cheese, bread and corn products using HPLC with atmospheric pressure ionization mass spectrometric detection. *Food Additives & Contaminants: Part A, 13*(3), 343-358.

Senyuva, H. Z., & Gilbert, J. Immunoaffinity column clean-up techniques in food analysis: A review. *Journal of Chromatography B, 878*(2), 115-132.

Sheibani, A., Tabrizchi, M., & Ghaziaskar, H. S. (2008). Determination of aflatoxins B1 and B2 using ion mobility spectrometry. *Talanta, 75*, 233-238.

Shim, W. B., Yang, Z. Y., Kim, J. S., Kim, J. Y., Kang, S. J., Woo, G. J., et al. (2007). Development of immunochromatography strip-test using nanocolloidal gold-antibody probe for the rapid detection of aflatoxin B1 in grain and feed samples. *Journal of Microbiology and Biotechnology, 17*, 1629-1637.

Stroka J., von Holst C., Anklam E., Reutter M. (2003). Immunoaffinity column cleanup with liquid chromatography using post-column bromination for determination of aflatoxin B1 in cattle feed: collaborative study.*J. AOAC Int.* 86, 1179.

Sulyok, M., Krska, R., Schuhmacher, R. (2010). Application of an LC-MS/MS based multi-mycotoxin method for the semi-quantitative determination of mycotoxins occurring in different types of food infected by moulds. *Food Chemistry, 119*, 408-416.

Sun, A. L., Qi, Q. A., Dong, Z. L., & Liang, K. Z. (2008). An electrochemical enzyme immunoassay for aflatoxin B1 based on bio-electrocatalytic reaction with room-temperature ionic liquid and nanoparticle-modified electrodes. *Sensing and Instrumentation for Food Quality and Safety, 2*(1), 43-50.

Sun, X. L., Zhao, X. L., Tang, J., Zhou, J., & Chu, F. S. (2005). Preparation of gold-labeled antibody probe and its use in immunochromatography assay for detection of aflatoxin B1. *International journal of food microbiology, 99*, 185-194.

Tan, Y., Chu, X., Shen, G. L., & Yu, R. Q. (2009). A signal-amplified electrochemical immunosensor for aflatoxin B1 determination in rice. *Analytical biochemistry, 387*(1), 82-86.

Tanaka, R., Yuhi, T., Nagatani, N., Endo, T., Kerman, K., Takamura, Y., et al. (2006). A novel enhancement assay for immunochromatographic test strips using gold nanoparticles. *Analytical and bioanalytical chemistry, 385*, 1414–1420.

Tang, L., Zeng, G. M., Shen, G. L., Li, Y. P., Zhang, Y., & Huang D. L. (2008). Rapid Detection of Picloram in Agricultural Field Samples Using a Disposable Immunomembrane-Based Electrochemical Sensor. *Environmental Science and Technology, 42*, 1207–1212.

Tassaneeyakul, W., Razzazi-Fazeli, E., Porasuphatana, S., Bohm, J. (2004). Contamination of aflatoxins in herbal medicinal products in Thailand. Mycopathologia, 158, 239-244.

Vahl, M., & J rgensen, K. (1998). Determination of aflatoxins in food using LC/MS/MS. *Zeitschrift für Lebensmitteluntersuchung und-Forschung A, 206*(4), 243-245.

Van Egmond, H.P.,Jonker, M.A. Reglamentos a nivel mundial para las micotoxinas en los alimentos y las raciones en el ano 2003. Food & Agriculture Org., 2004

Vatinno, R., Vuckovic, D., Zambonin, C. G., & Pawliszyn, J. (2008). Automated high-throughput method using solid-phase microextraction-liquid chromatography-tandem mass spectrometry for the determination of ochratoxin A in human urine. *Journal of Chromatography A, 1201*(2), 215-221.

Ventura, M., Gomez, A., Anaya, I., Díaz, J., Broto, F., Agut, M., & Comellas, L. (2004). Determination of aflatoxins B1, G1, B2 and G2 in medicinal herbs by liquid chromatography-tandem mass spectrometry. *Journal of Chromatography A, 1048*(1), 25-29.

Vig, A., Radoi, A., Mu oz-Berbel, X., Gyemant, G., & Marty, J. L. (2009). Impedimetric aflatoxin M1 immunosensor based on colloidal gold and silver electrodeposition. *Sensors and Actuators B: Chemical, 138*(1), 214-220.

Wang, L., & Gan, X. X. (2009). Biomolecule-functionalized magnetic nanoparticles for flow-through quartz crystal microbalance immunoassay of aflatoxin B1. *Bioprocess and biosystems engineering, 32*(1), 109-116.

Wang, S., Quan, Y., Lee, N., & Kennedy, I. R. (2006). Rapid determination of fumonisin B1 in food samples by enzyme-linked immunosorbent assay and colloidal gold immunoassay. *Journal of Agricultural and Food Chemistry, 54,* 2491-2495.

Wang, X. H., Liu, T., Xu, N., Zhang, Y., Wang, S. (2007). Enzyme-linked immunosorbent assay and colloidal gold immunoassay for ochratoxin A: investigation of analytical conditions and sample matrix on assay performance. *Analytical and bioanalytical chemistry, 389,* 903-911.

Wang, Y., Dostálek, J., & Knoll, W. (2009). Long range surface plasmon-enhanced fluorescence spectroscopy for the detection of aflatoxin M1 in milk. *Biosensors and Bioelectronics, 24*(7), 2264-2267.

Xu, Y., Huang, Z. B., He, Q. H., Deng, S. Z., Li, L. S., & Li, Y. P. (2010). Development of an immunochromatographic strip test for the rapid detection of deoxynivalenol in wheat and maize. *Food Chemistry, 119,* 834-839.

Yu, J. C. C., & Lai, E. P. C. Molecularly Imprinted Polymers for Ochratoxin A Extraction and Analysis. *Toxins, 2*(6), 1536-1553.

Zambonin, C. G., Monaci, L., & Aresta, A. (2001). Determination of cyclopiazonic acid in cheese samples using solid-phase microextraction and high performance liquid chromatography. *Food Chemistry, 75*(2), 249-254.

Zhang, D., Li, P, Zhang, Q., Zhang, W., Huang, Y., Ding, X., Jiang J. (2009). Production of ultrasensitive generic monoclonal antibodies against major aflatoxins using a modified two-step screening procedure. *Analy. Chim. Acta, 636:* 63-69

Zhang, D., Li, P.., Zhang, Q., & Zhang, W., (2011). Ultrasensitive nanogold probe-based immunochromatographic assay for simultaneous detection of total aflatoxins in peanuts. *Biosensors and Bioelectronics, 26,* 2877–2882.

Zhang, J, Li, P., Zhang, W., Zhang, Q., Ding, X., Chen, X., Wu, W., Zhang, X. (2009). Production and Characterization of Monoclonal Antibodies Against Aflatoxin G1. *Hybridoma, 28*(1): 67-70

Zheng, M. Z., Richard, J. L., & Binder, J. (2006). A review of rapid methods for the analysis of mycotoxins. *Mycopathologia, 161*(5), 261-273.

Zhou, Y., Pan, F. G., Li, Y. S., Zhang, Y. Y., Zhang, J. H., Lu, S. Y., et al. (2009). Colloidal gold probe-based immunochromatographic assay for the rapid detection of brevetoxins in fishery product samples. *Biosensors and Bioelectronics, 24,* 2744-2747.

Zllner, P., & Mayer-Helm, B. (2006). Trace mycotoxin analysis in complex biological and food matrices by liquid chromatography-atmospheric pressure ionisation mass spectrometry. *Journal of Chromatography A, 1136*(2), 123-169.

12

New Development in Aflatoxin Research: From Aquafeed to Marine Cells

Maria Pia Santacroce[1], Marcella Narracci[2], Maria Immacolata Acquaviva[2],
Rosa Anna Cavallo[2], Valentina Zacchino[1] and Gerardo Centoducati[1]
*[1]Department of Public Health and Animal Science, Division of Aquaculture,
Faculty of Veterinary Medicine, University of Bari,
[2]Institute for Coastal Marine Environment, National Research Council of Italy, Taranto,
Italy*

1. Introduction

Available data on the real impact of aflatoxins on farm aquatic species are very limited. Since long time, aflatoxin B1 (AFB_1) has been considered the most potent food-born hepatotoxicant, frequently found in animal feedstuff. At present, it has been reported as responsible agent in unforeseen outbreaks of fish mortality due to acute or chronic aflatoxicosis, mainly well documented in freshwater species. The lack of information on the incidence of aflatoxicosis in marine reared teleosts may be partially due to the difficulty in accurately diagnosing the disease in fish, as well as to the lack of specie-specific *in vitro* models for toxicity studies.

In this work: 1) we have verified that pelletted fish feed might be considered as sources of AFB_1 contamination in aquaculture due to the isolation and identification of blue eye fungi (*Aspergillus* spp., *Penicillium* spp.) in feed samples, as well as other several genera (*Fusarium, Cladosporium, Alternaria, Geotrichum, Mucor, Rizophus, Acremonium*); 2) we have performed an *in vitro* evaluation of AFB_1 potential cytotoxic on *Sparus aurata* hepatocyte primary cultures (SaHePs), using a multiple endpoint screening. Our results demonstrate that seabream hepatocytes are highly sensitive to AFB_1 exposure and especially indicate three distinct pathways of cytotoxic response: necrotic cell death, apoptotic cell death and uncontrolled cell proliferation; 3) we have compared the dose response curves obtained by measuring the bioluminescence of *Vibrio fischeri* upon AFB_1 exposure to those obtained from *in vitro* cell culture system. Results show equivalent and overlapping toxic responses with those from seabream hepatocytes.

2. Impact of Aflatoxins in aquatic species

Aflatoxins are the most potent natural toxic metabolites produced by toxinogenic strains belonging to molds of the *Aspergillus* genus contaminating foods, feed ingredients and products of animal origin. Since their discover in the '60s, the aflatoxins poisoning continue to represent a growing potential threat to human and animal health based on their carcinogenic, immunosuppressive and other severe adverse effects (Chavez-Sanchez et al. 1994; Halver 1969; Han et al., 2010; Jantrarotai & Lovell 1990; Sahoo et al., 1996; Santacroce et al., 2008). The real impact of aflatoxins on farm animals have been largely studied especially in mammalians or in the zootechnical field of terrestrian vertebrates. In contrast, very limited data on aquatic species are now available (Han et al., 2008; Lovell, 1992; Murjani, 2003; Pestka, 2007). Among all aflatoxins, aflatoxin B1 (AFB_1) is considered the most potent food-born hepatotoxicant frequently found in animal feedstuffs and responsible agent in unforeseen outbreaks of fish mortality attributed to aflatoxicosis, well documented in freshwater species since long time (Agag 2004; Cagauan et al., 2004; Santacroce et al., 2008). Aquatic species have shown dissimilar susceptibility to the hepatotoxic and carcinogenic effects of AFB_1 depending on the particular species . In fact, the fish susceptibility to AFB_1, largely studied for more than 50 years in USA and North Europe in freshwater fish and crustaceans, seems to be related with interspecies variations of AFB_1 biotransformation efficiency (Eaton & Groopman, 1994; Hendricks, 1994; Wales 1970). The increased use of plant origin ingredients in aquafeed formulations has intensified the potential onset for aflatoxicosis in fish farming systems due to the carryover of high loads of aflatoxin contamination by vegetable sources (Cagauan et al., 2004; Ellis et al., 2000; Fegan, 2005; Naylor et al., 2009; Spring, 2005). Aflatoxin production by the most toxinogenic strains can occur directly in the field, during insiling, feed formula preparation, and also during improper feed storage in the farm. On the other hand, the thermal treatments, applying high temperature pelleting procedures, even though destroy the mould but do not inactivate the heat-stable toxins present in spores and mycelium. Toxins accumulate in fish meal thus representing an high risk for the farmed species and then for the customer health and safety (IARC, 1993). As a result, the problem of aflatoxin contamination in aquaculture has amplified. Several studies revealed that AFB_1 residues can be retained in aquatic animal tissues, giving rise to potential public health risks after ingestion (Han et al., 2010; Messonnier et al., 2007; Puschner 2002; Tacon & Metian, 2008). Moreover, the presence of aflatoxins decrease the nutritional value of administrated feed in fish farm, both affecting the fish welfare status and the product quality (Hassan et al., 2010; Naylor et al., 2009). In intensive aquaculture, the features of administrated feed play a main rule being the major alimentary source involved with the fish growing and their nutritional requirements. The cases of acute intoxication by aflatoxin are almost rare and exceptional, while the chronic toxicity is the serious and most prevalent problem, because of AFB1 carcinogenicity upon long term microexposures. When moderate to high doses of aflatoxin are ingested, fish develop an acute intoxication, called acute aflatoxicosis, that generally gives rise to poor heath and fertility, loss in productivity, reduced weight gain, and immunosuppression (Stewart & Larson, 2002). Chronic aflatoxicosis occurs when low to moderate doses of aflatoxins are ingested over a long period of time. Generally, it is difficult to recognise or diagnose this condition because of its slow, subclinical trend. The majority of clinical signs is related to chronic status, such as impaired liver function, reduced feed efficiency, weight loss, increased susceptibility to secondary infectious diseases, necrosis and tumour development in liver and other organs, and increased mortality (Murjani, 2003). More insidious, pathological signs occur as a consequence of prolonged dietary exposure, causing genotoxic, tumorigenic and teratogenic, hormonal or neurotoxic effects in fish, as well as in humans. Chronic aflatoxicosis is of great concern in aquaculture systems, since it was found to be implicated both with a gradual decline of reared fish health status and with decreased stock quality. While considerable epidemiological data have been obtained on AFB_1 adverse effects in humans, farm animals and freshwater species, there is a substantial need to obtain such data especially on aquacultured euryaline fish (Santacroce et al., 2008).

The effect of AFB_1 on marine teleosts is quite unknown, although AFB_1 feed contamination is becoming of increasing interests in marine aquaculture (El-Sayed & Khalil, 2009). Even though the problem of aflatoxicosis in fish was discovered about 50 years ago, sudden outbreaks of fish mortality continue to be reported, suggesting that the problem is still misunderstood and that scarce preventive measures have been adopted (Santacroce et al., 2008). At present, there is a gap in information regarding differences in AFB_1 susceptibilities in marine-reared fish. This means that the real exposure risk to AFB_1 in such species is still not understood.

Objectives of this work were: 1) to verify that pelletted fish feed might be considered as sources of AFB_1 contamination in aquaculture, isolating and identifying the contaminating toxigenic moulds; 2) to perform an *in vitro* evaluation of AFB_1 cytotoxic potential on *S. aurata* hepatocyte primary cultures, using a multiple endpoint screening; 3) to compare the dose response curves obtained by measuring the bioluminescence of *V. fischeri* upon AFB_1 exposure to those obtained from *in vitro* cell culture system.

3. Case report in aquaculture: Contaminated feeds

Most toxigenic molds, able to produce toxins, belong to the *Aspergillus, Penicillium,* and *Fusarium* genera (Moss, 1998). Fungal life consists of two different steps, mold growth and mycotoxin production, each one requiring specific and restricted conditions. Secondary metabolites are produced by toxigenic molds at the end of the active growth and under favorable conditions; they can be collected both in spores and vegetative mycelium or secreted into the growth substrate (Moss, 1991). Such metabolites promote the competitive fungal survival, but are not necessary for the essential metabolic functions of the fungus; they are commonly associated with the sporulation process and usually require strictly environmental conditions (Sekiguchi & Gaucher, 1977). Within the same fungal species, toxigenic strains can produce different quantity of mycotoxin and different types of toxic secondary metabolites, even if they show the same metabolic activity and speed of growth. The ubiquitus nature and biosynthetic heterogeneity of fungi hardly favors mycotoxin contamination of feedstuff (Dragoni et al., 2000). In animal feed colonized by toxigenic fungi are commonly present several mycotoxins, often found unchanged after feed processing (Jackson et al., 1996), because of their highly stable chemical structure. Studies carried on mycotoxin contamination of feed and food have led to the identification of over 100 toxigenic molds and at least 300 mycotoxins (Miller & Trenholm, 1994; Sharma & Salunkhe, 1991).

Herein, we highlight the existence of fish feed contamination in samples of spoiled grain pellets taken from a sea bream farm, entirely covered by moulds and spores (Fig. 1). After the grain was removed from the bin, green and blue eye moulds appeared both on pellets and surfaces. Feed portions of 50 g were transferred to sterile glass beacker and mixed with 450 mL of Sabouraud broth. After vortexing for 30 minutes, the suspension was opportunely diluted (1:10 and 1:100) and 100 μL of suspension and each dilution were streaked, in triplicate, into Sabouraud agar and incubated at 25-30 °C. All the suspect colony types were selected, streaked onto Sabouraud agar to obtain pure cultures and screened for morphological characterization (Fig. 2).

The fungal species most frequently identified in the group of the blue eye fungi belonging to the taxa of Ascomycota were *Aspergillus* spp. (Fig. 3) and *Penicillium* spp. (Fig. 4). Numerous other fungi were isolated, and the genera, *Mucor* (Fig. 5), *Cladosporium* (Fig. 6), *Fusarium* (Fig. 7), *Geotrichum* (Fig. 8), *Alternaria, Rhizopus, Acremonium* were predominant among the moulds found. Acco rding to several authors, the identification of wrong storage conditions since the presence of blue eye group fungi, certainly indicates that the grain has been improperly stored.

Fig. 1. Fish feed contamination: spoiled pellets from a sea bream farm, entirely covered by moulds and spores

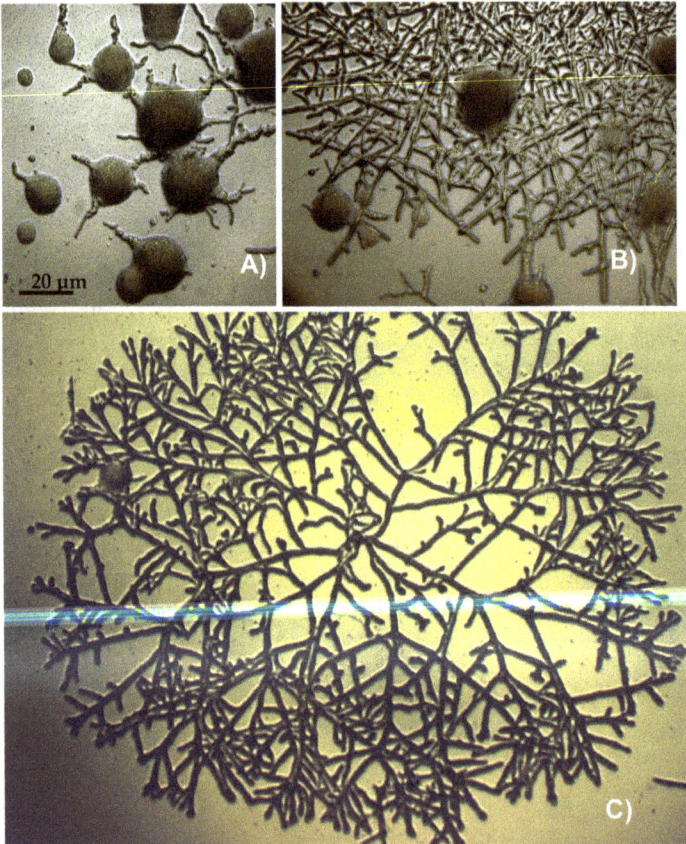

Fig. 2. Seed suspension on Sabouraud agar plate: A) Hyphal germination (x200); B) Hyphal elongation and branching (x200); C) Mycelium growth and mass of hyphae (x100)

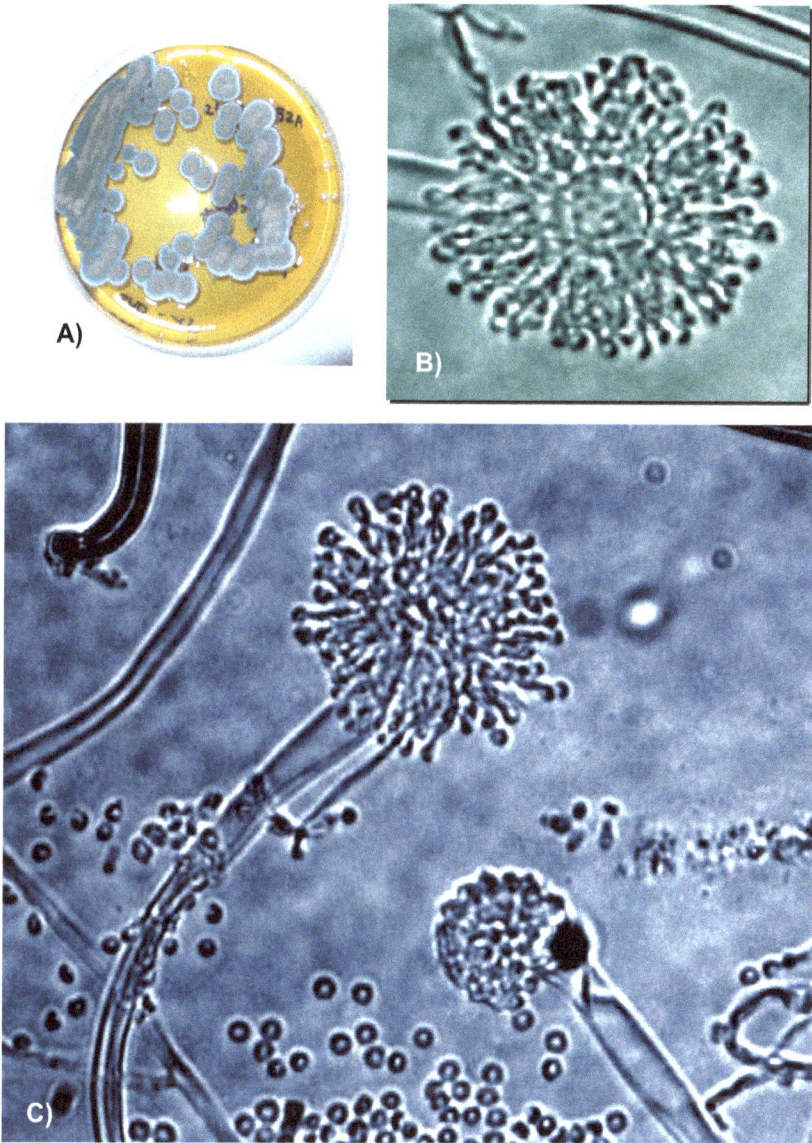

Fig. 3. *Aspergillus flavus* isolated from contaminated aquafeed administered to aquacultured seabream: A) mould growth on Sabouraud agar plate; B), C) Direct microscopy of a Scoth test on slide (x400)

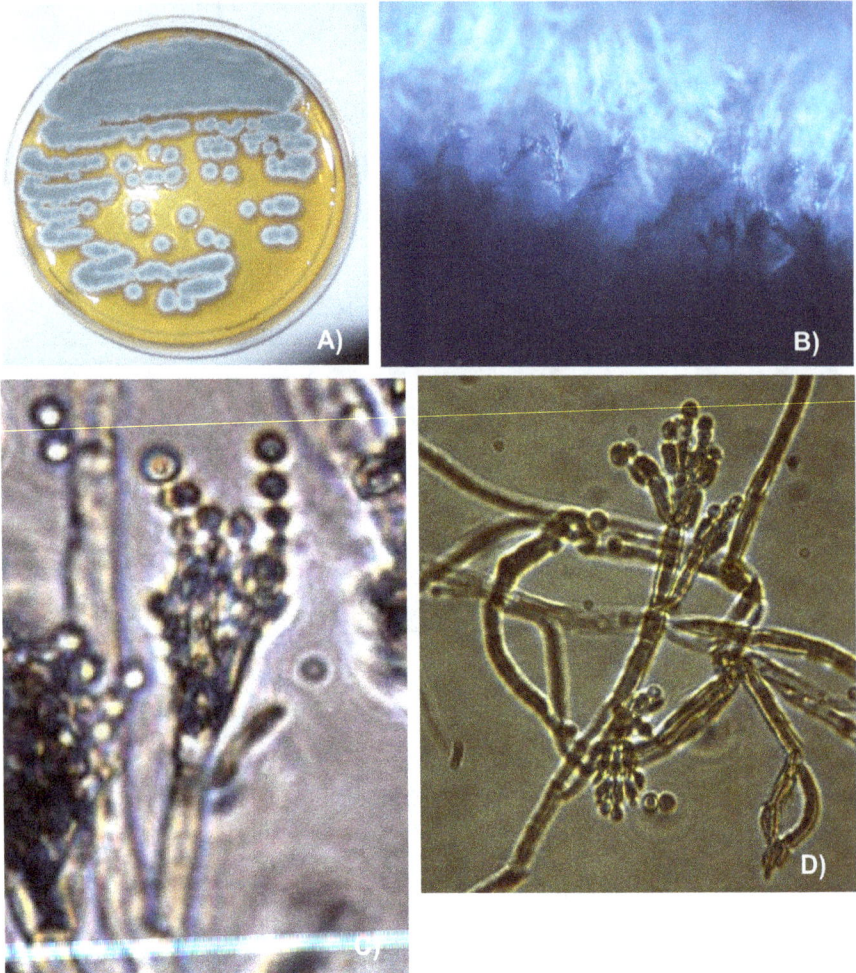

Fig. 4. *Penicillium crysogenum* isolated from contaminated aquafeed administered to aquacultured seabream: A) mould growth on Sabouraud agar plate; B) growth of aerial hyphae (x400); C) Conidial head two-stage branched (x600); D) Direct microscopy of a Scoth test on slide (x400)

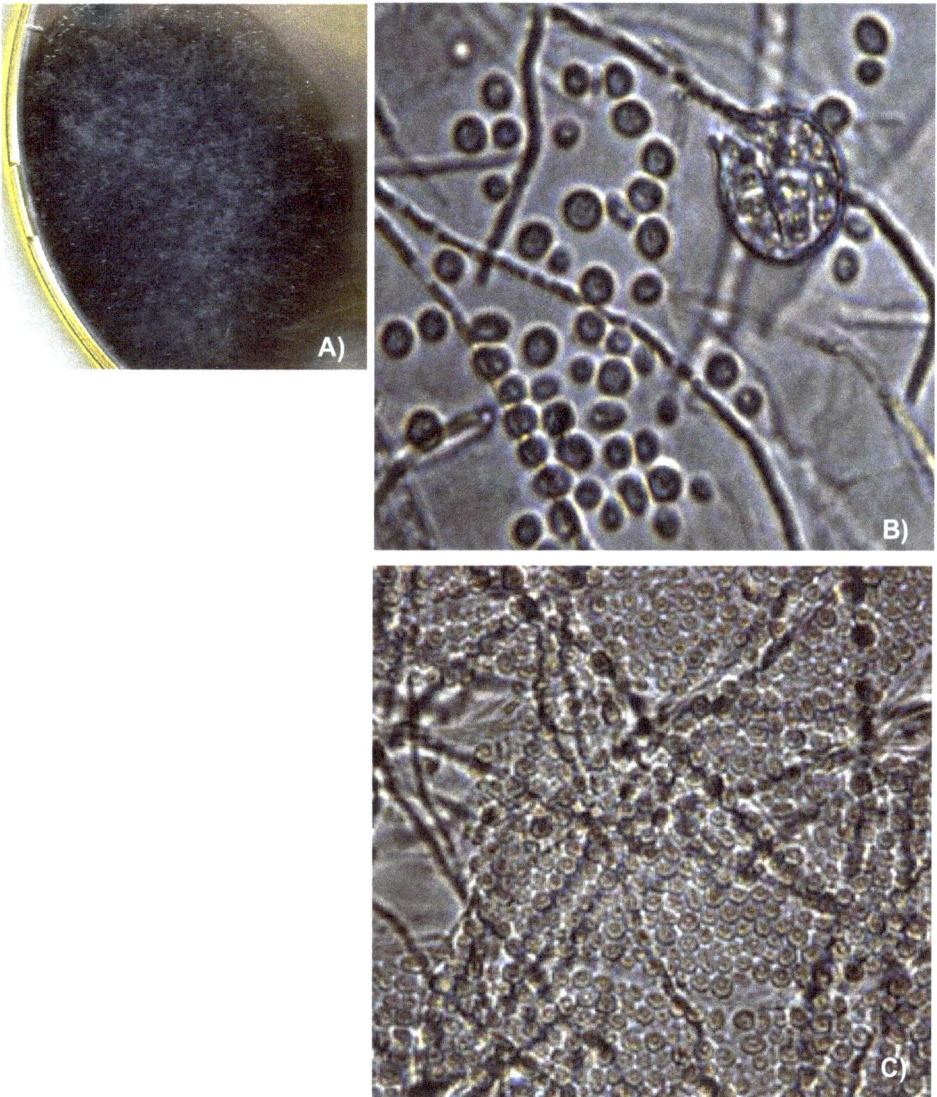

Fig. 5. *Mucor* spp. isolated from contaminated aquafeed administered to aquacultured seabream: A) mould growth on Sabouraud agar plate; B) dispersal of conidia from ascospores (x600); D) Direct microscopy of a Scoth test on slide (x400)

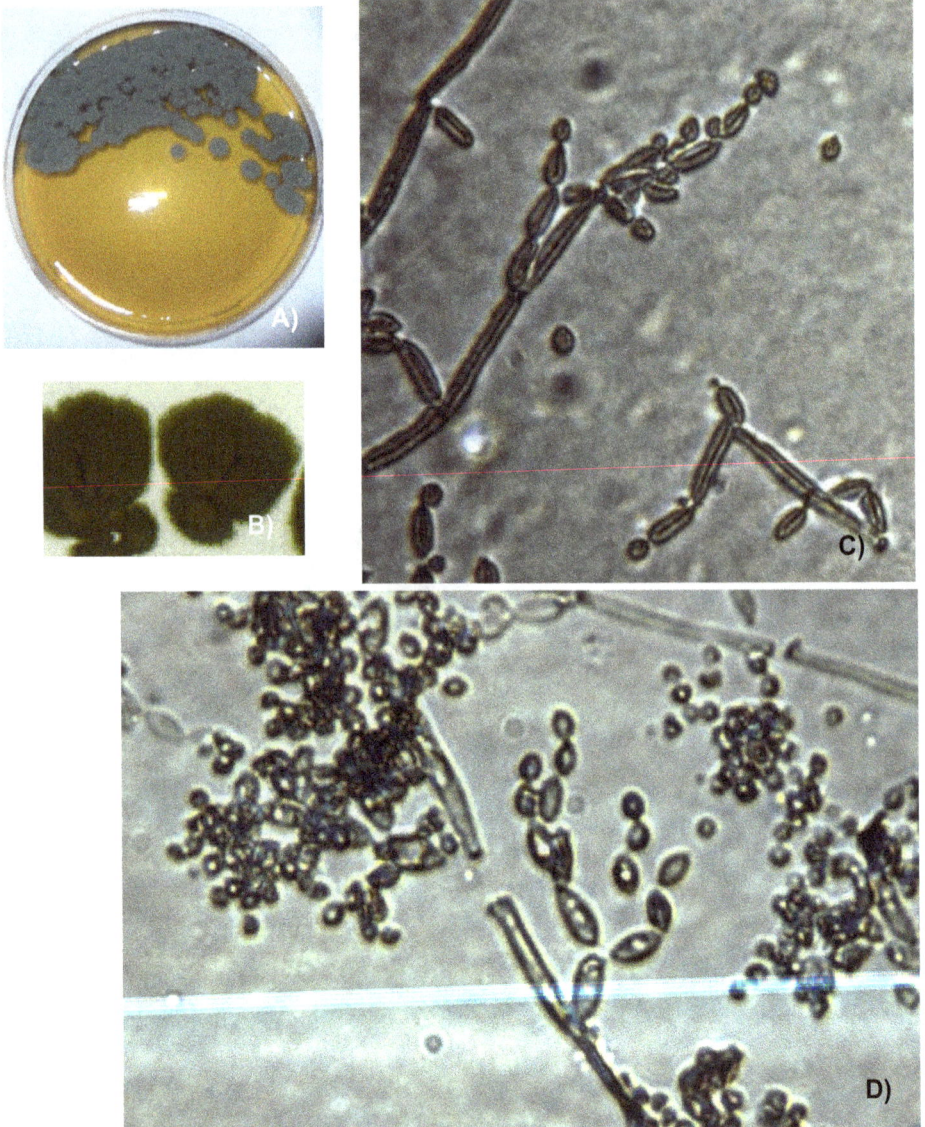

Fig. 6. *Cladoaporium* spp. isolated from contaminated aquafeed administered to aquacultured seabream: A) mould growth on Sabouraud agar plate; B) dispersal of conidia from ascospores (x400); D) Direct microscopy of a Scoth test on slide (x400)

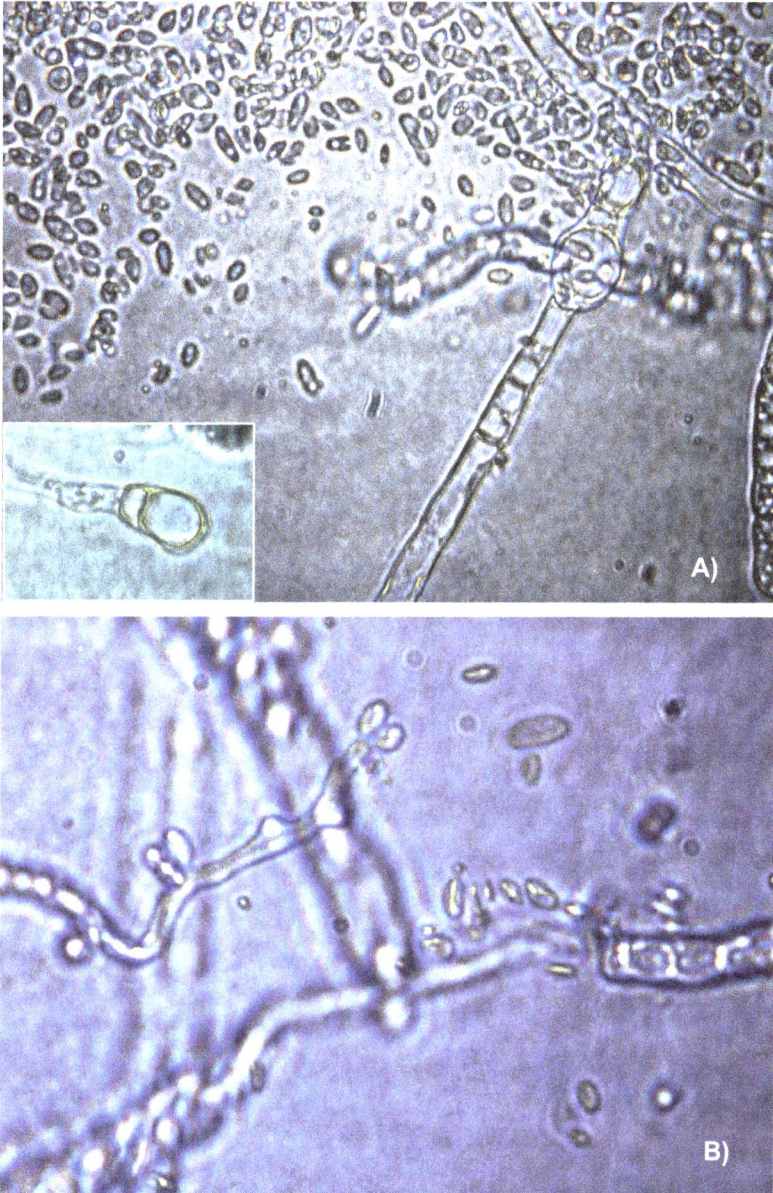

Fig. 7. *Fusarium* spp. isolated from contaminated aquafeed administered to aquacultured seabream: A) Direct microscopy: conidiophores with intercalary chlamydospores, and terminal clamydospore (left lower corner) (x600); D) Direct microscopy: conidiophores with terminal phialides and microconidial dispersal. (x600)

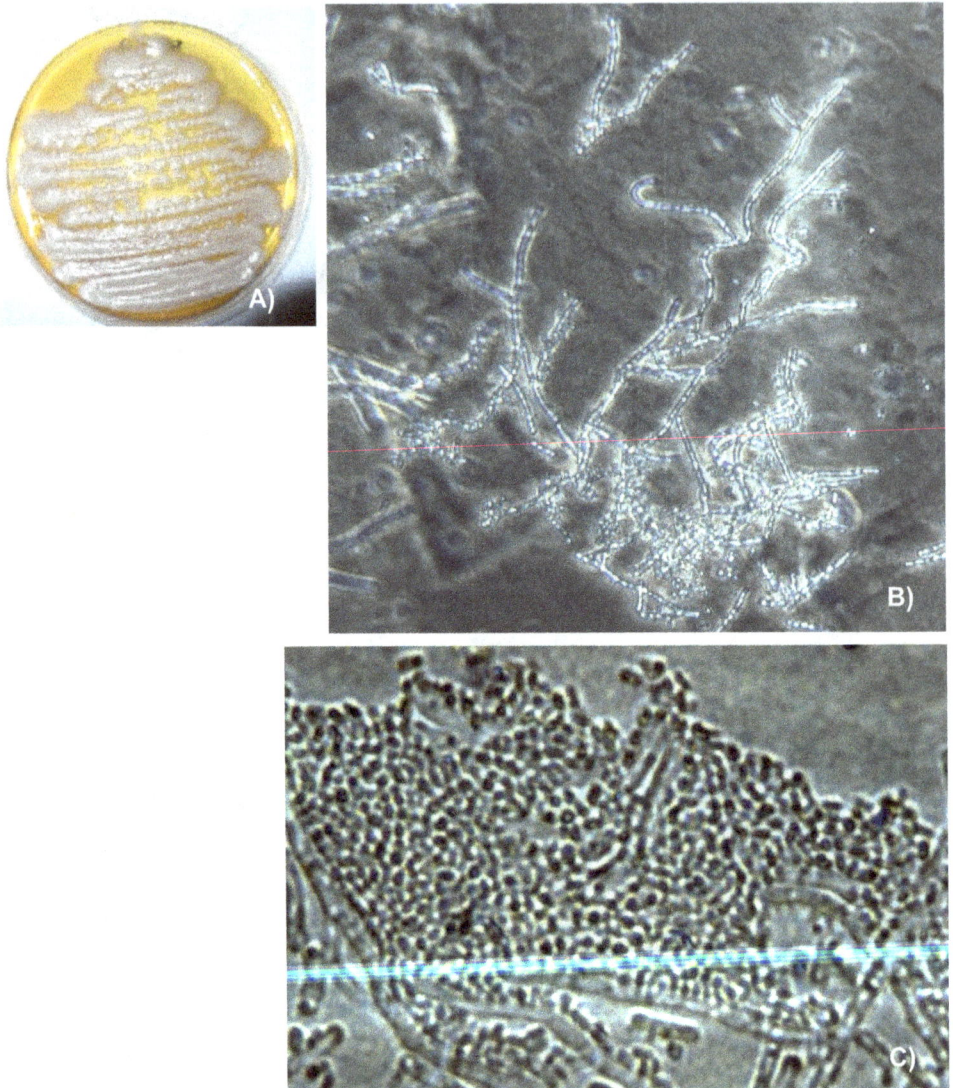

Fig. 8. *Geotrichum* sp. isolated from contaminated aquafeed administered to aquacultured seabream: A) mould growth on Sabouraud agar plate; B) (x200) fragmentation of undifferentiated hyphae; C) (x400). Direct microscopy: arthroconidia

4. New models of study: Primary cultures of marine teleostean hepatocytes

Based on the previous reported data, it may be assumed that the lack of information regarding the incidence of aflatoxicosis in marine reared species may be in part due to the difficulty in accurately diagnosing aflatoxicosis in fish, as well as to the lack of specie-specific *in vitro* models for toxicity studies. It is well known that animal cell cultures permit the comparison of species at a cellular level under equivalent conditions of toxicant exposure. Because of the insufficient availability of *in vitro* marine liver systems on a species-specific basis, comparison studies between AFB$_1$ cytotoxicity on freshwater and seawater fish cells have never been made. Although primary hepatocytes represent the most employed within all the *in vitro* liver models, an ideal hepatocyte system targeted for marine fish has not yet fully established. In order to overcome this issue, we have developed a new *in vitro* model which firstly describes the isolation and cultivation of hepatocytes from a marine Mediterranean teleost of great economic value, the gilt-head seabream (*Sparus aurata*), applying a method different from the ancient liver perfusion (Santacroce et al. 2010). In this method, seabream hepatocytes were quickly derived from the explanted liver, without any passage of liver perfusion *in vivo*, through several steps of mechanical separation, multiple enzymatic digestion and isopicnic cell purification. Previously works on isolation and cultivation of teleost hepatocytes were done following the two-step perfusion procedure, firstly described by Seglen (1976), followed by Mommsen et al. (1994) and then adopted from over forty years by other authors (Segner, 1998; Mommsen *et al.*, 1994; Pesonen & Andersson, 1991). The protocol involved a first perfusion of the liver with a Ca$_2$+-free balanced saline solution followed by a second perfusion with the digesting enzyme collagenase. The liver perfusion was the first important technique implemented for the preparation of primary hepatocytes (Guguen-Guillouzo, 1992). Firstly, the method was developed in rodents, (Berry & Friend, 1969; Seglen, 1976) and then improved to obtain hepatocytes from other animal sources like pig (Chen et al., 2002; Koebe & Schildberg, 1996), sheep (Clark & Vincent, 2000) and finally fish (Birnbaum et al., 1976; Blair et al. 1990; Braunbeck & Segner, 2000; Mommsen et al., 1994; Segner, 1998). Since the early 70s, this was the main method largely used for aquatic vertebrates, event though the main disadvantage of this procedure is the large size of animal (approximately 100 g/body weight) required to permit the *in vivo* abdomen insertion of perfusion devices connected to a peristaltic pump. Mitaka et al. (1995) proposed a selective separation method for obtaining and culturing small hepatocytes by using hyaluronic acid-attached carrier. This technique comprised the isolation of hepatocytes from the liver of adult rats by perfusion through portal vein according the method of Seglen (1976). Other alternative procedures established in freshwater fish involved cutting the liver into pieces and incubating in a solution of 0.5% of collagenase until full digestion (Mitaka & Kon, 2004). This procedure was outlined by Bouche et al. (1979) in carp but only after liver perfusion through the arteria coeliaca. In rare exceptions, alternative enzymes were used as single or mixture addition, but generally reported as components of the perfusion medium adopted during the liver perfusion (Braunbeck & Segner, 2000). Those used for the dispersion of fish hepatocytes were hyaluronydase, protease, elastase and nagarase (Mommsen et al., 1994) or trypsin, as reviewed by Braunbeck & Segner (2000). Only one study is reported in literature on the isolation of primary seabream hepatocytes, but still based on the liver perfusion technique according to Mommsen and colleagues (1994) (Bevelander et al., 2006). The liver was perfused via the heart for 15 min, then excised, cut in small pieces and incubated with a solution of collagenase (0.3 mg/mL) for 45 min. Unfortunately, those hepatocytes showed only a limited survival, up to 1 week, suggesting a limitation in cell viability. The above method is not applicable with fish of relative small size, like juveniles or small aquarium fish. Hence, the pressing necessity to improve the isolation and tissue culture techniques involving cells for a better implementation of biotechnological methods on marine species.

The method herein described for isolating seabream hepatocytes differs from those previously cited for the medium composition, the type of enzyme mixture and concentration, digestion time and temperature, time and speed of centrifugation steps, filter type and selectivity, isopicnic separation of hepatocytes from nucleated erythrocytes, cultivation conditions (temperature, CO_2 tension, refrigerate incubator). *S. aurata* juveniles (30 ± 4 g mean body weight, n = 45) were used for establishing hepatocyte primary cultures (SaHePs) according to the new procedure, suitable for fish of small size (Santacroce et al., 2010). Hepatocytes were isolated as described previously by tissue physical disaggregation combined with enzyme digestion, and purificated by several steps of centrifugation (Santacroce et al., 2010). Freshly isolated hepatocytes were tested for viability by Trypan blue, counted and seeded at a density of 30,000 cells/cm² in 96-well plates Falcon BD previously pre-coated with collagen I. Cells were cultured in a refrigerate incubator at 18°C in humidified atmosphere of 97% air/3% CO_2 (Fig. 9).

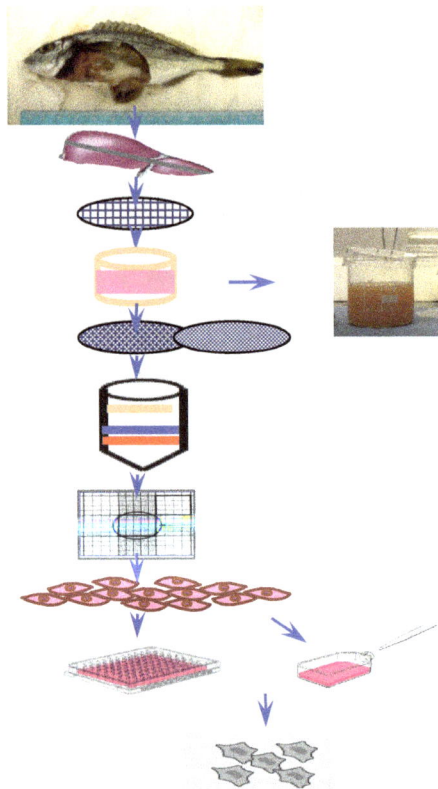

Fig. 9. Seabream hepatocyte primary culture protocol

The cells were grown in Leibovitz's L-15 medium with 2 mM L-glutamine, 10% FBS, 100 IU/mL penicilin, 100 μg/mL streptomycin, 100 μg/mL amphotericine, 50 μg/mL gentamycin, 1 mM Na Pyruvate, 5 mM D-Glu, 10 mM HEPES, 12 mM NaHCO$_3$, and supplemented with 20 mM NaCl, 0.05% ITS *plus* (insulin/transferrin/sodium selenite *plus* oleic acid/linoleic acid/BSA), 0.01 mM MEM non-essential amino acid, 0.01 mM MEM-vitamin mix, 0.1 mM ascorbic acid, and 0.01 μg/mL epidermal growth factor, 0.005 μg/mL hepatocyte growth factor. Cells were allowed to attach for 12 h, afterward fresh nutrient L-15 medium was added. After 24 h of incubation, the non-adherent cells were removed by washing wells twice with 1X PBS, then L-15 medium was changed every 24 h thereafter. Fig. 10 shows phase contrast images of *S. aurata* primary hepatocytes cultured on collagen I coated flasks during monolayer formation, development and differentiation. Starting from the third day to sixth, during the phase of monolayer development, SaHePs consisted basically of two cell types: small hepatocytes, represented by islands with tightly packed small cells, each containing multiple proliferating islets that grew inside, and large hepatocytes, formed by cords of large bright cells. Then, more multicellular islands appeared, and cords of large hepatocytes moved to fill the space between the islands of small hepatocytes. During development, most cells adopted morphological changes, showing a hepatocyte typical polygonal shape. In the second week from the seeding, monolayer reached about 70% of confluence. By the third week, cells formed a compact monolayer, fully differentiated. Morphologically, cells presented one or two nucleus with two large evident nucleolus, apparent cytoskeleton, and characteristic biliary canalicular structures in proximity of intercellular junction of two adjacent hepatocytes.

Fig. 11 shows immunofluorescence images of primary hepatocytes positive stained for several liver markers after 3 days seeding on to collagen I coated chamber slides: (A) Cytochrome P450, the major drug metabolizing enzyme located in the endoplasmatic reticulum; (B) Albumin, the most abundant protein synthesized by mature and functional hepatocytes; (C) CK-18, a special skeleton protein of hepatocyte; (D) production of extracellular matrix stained with Ab anti-CK-18-TRITC, (E) viable cells (green) and viable nuclei stained with DAPI (blue), (F) metabolic active cells in CFDA, carboxyfluorescein diacetate. G) and H) show the cell proliferative capacity: mitotic nuclei (white) are double immunolabelled with Ab anti PCNA, an intranuclear protein cell cycle dependent, which assists DNA polymerase delta during DNA replication, hence its expression is considered a marker of DNA synthesis, and PI, a nuclear counterstaining; resting nuclei are red, and viable cells are green (CFDA- FITC).

The whole microscopy analysis was performed, for brightfield and fluorescence, by a Motic AE31 Epi-Fluorescent Inverted Microscope, equipped with DAPI/TRITC/FITC fluorescence filter cube set. Digital image capture was performed by Moticam 3000C Cooled CCD digital color camera (3.3 Megapixel, 1/2″ CCD), capture system in origin Live Cam 1.0 (32-32) and Motic Images Advanced (V. 3.2) acquisition software (Motic, Seneco, Milan, Italy). Image analysis and assemblage was performed with Motic Images Advanced (V. 3.2) (Motic, Seneco, Milan, Italy) and Adobe Photoshop 8.0 (Adobe, Inc.).

5. AFB₁ exposure in two *in vitro* systems

The nature and the degree of possible harmful effects produced on living species by toxicants can be evaluated using both analytic laboratory tests and bioassays. The information obtained by these tests are useful for implementing environmental risk assessment. Although analytic chemistry is sensitive, the biological models offer more information about chemical damage. In the last years, several biological models were used in both short- and long-term tests implementing various test organisms. Short-term tests are based on the assessment of quickly measurable parameters, such as bioluminescence in Microtox® system (Bulich, 1986), extensively applied. Long-term tests rate parameters such as cell viability or growth using as model human (Delmas et al., 2000; Gaubin et al., 2000), duckweed (Ince et al., 1999), or fish (Santacroce et al., 2010) cell lines.

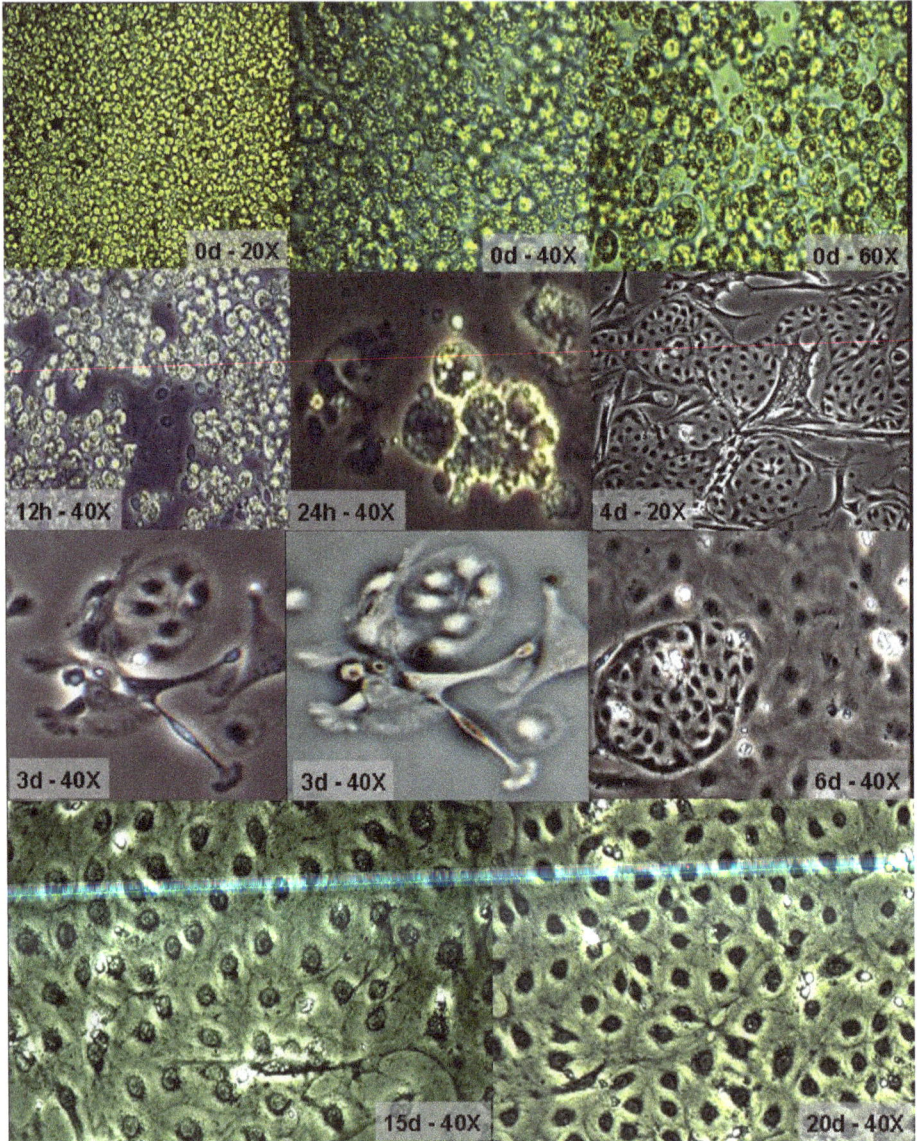

Fig. 10. *S. aurata* primary hepatocytes culture

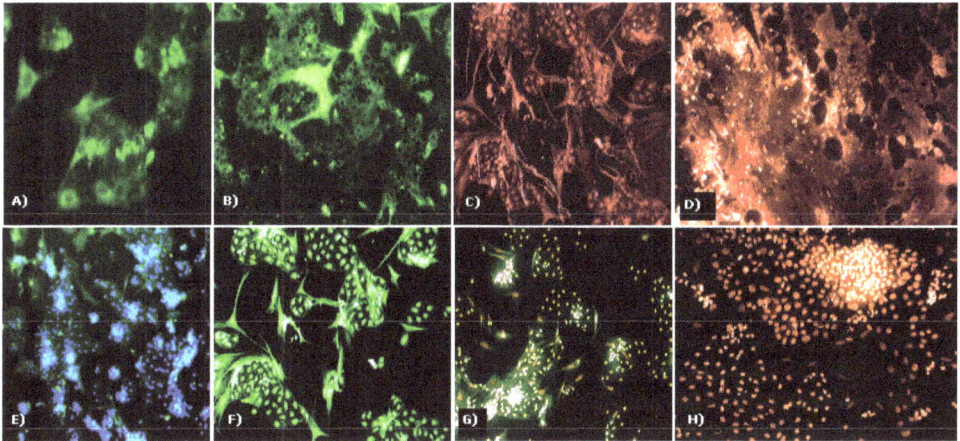

Fig. 11. Immunocytochemistry and morphology of seabream hepatocytes after 3 days seeding in Collagen I coated chamber slides. Immunofluorescence images show primary hepatocytes positive stained for several liver markers: (A) Cyp1A1/2-FITC, (B) Ab anti-Albumin-FITC, (C) Ab anti-CK-18-TRITC, (D) production of extracellular matrix stained with Ab anti-CK-18-TRITC; (E) viable cells (green) and viable nuclei counterstained with DAPI (blue) FITC/DAPI merged; (F) Metabolic active cells in CFDA (FITC); G) and H) show results pertaining to proliferative capacity: mitotic nuclei (white) are double immunolabelled with Ab anti PCNA-FITC and PI-TRITC, resting nuclei are red, and viable cells are green (CFDA-FITC), (G) CFDA-FITC/PI-TRITC merged; (H) Ab anti PCNA-FITC and PI-TRITC

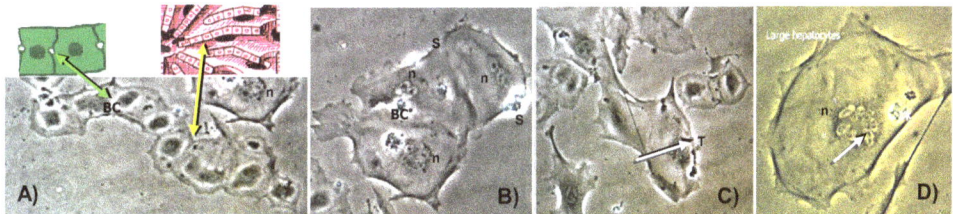

Fig. 12. Phase contrast images of primary primary seabream hepatocytes cultured at 18°C, 3% CO_2 on Collagen I coated T-25 flasks showing the apical, canalicular membrane with canaliculi resembling structures: a) cord of hepatocytes in culture; b) binucleated (n) hepatocytes with biliar canaliculi (BC) and granular secretions (S); c) transcytosis mechanism (T); d) mature polyploid hepatocyte highly differentiated

Based on the serious effects that aflatoxins can have on farm management and human health, the knowledge of their detection, toxicity, biosynthesis, and regulation is necessary to give proper responses to aflatoxin intoxication (Do & Choi, 2007).

5.1 AFB₁ effects on *Sparus aurata* hepatocytes

An *in vitro* evaluation of AFB_1 cytotoxic potential on SaHePs was carried out as second goal of this study. We performed a toxicity assessment by using a multiple endpoint screening. The work was based on series of *in vitro* cytotoxicity and functional assays in order to: provide new information on the toxic properties of AFB$_1$ at cellular level, characterize the type and degree of damage, the threshold hazard dose for reared seabream, and the boundary between acute and chronic toxicity. SaHePs were treated with a wide range of AFB$_1$ concentrations (from 0.25 mg/mL to 0.001 pg/mL) for 24, 48 and 72 hours, thus mimicking acute and chronic conditions. After each exposure, hepatocytes were examined for morphologic alterations, viability and citotoxicity, and apoptosis induction. The cytotoxic activity of AFB$_1$ was characterized by measuring two different viability endpoints, such as the MTT assay, as marker of cellular metabolic activity, to check the mitochondrial dehydrogenase activity using 3-(4,5-dimethyl-thiazol-2yl)-2,5-diphenyltetrazolium bromide (MTT) as substrate, and the neutral red (NR) retention assay to check lysosomal function upon AFB$_1$ exposure. Finally, the release of the cytoplasmatic enzyme lactate dehydrogenase (LDH) was performed as marker of lethality to check the membrane integrity. For each treatment time, the cytotoxic effect was determined as the half-maximal inhibiting concentration (IC$_{50}$) resulting in 50% of reduction in cell viability. The IC$_{50}$ values were determined fitting data to a four-parameter logistic model by using a Hill function non-linear regression analysis with the GraphPad Prism v.5.00 software package. Apoptosis was evaluated by assessing the phosphatidylserine (PS) exposition in the outer leaflet of plasma membrane at the cell surface of dying apoptotic, using the Annexin V-Cy3.18 binding in fluorescence microscopy. This assay allowed to identify the subacute cytotoxicity by differentiating early apoptotic cells from viable or necrotic ones. Tumorigenesis was evaluated by the proliferating cell nuclear antigen (PCNA) labelling, an intranuclear protein cell cycle dependent considered as marker of DNA synthesis. Results showed that AFB$_1$ exhibited dose- and time-dependent cytotoxic effect, the IC$_{50}$ being inversely related to the exposure time (Table 1). Dose-response curves obtained after 24, 48 and 72 h revealed that prolonged exposure times lead to a significant increase of the toxic potency of AFB$_1$ (Fig. 13).

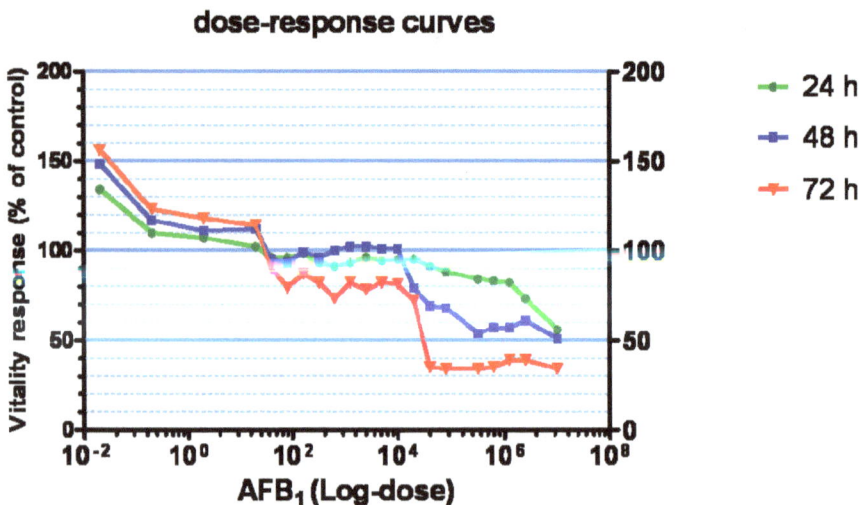

Fig. 13. Concentration-response curves for AFB$_1$ exposures on SaHePs

Although results showed that the viability endpoints used (NR, MTT) for measuring the AFB_1 cytotoxic potential were comparable, the IC_{50} value of the MTT was a more sensitive parameter of cytotoxicity (Table 1). In fact, the three cytotoxicity endpoints have been combined into a single value by applying the equation of Castano et al. (1994) which resulted strictly close to the MTT value, CI 0.067 µg/mL versus 0.609 µg/mL.

Assays	IC_{50}-24h	IC_{50}-48h	IC_{50}-72h
NR	3 µg/mL	0.3 µg/mL	0.03 µg/mL
MTT	5 µg/mL	0.6 µg/mL	0.06 µg/mL
LDH	4 µg/mL	3 µg/mL	2 µg/mL

Table 1. IC_{50} values at 24, 48, 72 h of NR, MTT and LDH assays

The maximum lethality response was assessed after 72 h exposure at 1.95 µg/mL (25% of metabolic activity), afterward cell death reached a plateau in almost 70-80 % of hepatocytes without any further damage recover up to 250 µg/mL. The dose-response curve at 72 h had a higher threshold but a steeper slope than 48 and 24 h. The release of LDH allowed to monitor primary necrosis over treatment times and to define the acute toxicity boundary. The threshold dose level (LOEC), where toxicity first appears, was estimated at 10 ng/mL, whereas the no-observable-adverse-effect-concentration (NOEC) was at 5 ng/mL. However, at doses within this apparent safe level, and approaching the baseline, cytotoxic effects and delayed secondary cell death were observed ranging from 0.02 µg/mL to 0.005 µg/mL, with signs of cell suffering. In this range cell vitality decreased in a time dependent manner, since about 0-5% of cells death was registered after 24 and 48 h, while at 72 h cell survival lowered up to a maximum of 20 % (Fig. 14).

In order to distinguish whether such increase in cell death after 72 h was due to primary necrosis or delayed secondary cell death, an immunocytochemical analysis was performed using direct immunofluorescence with the Annexin V-Cy3.18 staining. Apoptosis marker response confirmed that this apparent safe level hid a delayed mortality for apoptosis induction, even observable up to dose of 0.2 ng/mL. Figure 15 shows hepatocytes cultured in a 4-well slide (500 µL/well) exposed to the toxin for 24 h at doses of: 0 µg/mL (A) as normal control (annexin V-/6-CFDA+) with living cells marked in green; 1.9 µg/mL (B) (annexin V+/6-CFDA-) with necrotic cells in red; 0.02 µg/mL (C) (annexin V+, 6-CFDA+) apoptotic cells in orange; and 0.1 ng/mL (D) (annexin V+, 6-CFDA+) early apoptotic cells in green-yellow. At this latter dosage, cell membranes are still undamaged, even though cells are genotoxically compromised and designated to death by exposure time. Figure 15 (E) shows the typical apoptotic morphological changes upon exposure to 0.1 ng/mL for 72 h, characterized by detachment, swelling to dense rounded mass, cell shrinkage, apoptotic bodies cluster formation "popcorn like" and cell death with final lysis stage.

Below the non effect zone (0.005 µg/mL to 0.2 ng/mL), the response started to deviate up and down from the control baseline, identifying the hormesis zone. At lower doses, an increase in the occurrence of tumorigenic transformed cells was observed by PCNA labelling (Fig. 16 A, B). Presumably, the genotoxic DNA lesions, induced during the apoptotic pathway, may determine a reduced efficiency in the cell cycle check point, so contributing the development of transformed cellular phenotypes and tumours. In fact, the persistence of mutant cells, which evade the apoptosis, leaded to an increased number and dimension of tumoral foci over time as doses lowered. High proliferation was registered after 72 h of exposure when AFB_1 was ranging from 0.2 pg to 0.001 pg/mL. This was confirmed by the number of mitotic cells positives for immunofluorescent labeling of DNA (Fig. 16 A, B, C).

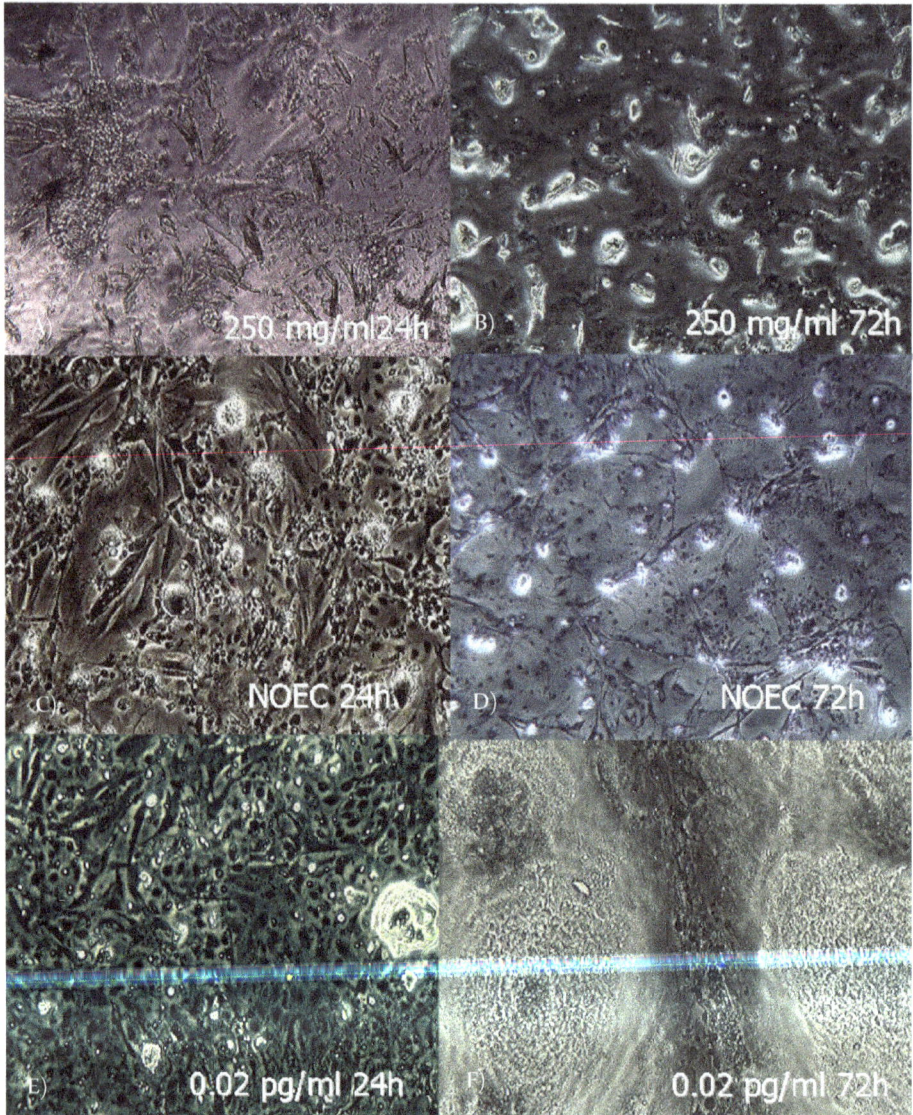

Fig. 14. AFB_1 damage on SaHePs after 24 h and 72 h of AFB_1 exposure (original magnification x200). A) and B) SaHePs exposed at 250 mg/mL; C) and D) SaHePs exposed at 5ng/mL; E) SaHePs exposed at 0.02 pg/mL after 24 h: cell shrinkage, pyknosis; F) SaHePs exposed at 0.02 pg/mL after 72 h: extensive cell proliferation with the loss of contact inhibition, and formation of multicellular overgrowth nodules (tumoral foci).

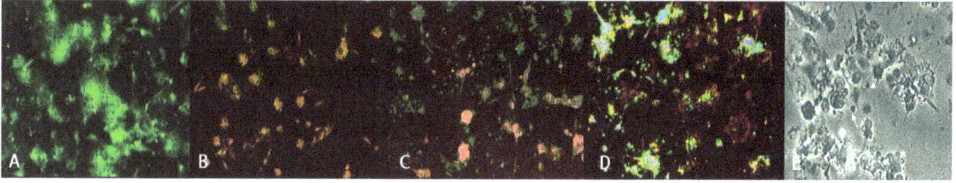

Fig. 15. Hepatocytes cultured in a 4-well slide exposed to AFB$_1$ for 24 h (original magnification x200)

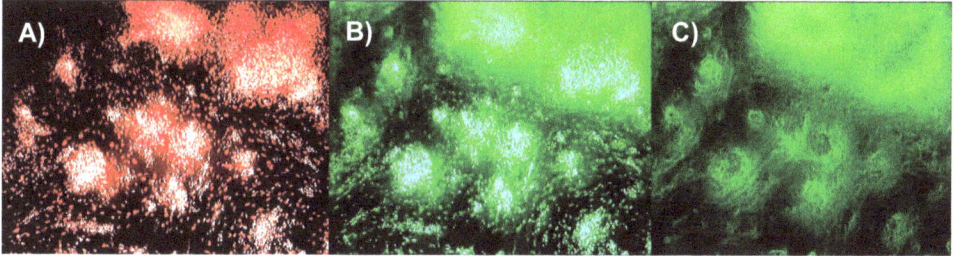

Fig. 16. SaHePs exposed to AFB$_1$ at dose of 0.001 pg/mL after 72 h. Tumoral foci formation: mitotic nuclei (white) are double immunolabelled with A) Ab anti PCNA-FITC and counterstained with PI-TRITC, resting nuclei are red; in B) and C) viable cells are green, B) PCNA-CFDA-FITC (merged); C) CFDA-FITC

High doses appeared to lead to a necrotic cell death due to mitochondrial impairment and membrane leakage. Low doses appeared to inhibit both apoptosis and cell cycle check point regulation leading to aberrant cell proliferation. Proceeding from high to low doses an hormetic zone was observed along the curve approaching the baseline which preludes to neoplastic transformation *in vitro*.

Sublethal and subcytotoxic concentrations of AFB$_1$ trigger apoptosis prior to induce necrosis, as assessed by the occurrence of a damage which is not recoverable, but permanent, even if the toxic insult is removed. Our results indicate almost three distinct pathways of cytotoxic response in AFB$_1$ treated seabream hepatocytes: necrotic cell death, apoptotic cell death, and uncontrolled cell proliferation. Such findings demonstrate that seabream hepatocytes are highly sensitive to AFB$_1$ exposure.

5.2 AFB$_1$ effects on *Vibrio fischeri*

In the last years, several biological models were used in both short- and long-term tests implementing various test organisms. The toxicological studies need for simple, inexpensive, rapid and sensitive test in order to screen an increasing number of chemicals and assess their acute and chronic effects (Fargasová, 1994; Ghosh et al., 1996; Radix et al., 2000). Bacteria are considered test organisms that offer good response to these necessity (Ghosh et al. 1996). Among the available bacterial assays, the standard Microtox® system is the most popular due to its rapidity, sensitivity, reproducibility, as well as cheap costs (Kwan & Dutka, 1990). This system measures the decrease of light emission by *Vibrio fischeri* (NRRL B-11177) after being exposed to chemicals, and it was successfully used for the toxicity evaluation of a large number of substances (Arufe et al., 2004; Fulladosa et al., 2007), contaminated water (Fernandez et al., 1995) and sediments (Narracci et al., 2009). This method is more sensitive than other acute toxicity tests (Weideborg et al., 1997) and can be used for the prediction of chemical toxicity in other aquatic organisms (Chen & Que Hee, 1995; Zhao et al., 1995).

V. fischeri is a Gram-negative, rod-shaped, flagellate, heterotrophic bacterium recovered in marine ecosystem, with bioluminescence capability (Fig. 17). Bioluminescence is produced by bacterial luciferin-luciferase system: there is a substrate (luciferin) oxidation in presence of enzyme (luciferase). Several factors, both external and proper of bacteria, are involved in the induction or inhibition of the enzymatic transcription system in the light emission.

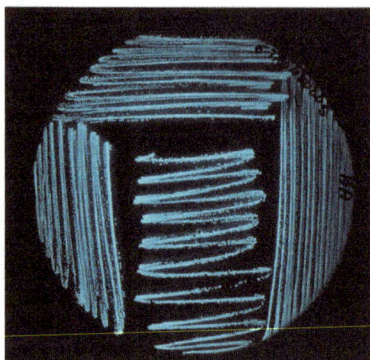

Fig. 17. *Vibrio fisheri* on Sea Water Complete (SWC) agar

The end point of Microtox® system is based on the evaluation of the light developed by this bacterium as an end product of its respiration. Any inhibition of cellular activity causes a change in the respiration rate and a corresponding variation of bioluminescence. Therefore, light emission can be considered as a signal of " health status": a toxic chemical can inhibit one of the several enzymes directly or indirectly involved in bioluminescence, leading to a gradual reduction of the light in a dose-dependent manner.

We built dose response curves measuring the bioluminescence emitted by *V. fischeri* upon AFB_1 exposure, and compared this data to those obtained from *in vitro* cell culture system, to lastly correlate the *in vitro* basal cytotoxicity data (ICs) with the validated EC_{50} value tested by the Microtox® system.

Toxicity screening was carried out by a Microtox® Model 500 Analyzer, equipped with a 30 well temperature controlled incubator, one reaction and one read well and interfaced with a PC equipped with the Microtox® Omni 1.16 software for Windows 98 for acquisition and data handling. The basic protocol employed a non-toxic control (blank) and four serial dilutions of the original sample. Reagent consisted of living luminescent bacteria grown in optimal conditions, harvested and lyophilized, and rehydrated with Reconstitution Solution to obtain a suspension of organisms for the test execution. The system measures bacterial light emission of a sample and compares it to the light emission of a control. The difference in light output is the effect of the sample on organisms.

For the toxic evaluation of AFB_1 was used the Microtox® Basic Test (BT) according to standard operating procedure (Azur, 1994). *V. fischeri* were exposed to a concentration range of AFB_1 from 0.1 pg/mL to 10 μg/mL. AFB_1 was diluted using diluent reagent for Microtox® and the Osmotic Adjusting Solution (OAS), necessary to correct the osmotic pressure of the sample to about 2% NaCl. The bacterial light emission was measured and compared to a control after 5, 15, 30 min and 3.5 h; the incubation temperature was of 15°C. An apparent time dependent response was observed in the range 2,5-10 µg/mL, where a decrease in the light emission corresponded to a toxic evaluation from high to very high, and in the dose range 0.1-0.5 pg/mL, where a change from a non toxic assessment to a stimulatory effect was observed (Fig. 18). A clear biostimulation was also evident in the range 0.312-0.468 µg/mL, whereas a transition from stimulatory effect to non toxic evaluation was detected at 0.005 and 0.006 µg/mL. Moreover, the increase of AFB_1 concentration determined a progressive decrease in bioluminescence which was related to an enhance of toxic degree, even reaching a very high level of toxicity. This linear trend was observed starting from a dose of 1.25 µg/mL. At lower AFB_1 concentrations an alternation between non toxic assessment and biostimulation was observed. Finally, the biostimulatory effect was clearly detected at very low concentrations (0.1-0.5 pg/mL), with a time dependent increase, while the maximum biostimulation was reached for all the time exposures at 0.312 µg/mL. The mean value of EC_{50} was of 2.53 µg/mL.

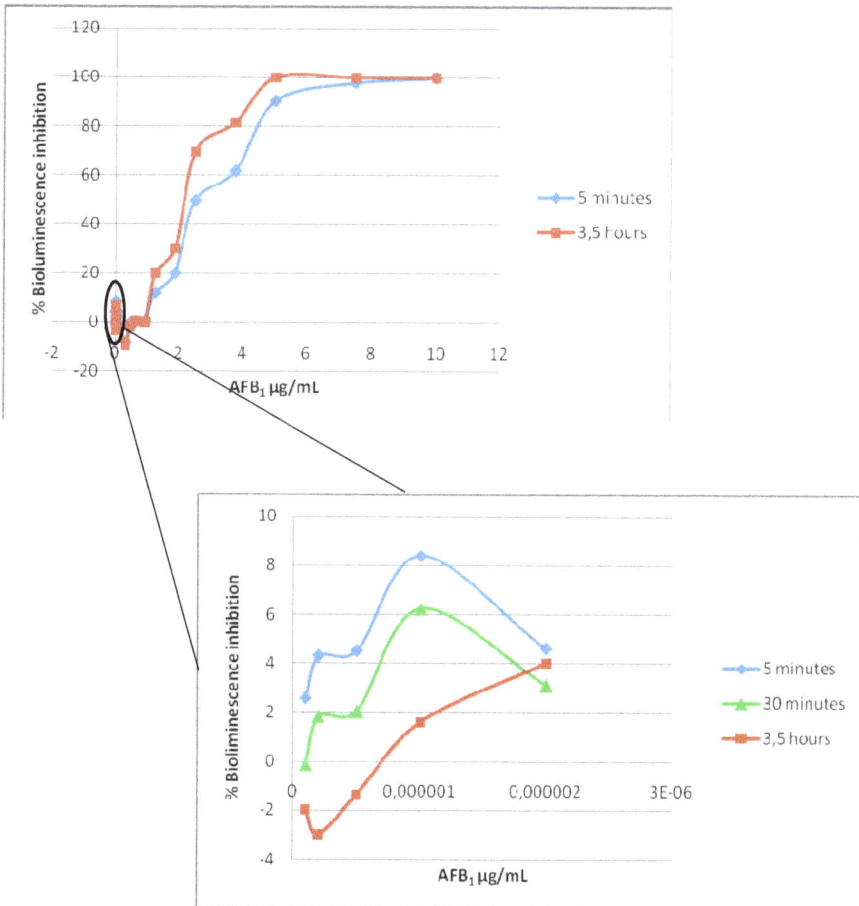

Fig. 18. Effects on bioluminescence of AFB_1 on *Vibrio fisheri* using Microtox®. Evidenced response at 0.1-2 pg/µL AFB_1 concentration.

These results shows that low concentrations of AFB_1 are able to increase luminescence intensity of bacteria compared to control, while higher concentrations are correlated with a toxic evaluation. Such evidence may be related to hormesis, which is defined as a stimulating and beneficial effect of substances at very low concentrations and harmful at high doses, or generally, a process with a biphasic trend that changes both in magnitude and in sign (Calabrese, 1999, 2002; Murado & Vázquez, 2007).

6. Conclusion

Comparing the results obtained by the two *in vitro* systems, the toxic responses are equivalent and overlapping. Therefore, this work could be considered a useful starting point for the design of new test batteries for the assessment of potentially toxic substances in aquafeed. Using an integrated approach of *in vitro* trials, during the early stages of exposure studies, can benefit *in vivo* experiments for acute and chronic exposure, determining a smaller number of better designed and targeted trials, and consequently reducing the number of sacrificed animals.

Based on the scarcity of published reports on the AFB_1 toxicity in aquacultured euryaline fish, we conclude that further research is needed in order to realize a quantitative comparison with other species. In order to provide a high level of public health protection, it would be useful to: a) investigate the bioaccumulation of AFB_1 and its metabolites in aquatic organisms of the food chain; b) investigate the quantitative correlation between AFB_1 levels found in aquafeed and the resulting residues of metabolites in fish flesh for human consumption. Overall, the finding herein presented will improve information on hazard identification and damage characterizations, providing new insights to investigate the real impact of aflatoxin B_1 on marine farmed teleosts.

Furthermore, the *in vitro* cytotoxicity cell culture based assay paralleled with the Microtox® system might provide a useful database of IC_{50}/EC_{50} values, functional to calculate or mathematically ipothesize the toxic levels of AFB_1 in other euryhaline species with unknown sensitivity. This is important for understanding the relative potencies of toxicants by using two different *in vitro* systems, and in different species.

7. References

Agag, B.I. (2004). Mycotoxins in foods and feeds Ass. *Univ Bull Environ Res*, Vol. 7, pp. 173–206

Arufe, M.J., Arellano, J., Moreno, M.J. & Sarasquete, C. (2004). Toxicity of a commercial herbicide containing terbutryn and triasulfuron to seabream (*Sparus aurata* L.) larvae: a comparison with the Microtox test. *Ecotoxicology and Environmental Safety*, Vol. 59, pp. 209-216, ISSN 0147-6513

Azur Environmental 1994. Microtox® M500 Manual (A toxicity testing handbook). Carlsbad, CA, USA

Berry, M.N., & Friend, D.S. (1969). High-yield preparation of isolated rat liver parenchymal cells. *The Journal of Cell Biology*, Vol. 43, pp. 506-520, ISSN 0021-9525

Bevelander, G.S., Hang, X., Abbink, W., Spanings, T., Canario, A.V.M., & Flik, G. (2006). PTHrP potentiating estradiol-induced vitellogenesis in sea bream (*Sparus auratus,* L.). *General and Comparative Endocrinology*, Vol. 149, pp. 159-165, ISSN 0016-6480

Birnbaum, M.J., Schulta, J., & Fain, J.N. (1976). Hormone-stimulated glycogenolysis in isolated goldfish hepatocytes. *American Journal of Physiology*, Vol. 231 pp. 191–197, ISSN 0363-6135

Blair, J.B., Miller, M.R., Pack, D., Barnes, R., The, S.J., & Hinton, D.E. (1990). Isolated trout liver cells: establishing short-term primary cultures exhibiting cell-to-cell interactions. *In Vitro Cellular & Developmental Biology*, Vol. 26, pp. 237–249, ISSN 1071-2690

Bouche. (1979). Isolation of carp hepatocytes by centrifugation on a discontinuous Ficoll gradient. A biochemical and ultrastructral study. *Biologie cellulaire*, Vol. 36, No 1, pp. 17-24, ISSN 0399-0311

Braunbeck, T., & Segner, H. (2000). Isolation and cultivation of teleost hepatocytes. In: *The hepatocytes review,* Berry, M.N. & Edwards, A.M., pp 49–71, Kluwer Academic Publishers, ISBN 0-7923-6177-6, Dordrecht, NL

Bulich, A. (1986). Bioluminescence assays. In: *Toxicity testing using microorganisms*, Bitton G, Dutka B, Vol. 1, pp 27–55, ISBN 0849352568, CRC, Boca Raton, FL

Cagauan, A.G., Tayaban, R.H., Somga, J., & Bartolome, R.M. (2004). Effect of aflatoxin-contaminated feeds in Nile tilapia (Oreochromis niloticus L.). *Proceedings of the 6th International Symposium on Tilapia in Aquaculture (ISTA 6).* Section: Health Management and Diseases, Manila, Philippines. 12–16 September

Calabrese, E. J. (2002). Hormesis: Changing view of the dose response, a personal account of the history and current status. *Mutation Research*, Vol. 511, pp. 181–189, ISSN 1383-5718

Calabrese, E.J. (1999). Evidence that hormesis represents an overcompensation response to a disruption in homeostasis. *Ecotoxicology and Environmental Safety*, Vol. 42, pp. 135-137, ISSN 0147-6513

Castaño, A., Vega, M., Blazquez, T., & Tarazona, J.V. (1994). Biological alternatives to chemical identification for the ecotoxicological assessment of industrial effluents: The RTG-2 in vitro cytotoxicity test. *Environmental Toxicology and Chemistry*, Vol. 13, No. 10, pp. 1607–1611, ISSN 0730-7268

Chavez-Sanchez, M.C., Martınez Palacios, C.A. & Osorio Moreno, I. (1994). Pathological effects of feeding young *Oreochromis niloticus* diets supplemented with different levels of aflatoxin B1. *Aquaculture*, Vol. 127, pp. 49 – 60, ISSN 0044-8486

Chen, H.F. & Que Hee, S.S. (1995). Ketone EC50 values in the Microtox_ test. *Ecotoxicology and Environmental Safety*, Vol. 30, pp. 120–123, ISSN 0147-6513

Chen, Z., Ding, Y., & Zhang, H. (2002). Morphology, viability and functions of suckling pig hepatocytes cultured in serum-free medium at high density. *Digestive surgery*, Vol. 19, pp. 184-191, ISSN 0253-4886

Clark, M.G., & Vincent, M.A. (2000). Preparation and properties of isolated hepatocytes from sheep. In: *The Hepatocyte Review*, Berry, M.N. & Edwards, pp. 27-36, Kluwer Academic Publishers, ISBN 0-7923-6177-6, Dordrecht, NL

Delmas, F., Villaescus, I., Woo, N.Y.S., Soleilhavoup, J.P., & Murat J.C. (2000). A cellular method for the evaluation of the noxiousness of inorganic pollutants in industrial wastes: Calculation of a safety index for monitoring sludge discharge. *Ecotoxicology and Environmental Safety,* Vol. 45, pp.260–265, ISSN 0147-6513

Do, J.H. & Choi. D.K. (2007). Aflatoxins: detection, toxicity and biosynthesis. *Biotechnology and Bioprocess Engineering*, Vol. 12, pp. 585-593, ISSN 1226-8372

Dragoni, I., Cantoni, C., Papa, A. & Vallone L. (2000). *Muffe, alimenti e micotossicosi.* Citta`Studi Edizioni, ISBN 88-251-7187-0, UTET Libreria srl, Milano, Italia.

Eaton, D.L. & Groopman, J.D. (Eds.). (1994). *The toxicology of aflatoxins: human health, veterinary, and agricultural significance*, Academic Press, San Diego

Ellis, R.W., Clements, M., Tibbetts, A., & Winfree, R. (2000). Reduction of the bioavailability of 20 mg/kg aflatoxin in trout feed containing clay. *Aquaculture*, Vol. 183, pp. 179–188, ISSN 0044-8486

El-Sayed, Y.S., & Khalil, R.H. (2009). Toxicity, biochemical effects and residue of aflatoxin B_1 in marine water-reared sea bass (*Dicentrarchus labrax* L.). *Food and Chemical Toxicology*, Vol. 47, pp. 1606–1609, ISSN 0278-6915

Fargasová, A. (1994). Comparative toxicity of five metals on various biological subjects. *Bulletin of Environmental Contamination and Toxicology*, Vol. 53, pp. 317-324, ISSN 0007-4861

Fegan, D. (2005). *Mycotoxins: the hidden menace?* http://www.alltech.com/

Fernández, A., Tejedor, C., Cabrera, F. & Chordi, A. (1995) Assessment of toxicity of river water and effluents by the bioluminescence assay using *Photobacterium phosphoreum*. *Water Research*, Vol. 29, pp. 1281–1286., ISSN 0043-1354

Fulladosa, E., Murat, J.C., Bollinger, J.C. & Villaescusa, I. (2007). Adverse effects of organic arsenical compounds towards *Vibrio fischeri* bacteria. *Science of the Total Environment*, Vol. 377, pp. 207-213, ISSN 0048-9697

Gaubin, Y., Vaissade, F., Croute, F., Beau, B., Soleilhavoup, J.P. & Murat, J.C. (2000). Implication of free radicals and glutathione in the mechanism of cadmium-induced expression of stress proteins in the A549 cell-line. *Biochimica et Biophysica Acta,* Vol. 1495, pp.4–13, ISSN 01674889

Ghosh, S.K., Doctor, P.B. & Kulkarni, P.K. (1996). Toxicity of zinc in three microbial test systems. *Environmental Toxicology and Water Quality*, Vol. 11, pp. 13-19, ISSN 1053-4725

Guguen-Guillouzo, C. (1992). Isolation and culture of animal and human hepatocytes. In: *Culture of Epithelial Cells*, Freshney, R.I., Vol. 1, pp. 197-223, Alan R. Liss Inc., ISBN 0-471-40121-8, Glasgow

Halver, J.E. (1969). Aflatoxicosis and trout hepatoma. In: *Aflatoxin*, Goldblatt L.A., pp. 265-306, Academic press, New York

Han, D., Xie, S., Zhu, X., Yang, Y. & Guo, Z. (2010). Growth and hepatopancreas performances of gibel carp fed diets containing low levels of aflatoxin B1. *Aquaculture Nutrition*, Vol. 16, pp. 335-342, ISSN 1353-5773

Han, X.Y., Huang, Q.C., Li, W.F., & Xu, Z.R. (2008). Changes in growth performance, digestive enzyme activities and nutrient digestibility of cherry valley ducks in response to aflatoxin B_1 levels. *Livestock Science*, Vol. 119, pp. 216–220, ISSN 1871-1413

Hassan, A.M., Kenawy, A.M., Abbas, W.T., & Abdel-Wahhab, M.A. (2010). Prevention of cytogenetic, histochemical and biochemical alterations in *Oreochromis niloticus* by dietary supplement of sorbent materials. *Ecotoxicology and Environmental Safety*, Vol. 73, pp. 1890–1895, ISSN 0147-6513

Hendricks, J.D. (1994). Carcinogenicity of aflatoxins in non mammalian organisms. In: *The toxicology of aflatoxins: human health, veterinary, and agricultural significance*, Eaton D.L., & Groopman J.D., pp. 103–136, Academic Press, New York

IARC. (1993). Some naturally occurring substances, food items and constituents, heterocyclic aromatic amines, and mycotoxins. *IARC Monographs on the Evaluation of Carcinogenic Risks to Humans*, Vol. 56, pp. 489–521, ISSN 1017-1606

Ince, N.H., Dirilgen, N., Apikyan, I.G., Tezcanli, G. & TstUn ,B. (1999). Assessment of toxic interactions of heavymetals in binary mixtures: A statistical approach. *Archives of Environmental Contamination and Toxicology*, Vol. 36, pp.365–372, ISSN 0090-4341

Jackson, L.S., Hlywka, J.J., Senthil, K.R. & Bullerman, L.B. (1996). Effect of thermal processing on the stability of fumonisins. *Advances in Experimental Medicine and Biology*, Vol. 392, pp. 345–353, ISSN 0065-2598

Jantrarotai, W. & Lovell, R.T. (1990). Subchronic toxicity of dietary aflatoxin B1 to channel catfish. *Journal of Aquatic Animals and Health*, Vol. 2, pp. 248 – 254, ISSN 0899-7659

Koebe, H.G., & Schildberg, F.W. (1996). Isolation of porcine hepatocytes from slaughterhouse organs. *The International Journal of Artificial Organs*, Vol. 19, pp. 53-60, ISSN 0391-3988

Kwan, K.K. & Dutka, B.J. (1990). Simple two step sediment extraction procedure for use in genotoxicity and toxicity bioassays. *Toxicity Assessment*, Vol. 5, pp. 395-404, ISSN 0884-8181

Lovell, R.T. (1992). Mycotoxins: hazardous to farmed fish. *Feed International*, Vol. 13, pp. 24–28

Meissonnier, G.M., Laffitte, J., Loiseau, N., Benoit, E., Raymond, I., Pinton, P., Cossalter, A.M., Bertin, G., Oswald, I.P., & Galtier, P. (2007). Selective impairment of drug metabolizing enzymes in pig liver during subchronic dietary exposure to aflatoxin B1. *Food and Chemical Toxicology*, Vol. 45, pp. 2145-2154, ISSN 0278-6915

Miller, J.D. & Trenholm, H.L. (1994). Mycotoxins in grain: compounds other than aflatoxins. Eagan Press, ISBN 0-9624407-5-2, St. Paul, MN

Mitaka, & Kon. (2004). EP 1 783 209 A1. *In*: EPC Bulletin 2007/19 - Selective Culture Method And Separation Method For Small Hepatocytes With The Use Of Hyaluronic Acid.

Mitaka, T., Kojima, T., Mizuguchi, T., & Mochizuki, Y. (1995). Growth and Maturation of Small Hepatocytes Isolated from Adult Rat Liver. *Biochemical and Biophysical Research Communications*, Vol. 214, pp. 310-317, ISSN 0006-291X

Mommsen, T.P., Moon, T.W., & Walsh, P.J. (1994). Hepatocytes: isolation, maintenance and utilization. In: *Biochemistry and Molecular Biology of Fishes*, Hochachka P.W., & Mommsen T.P., pp.356-373, Elsevier, Amsterdam

Moss, M.O. (1991). The environmental factors controlling mycotoxin formation, In: *Mycotoxins and animal foods*, Smith JE & Anderson RA, pp. 37–56 CRC Press, ISBN 0-8493-4904, Boca Raton, FL

Moss, M.O. (1998). Recent studies on mycotoxins. *Journal of Applied Microbiology Symposium Supplement*, Vol. 84, pp. 62–76, ISSN 1364-5072

Murado, M.A. & Vázquez, J.A. (2007). The notion of hormesis and the dose-response theory: a unified approach. *Journal of Theoretical Biology*, Vol. 244, pp. 489-499, ISSN 00225193

Murjani, G. (2003). Chronic aflatoxicosis in fish and its relevance.to human health. *Central Institute of Freshwater Aquaculture*, India

Narracci, M., Cavallo, R.A, Acquaviva, M.I., Prato, E. & Biandolino, F. (2009). A test battery approach for ecotoxicological characterization of Mar Piccolo sediments in Taranto (Ionian Sea, Southern Italy). *Environmental Monitoring and Assessment*, Vol. 148, No. 1, pp. 307-314, ISSN 0167-6369

Naylor, R.L., Hardy, R.W., Bureau, D.P., Chiu, A., Elliott, M., Farrell, A.P., Forster, I., Gatlin, D.M., Goldburg, R.J., Hua, K., & Nichols, P.D. (2009). Feeding aquaculture in an era of finite resources. *Proceedings of the National Academy of Sciences*, Vol. 106, pp. 15103–15110, ISSN 0027-8424

Pesonen, M., & Andersson, T. (1991). Characterization and induction of xenobiotic metabolizing enzyme activities in a primary culture of rainbow trout hepatocytes. *Xenobiotica,*Vol. 21, pp. 263–271, ISSN 0049-8254

Pestka, J.J. (2007). Deoxynivalenol: toxicity, mechanisms and animal health risks. *Animal Feed Science and Technology*, Vol. 137, pp. 283–298, ISSN : 0377-8401

Puschner, B. (2002). Mycotoxins. *Veterinary Clinics of North America Small Animal Practice*, Vol. 32, pp. 409– 419, ISSN 0195-5616

Radix, P., Léonard, M., Papantoniou, C., Roman, G., Saouter, E., Gallotti-Schmitt, S., Thiébaud, H. & Vasseur, P. (2000). Comparison of four chronic toxicity tests using algae, bacteria, and invertebrates assessed with sixteen chemicals. *Ecotoxicology and Environmental Safety*, Vol. 47, pp. 186-194, ISSN 0147-6513

Sahoo, P.K., Chattopadhyay, S.K. & Sikdar, A. (1996). Immunosuppressive effects of induced aflatoxicosis in rabbits. *Journal of Applied Animal Research*, Vol. 9, pp. 17 – 26, ISSN 0971-2119

Santacroce, M.P., Conversano, M.C., Casalino, E., Lai, O., Zizzadoro, C., Centoducati, G. & Crescenzo, G. (2008). Aflatoxins in aquatic species: metabolism, toxicity and perspectives. *Reviews of Fish Biology and Fisheries*, Vol. 18, pp. 99 – 130, ISSN 0960-3166

Santacroce, M.P., Zacchino, V., Casalino, E., Merra, E., Tateo, A., De Palo, P., Crescenzo, G. & Centoducati, G. (2010). Expression of a highly differentiated phenotype and hepatic functionality markers in gilthead seabream (*Sparus aurata* L.) long-cultured hepatocytes: first morphological and functional in vitro characterization. *Reviews in Fish Biology and Fisheries*, ISSN 0960-3166

Seglen, P.O. (1976). Preparation of isolated rat liver cells. *Methods in Cell Biology*, Vol. 13, pp. 31-83, ISSN 0091-679X

Segner, H. (1998). Isolation and primary culture of teleost hepatocytes. *Comparative biochemistry and physiology*, Vol. 120, pp. 71–81, ISSN 0300-9629

Sekiguchi, J. & Gaucher, G.M. (1977). Conidiogenesis and secondary metabolism in *Penicillium urticae*. *Applied and Environmental Microbiology*, Vol. 33, pp. 147–158, ISSN 0099-2240

Sharma, R.P. & Salunkhe, D.K. (1991). *Mycotoxins and phytotoxins*. CRC Press, Boca Raton, FL

Spring, P. (2005). Mycotoxins – a rising threat to aquaculture. *Feedmix*, Vol. 13, pp. 5

Stewart, D., & Larson, E. (2002). *Aflatoxicosis in wildlife*. Information Sheet 1582. Mississippi State Univ. Extension Service, Cooperating with U.S. Dept. of Agriculture

Tacon, A.G.J. & Metian, M. (2008). The Role of the Food and Agriculture organization and the Codex Alimentarius. *Annals of the New York Academy of Sciences*, Vol. 1140, pp. 50–59, ISSN 0077-8923

Wales, J.H. (1970). Hepatoma in rainbow trout. *Proceedings of A symposium on diseases of fishes and shellfishes*, Snieszko, S.F., No. 5, pp. 351-365, American Fisheries Society, Washington, D.C.

Weideborg, M., Vik, E.A., Ofjord, G.D. & Kjonno, O. (1997). Comparison of three marine screening tests and four Oslo and Paris Commission procedures to evaluate toxicity of offshore chemicals. *Environmental Toxicology & Chemistry*, Vol. 16, pp. 384–389, ISSN 0730-7268

Zhao, Y.H., He, Y.B. & Wang, L.S. (1995). Predicting toxicities of substituted aromatic hydrocarbons to fish by *Daphnia magna* or *Photobacterium phosphoreum*. *Toxicological and Environmental Chemistry*, Vol. 51, pp. 191–195, ISSN 0277-2248

Part 3

Approaches for Prevention and Control of Aflatoxins on Crops and on Different Foods

13

Comparative Evaluation of Different Techniques for Aflatoxin Detoxification in Poultry Feed and Its Effect on Broiler Performance

T. Mahmood[1], T.N. Pasha[1] and F.M. Khattak[2]
[1]Department of Food and Nutrition,
University of Veterinary and Animal Sciences, Lahore,
[2]Avian Science Research Centre, SAC, Edinburgh,
[1]Pakistan
[2]UK

1. Introduction

Aflatoxins (AF), the toxic secondary metabolites produced by *Aspergillus flavus* and *Aspergillus parasiticus*, are a major concern in the poultry production. AF metabolites are stable and fairly resistant compounds to degradation (Dalvi, 1986; Park, 2002; Desphande, 2002; Lesson *et al.*, 1995; Feuell, 1996). These metabolites are usually produced during the growth of the *Aspergillus flavus*, *Aspergillus parasitcus* and *Aspergillus nominus* on certain foods and feedstuffs under favourable conditions of moisture, temperature and aeration (Goto *et al.*, 1997; Dutta and Das, 2001). Their toxicity depends on several factors including its concentration, the duration of exposure, the species, sex, age, and health status of animals (Jewers, 1990). Contamination of AF in feed causes aflatoxicosis in poultry that is characterised by reduced feed intake, decreased weight gain, poor feed utilization (Tedesco *et al.*, 2004; Bailey *et al.*, 2006; Shi *et al.*, 2006, 2009), increased susceptibility to environmental and microbial stresses, and increased mortality (Leeson *et al.*, 1995). AF can also cause productive deterioration which is associated with changes in biochemical and hematological parameters (Denli *et al.*, 2004; Basmacioglu *et al.*, 2005; Bintvihok and Kositcharoenkul, 2006), liver and kidney abnormalities, and impaired immunity, which is able to enhance susceptibility to some environmental and infectious agents (Ibrahim *et al.*, 2000; Oguz *et al.*, 2003). AF has been reported to have effect on metabolism in poultry by decreasing the activities of several enzymes that are important in the digestion of starch, proteins, lipids and nucleic acids. Consequently, the activities of serum glutamate pyruvatate transaminase, serum gluatamate oxaloacetate tranferase and γ-glutamyl transferase are increased, primarily indicating hepatic damage (Devegowda and Murthy, 2005). AF is also known to interfere with metabolism of vitamin D, iron and copper and can cause leg weakness (Khajarern and Khajarern, 1999). Severe economic losses have been reported in the poultry industry due to aflatoxicosis (Kubena *et al.*, 1991, 1995). Ultimately, the transmission of AF and its metabolites from feed to animal edible tissues and products, such as liver and eggs, becomes a potential hazard for human health.

The occurrence of mycotoxin in nature is considered a global problem. However, in certain regions of the world, some of the mycotoxins are produced more commonly than others. Several *invitro* and *invivo* studies conducted in India, Pakistan, Egypt & South Africa suggested that AF are often present in substantial levels in mixed feed & ingredients (Devegowda and Murthy, 2005). Although, AF in feed and food is considered to be a major concern in warm and humid climatic regions of the world, however, caution must be exercised even in colder regions, when using feedstuffs imported from warm and humid countries.

With increasing knowledge and awareness of AF as a potent source of health hazards to both man and farm animals, producers, researchers and government organizations are making great effort to develop effective preventive management and decontamination technologies to minimise the toxic effects of AF content in foods and feedstuffs. In order to reduce the toxic and economic impact of mycotoxins, established regulations and legislative limits have been set for AF in poultry feed. Many countries follow a maximum permissible level of 20ppb for AF in poultry feed (CAST, 2003; FAO, 1995).

Appropriate pre and post-harvest contamination can be reduced by using appropriate agricultural practices. However, the contamination is often unavoidable and still remains a serious problem associated with many important agricultural commodities, which emphasizes the need for a suitable process to inactivate the toxin. Besides the preventive management, several approaches have been employed including physical (feed mill techniques, blending, extraction, irradiation, and heating), chemical (acids, bases, alkali treatments and oxidizing agents) biological treatments (certain species of fungi and bacteria) and solvent extraction to detoxify AF in contaminated feeds and feedstuffs (Coker *et al.*, 1986; Piva *et al.*, 1995; Parlat *et al.*, 1999). Since the beginning of 1990s, the adsorbent-based studies have also been reported to be effective in removing AF from contaminated feed and minimise the toxicity of AF in poultry (Ibrahim *et al.*, 2000). Among several adsorbents commercially available in the market, Zeolites (Miazzo *et al.*, 2000), bentonites (Rosa *et al.*, 2001, Pasha et al, 2007, 2008) and clinoptilolite (CLI), (Oguz and Kurtoglu, 2000; Oguz *et al.*, 2000 a, b), were preferred because of their high binding capacities for AF and their reducing effect on AF-absorption from the gastrointestinal tract.

All these methods cannot be used in practical feed manufacturing, because of the limitation of the nutrients decomposition, non availability of commercial methods and their residual effects. The increasing number of reports on detoxification of AF in poultry feed using different techniques has given rise to a demand for practical and economical detoxification procedures. Some of the physical treatments are reported to be relatively costly and may also remove or destroy essential nutrients in feed. Whereas, chemical methods are considered to be time consuming, expensive as they mostly require suitable reaction facilities, and are reported to have deteriorating residual effects on animal health (Coker, 1979; Coker *et al.*, 1985). Certain legal implications are also associated with the use of different detoxifying methods. For example, European community (EC) is in favour of use of physical decontamination processes and sorting procedures. However, neither the use of chemical decontamination processes, nor the mixing of batches with the aim of decreasing the level of contamination below the maximum tolerable level is legal within the European Union (Avantaggiato *et al*, 2005). Although several mycotoxin detoxitoxifying or adsorbing techniques have been assessed independently however, limited information is available on the comparison of different techniques. To further understand the mechanisms of aflatoxin and detoxification of poultry feed by heat treatment (extrusion) and added adsorbents

(Sodium bentonite or Mycofix®Plus), a study was conducted to compare different detoxification techniques and to further investigate its effect on broiler performance.

2. Materials and methods

2.1 Birds and diet
Two hundred day-old commercial broiler chicks were randomly distributed to 5 dietary treatments with 4 replicates of 10 chicks each. During the first 21 days, all birds were fed on diet 1 which was the basal starter ration without any aflatoxin contamination (AF) and detoxifying treatment (DT). The ingredient composition of the basal diet is presented in Table 1. Experimental diets were prepared by replacing maize with contaminated maize having 70 ppb AF (Treatment 2) and were subjected to different DT (Treatment 3 to 5).

Ingredient	Starter (%)	Grower (%)
Maize	50.70	60.00
Rice tips	10.00	10.00
Corn gluten meal 60 %	5.00	3.00
Soybean meal	10.00	6.00
Guar meal	5.00	5.00
Cotton seed meal	7.00	7.00
Rape seed meal	7.00	7.00
Fish meal	3.00	
Molasses	1.00	
Di-calcium Phosphate	1.00	1.00
Lysine	0.30	0.35
Calculated nutrient composition		
M E (kcal/kg)	3000	3100
Crude protein	22.00	19.00
Crude fiber	4.53	4.48
Lysine	1.08	0.93
Methionine	0.49	0.45
Cystine	0.27	0.27
Met+Cys	0.88	0.80
Linoliec acid	1.21	1.36
Calcium	1.00	0.92
Phosphorous total	0.75	0.69
Phosphorus available	0.44	0.41

Table 1. Ingredient and nutrient composition of basal starter and grower diet of broilers.

On day 21, birds were fed on one of five experimental diets. Experimental diets were fed from day 21 to day 42 of the trial. Feed and water was available on *ad libitum* basis. All the birds were vaccinated against Newcastle Disease (N.D.), with Lasota strain eye droppings at day 7 and with oil based vaccine (intra muscular) at day 21 of the experiment.

2.2 Experimental design
The experimental design consists of five dietary treatments; 1 (0 ppb AF & no DT); 2 (70 ppb AF & no DT); 3 (70 ppb AF & DT by Extrusion); 4 (70 ppb & DT by Mycofix®Plus); 5 (70 ppb & DT by Sodium bentonite).

2.3 Aflatoxin production and analysis

Aflatoxins were produced by the inoculation of fungus on corn as described by (Lillehoj, *et al.*, 1974) with some amendments. Fermentation was carried out in 1-liter Erlenmeyer flasks containing 50g of whole corn kernels. 25 ml distilled water was added to the corn (50 g) in the Erlenmeyer flasks, and the mixture was allowed to stand for 2 hrs with frequent shaking. The flasks were tightly plugged with cotton and autoclaved at 121ºC at 15 psi for 15 min and cooled at room temperature. They were then inoculated with 3ml spore suspension in a sterile environment, placed on an orbital shaker at 200rpm and incubated at 28 °C. At 24 and 48hr, sterile water (3-5ml) was added in the flask, quantity of water adjusted in a manner that individual kernels do not adhere with each other.

If the corn did pack in clumps, the material was loosened by vigorous shaking, and if required, clumps were smashed with the help of a sterile rod within sterile environment to make sure that individual kernel should be kept free from others. On 7-8d the flasks were again autoclaved at 121ºC at 15 psi for 15 min, and placed in a hot oven at 60 ºC for 24hr till all the moisture was removed.

The AF containing corn Kernels were grinded to powdered form and was quantitatively evaluated by direct competitive enzyme link immuno-sorbent assay (ELISA) described by Barabolak (1977) using (RIDASCREEN® FAST Aflatoxin Total) kit and mixed in feed according to the calculation to get the desirable level of aflatoxin (70 ppb) in the feed. The prepared experimental diets were analysed again using ELISA technique to confirm the AF levels.

2.4 Methods used for detoxification of AF
2.4.1 By extrusion

The AF containing corn was passed through the extruder, following the procedure described by Grehainge *et al*, 1983. The corn in the experimental starter and grower diets was replaced by extruded corn (Treatment 3).

2.4.2 By Mycofix

A commercially available Mycofix® plus was added in the experimental diet at the recommended dose rate of 0.5 Kg per ton of feed (Treatment 4).

2.4.3 By Sodium Bentonite (SB)

The source and composition of SB used was same as described previously (Pasha et al, 2007). The supplemental SB was added in the experimental diet at 1% of the feed (Treatment 5).

2.5 Sampling

Body weights and feed intake per pen was recorded weekly and mortality was recorded as it occurred. At day 28, 35 and 42 of bird's age, five birds per replicate were randomly selected for estimation of antibody titers against Newcastle disease (ND). The blood samples (3 ml) were collected from wing vein. The blood serum was separated and analysed by Haemagglutination inhibition (HI) method described by Sever (1962). After blood collection, birds were humanely killed and bursa of Fabricius was removed and weighed.

3. Statistical analysis

The results (pen means) were subjected to one–way ANOVA as a complete randomized design (CRD) using Genstat 11 for window. Treatment means were compared by the Tukey's test and statistical significance was accepted at $P < 0.05$

4. Results and discussion

4.1 Growth performance

The body weight gain, feed intake and FCR values showed no differences between different treatments (Tables 2). This lack of difference in performance between treatments could either be due to lower level of AF (70ppb) or shorter administrative period (day 21 to 42 of age). Similarly no change in production parameters have been reported previously in several studies (Ouz et al., 2000b, 2003; Ortatatli et al., 2005; Magnoli et al., 2008a,b, 2011) when birds were fed diets low in toxin (50 to 100 ppb AF) for a period of 46 days.

The results from the present study are not in agreement with other studies where significantly reduced body weights were observed when birds were exposed to higher dietary AF (400 and 600 ppb AF). The depression in growth upon feeding AF was attributed to reduced protein and energy utilization (Smith and Hamilton, 1970; Lanza et al. 1980; Doerr et al., 1983; Dalvi and Ademoyero, 1984; Verma et al., 2002) which impaired nutrient absorption and reduced pancreatic digestive enzyme production (Osborne and Hamilton, 1981) and consequently reduced appetite (Sharline et al, 1980). Similarly, significant depressions in body weight gain were also recorded in broilers given diets containing 1 and 2 mg/kg of AF (1000 to 2000 ppb) at 4 and 7 weeks of age. Several studies have also shown that dietary AF adversely affected the growth of broilers in a dose-dependent manner (Johri and Sadagopan, 1989; Espada et al., 1992; Beura et al., 1993). A similar reductions in weight gain were also observed in broilers at dietary AF contents of 0.75 mg/kg (750 ppb) and above and these depression in body weight in toxin fed groups were reported to be dose dependent (Reddy et al., 1984).

Treatment	DESCRIPTION	Weight Gain (g)	Feed Intake (g)	FCR
1	(0 ppb AF & no DT)	1566	3221	2.05
2.	(70 ppb AF & no DT)	1527	3215	2.10
3	70 ppb AF + DT by extrusion	1533	3249	2.11
4	70 ppb AF + DT with Mycofix®Plus	1544	3233	2.09
5	70 ppb AF + DT with Sodium bentonite	1564	3244	2.07
	SED	24.94	24.51	0.048
	P-value	0.444	0.608	1.00

Means within a column were not different (P < 0.05). Tukeys T test was used for means separation; SED – standard error of the difference.

Table 2. Weight gain, feed intake, and FCR of birds fed different experimental diets.

In contrast to our results, reduced feed intake and poor feed efficiency in broilers has also been reported in birds fed diets containing AF at 2, 4 and 6 weeks of age when level of dietary AF was higher than 100 ppb (Sharline et al. 1980; Huff and Doerr, 1981; Nandkumar et al. 1984; Rajasekhar Reddy et al. 1982; Johri and Majmudar, 1990; Verma et al., 2004). These authors have suggested that the reduced appetite during aflatoxicosis could be due to impaired liver metabolism caused by the liver damage. The reason for these differences in performance compared to present study could be due to different type, age and dose of AF. It is likely that juvenile birds may respond differently to the same dose of AF in diet as their physiological needs and capacity to absorb is higher compared to older birds.

It is suggested that extrusion-cooking is an efficient process used for eliminating some of the naturally occurring food toxins. The process involves high temperature (up to 250°C), short time (usually 1-2 min), high pressures (up to 25 mPa) and low water contents (below 30) and has been used to eliminate some of the toxins in food and feed ingredients (Harper, 1989; Fast, 1991; Kohlwey et al. 1995;). It has also been used as a kind of bioreactor to decontaminate AF. In spite the fact that feed moisture, barrel temperature and the die diameter are identified as different variables that can influence AF reduction during extrusion cooking, a reduction in total AF content (up to 84%) has been reported when artificially contaminated peanut meal was detoxified using extrusion cooking (Grehaigne et al., 1983; Cheftel, 1986, 1989). Similarly, commercial adsorbents are also reported to bind the aflatoxins in feed and prevent its absorption in animal gastrointestinal tract (Ramos and Hernandez, 1997). Numerous studies have shown the effectiveness of these agents to bind aflatoxins in vitro (Huff et al., 1992; Diaz et al., 2002). Mycofix, is one of the adsorbent that can be added in poultry feed and is claimed to neutralize moderate levels of aflatoxin (up to 2500-3500 ppb) in poultry feed. Mycofix deactivates aflatoxin with its polar functional group, due to AF fixation to adsorbing components in Mycofix, with stable binding capacity. Adsorption starts in the oral cavity during salivation and continues in stomach and gut. The fixed mycotoxin being unable to enter the blood and subsequently excreted in faeces after 98% adsorption of AF by Mycofix (Biomin®, 2000). Similarly, incorporation of SB in poultry diet is another proved adsorbent, to have high AF binding capacities both in vitro (Magnoli et al., 2008a) and in vivo (Rosa et al., 2001; Magnoli et al., 2008b). Bentonites are basically clays with strong colloidal properties that absorb water rapidly, which results in swelling and a manifold increase in volume, giving rise to a thixotropic, gelatinous substance (Bailey et al, 1998). Bentonites are composed of hydrated aluminosilicates of sodium (Na), potassium (K), calcium (Ca), and occasionally iron, magnesium, zinc, nickel, etc. They have a high negative charge and are balanced by cations such as Mg, K, and Na, therefore, they do not react with food/feed ingredients and act as inert material due to their neutral pH or slightly alkaline nature. However, the adsorption ability of these clays varies from one geological deposit to another.

In the current study, neither extrusion nor any of the absorbent (Mycofix®Plus and SB) resulted in any significant improvement in birds performance. The reason for lack of significant effects for DT methods used could probably be due to the performance of the birds on the AF containing diets (treatment 2). If the diet containing 70ppb AF (Treatment 2) had negatively influenced performance, it would be expected that DT methods used would restore or improve production. Therefore, this result does not imply that the DT methods used are not effective but rather indicate that birds exposed to higher levels of AF in diets are more likely to be benefited from the detoxified feed.

4.2 Antibody titre

The means of antibody titre (HA) against Newcastle disease (ND) showed no difference (P>0.05) between treatments when analysed at 21, 28, 35 and 42 day of the trial (Table 3).

The presence of AF in the feed is reported to decrease vaccinal immunity and may therefore lead to the occurrence of disease even in properly vaccinated flocks (Lesson et al., 1995). Aflatoxins have been associated to have immunosuppressive effect due to direct inhibition of protein synthesis, including those with specific functions such as immunoglobulins IgG, IgA, inhibition of migration of macrophages, interferance with the haemolytic activity of complement, reduction in the number of lymphocytes through its toxic effect on the Bursa of Fabricius and impairment of cytokines formation by lymphocytes (Tung *et al.*, 1975; Creppy *et al.*, 1979, Devegowda and Murthy, 2005). In present study, no difference (P>0.05) in HA titres was observed when treatment with AF (treatment 2) was compared with all other treatment groups suggests that birds exposed to 70ppb AF in diet do not show any signs of immunosuppression. Similarly, Gabal and Azzam, (1998) suggested that prolonged administration of AF at the low levels do not markedly change the hematological and serological parameters of broiler chickens, but may cause relevant lesions in liver and renal tissues. Moreover, the metabolism of broilers seems to be more adapted to high concentrations of aflatoxin in the feed when administered from 21 to 42 d of age, when compared with data reported from similar experiments conducted with broilers aging 1 to 21 d and with other species such as turkey poults (Pier *et al.*, 1979; Campbell *et al.*, 1988; Gabal and Azzam, 1998).

Treatment	Description	HA		
		Day 28	Day 35	Day 42
1	0 ppb AF & no DT	14.9	257.6	184.0
2	70 ppb AF & no DT	13.7	222.9	138.9
3	70 ppb AF + DT by extrusion	17.1	268.9	168.9
4	70 ppb AF + DT with Mycofix®Plus	13.9	237.2	168.9
5	70 ppb AF + DT with Sodium bentonite	16.5	245.3	184.4
	SED	4.65	23.6	17.65
	P-Value	0.924	0.396	0.143

Means within a column were not different (P < 0.05). Tukey's test was used for means separation; SED – standard error of the difference.

Table 3. Effects of experimental diets on haemagglutination titres (HA) against ND at day 28, 35 and 42 of the trial.

In contrast to our study dietary AF has been reported to cause vaccine failure as indicated by significantly reduced (P<0.05) antibody titres against Newcastle disease vaccine when birds were fed diets containing 2000 to 3000 ppb AF (Rathore *et al.*, 1987; Mangat *et al.*, 1988; Viridi *et al.*, 1989; Ghosh *et al.*, 1990; Bakshi, 1991; Mohiuddin and Reddy, 1993; Sharma, 1993; Mohiudin, 1993). Similarly, Azzam and Gabal (1998) reported that even low levels of dietary AF (200 ppb) can cause reduction in antibody titers to vaccines for Newcastle disease, infectious bronchitis, and infectious bursal disease in layers, when fed for a longer period (40 weeks). This difference in response to HA titres results could be attributed to the higher inclusion levels of dietary AF (200 to 3000 ppb) used in these studies.

The use of feed adsorbents is considered the most promising and economical approach for reducing mycotoxicosis in animals. The beneficial effect of Mycofix® has been reported to ameliorate the negative effect of AF on IBDV antibody titres and the effects are attributed to the presence of phytogenic substances, a hepatoprotective flavolignins (silymarin) in Mycofix, which prevents toxins from entering the liver cell membranes, and as it contains the terpenoid complexes, which reduce inflammations and protect the mucous membranes (Biomin®, 2000). Similarly, Ibrahim et al., (2000) reported that SB is also effective in ameliorating the suppressive effect of AF on the HI-titer in chicks vaccinated against Newcastle disease and the best result was obtained when SB was added at a rate of 0.4% of feed to the AF-containing diets. This effect was attributed to the role of SB as a sequestering agent against AF present in the diet through reducing its bioavailability in the gastrointestinal tract (Araba and Wyatt, 1991). However, in the present study, no differences (P>0.05) in ELISA titres were observed when birds fed AF diet (Treatment 2) were compared with all other treatment groups. This result further support our growth performance results and indicates that diets low in AF (70 ppb) do not depress broiler growth and vaccinal immunity.

4.3 Bursal body weight ratio (BBR)

The sensitivity of the immune system to mycotoxin-induced immunosuppression arises from the vulnerability of the continually proliferating and differentiating cells that participate in immune mediated activities and regulate the complex communication network between cellular and humoral components. AF are reported to inhibits the histological development and functional maturation of lymphoid organs (Celik *et al.*, 2000). Morphological evidence to explain the immunosuppressive effects of AF (2500 ppb) was documented by Celik *et al.* (2000) in broiler chickens after 21 days of feeding and the major signs were reduction in the weights of lymphoid organs including bursa of Fabricius, spleen and thymus. Similarly, Verma *et al.*, (2004) reported a significant decrease in the relative weight of the bursa of Fabricius when birds were exposed to diets having 2000 ppb AF. Similar reduction in BBR and moderate histopathological changes have been reported in broilers (Giambrone *et al.*, 1985; Marquez and Hernandez, 1995), laying hens (Dafalla *et al.*, 1987), ducks (Sell *et al.*, 1998; Khajarern and Khajarern, 1999) and wild turkeys (Quist *et al.*, 2000) when birds were fed diets having various levels of AF (100 to 500 ppb). In addition, vacuolation of liver cells and cellular depletion in the follicle medulla of the bursa Fabricii has been reported to be produced as an indication of aflatoxicosis by feeding lower levels of AF (100 ppb) over a long-term period of 42 days (Espada *et al.*, 1992). In contrast, the present study indicated no significant difference (P>0.05) in BBR between different treatment groups (Table 4). However, it cannot be concluded from the present investigation whether 70 ppb AF level in broiler diet can cause aflatoxicosis in broilers, as no significant difference (P>0.05) was observed when different response parameters tested were compared to the those of AF contaminated diet. This difference in results probably could be due to differences in age or genetic strain of birds, nutritional status, and source of mycotoxins, exposure time, vaccination schedule, serologic technique and management practices used in these studies.

Treatment	Description		BBR
1	0 ppb AF & no DT		1.69
2	70 ppb AF & no DT		1.54
3	70 ppb AF + DT by extrusion		1.77
4	70 ppb AF + DT with Mycofix®Plus		1.60
5	70 ppb AF + DT with Sodium bentonite		1.70
		SED	0.77
		P-value	0.998

Means within a column were not different (P < 0.05). Tukeys T test was used for means separation; SED – standard error of the difference..

Table 4. Effect of experimental diets on the average mean bursal body weight ratio (BBR) of birds at 42 days of age.

5. Conclusion

The manifestation and magnitude of a AF related response depends upon dose of AF, time period that the determined dose is exposed to the animal and interactions (such as age of animal, nutritional status at the time of AF exposure, presence of multiple mycotoxins in the diets etc). Prevention and control of AF in the poultry production chain requires the knowledge and consideration of all factors influencing mycotoxin formation in the field and during the storage of feedstuffs. The results from the current study demonstrated that growth performance and immune response was not depressed when broilers aged 21 to 42 days were exposed to diets containing 70ppb AF. However, methods of DT compared did not result in any significant improvement (P>0.05) in any of the response parameter. Further studies are recommended to evaluate the efficacy of the detoxifying agents by using a factorial designs that include a non-contaminated diet and a contaminated diet, both with and without DT.

6. References

Araba, M & Wyatt, RD. (1991). Effects of sodium bentonite hydrated aluminosilicate (Novasil) and charcoal on aflatoxicosis in broiler chickens. *Poultry Science*, 1(70): 6-11.

Avantaggiato, G., Solfrizzo, M., Visconti, A. (2005). Recent advances on the use of adsorbent materials for detoxification of Fusarium mycotoxins. *Food additives and Contaminants*, 22: 379-388.

Azzam A. H. & Gabal M. A. (1998). Aflatoxin and immunity in layer hens. *Avian Pathology*; 27:570-577.

Bailey, C. A., Latimer, G. W, A., Barr, C., Wigle W. L., Haq, A. U., Balthrop, J. E. & Kubena, L. F. (2006). Efficacy of montmorillonite clay (NovaSil PLUS) for protecting full-term broilers from aflatoxicosis. *Journal of Applied Poultry Science*, 15:198–206

Bailey, R.H., Kubena, L.F., Harvey, R.B., Buckley, S.A. & Rottinghaus, G.E. (1998). Efficacy of various inorganic sorbents to reduce the toxicity of aflatoxin and T-2 toxin in broiler chickens. *Poultry Science*, 77, 1630–1632.

Bakshi, C.S. (1991). Studies on the effect of graded dietary levels of aflatoxin on immunity in commercial broilers. *M.V.Sc. Thesis*, Indian Veterinary Research Institute, Izatnagar, U.P., India.

Barabolak, R.J. (1977). Improved procedure for quantitative determination of aflatoxin in corn and wet milled corn products. *AOAC*, 60 (2): 308.

Basmacioglu, H., Oguz, H., Ergul, M., Col, R. &. Birdane, Y. O. (2005). Effect of dietary esterified glucomannan on performance, serum biochemistry and haematology in broilers exposed to aflatoxin. *Czech Journal of Animal Science*, 50: 31–39.

Beura, C.K., Johri, T.S., Sadagopan, V.R. & Panda, B.K. (1993). Interaction of dietary protein level on dose response relationship during aflatoxicosis in commercial broilers. I. Physical responses livability and nutrient retention. *Indian Journal of Poultry Science*, 28: 170−177.

Bintvihok, A., and. Kositcharoenkul, S. (2006). Effect of dietary calcium propionate on performance, hepatic enzyme activities and aflatoxin residues in broilers fed a diet containing low levels of aflatoxin B_1. *Toxicon*, 47:41–46

Biomin®. (2000) Mycofix® Plus 3.0. A modular system to deactivate mycotoxins. *Biomin,* GTI GmbH. Herzogenburg, Austria.

Campbell, M.L., May, D., Huff, W.E. & Doer, J.A. (1988). Evaluation of immunity of young broiler chickens during simultaneous aflatoxicosis and ochratoxicosis. *Poultry Science,* 62: 2138-2144

Celik, I., Oguz, H., Demet, O., Donmez, H.H., Boydak, M. and Sur, E. (2000). Efficacy of polyvinylpolypyrrolidone in reducing the immunotoxicity of aflatoxin in growing broilers. *British Poultry Science,* 41: 430-439

Cheftel, J. C. (1986). Nutritional effects of extrusion-cooking. *Food Chemistry.* 20. 262-285.

Cheftel, J. C. (1989). Extrusion cooking and food safety. In *Extrusion Cooking* Eds. Mercier, C., Linko, P. and Harper, J.M. pp. 435±462. St Paul, Minnesota, USA: American Association of Cereal Chemists.

Coker, R.D. (1979). Aflatoxin : past, present and future.- *Tropical* Science, 21: 143-162.

Coker, R.D., Jewers, K. & Jones, B.D. (1986). The treatment of aflatoxin contaminated commodities. *International Biodeterioration,* 22S, pp. 103-1 08.

Coker, R.D., Jewers, K., Jones, B.D. (1985). The destruction of aflatoxin by ammonia : practical possibilities. *Tropical Science,* 25: 139-154.

Council for Agricultural Science and Technology (CAST). (2003). *Mycotoxins: Risks in plant, animal and human systems.* Task Force Report No 139. Ames, Iowa, USA.

Creppy, E.E., Lugnier, A.A.J., Fasiolo, F., Heller, K., Roschenthaler, R. & Dirheimer, G. (1979). In vitro inhibition of yeast phenylalanyl-t-RNA synthatase by ochratoxin A. *Chemical and Biological Interactions,* 24: 257-261.

Dafalla, R., Yagi, A.I. & Adam, S.E.I. (1987). Experimental aflatoxicosis in hybro-type chicks; sequential changes in growth and serum constituents and histopathological changes. *Veterinary and Human Toxicology* 29: 222–225.

Dalvi, R. R. & A. A. Ademoyero. (1984). Toxic effects of aflatoxin B_1 in chickens given feed contaminated with *Aspergillus flavus* and reduction of the toxicity by activated charcoal and some chemical agents. *Avian Disease,* 28: 61-69

Dalvi, R.R. (1986). An overview of aflatoxicosis of poultry. Its characteristics, prevention and reduction. *Veterinary Research and Communication,* 10: 429-443.

Denli, M., Okan, F. &. Doran, F. (2004). Effect of conjugated linoleic acid (CLA) on the performance and serum variables of broiler chickens intoxicated with aflatoxin B_1. *South African Journal Of Animal Sciences,* 34: 97–103

Desphande, S. S. (2002). Fungal toxins. In S. S. Desphande (Ed.), *Handbook of food toxicology,* New York: Marcel Decker. 387-456.

Devegowda, G & Murthy, T.N.K. (2005). Mycotoxins: Their effects in poultry and some practical solutions. In. *The Mycotoxin Blue book,* Ed: Diaz, D.E. pp: 25-57. Nottingham University press, United Kingdom:

Diaz, D. E., Hagler, W. M, Hopkins, B. A. & Whitlow, L. M. (2002). Aflatoxin binders I: In vitro binding assay for aflatoxin B_1 by several potential sequestering agents. *Mycopathologia,* 156: 223–226

Doerr, J.A., Huff, W.E., Webeck, C.J., Chhaloupka, G.W., May, J.D. & Merkley, J.W. (1983) Effects of low level chronic aflatoxicosis in broiler chickens. *Poultry Science,* 62: 1971 – 1977.

Dutta, T.K. & Das, P.I. (2001). Isolation of aflatoxigenic strains of aspergillus and detection of aflatoxin B1 from feeds in India, *Mycopathologia,* 151: 29-33.

Espada, Y., Domingo, M., Gomez, J. & Calvo, M.A. (1992). Pathological lesions following an experimental intoxication with aflatoxin B1 in broiler chickens. *Research in Veterinary Science*, 53: 275–279.

Fast, R.B. (1991) Manufacturing technology of ready-to-eat cereals. In *Breakfast Cereal and How They Are Made,* Eds. Fast, R.B. and Caldwell, E.F. pp. 15±42. St Paul, Minnesota, USA: American Association of Cereal Chemists.

Feuell, A. (1996). Aflatoxins in groundnuts. Part 9: Problems of detoxification. *Tropical Science*, 8: 61-70.

Food and Agricultural Organization.(1995). Worldwide Regulations for Mycotoxins. *FAO Food and Nutrition paper 64,* FAO, Vialedella Terme di Caracalla, 00100, Rome, Italy.pp.43.

Gabal, M. A. and Azzam, A. H. (1998). Interaction of aflatoxin in the feed and immunization against selected infectious diseases in poultry. II. Effect on one-day-old layer chicks simultaneously vaccinated against Newcastle disease, infectious bronchitis and infectious bursal disease', *Avian Pathology*, 27(3): 290 − 295

GenStat. 11th edn for windows. Hemel Hempstead, UK: VSN International Ltd.

Ghosh, R.C., Chauhan, H.V.S. & Roy, S. (1990). Immunosuppression in broilers under experimental aflatoxicosis. *British Veterinary Journal*, 146: 457 − 462.

Giambrone, J.J., Diener, U.L., Davis, N.D., Panangala, V.S. & Hoerr, F.J. (1985). Effects of aflatoxin on young turkeys and broilers chickens. *Poultry Science*, 64, 1678–1684.

Goto, T., Peterson, S. W., Ito, Y. & Wilkins, D. T. (1997). Mycotoxin producing ability of *Aspergillus tamari. Mycotoxins*, 44: 17-20.

Grehaigne, B., Chouvel, H., Pina, M., Graille, J. & Cheftel, J.C. (1983). Extrusion cooking of aflatoxin containing peanut meal with and without addition of ammonium hydroxide. *Lebensmittel-Wissenschaft und-Technologie*, 16: 317–322.

Harper, J.M. (1989). Food extruders and their applications. In *Extrusion Cooking,* Ed. Mercier, C., Linko, P. and Harper, J.M. pp. 1±16. St Paul, Minnesota, USA: American Association of Cereal Chemists.

Huff, W.E. & Doerr, J.A. (1981). Synergism between aflatoxin and ochratoxin A in broiler chickens. *Poultry Science*, 60: 550 − 555.

Huff, W.E., Kubena, L.F., Harvey, R.B. & Phillips, T.D. (1992). Efficacy of hydrated sodium alluminosilicate to reduce the individual and combined toxicity of aflatoxin and ochratoxin A. *Poultry Science*, 71: 64 − 69.

Ibrahim, I.K., Shareef, A.M., Al-Joubory, K.M.T. (2000). Ameliorative effects of sodium bentonite on phagocytosis and Newcastle disease antibody formation in broiler chickens during aflatoxicosis. *Research in Veterinary Science,* 69: 119–122.

Jewers, K. (1990). Mycotoxins and their effect on poultry production. Options *Méditerranéennes*, Sér. A/No. 7:195–202.

Johri, T.S. & Majumdar, S. (1990). Effect of methionine, choline, BHT, supplemented aflatoxic diets, *Proc. XIII National Symposium of Indian Poultry Science.*, Assoc, held at Bombay on 20-22 Dec., 1990.

Johri, T.S. & Sadagopan, V.R. (1989). Aflatoxin occurrence in feedstuffs and its effect on poultry production. *Journal of Toxicology and Toxin Reviews*, 8: 281 − 287.

Khajarern, J. & Khajarern, S. (1999). Positive effects of Mycosorb against aflatoxicosis in ducklings and broilers. In: Poster presentation at *Alltech's 15th Annual Symposium on Biotechnology in the Feed Industry*, Lexington, KY.

Kohlwey, D.E., Kendall, J.H.& Mohindra, R.B. (1995). Using the physical properties of rice as a guide formulation. *Cereal Food World*, 40: 728-732.

Kubena, L.F., Edrington,T.S., Kamps-Holtzapple, Harvey, C. R., Elissalde, M.H. & Rottinghaus, G.E. (1995). Effect of feeding fuminosin B1 present in *Fusarium moniliforme* culture material and aflatoxin singly and in combination to turkey poults. *Poultry Science*, 74: 1295-1303.

Kubena, L.F., Huff, W.E., Harvey, R.B., Yersin, A.G., Elissad, M.H., Witzel, D.A., Giroir, L.E., Philips, T.D. & Petersen, H.D. (1991). Effect of a hydrated sodium calcium aluminosilicate on growing turkey poults during aflatoxicosis. *Poultry Science*, 70: 1823-1930.

Lanza, G.M., Washburn K.W. & Wyatt, R.D. (1980). Variation with age in response of broilers to aflatoxin. *Poultry Science*, 59(2): 282-288.

Leeson, S., Diaz, G., & Summers, J.D. (1995). Aflatoxins. In: Leeson, S., Diaz, G., Summers, J.D. (Eds.), *Poultry Metabolic Disorders and Mycotoxins*. pp. 248–279. University Books, Canada, Ont.

Lillehoj, E. B., Garcia, W. J. & Lambrow M. (1974). *Aspergillus flavus* infection and aflatoxin production in corn: influence of trace elements. *Applied Microbiology*, 28 (5): 763-767.

Magnoli, A.P., Cavagglieri, L., Magnoli, C., Monge, J.C., Miazzo, R., Peralta, M. F., Salvano, M., Rosa, R.C.A., Dalcero, A. &. Chiacchiera, S.M. (2008a). Bentonite performance on broiler chickens fed with diets containing natural levels of aflatoxin B_1. *Revue de Medecine Veterinaire*, 30: 55–60.

Magnoli, A.P., Monge, M.P., Miazzo, R.D. , Cavaglieri, L.R., Magnoli, C.E., Merkis C.I., Cristofolini, A.L., Dalcero, A.M., &. Chiacchiera, S.M. (2011). Monensin affects the aflatoxin-binding ability of a sodium bentonite. *Poultry Science*, 90: 48-52.

Magnoli, A.P., Tallone, L., Rosa, R.C.A., Dalcero A.M., Chiacchiera S.M., &. Torres Sanchez R.M. (2008b). Structural characteristic of commercial bentonites used as detoxifier of broiler feed contamination with aflatoxin. *Applied Clay Science*, 40: 63–71.

Mangat, A.P.S., Gill, B.S. & Maiti, N.K. (1988). Immunosuppressive effect of low level of aflatoxin against Ranikhet disease in chicken. *Indian Journal of Comparative Microbiology, Immunology and Infectious Diseases*, 10: 25 – 30.

Marquez, R.N.M. & Hernandez, T.R. (1995). Aflatoxin adsorbent capacity of two Mexican aluminosilicates in experimentally contaminated chick diet. *Food Additives and Contaminants*, 12: 431–433.

Miazzo, R., Rosa, C.A., De Queiroz Carvalho, E.C., Magnoli, C., Chiacchiera, S.M., Palacio, G., Saenz, M., Kikot, A., Basaldella, E. & Dalcero, A. (2000). Efficacy of synthetic zeolite to reduce the toxicity of aflatoxin in broiler chicks. *Poultry Science*, 79: 1–6.

Mohiuddin, S.M. & Reddy, V.M. (1993). Immunosuppressive effect of aflatoxin in bursectomised chicks against Ranikhet disease vaccine. *Indian Journal of Animal Sciences*, 63: 279 – 280.

Mohiudin, S M. (1993). Effects of aflatoxin on immune response in viral diseases. *Poultry adviser*, 26: 63-66

Nandakumar Reddy, D. Rao, P.V., Reddy, V.R. & Yadgiri, B. (1984). Effect of selected levels of dietary aflatoxin on the performance of broiler chicken. Indian *Journal of Animal Science*. 54(1): 68-73.

Oguz, H. & Kurtoglu, V. (2000). Effect of clinoptilolite on fattening performance of broiler chickens during experimental aflatoxicosis. *British Poultry Science*, 41: 512–517.

Oguz, H., Hadimli, H.H., Kurtoglu, V., & Erganis, O. (2003). Evaluation of humoral immunity of broilers during chronic aflatoxin (50 and 100 ppb) and clinoptilolite exposure. *Revue de Medicine Veterinaire* 154: 483–486.

Oguz, H., Kecec, T., Birdane, Y.O., Onder, F. & Kurtoglu, V. (2000a). Effect of clinoptilolite on serum biochemical and haematological characters of broiler chickens during experimental aflatoxicosis. *Research in Veterinary Science*, 69: 89–93.

Oguz, H., Kurtoglu, V. & Coskun, B. (2000b). Preventive efficacy of clinoptilolite in broilers during chronic aflatoxin (50 and 100 ppb) exposure. *Research in Veterinary Science*, 69: 197–201.

Ortatatli, M., Ouz, H., Hatipoglu, F. & Karaman, M. (2005). Evaluation of pathological changes in broilers during chronic aflatoxin (50 and 100 ppb) and clinoptilolite exposure. *Research Veterinary Science*, 78: 61–68.

Osborne, D.J. & Hamilton, P.B. (1981). Decreased pancreatic digestive enzymes during aflatoxicosis. *Poultry* Science, 60: 1818-1821.

Park, DL. (2002). Mycotoxins and food safety. In: *Advances in experimental medicine and biology* Eds: DeVries JW; Trucksess MW; Jackson LS 504: 173-179

Parlat, S.S., Yildiz, A.O. & Oguz, H. (1999). Effect of clinoptilolite on fattening performance of Japanese quail (Coturnix coturnix japonica) during experimental aflatoxicosis. *British Poultry Science*, 40: 495–500.

Pasha, T.N., Farooq M.U., Khattak, F.M.,. Jabbar ,M.A, &. Khan. A.D. (2007). Effectiveness of sodium bentonite and two commercial products as aflatoxin absorbents in diets for broiler chickens. *Animal Feed Science & Technology*, 132:103–110.

Pasha, T.N., Mahmood, A., Khattak, F.M., Jabbar ,M.A, &. Khan. A.D. (2008). The effect of feed supplementation with different sodium bentonite treatments on broiler performance. *Turkish Journal of Veterinary and Animal Sciences*, 32: 245-248.

Pier, A.C., Richard, J.L. & Thurston, J.R. (1979). The influence of mycotoxins on resistance and immunity. In *Interaction of Mycotoxins in Animal Production*, (pp. 96-117). Washington, DC: National Academy of Sciences.

Piva, G., Galvano, F.P.F., Pietri, A. & Piva, A. (1995). Detoxification methods of aflatoxins. *A Review of Nutrition Research*.15: 767–776.

Quist, C.F., Bounous, D.I., Kilburn, J.V., Nettles, V.F. & Wyatt, R.D. (2000.) The effect of dietary aflatoxin on wild turkey poults. *Journal of Wildlife Disease*, 36: 236–444.

Rajashekhara Reddy, A., Reddy, V.R., Rao, P.V. & Yadgiri, B. (1982). Effect of experimentally induced aflatoxicosis on the performance of commercial chicks. *Indian Journal of Animal Science*, 52: 405-410.

Ramos, A. J., and E. Hernandez. (1997). Prevention of aflatoxicosis in farm animals by means hydrated sodium calcium aluminosilication addition to feedstuffs. A review. *Animal. Feed Science & Technology*, 65: 197–206.

Rathore, B.S., Verma, K.C., Singh, S.D. & Khera, S.S. (1987) Epidemiological studies on Ranikhet disease vaccine failure in chickens. *Indian Journal of Comparative Microbiology, Immunology and Infectious Diseases*, 8: 175−177.

Reddy, P.S., Reddy, C.V., Reddy, V.R. & Rao, P.V. (1984) Occurrence of aflatoxin in some feed ingredients in three geographical regions of Andhra Pradesh. *Indian Journal of Animal Science*, 54: 235−238.

Rosa, C.A., Miazzo, R., Magnoli, C., Salvano, M., Chiac, S.M., Ferrero, S., Saenz, M., Carvalho, E.C. & Dalcero, A. (2001). Evaluation of the efficacy of bentonite from the south of Argentina to ameliorate the toxic effects of aflatoxin in broilers. *Poultry Science*, 80: 139–144.

Sell, S., Xu, K.L., Huff, W.E., Kubena, L.F., Harvey, R.B. & Dunsford, H.A. (1998). Aflatoxin exposure produces serum a fetoprotein elevations and marked oval cell proliferation in young male Pekin ducklings. *Pathology* 30: 34–39.

Sever, J. L. (1962). Application of a microtechnique to viral serological investigations. *Journal of Immunology*, 88; 320-329.

Sharline, K.S.B., B.J. Howarth & R.D. Wyatt. (1980). Effect of dietary aflatoxin on reproductive chicks. *Poultry Science*, 72: 651-657.

Sharma R. P. (1993). Immuntoxicity of mycotoxins. *Journal of Dairy Science*, 76: 892- 897.

Shi, Y. H., Xu, Z. R., Feng, J. L. &. Wang, C. Z. (2006). Efficacy of modified montmorillonite nanocomposite to reduce the toxicity of aflatoxin in broiler chicks. *Animal Feed Science Technology*. 129: 138–148

Shi, Y., Xu, Z., Sun, Y., Wang, C.&. Feng, J. (2009). Effects of two different types of motmorillonite on growth performance and serum profiles of broiler chicks during aflatoxicosis. *Turkish Journal of Veterinary and Animal Science*, 33: 15–20.

Smith, J.W. & Hamilton, P.B. (1970). Aflatoxicosis in broiler chickens. *Poultry Science*, 49: 207−215.

Tedesco, D., Steidler S., Galletti, S., Taneni, M., Sonzogni O.& Ravarotto, L.(2004). Efficacy of silymarin-phospholipid complex in reducing the toxicity of aflatoxin B_1 in broiler chicks. *Poultry Science*, 83:1839–1843

Tung, H.T., Wyatt, R.D., Thaxton, P. & Hamilton, P.B. (1975) Concentration of serum proteins during aflatoxicosis. *Toxicology and Applied Pharmacology*, 34: 320−326.

Verma, J., Johri, T. S. , Swain, B. K. and Ameena, S. (2004) 'Effect of graded levels of aflatoxin, ochratoxin and their combinations on the performance and immune response of broilers. *British Poultry Science*, 45 (4):512-518.

Verma, J., B. K. Swain, and T. S. Johri. (2002). Effect of various levels of aflatoxin and ochratoxin A and combinations thereof on protein and energy utilisation in broilers. *Journal of Science Food and Agriculture*, 82:1412–1417.[

Virdi, J.S., Tiwari, R.P., Saxena, M., Khanna, V., Singh, G., Saini, S.S. & Vadehra, D.V. (1989) Effect of aflatoxin on immune system of the chick. *Journal of Applied Toxicology*, 9: 271-275.

14

Phytoinhibition of Growth and Aflatoxin Biosynthesis in Toxigenic Fungi

Abdolamir Allameh[1], Tahereh Ziglari[2] and Iraj Rasooli[3]
[1]Department of Biochemistry, Faculty of Medical Sciences,
Tarbiat Modares University, Tehran,
[2]Faculty of Medicine, Islamic Azad University, Qeshm International Branch,
Qeshm Island
[3]Faculty of Basic Sciences, Shahed University, Tehran,
I. R. Iran

1. Introduction

Aflatoxins are primarily produced by the fungi *Aspergillus flavus* and *Aspergillus parasiticus*, which contaminate a wide variety of food and feed commodities including maize, oilseeds, spices, groundnuts, tree nuts, milk and dried fruits [Strosnider et al., 2006].

Presence of aflatoxins in food chain is associated with decrease in quality and quantity of food and feed materials. In addition, consumption of aflatoxin-contaminated products can pose a risk of development of various diseases in human and animals. Aflatoxins are produced in toxigenic fungi after undergoing biosynthesis pathway involving several enzymes and reactions. Upon consumption of aflatoxin contaminated products by human and animals, the toxin undergoes metabolism via cytochrome P450 enzymes in the liver. Aflatoxin metabolism in mammalian organs is a committed process and different metabolites are produced which can exert adverse effects of toxic metabolites. Aflatoxin epoxide (8,9-epoxide) is the major toxic metabolite which can bind to DNA and induce hepatocellular carcinoms. The extent of aflatoxin toxicity and carcinogenicity in human and animals depends on several factors including the metabolic capacity of the organism. Aflatoxin contamination of food products is associated with health and socioeconomic costs which is difficult to valuate in the developing countries. Moreover, the current regulations do little to help reduce aflatoxin and related health effects. Therefore the focus should be on promoting the adaptation of strategies that can control aflatoxin and its associated health risks. According to Wu and Khlangwiset (2010), interventions to reduce aflatoxin-induced illness can be grouped into three categories; agricultural, dietary and clinical. Agricultural interventions are methods that can be applied either in the field (preharvest) or in drying, storage and transportation (postharvest) to reduce aflatoxin levels in food. The dietary and clinical interventioans are considered as secondary interventions by which the aflatoxin-related illness can be reduced. These two types interventions are associated with advantages and disadvantages.

Due to concern for the potential effects of aflatoxins on human health, most countries have legislation that restricts marketing of aflatoxin-contaminated grains [Van Egmond, 1989]. The United States Food and Drug Administration has set an aflatoxin limit of 20 µg/kg for foods and for most feeds and feed ingredients. The European Union has enacted a very stringent aflatoxin tolerance threshold of 2 μg/kg aflatoxin B1 and 4 μg/kg total aflatoxins for nuts and cereals for human consumption [Bankole and Adebanjo, 2003].

The objective of the present article is to review different approaches by which aflatoxins can be reduced or eliminated in the food chain. The feasibility and the safety of aflatoxin detoxification process in food materials depend on different factors. The safety issue of food products that undergo detoxification treatment could be improved by using phytochemical agents with potential antimicrobial activities.

One of the characteristics of aflatoxin inactivation processes is that it should destroy the mycelia and spores of the toxic fungi, which may proliferate under favorable condition. The pH and moisture content of the foodstuffs have been reported as the main abiotic factors affecting the fungal infestation. [Prakash et al., 2011]. The chemical profile of the substrate may also play a major role in the growth and proliferation of moulds on the foodstuffs as has been emphasized by Singh et al. (2008).

Because of the toxic and carcinogenic potential of aflatoxins, much emphasis has been focused on the control or elimination of these fungi and/or their toxic metabolites in food grains and livestock feeds. Cultural practices, such as adjustments of sowing and harvesting time can be effective to a certain extent in preventing pre-harvest aflatoxin contamination. However, in case of inappropriate storage conditions, the fungi can invade the grains causing serious damage and toxin accumulation in the grains. Though some of the fungicides are effective in preventing the growth of *Aspergillus flavus* in storage especially as a fumigant [Paster et al., 1995], consumer concerns about possible risks associated with the use of fungicides have resulted in an intensive search for safer and more effective control options that pose minimal risk to human health and the environment. [Velazhahan et al., 2010,].

Whichever decontamination strategy is used, it must meet some basic criteria [Park, 1993; Beaver, 1991; Pomeranz et al., 1990]:

• The mycotoxin must be inactivated (destroyed) by transformation to non-toxic compounds.
• Fungal spores and mycelia should be destroyed, so that new toxins are not formed.
• The food or feed material should retain its nutritive value and remain palatable.
• The physical properties of raw material should not change significantly.
• It must be economically feasible (the cost of decontamination should be less than the value of contaminated commodity).

Principally there are three possibilities to avoid harmful effects of contamination of food and feed caused by mycotoxins:

• Prevention of contamination,
• Decontamination of mycotoxin-containing food and feed,
• Inhibition of absorption of mycotoxin in consumed food in the digestive tract [Bata et al 1999].

2. Mycology

It is well established that not all molds are toxigenic and not all secondary metabolites from molds are toxic. About 300 different secondary metabolites are known [Bhatnagar et al., 2002], however only a few of them play a role as contaminants in food. These are especially aflatoxins, trichothecenes, fumonisins, ochratoxin A and patulin [Bennett et al., 2003]. For all of them statutory limits have been set or are under discussion within the European Union. The most important fungal genera, which produce these mycotoxins, belong to the genera *Aspergillus, Penicillium* or *Fusarium*. Aflatoxin is produced mainly by toxigenic strains of *Aspergillus flavus, Aspergillus.parasiticus and Aspergillus nomius* [Beck et al., 1990; Karolewiez and Geisen, 2005; Kimura et al., 2003; O'Callaghan et al., 2003; Proctor et al., 2003; Penalva and Arst Jr. et al., 2002; Yu et al., 2004a]. Accumulation of aflatoxin B has been reported from members of three diffrent groups of *Aspergilli*: *Aspergillus* section Flavi: *Aspergillus flavus, Aspergillus flavus var. parvisclerotigenus, Aspergillus parasiticus, Aspergillus toxicarius, Aspergillus nomius, Aspergillus pseudotamarii, Aspergillus zhaoqingensis, Aspergillus bombycis. Aspergillus* section Nidulantes: *Emericella astellata and Emericella venezuelensis. Aspergillus* section Ochraceorosei: *Aspergillus ochraceoroseus* and *Aspergillus rambellii*. G type aflatoxins have only been found in some of the spices in *Aspergillus* section Flavi, while B type aflatoxins are common in all three groups. However it is a well known fact that the presence of a mycotoxigenic fungus in a food sample does not ultimately indicate the production of the respective mycotoxin. The biosynthesis of secondary metabolites, like the mycotoxins, is tightly regulated depending on environmental conditions like substrate, pH, water activity or temperature [Hope et al., 2005]. These facts may suggest, that for a complete assessment of the mycotoxicological status of a food, not merely the detection of a putative mycotoxin producing fungus is important, but the knowledge about the ability of the fungus to activate mycotoxin biosynthesis genes under the environmental conditions suitable for the food chain. [Schmidt-Heydt et al., 2007]. Factors contributing to the presence or production of mycotoxins in foods or feeds include storage, environmental, and ecological conditions.

2.1 Biosynthesis
The structure of aflatoxins consists of a coumarin nucleus attached to a bifuran and either pentanone (aflatoxin B1 and aflatoxin B2) or a six-membered lactone (aflatoxin G1 and aflatoxin G2). Aflatoxin B1, B2, G1, and G2 are the four main naturally-occurring aflatoxins, among which aflatoxin B1 ($C_{17}H_{12}O_6$) is known to be the most significant in terms of animal and human health risk [Pier, 1992; Coulombe 1993, Bluma et al., 2008].

Aflatoxins belong to the polyketide class of secondary metabolites produced by toxigenic strais of *Aspergillus flavus* and *Aspergillus parasiticus,* and are synthesized by enzymes encoded within a large gene cluster [Yabe and Nakajima 2004; Yu et al., 2004b]. As shown in figure 1, the initial step in the generation of the polyketide backbone of aflatoxins is proposed to involve polymerization of acetate and nine malonate units (with a loss of CO_2) by a polyketide synthetase in a manner analogous to fatty acid biosynthesis [Dutton, 1988; Bhatnagar et al., 1992]. Aflatoxin synthesis is controlled by different enzymes which are expressed through gene expression processes. Genetic studies on aflatoxin biosynthesis in *Aspergillus flavus* and *Aspergillus parasiticus* led to the cloning of 25 clustered genes within a 70 kb DNA region responsible for the enzymatic conversions in the aflatoxin biosynthetic pathway. Regulatory elements such as aflR and aflS (aflJ), nutritional and environmental factors, fungal developmental and sporulation were also found to affect aflatoxin formation. In *Aspergillus flavus* there are eight chromosomes with an estimated genome size of about 33–36 Mbp that harbor an estimated 12,000 functional genes [Yu et al., 2004b].

Many inhibitors of aflatoxin biosynthesis may act at three levels: (1) Modulate environmental and physiological factors affecting aflatoxin biosynthesis, (2) inhibit signaling circuits upstream of the biosynthetic pathway, or (3) directly inhibit gene expression or enzyme activity in the pathway. The known inhibitory compounds either alter known environmental and physiological modulators of aflatoxin biosynthesis or they alter signal transduction pathways in the upstream regulatory network [Holmes et al., 2008]. Each step in gene expression, transcription, RNA transport and processing, translation, protein processing and localization can be inhibited by natural plant products or other agents [Trail et al., 1995].

2.2 Toxicity and detoxification of aflatoxin in mammalian organs

Aflatoxins, which are known to be potent mutagenic, carcinogenic, teratogenic, hepatotoxic, immunosuppressive, also inhibit several metabolic systems [International Agency for Research on Cancer, 1993] and causing damages such as toxic hepatitis, hemorrhage, and edema [Santos et al., 2001]. Aflatoxins have been detected in cereal grains, oil seeds, fermented beverages made from grains, milk, cheese, meat, nut products, fruit juice and numerous other agricultural commodities [Bullerman, 1986].

As shown in figure-2, aflatoxin B1 undergoes metabolism in mammalian liver leading to the formation of metabolites such as aflatoxin B1-epoxide and hydroxylated metabolites (aflatoxin M1, aflatoxin P1, aflatoxin Q1, and aflatoxicol). The metabolites produced in phase-I undergo phase II biotransformation involving the enzymes glutathione S-transferase (GST), β-glucuronidase, and/or sulfate transferase which produce conjugates of aflatoxin B1-glutathione, aflatoxin B1-glucuronide, and aflatoxin B1-sulfate, respectively.

Aflatoxin B1 caused damage by two different ways in the cells. Firstly, it is activated to aflatoxin B1-8, 9-epoxide and forms adduct primarily at N7 position of guanine and is responsible for its mutagenic and carcinogenic effects [Wang and Groopman, 1999; Denissenko et al., 1999]. Secondly, aflatoxins especially aflatoxin B1, produce reactive oxygen species such as superoxide radical anion, hydrogen peroxide and lipid hydroperoxides; though these do not appear to interact with DNA, but they are precursors to the hydroxyl radical. The hydrox radicals interact with DNA which may cause mutations [Halliwell and Gutteridge, 1999]. The major conjugate of aflatoxin B1-epoxide identified is the aflatoxin-B1-glutathione conjugate [Monroe & Eaton, 1987].

To control the level of reactive oxygen species and to protect cells under stress conditions, living tissues contain enzyme systems such as, superoxide dismutase, glutathione peroxidase, and catalase as well as antioxidant substances. The effect of reactive oxygen species is balanced by the antioxidant action of non-enzymatic antioxidants, as well as by antioxidant enzymes. Such antioxidant defenses are extremely important as they represent the direct removal of free radicals (prooxidants), thus, providing maximal protection for biological sites [Valko et al., 2006].

It is worth mentioning that glutathione conjugation system is present in aflatoxin-producing fungi which can facilitate detoxification of the toxic metabolite of aflatoxin from the mycelia. Aflatoxin B1-glutathione conjugation in the toxigenic fungi depends on the levels of fungal glutathione and glutathione S-transferase which are inducible in fungi cultured in presence of classic inducers. Likewise the fungi may express enzymes which are involved in inactivation of free radicals as a result of metabolic functions [Saxena et al. 1989; Ziglari et al. 2008].

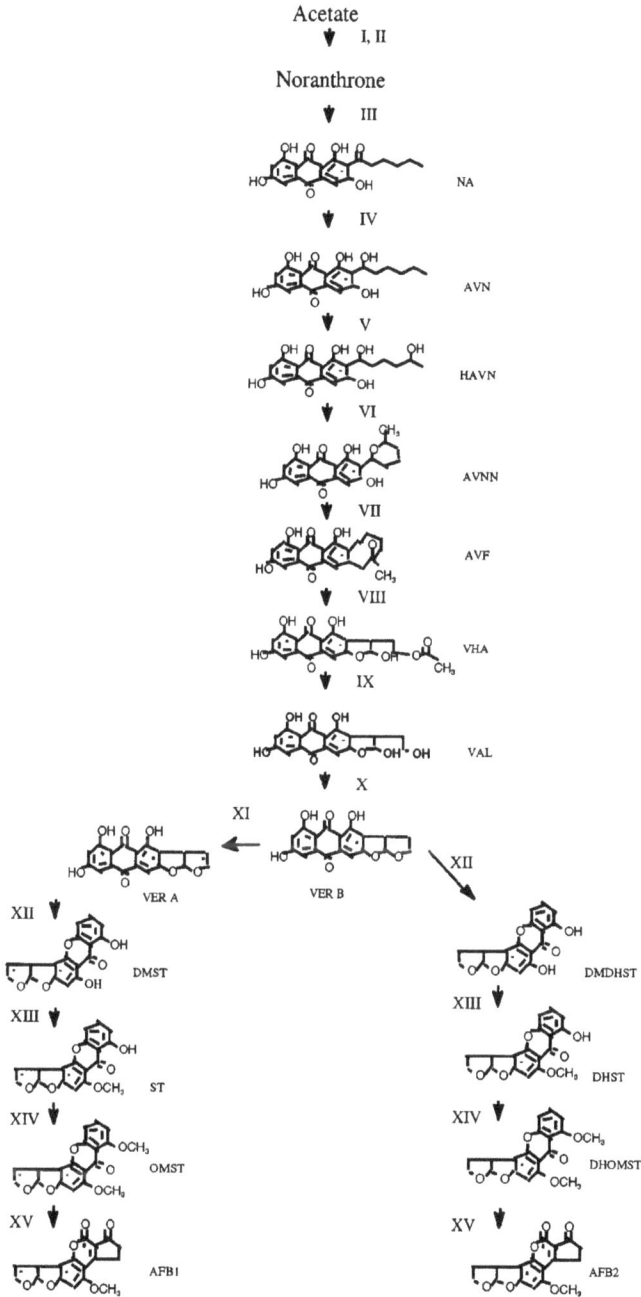

Fig. 1. Biosynthesis of aflatoxins in toxigenic strains of *Aspergillus flavus* and *Aspergillus parasiticus*. Adopted from Payne and Brown, 1998. The numbers show the metabolites formed during the biosynthesis.

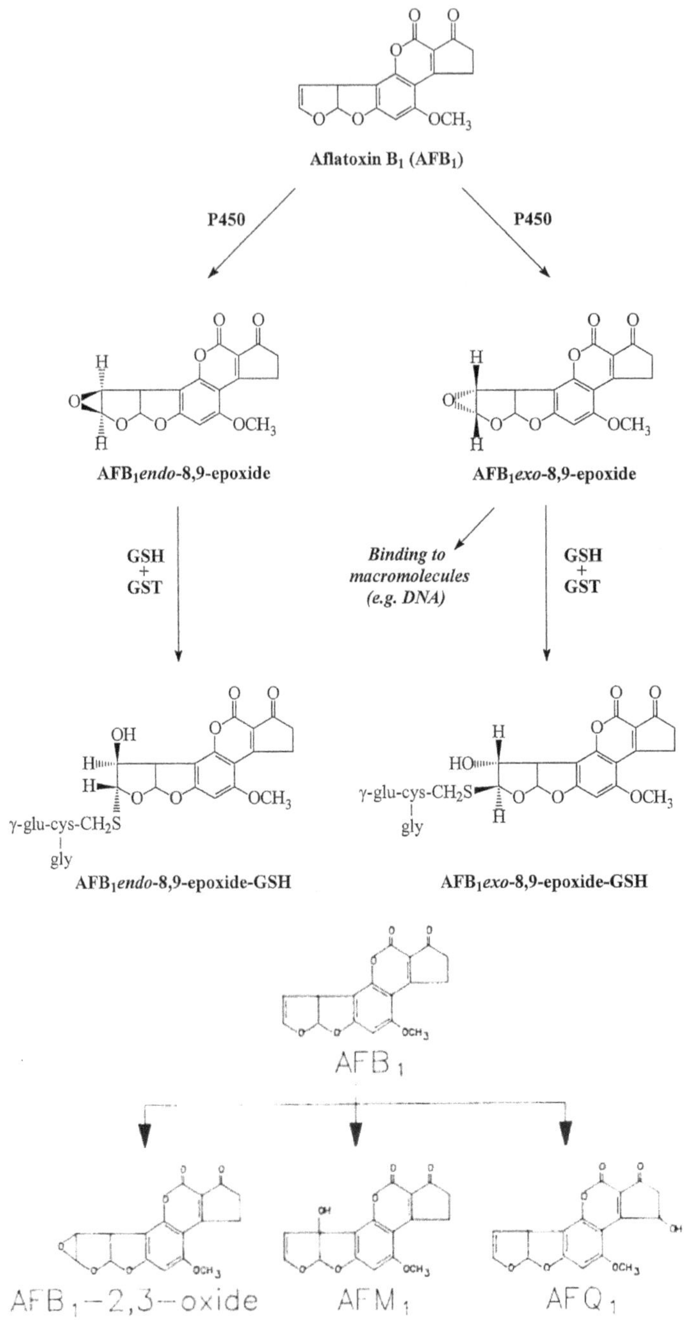

Fig. 2. Biotransformation aflatoxin B1 by phase I and phase II xenobiotic metabolizing enzymes in mammalian cells. [Daniels et al., 1990]

3. Aflatoxin elimination methods

Contamination of food commodities with aflatoxin resulting from fungal attack can occur before, after and during the harvest and storage operations. The enormous health and economic significance of food and feed contaminants have become steadily clearer since the 1960s, when mycotoxin was first discovered.

There are two strategies to reduce the levels of aflatoxins in food and feed materials. The first approach is to control or prevent food contamination in aflatoxin producing fungi. This strategy so called preharvest strategy is relatively easy and can be implicated during cultivation and harvesting. However, the postharvest strategy which deals with elimination of aflatoxins and aflatoxin-producing fungi in the agricultural products appears to be more complicated and not recommended for human consumption. The complexity of elimination methods varies depending on the quantity and the nature of the food material. The safety of the products undergoing elimination is currently suggested for animal feed and their use for human consumption is not recommended. Inactivation or removal of aflatoxins in food and feed commodities at a large scale can recycle a major part of the protection for animal consumption. Chemical and physical treatment are currently major procedures used at large scale. However, the safety of these methods can be increased by using phytochemical agents to reduce aflatoxin production in food and feed products. Phytochemical agents which may directly or indirectly enter to human food may not pose a toxicity threat to humans. Hence, replacement of phytochemicals is in the benefit of human safety.

3.1 Physical control

Various physical techniques have been devised to remove, destroy or suppress the toxicity of the mycotoxins. These techniques include physical removal of the contaminated portions of the foodstuffs, treatment with heat, and radiation in order to convert the toxins into relatively innocuous compounds or the addition of adjuvants to suppress or otherwise mask the ill effects of toxins [Park et al., 2007]. Many physical methods such as microwave heating and treatments with ozone (ozonation) have been recommended for detoxification of aflatoxin contaminated food [Farag et al., 1996; Xu, 1999; Prudente and King, 2002, Inan et al., 2007]. Washing the grain, heating and drying are other traditional methods used to reduce mycotoxins in food products. Physical method is believed to be the most effective method for the reduction of mycotoxins in contaminated commodities. However, a technique such as, gamma radiation is limited due to high cost of equipment [Jalili et al., 2010].

3.1.1 Drying and roasting of food products

Traditionally food and feed materials are dried using sunlight. The level of aflatoxins was reduced to over 40% by roasting and heating peanuts [Rustom, 1997]. Buser and Abbas reported that an extrusion process is able to decrease the level of aflatoxins to 33%. [Hwang et al., 2006]. The influence of temperature and pressure was also examined in an extruder which is a bioreactor that transforms cereals, under high temperature and pressure. As a hydrated powder, the crude material feeds the extruder, undergoing chemical and physical transformations because of the thermal effect and severe shear stress [Chiruvella et al., 1996]. Thus, extrusion-cooking is an attractive process for continuous food/feed processing, and has been developed extensively in recent years as an efficient manufacturing process. High temperature/short time extrusion-cooking is commonly used in the industry to directly produce expanded products such as snack foods, breakfast cereals and pet foods [Miller, 1990; Moore, 1994; Rokey, 1994; Rahman, 1995]. For temperatures between 140 and 200°C and moisture content ranging from 170 to 270 g/kg, reductions in aflatoxin levels between 50 and 75% were obtained. Cazzaniga et al. (2001) reported that extrusion of maize flour with low levels of aflatoxin B1 (50 ng/g) was partially successful (10–25%) for the decontamination of aflatoxins with metabisulphite addition (1%) at temperatures of 150 and 180 °C, respectively. [Mendez-Albores et al., 2009]. The level of aflatoxin B1 in dried wheat was decreased to 50% and 90% by heating at 150 and 200°C, respectively. However, the reduction of aflatoxin B1 in wet wheat in which water (10%) was intentionally added was higher by heating than in dried wheat. The reduction of aflatoxin B1 was increased by 8% and 23% in wet United States wheat (soft red white wheat) and Korean wheat (Anbaekmil) compared to dry United States and Korean wheat, respectively, through heat treatment. Traditional processing used in Korean foods such as Sujebi (a soup with wheat flakes) and steamed bread caused 71% and 43% decrease in aflatoxin B1 content. Reduction of aflatoxin B1 toxicity was directly proportional to washing time in both Korean and United States wheat. [Hwang, et al., 2006].

3.1.2 Irradiation

It has been shown that gamma ray treatment of food products is effective in reducing mycotoxin concentration in different foods. It was found that by increasing the gamma doses from 10 to 60 kGy, mycotoxin reduction significantly increased, however, there was no reduction in the mycotoxin content at doses less than 10 kGy. In a related study, doses of 15, 20, 25 and 30 kGy were used to destroy aflatoxin B1 in peanut sample by 55–74% [Prado et al., 2003]. Aflatoxin B2 and aflatoxin G2 in all of the treatments showed lower reduction comparing with other mycotoxins. According to Aziz and Youssef (2002) a dose of 20 kGy was sufficient for complete destruction of aflatoxin B1 in peanut, yellow corn, wheat and cotton seed meal. Ghanem et al. (2008) demonstrated that aflatoxin B1 degradation in food samples was inversely related to the oil content in irradiated samples. It has been suggested that water content is an important factor in the destruction of mycotoxin by gamma rays, since the radiolysis of water leads to the formation of highly reactive free radicals that can readily attack aflatoxin B1 at the terminal furan ring and produce molecules with lower biological activity [Jalili et al., 2010]. EL-Bazza et al. (1996) reported that gamma irradiation dose level of 3.0 kGy proved to be effective in decontamination of Carum carvi and Matricaria chamomilla samples from fungi, while, Ammi visnaga and Artimisia judica samples could only be decontaminated at a higher dose level of 4.0 kGy. A gamma radiation dose of 6.0 kGy was found to be sufficient to free the tested seed samples from fungi [El-Bazza et al., 2001]. The adverse effects of gamma radiation on food quality have been demonstrated by treating Ashanti pepper with the optimal gamma radiation dose. Both ground and whole forms of Ashanti pepper were subjected to 0, 2.5, 5.0, 7.5 and 10 kGy doses of gamma rays from a 60Co source, showed that, the 2.5-kGy doses reduced the fungal and bacterial load by 2 log cycles and 7.5 kGy eliminated the fungal population. A dose of 10 kGy was required to decontaminate the samples irrespective of sample form, although grinding and not irradiation affected the essential oil composition of the spice [Onyenekwe et al., 1997]. Different combinations of temperature and pressure on the influence of gamma radiation have been also studied. Combination treatment of heat and irradiation (3.5 kGy) reduced the Aspergillus flavus spore inoculum size by about 4 log cycles and yielded the highest amount (41.1 µg/ml) of aflatoxin B1 in Maize Meal broth supplemented with 2% glucose and 2% peptone (AMMB). However, moist heat treatment of spores receiving the same dose (3.5 kGy) reduced toxin formation by 25%. Aflatoxin B1 formation by Aspergillus flavus spores incubated in AMMB was completely prevented by a combination treatment of moist heat and 4.0 kGy of gamma irradiation. A similar treatment attenuated aflatoxins B2, G1 and G2 production which were formed with B1 by Aspergillus flavus NRRL 5906 [Odamtten et al., 1986].

The self-designed microwave-induced argon plasma system [Park et al., 2007] is a technique required much less exposure time for mycotoxin degradation than other methods, such as visible or UV light and gamma ray. The UV irradiation and etching by plasma may be responsible for degrading and removing the mycotoxins. This plasma system has many advantages, such as increased ionization by reactive species and relatively high intensity of UV light (75–102 mW/cm2), low average temperature (75–130°C) and easy operation. In summary, the mycotoxins, aflatoxin B1, deoxynivalenol, and nivalenol were completely removed after 5 seconds of plasma treatment. Moreover, the cytotoxicity of mycotoxins was significantly reduced with progress in the treatment time [Park et al., 2007].

3.1.3 Bioabsorption of aflatoxins in animal feed

One of the most important approaches aimed at reducing the risk of aflatoxicosis or in limiting decrease in animal performance and toxic metabolite carry-over in milk, meat and eggs, is the use of clays in contaminated feeds to reduce aflatoxin absorption in the intestine. Some in vitro tests [Philips et al., 1988] showed that various absorbing materials such as alumina, silica and aluminosilicate are capable of binding aflatoxin in solution. Extraction using various solvents at different temperatures and pH showed a release which varied in intensity in function of the type of material used. It has been demonstrated that the hydrated sodium calcium aluminosilicates (HSCAS) were particularly efficacious in binding aflatoxin. Analogous detoxification trials have been performed using zeolites, bentonites and modified phylloaluminosilicates. A micronized zeolite [Stankov et al., 1992] was tested as an aflatoxin sorbent in feeds for weaning piglets and it induced a marked reduction in mortality rate and increase of feed consumption and body weight. In contrast, a study on dairy cows [Pietri et al., 1993] did not detect any zeolite induced reducing action on carry-over, while a test on broilers of domestic fowl [Sova et al., 1991] showed the total absence of beneficial effects determined by addition of zeolite. In fact, in synthetic zeolites, as opposed to natural ones, the pore size distribution varies very little and is generally concentrated within a narrow diameter range. If the size of the pores is compatible with those of the aflatoxin molecules, conspicuous adsorption occurs. In contrast, adsorption can be easily nil because no intermediate sized pores are present. A test on bentonite as an aflatoxin sorbent conducted on dairy cows [Veldman, 1992] revealed a 33% carryover reduction; while in vitro trials on trout feed [Winfree & Allred 1992] achieved adsorption of 70% the aflatoxin B, present in the feed. An in vitro test [Sison., 1992] demonstrated the efficacy of a commercial product (Mycobond) made of chemically modified phylloaluminesilicate combined with multilayered montmorillonite and detected the formation of an inert, and stable complex capable of preventing absorption of mycotoxins in the intestine [Piva et al., 1995].

3.1.4 Other methods
Practical methods to degrade mycotoxins using ozone gas (O3) have been limited due to low O3 production capabilities of conventional systems and their associated costs. Recent advances in electrochemistry (i.e. proton-exchange membrane and electrolysis technologies) have made available a novel and continuous source of O3 gas up to 20% by weight. It is possible that the rapid delivery of high concentrations of O3 will result in mycotoxin degradation in contaminated grains-with minimal destruction of nutrients. Results indicated that aflatoxin B1 and aflatoxin G1 were rapidly degraded using 2% O3, while aflatoxin B2 and aflatoxin G2 were more resistant to oxidation and required higher levels of O3 (20%) for rapid degradation [McKenzie et al., 1997]. Ozonation, an oxidation method, has been developed for the detoxification of aflatoxins in foods [Samarajeewa et al., 1990]. Ozone, or triatomic oxygen, is a powerful disinfectant and oxidising agent [McKenzie et al., 1997]. It reacts across the 8, 9-double bond of the furan ring of aflatoxin through electrophilic attack, causing the formation of primary ozonides followed by rearrangement into monozonide derivatives such as aldehydes, ketones and organic acids [Proctor et al., 2004]. As a disinfectant, ozone is 1.5 times stronger than chlorine and is effective over a much wider spectrum of micro-organisms [Xu, 1999, Maeba et al., 1988] have confirmed the destruction and detoxification of aflatoxins B1 and G1 with ozone. Aflatoxin B1 and G1 were sensitive to ozone and degraded with 1.1 mg/l of ozone in 5 min in model experiments. The reductions of content of aflatoxin B1 levels in flaked and chopped red peppers around 80% and 93% after exposures to 33 mg/l ozone and 66 mg/l ozone for 60 min, respectively was shown. [Inan et al., 2007].

3.2 Chemical control
A number of chemicals have been investigated for their ability to destroy, transform, or inactivate aflatoxin [Dollear, 1969; Mann et al., 1970]. Developing measures to control mycotoxin contamination is a high priority for the food and animal feed industries. The most reliable method to prevent mycotoxicosis is to avoid the use of contaminated materials to disinfect fungi and to inactivate mycotoxin. Most of the chemical treatments proposed are not necessarily practical, however, because they not only decompose aflatoxin but also deplete the quality of the food and feed materials themselves. The chemical used for elimination of aflatoxins are mainly antifungal agents, but they can also be exclusively used for inhibition of aflatoxin biosynthesis and destruction of the toxins.

3.2.1 Antifungal agents
So far a large number of compounds have been found to inhibit aflatoxin production. Most of them appear to do so by inhibiting fungal growth. For example, some surfactants have suppressed the growth of *Aspergillus flavus* and aflatoxin synthesis. [Bata et al., 1999]. Among the chemical compounds screened, propionic acid (0.1–0.5%), ammonia (0.5%), copper sulphate (0.5–1%) and benzoic acid (0.1–0.5%) completely inhibited *A. parasiticus* growth. It has been shown that sodium benzoate has antimicrobial effect on the growth, survival and aflatoxin production of *Aspergillus niger*, *Aspergillus flavus* and *Aspergillus fumigatus* in packaged garri (2 kg/pack) during storage at ambient temperature (30±2 °C) [Ogiehor et al., 2004]. Sodium hypochlorite (0.1–0.5%) exhibited high anti-fungal property (68–84%). Urea (0.1–0.5%), citric acid (0.2–0.5%) and sodium propionate (0.1–0.5%) were moderate in inhibiting fungal growth. Citric acid below 0.2% had poor anti-fungal effect [Gowda et al., 2004]. Ammonia at 0.2% level and copper sulphate below 0.08% level had moderate anti-fungal activity (60 and 36%, respectively).

3.2.2 Inhibition of aflatoxin biosynthesis and degredation of the toxin

Two extensively studied inhibitors of aflatoxin synthesis are dichlorvos (an organophosphate insecticide) and caffeine [Hsieh, 1973]. A large number of chemicals can react with aflatoxins and convert them to less toxic and mutagenic compounds. These chemicals include acids, bases, oxidizing agents, bisulphites and gases [Dollear et al., 1968; Mann et al., 1970; Mendez-Albores, 2007]. The components of the neem tree (Azadirachta indica) is well known for its interference in aflatoxin biosynthesis with very little action on the fungal mycelia [Bhatnagar et al., 1988., Allameh et al., 2001] . There are evidences which show that neem leave extracts exclusively inhibit aflatoxin biosynthesis in toxigenic fungi without a major change in the mycelia. Information about the effectiveness and mode of action of the neem components will be discussed under section "3-3-3 Interference of natural products in aflatoxin biosynthesis" of this chapter.

In addition, most of the antifungal agents can also inhibit aflatoxin biosynthesis and cause destruction of aflatoxin structures. For example, ammoniation process is believed to detoxify aflatoxins in various raw materials with high efficiency [Buser and Abbas, 2002]. Alkaline compouds, such as ammonia, sodium- and calcium hydroxide etc, were used particularly for destruction of aflatoxin (for a review, see Samaraeva et al., (1990). Elimination of aflatoxins in feed by ammonia treatment is one of the approaches to reduce aflatoxicosis. After replacing the aflatoxin-containing maize with ammoniated grains (1%v/w) in diet of one-day-old broiler chicks, the mortality rate significantly decreased [Allameh et al., 2005]. Treatment with ammonia in the gaseous phase, in solution, or with substances capable of releasing it, achieved optimum results in detoxifying peanut, cotton and corn meals. It was observed that the aflatoxin B1 molecular structure is irreversibly altered when exposure to ammonia lasts long enough. In contrast, if exposure is not sufficiently protracted, the molecule can revert to its original state [Piva et al., 1995]. Our experience showed that ammonia vapors reduce aflatoxins after destruction of fungal mycelia and spores of the toxic fungi [Namazi et al., 2002]. Efficient detoxification of aflatoxin-contaminated groundnut meal with ammonia during pelleting using 5% NH_3 and 200 g water/kg during a 10-day period further confirm the effectiveness of ammonia [Thiesen, 1977].

Sodium bisulfite treatment is a common aflatoxin B1 detoxification method [Moerck et al., 1980; Sommartya et al., 1988; Hagler et al., 1982]. Although it is less efficacious than ammonia detoxification it overcomes some of the typical disadvantages of ammonia methods and also has much lower costs. The main reaction product has been isolated and identified as a sulfonate, called 15α-sodium sulfonate or aflatoxin B1S (aflatoxin B1S) [Hagler et al., 1983; Yagen et al., 1989], which forms by insertion of NaHSO, at the double bond of fitrofuranic ring, depriving the aflatoxin B1 molecule of main DNA molecule reaction site, thus reducing its mutagenic potential. The efficacy of nixtamalization, a traditional practice widely used in South America to prepare typical corn tortillas consisting of cooking the corn in boiling water supplemented with calcium hydroxide has been reviewed [Piva et al., 1995].

Formaldehyde is a compound which is moderately efficacious in attacking and neutralizing the aflatoxin B1 molecule, even if no data on its reaction mechanism are available. Studies showed its enhanced efficacy in association with ammonia [Frayssinet et al., 1972] and calcium hydroxide [Codifer et al. 1976]. In contaminated milk samples addition of 0.5% formaldehyde could reduce 1.1 µg aflatoxin M, to 0.05 µg [Heimbecher et al., 1988]. Bleaching of flour with chlorine in a commercial mill resulted in a 10% reduction of deoxynivalenol content. Aqueous sodium bisulfate caused the greatest reduction in mycotoxin levels [Bata and Laszitity, 1999].

Aqueous citric acid exhibits detoxifying activity in aflatoxin B1-contaminated feeds and protects animals from chronic aflatoxin toxicity [Mendez-Albores, 2007]. The aqueous citric acid had detoxification activity in treating aflatoxin contaminated maize [Méndez-Albores et al., 2005]. The detoxification of aflatoxin B1 initially involves the formation of the β-keto acid structure, catalyzed by the acidic medium, followed by hydrolysis of the lactone ring yielding aflatoxin D1 (a nonfluorescent compound, which exhibits phenolic properties and lacks the lactone group derived from the decarboxylation of the lactone ring-opened form of aflatoxin B1); and to a lesser extent, a second compound (a nonfluorescent phenol, commonly known as aflatoxin D2), which retains the difurane moiety but lacks both the lactone carbonyl and the cyclopentenone ring, characteristic of the aflatoxin B1 molecule The addition of different amounts of citric acid in the milled sorghum resulted in a moderated improvement in the extent of detoxification when using concentrations of 0.5–2 N [Mendez-Albores et al., 2009].

Anti-aflatoxigenic activity of certain chemicals such as eugenol [Jayashree and Subramanyam, 1999] and hydrolysable tannins [Mahoney and Molyneux, 2004] as well as some plant components [Joseph et al., 2005] is due to their antioxidant capacities. Epoxides, which can lead to lipid peroxidation of fungal cells, stimulated aflatoxin biosynthesis [Fanelli et al., 1983]. Inhibition of aflatoxin B1 production by 2-chloroethyl phosphoric acid revelaed that this compound can greatly reduce the expression of two aflatoxin biosynthetic genes, aflR and AflD, indicating that ethylene-related inhibition in aflatoxin biosynthesis is partly due to transcriptional inhibition of aflatoxin biosynthetic genes [Huang et al., 2009].

3.3 Biological control

Biological factors possess antimicrobial properties can be classified based on their source and the mechanism of action. Certain bacteria, fungi and yeast have been identified for their potential action of aflatoxin producing fungi. The mechanisms of action of biological agents to control aflatoxins is mainly through, biodegradation of the secondary metabolites and antifungal activity. Great successes in reducing aflatoxin contamination have been achieved by application of nontoxigenic strains of Aspergillus *flavus* and Aspergillus *parasiticus* in fields of cotton, peanut, maize and pistachio. According to Yin and co-workers (2008), the nontoxigenic strains applied to soil can occupy the same niches as the natural occurring toxigenic strains. Therefore, they may be capable of competing and displacing toxigenic strains.

3.3.1 Biodegradation of aflatoxins

Inactivation of aflatoxin by physical and chemical methods has not yet proved to be effective and economically feasible [Mishra and Das, 2003]. Microorganisms, especially bacteria, have been studied for their potential to either degrade mycotoxins or reduce their bioavailability. In recent years scientists focused on identification and application of natural products for inactivation of aflatoxins. It has been suggested that the biological detoxification offers an attractive alternative for eliminating toxins and safe-guarding the quality of food and feed. Ciegler et al. (1979) screened over 1000 microorganisms for the ability to degrade aflatoxins. Only one bacterium, *Flavobacterium aurantiacum* B-184, was able to irreversibly remove aflatoxin from solutions. The early investigations showed that pH and temperature influenced the uptake of the toxin by the cells. The first important question which must be answered is whether *Flavobacterium aurantiacum* actually degrades the aflatoxin or whether the disapperance of the toxin resulted from adsorption to the cells.

Detoxification of aflatoxin B1 by *Enterococcus faecium* is probably due to the binding of the mycotoxin to the bacterial cell wall components, a mechanism which has also been postulated by other studies [Haskard et al., 2001]. Bacterial cell wall peptidoglycans and polysaccharides have been suggested to be responsible components for the mycotoxin binding by bacteria [Hosono et al., 1988].

It has been demonstrated that *Bacillus subtilis* could reduce the aflatoxin quantity in co-culture with *Aspergillus* producing aflatoxin. Perhaps, likely, the *Bacillus subtilis* metabolites inhibit both spore germination and hyphal elongation, which induces the decrease of fungal development and consequent reduction of the aflatoxin production.

According to Teniola and co-workers (2005), *Rhodococcus erythropolis* and *Mycobacterium fluoranthenivorans* are able to degrade aflatoxin B1 more effectively and within a shorter time than the two *Nocardia corynebacterioides* strains. It was particularly interesting to notice up to 70% aflatoxin B1 elimination within 1 h of applying cell free extracts from the two strains, and >90% degradation was observed within 4 h. There was no detectable aflatoxin B1 from any strain after 24 h, with the exception of *Nocardia corynebacterioides* DSM 12676 (formerly *Flavobacterium aurantiacum*). These results are similar to the observations of Smiley and Draughon (2000), who showed that about 74.5% aflatoxin B1 degradation by the bacterial cell free extract obtained by lysozyme treatment after 24 h of incubation. It has been observed a diminishing aflatoxin B1 degradation which was attributed to the effects of heat treatment and incorporation of proteinase K into their extract. Liquid cultures of *Rhodococcus erythropolis* were also able to degrade aflatoxin B1 very effectively. Optimal degradation by the four isolates occurred at 30 °C which makes them applicable in food in the tropical environment like West Africa [Teniola et al., 2005]. It has been demonstrated that the *Bacillus subtilis* could reduce the aflatoxin levels directly without affecting the fungal development. The probiotic activity of bacteria depends on the bacterial strain and the density of bacteria used.

Some of the species of bacteria and fungi have been shown to enzymatically degrade mycotoxins (Bata and Lasztity, 1999]. However, question remains on the toxicity of products of enzymatic degradation and undesired effects of fermentation with non-native microorganisms on quality of food.

Isolates of yeasts belonging to different species including *Saccharomyces cerevisiae* and *Candida krusei* were tested for aflatoxin binding, some of the isolates from West African maize were found to bind more than 60%(w/w) of the added toxins. Most of the yeast strains bound more than 15% (w/w) of aflatoxin B1 and the toxin binding was highly strain specific. There are many reports on use of physically separated yeast cell walls obtained from brewery as feed additive in poultry diet resulting in amelioration of toxic effects of aflatoxins (Santin et al., 2003). When dried yeast and yeast cell walls were added to rat diet along with aflatoxin B1, a significant reduction in the toxicity was observed (Baptista et al., 2004). In an *in vitro* study with the cell wall material, there was a dose-dependent binding of as much as 77% (w/w) and modified mannan-oligosaccharides derived from the *Saccharomyces cerevisiae* cell resulted in as much as 95% (w/w) binding (Devegowda et al., 1996). Available experimental supports suggest the role of both peptidoglycon and polysaccharides in toxin binding (Zhang & Ohta, 1991). Based on some of the studies reported, it is confirmed that removal of mycotoxins is by adhesion to cell wall components rather than by covalent binding or by metabolism, as the dead cells do not lose binding ability (Baptista et al., 2004; Santin et al., 2003). Reported literature indicates that mannan components of cell wall play a major role in aflatoxin binding by *Saccharamyces cerevisiae* (Devegowda et al., 1996). Animal feeding experiments with whole yeast and yeast cell wall [Santin et al., 2003) show that addition of *Saccharomyces cerevisiae* in the diet resulted in reduced mycotoxin toxicities, indicating possible stability of the yeast-mycotoxin complex through the gastrointestinal tract. Similarly, Gratz et al., 2004 showed that pre-exposure of cells of *Lactobacillus rhamnosus* strain GG to aflatoxin B1 reduces its binding with intestinal mucus, resulting in faster removal. (Shetty et al., 2006).

In recent years it became clear that fungi play a major role in the degradation of aflatoxin B1. Fungi have been implicated in aflatoxin B1 degradation include zygomycetous fungi (*Rhizopus sp.* and *Mucor sp.*), ascomycetous fungi (*Aspergillus niger* and *Trichoderma sp.*), plant pathogens (*Phoma sp.* and *Alternaria sp.*), as well as basidiomycetous fungi (*Armillariella tabescens* and other white rot fungi) (Leonowicz et al., 1999). When the degradation of polyphenolic xenobiotics are considered, fungi is considered as one of the major groups responsible for their degradation, presumably due to the large repertoire of extracellular enzymes produced by these fungi (Arora and Sharma, 2009). Treatment of aflatoxin B1 with laccase enzyme produced by white rot fungi in unconcentrated culture filtrates, pure fungal laccase as well as with recombinant laccase enzymes decreased the fluorescence properties of the aflatoxin B1 molecule as determined with HPLC. It has been shown that treatment of aflatoxin B1 with fungal laccase enzymes targets and changes the double bond of the furofuran ring of the aflatoxin B1 molecule causing changes in aflatoxin fluorescence and mutagenicity properties (Alberts et al., 2009).

The use a yeast strain, *Pichia anomala*, to reduce spore production of *Aspergillus flavus* on pistachio nut fruits, leaves, and flowers has also been reported (Hua, 2004). Another approach involves competitive exclusion of toxigenic strains with a nonaflatoxigenic isolate. Using nonaflatoxigenic *Aspergillus flavus* isolates to competitively exclude toxigenic *Aspergillus flavus* isolates in agricultural fields has become an adopted approach to reduce aflatoxin contamination. From screening subgroups of nonaflatoxigenic *Aspergillus flavus*, an *Aspergillus flavus* isolate, (TX9-8), has been identified, which competed well with three *Aspergillus flavus* isolates producing low, intermediate, and high levels of aflatoxins, respectively. TX9-8 has a defective polyketide synthase gene (pksA), which is necessary for aflatoxin biosynthesis. Co-inoculating TX9-8 at the same time with large sclerotial (L strain) *Aspergillus flavus* isolates at a ratio of 1:1 or 1:10 (TX9-8: toxigenic) prevented aflatoxin accumulation. The intervention of TX9-8 on small sclerotial (S strain) *Aspergillus flavus* isolates varied and depended on isolate and ratio of co-inoculation. At a ratio of 1:1 TX9-8 prevented aflatoxin accumulation by *Aspergillus flavus* CA28 and caused a 10-fold decrease in aflatoxin accumulation by *Aspergillus flavus* CA43. No decrease in aflatoxin accumulation was apparent when TX9-8 was inoculated 24 h after toxigenic L- or S strain *Aspergillus flavus* isolates started growing. The competitive effect is likely due to TX9-8 outgrowing toxigenic *Aspergillus flavus* isolates. [Chang, Hua 2007].

According to the literature, *Armillariella tabescen* (Scop. ex Fr.) *Sing.*, is a non-toxic, edible fungus possesses detoxification activity towards aflatoxin B1 contaminated media. The detoxification activity of the extracts obtained from *Armillariella tabescen* mycelium pellets is assigned to the enzymes in the active extracts [Liu et al., 1998]. There are also other microorganisms such as soil or water bacteria, fungi, and protozoa and specific enzymes isolated from microbial systems can degrade aflatoxin group members with varied efficiency to less- or nontoxic products [Wu et al., 2009].

3.3.2 Antifungal agents

Antifungal agents with natural sources, which prevent the contamination of food by controlling the growth of *Aspergillus flavus* and *Aspergillus parasiticus*, is probably the most rational to prevent the growth of toxic fungi during storage.Inactivation of aflatoxin by physical and chemical methods has not yet proved to be effective and economically feasible (Mishra and Das, 2003). In recent years scientists focused on identification and application of natural products for inactivation of aflatoxins. It has been suggested that the biological detoxification offers an attractive alternative for eliminating toxins and safe-guarding the quality of food and feed.

Essential oils with antimicrobial properties are probably promising for growth inhibition of potentially toxigenic fungi. However, limited studies carried out on the mechanism of action of essential oils on fungal mycelia growth show that probably the cell wall and cell membrane are the main targets of the oil compartments. The plasma membrane of *Aspergillus parasiticus*, in the presence of thyme essential oils at 250 ppm, was seen to be irregular, dissociated from the cell wall, invaginated and associated with the formation of lomasomes. These lomasomes are usually found in fungi treated with imidazole components. The marked action of oil components might have conferred lipophilic properties and the ability to penetrate the plasma membrane. It has been shown that essential oil derived from Hyssopus officinalis affected the wall synthesis of *Aspergillus fumigatus*. The presence of the oil in the culture medium induced marked changes in the content of galactose and galactosamine. The alterations were related to changes in the structure of the cells. Such modifications induced by essential oils may be related to the interference of essential oil components with enzymatic reactions of wall synthesis, which affects fungal morphogenesis and growth [Ghfir et al., 1997].

Kurita et al. (1981) suggested that the antifungal activity of essential oil components, particularly aliphatic aldehydes, might be due to their ability to form charge transfer complexes with electron donors in the fungus cell [Rasooli et al., 2005]. The action of the oils on the integrity of nuclear membrane has not been ruled out. Changes in ultrastructure of the aflatoxin-roducing fungi treated with neem leaf extracts showed that the mycelia membrane is very susceptible to this treatment [Allameh et al. 2002].

3.3.3 Interference of natural products in aflatoxin biosynthesis

Numerous studies have been conducted to determine the effects of various food additives, preservatives, chemical, and environmental condition to effectively inhibit fungal growth and aflatoxin production. Despite the efficiency of chemicals in removal of aflatoxin-producing fungi and aflatoxins, the residues of chemicals can pose serious hazards to human and animal health. Meanwhile considerable pressure from consumers to reduce or eliminate chemically synthesized additives in their foods has led to a renewal of scientific interest in natural substances [Nychas, 1995; Bluma et al., 2008]. Some studies have concluded that whole essential oils have a greater antibacterial activity than the major components mixed, which suggests that the minor components are critical to the activity and may have a synergistic effect or potentiating influence. Among the thousands of naturally occurring constituents so far identified in plants and exhibiting a long history of safe use, there are none that pose, or reasonably might be expected to pose a significant risk to human health at current low levels of intake when used as flavoring substances [Rasooli et al., 2007]. Numerous diverse compounds and extracts containing effects inhibitory to aflatoxin biosynthesis have been reported. The most of these inhibitors are plant-derived such as phenylpropanoids, terpenoids and alkaloids [Holmes et al., 2008]. Most plants produce antimicrobial secondary metabolites, either as part of their normal program of growth and development or in response to pathogen attack or stress. A novel way to reduce the proliferation of microorganism and/or their toxins production is the use of essential oils, which are mixtures of different lipophilic and volatile substances, such as monoterpenes, sesquiterpenes, and/or phenylpropanoids, and have a pleasant odor. Furthermore, they are considered to be part of the preformed defense system of higher plants [Reichling et al., 2009]. They are usually obtained by steaming or hydro-distillation which was first developed in the middle Ages by the Arabs. Essential oils can contain about 20-60 components in quite different concentrations. They are characterized by two or three major components at fairly high concentrations (20-70%) compared to others present in trace amounts [Alpsoy, 2010]. A range of synthetic preservatives are being used to prevent the growth of food spoiling microbes causing different food borne diseases. However, most of them have been reported to cause different side effects after application. They are also responsible for the enhancement of reactive oxygen species molecules causing oxidative diseases by damaging the proteins, lipids, nucleic acids and more importantly stimulation of aflatoxin biosynthesis [Prakash, et al., 2011]. Until now, many studies have revealed that Aspergillus growth was completely inhibited by many plants essential oils. The effects of essential oils of 58 plant species were examined on the development of Aspergillus flavus and/or Aspergillus parasiticus by Alpsoy et al (2010). Different concentrations of the essential oils was found to inhibit the development of Aspergillus species. It is possible to use a combination of essential oils to increase their effects on fungal growth and aflatoxin production. The antifungal efficacy of plant essential oils varies depending on the concentration and composition of the oils. The inhibitory effects of the components of essential oil on growth rate of Aspergillus flavus and Aspergillus parasiticus has also been reported. Some essential oils and other extracts (vitamins, riboflavin, carotenoids, beta-carotene, alfa-carotene, lycopene, ascorbic acid, curcumin, several flavonoids, phenolic compouds and synthetic phenolic compounds) of plants could potentially provide protection against aflatoxins especially aflatoxin B1 [Rasooli et al. 2004; Rasooli and Owlia, 2005; Rasooli et al., 2008; Bluma and Etcheverry, 2008]. Phenolic compounds such as acetocyringone, syringaldehyde and sinapinic acid not only inhibited aflatoxin B1 biosynthesis, but also reduced production of intermediate metabolites namely, norsolinic acid. It was observed that the oils of cassia, clove, star-anise, geranium and basil inhibited the mycelial growth of established seed-borne infections of Aspergillus flavus, Curvularia pallescens and Chaetomium indicum as well as preventing infection following inoculation with Aspergillus flavus, Aspergillus glaucus, Aspergillus niger and Aspergillus sydowi. These oils also preserved the grain from natural Aspergillus flavus infection during the experimental period. Many natural compounds found in dietary plants, such as extracts of herbs and fruit extracts, possess antimicrobial activities against Aspergillus parasiticus [Soliman and Badeaa, 2002; Rasooli & Owlia, 2005]. Also spices and herbs, such as cloves, anise and star anise seeds, basil, cinnamon, marigold and spearmint [Soliman and Badeaa, 2002], garlic and onion, thyme [Rasooli & Owlia, 2005; Rasooli et al., 2006a], cassia and sweet basil [Atanda et al., 2007] have been reported to inhibit toxigenic and foodborne moulds[Rasooli et al., 2007]. Alderman and Marth (1976), who tested the antimicrobial activity of citrus oils and D-limonene- the major constituent of citrus oils against an aflatoxin-producing strain of Aspergillus parasiticus, found that citrus oils at a concentration of 3000-3500 ppm, suppressed mold growth and aflatoxin production. Likewise, growth and aflatoxin formation in the basal medium was depressed by orange and lemon oils at a concentration of 1.6% through 10 days incubation [Lillehoj & Zuber, 1974; Bullerman et al., 1977]. The essential oils of anise, caraway and fennel showed inhibitory effects on the four tested fungi, Aspergillus flavus, Aspergillus parasiticus, Aspergillus ochraceus and Fusarium moniliforme, at various concentrations [Soliman and Badeaa, 2002]. The inhibitory effect of the oil is proportional to its concentration, and the anise essential oil has more inhibitory effect than the other two members of the Umbellifereae family, caraway and fennel. The antifungal activity of these oils can be assigned to their active components, such as anithole. Caraway and spearmint belong to different families but they contain carfone as a main component of their essential oils, which may be responsible for their antifungal activity. The essential oils of members of the Labiateae family have inhibitory effects on the Aspergillus flavus, Aspergillus parasiicus, Aspergillus ochraceous and Fusarium moniliforme. Thyme oil was more toxic to the four pathogenic fungi than spearmint and basil (two members of the Labiateae family). The antifungal activity of thyme, spearmint and basil was also demonstrated by Montes-Belmont and Carvajall (1998) and Basilico and Basilico (1999) on the toxigenic fungi Aspergillus flavus, Aspergillus parasiticus, Aspergillus ochraceus, Aspergillus fumigatus and Fusarium spp. The antifungal activity of mint, basil and thyme on some pathogenic fungi, including Aspergillus flavus and Aspergillus parasiticus has also been reported which suggested their inhibitory effects on the sporulation of fungi and aflatoxin production [Ela et al., 1996; Inouye et al., 1998; Inouye et al., 2000; El-Maraghy, 1995; Dube et al., 1989]. These effects could be related to several components known to have biological activities, such as α-pinene, β-pinene in thyme, basil and spearmint, respectively. In addition, thyme oil contained thymol and p-cymene as the most prevalent components. The major substances for basil oil were ocimene and methyl chavecol. The essential oils extracted from some medicinal plants belonging to the family Compositeae have fungistatic activity against all toxigenic fungi. The fungistatic effects of chamomile and hazanbul essential oils, marigold and quyssum ghafath essential oil (family: Rosaceae, Species: agrimonia eupatoria) on the growth of Aspergillus flavus, A. parasiticus, Aspergillus ochraceus and Fusarium moniliforme showed differential effects on growth of these fungi. In a similar study with essential oil of cinnamon it was shown that this oil can completely inhibit the growth of the test fungi. The three components of cinnamon that have been identified as the agents active against moulds are cinnamic aldehyde [Bullerman, 1974], O-methoxycinnamaldehyde [Morozumi, 1978] and carfone [Dwividi and Dubey, 1993]. Many previous studies had verified cinnamon oil as a fungistatic agent against many toxigenic fungi and proved its highly fungicidal activity [Sinha et al., 1993]. Morover, the effect of essential oils (thyme, cinnamon, anise and spearmint) in inhibition of toxin production in inoculated wheat further confirm the applications of essential oils [Soliman and Badeaa, 2002]. White wood, cinnamon and lavender significantly inhibited the growth of Aspergillus flavus IMI 242684 [Thanaboripat et al, 2007]. The major constituents of white wood oil was found to be monoterpene compounds such as terpinolene (24.74%) and γ-terpinene (22.84%) [Brophy et al., 2002]. There has been speculation on the contribution of the terpene fraction of the oils to their antimicrobial activity [Conner, 1993]. A number of compounds and substances have been found to effectively inhibit fungal growth and aflatoxin production, while others have stimulatory properties [Zaika & Buchanan, 1987]. It is worth mentioning that low concentrations of test compounds often stimulate fungal growth and/or toxin production, while higher concentrations may completely inhibit the fungal growth. For instance, clove oil at 50 and 100 µg/ml and cinnamon oil at 50 µg/ml stimulated the growth of Aspergillus flavus in liquid media whereas higher concentrations reduced the mycelial growth. Essential oils from different sources such as those extracted from cinnamon (Cinnamomum zeylanicum), peppermint (Mentha piperita), basil (Ocimum basillicum), origanum (Origanum vulgare), the flavoring herb epazote (Teloxys ambrosioides), clove (Syzygium aromaticum) and thyme (Thymus vulgaris) were also proved to completely inhibit Aspergillus flavus growth on maize kernels.

The concentration of essential oils used for fungal inhibition studies varies depending on different factors. For instance, five different oils namely; geraniol, nerol, citronellol (aliphatic oils), cinnamaldehyde (aromatic aldehyde) and thymol (phenolic ketone), used at concentration of 100 ppm could completely suppress growth of *Aspergillus flavus* and consequently prevented aflatoxin synthesis in liquid medium. The hydrosols of anise, cumin, fennel, mint, picking herb, oregano, savory and thyme showed a strong inhibitory effect on mycelial growth of *Aspergillus parasiticus* NRRL 2999 [Sinha et al., 1993].

Inhibition of *Aspergillus parasiticus* growth and its aflatoxin production in presence of the essential oils extracted from two varieties of thyme i.e. Thymus eriocalyx and Thymus x-porlock were studied. The oils from the above plants were found to be strongly fungicidal and inhibitory to aflatoxin production. The oils analyzed by GC and GC/MS lead to identification of 18 and 19 components in Thymus eriocalyx and Thymus x-porlock oils respectively. The profile of the oil components from Thymus eriocalyx was similar to that of Thymus x-porlock in almost all the compounds but at different concentrations. The major components of Thymus eriocalyx and Thymus x-porlock oils were thymol β-phellandrene and cis-sabinene hydroxide respectively. Thymus eriocalyx oil exerted higher antifungal as well as antitoxic effects than that of Thymus x-porlock. This difference in antifungal and aflatoxin inhibition efficacy of thymus essential oils may be attributable to the oil compositions. Thymus eriocalyx oil contains higher (more than 2-fold) thymol than T. x-porlock oil [Rasooli et al., 2006b].

Growth of *Aspergillus parasiticus* NRRL 2999 was also reported to be completely inhibited by thyme (wild), thyme (black), oregano and savory extracts at the 2% level in Czapek-Dox Agar [Ozcan, 1998]. Inhibition of growth in phytopathogenic fungi such as *Rhizoctonia solani, Pythium ultimum* var. *ultimum, Fusarium solani and Colletotrichum lindemuthianum* inhibition was reported to be associated with the degeneration of fungal hyphae after treatment with Thymus vulgaris L., Lavandula R.C., and Mentha piperita L. essential oils with the oil of thyme being more effective than that of lavender or mint [Zambonelli et al., 1996].

The effectiveness of *Thymus kotschyanus* and *Zataria multiflora* Boiss. on the growth of the *Aspergillus parasiticus* strain and aflatoxin production are probably due to major substances such as thymol and carvacrol showing antifungal effects [Pinto et al., 2006] and completely suppressing aflatoxin synthesis [Mahmoud, 1994]. It is well known that a phenolic-OH group is very reactive and can easily form hydrogen bonds with the active sites of enzymes [Farag et al., 1989; Rasooli et al., 2009]. Based on the antifungal potential of essential oils derived from *Thymus vulgaris* L., *Thymus tosevii* L., *Mentha spicata* L., and *Mentha piperita* L.

(Labiatae) it has been suggested that these products could be used as natural preservatives and fungicides [Sokovic et al., 2009].

It appears that carvone has better antifungal properties because of its high water solubility. One of the reasons for lower antifungal activity of *Mentha piperita* essential oil could be due to the large amount of menthyl acetate, which is probably responsible for causing a decrease in antifungal properties [Griffin et al., 2000; Sokovic et al., 2009].

Essential oil-related inhibition in mycelial growth was observed to be associated with significantly decreased levels of aflatoxin production. Exposure of toxigenic Aspergillus parasiticus to neem leaf aqueous extract resulted in the inhibition of aflatoxin production not fungal growth, while exposue of fungus to essential oils from Thymus species caused inhibition in both fungal growth and aflatoxin synthesis [Rasooli & Razzaghi-Abyaneh 2004; Razzaghi-Abyaneh et al., 2005b]. Bhatnagar and McCormic (1988) have demonstrated that addition of neem leaf extract above 10% (v/v) effectively inhibited aflatoxin production by *Aspergillus parasiticus* and *Aspergillus flavus*. Neem oil at 0.5% had moderate antifungal activity (84% reduction versus control). Neem oil at below 0.2%, neem seed cake at above 0.5–10% was moderate in preventing fungal growth i.e. 25–52%. A major feature of the neem leaf extracts is that when added to growth media did not affect the fungal growth, but it could essentially block aflatoxin biosynthesis at concentrations greater than 10%. These results were further confirmed in our laboratory by showing that different concentrations of aqueous neem leaf extract inhibited fungal growth and aflatoxin production by *Aspergillus parasiticus* (NRRL 2999). The inhibition of aflatoxin synthesis by neem extracts was found to be time- and dose-dependent [Ghorbanian et al.,2007]. The maximum inhibitory effect was 80-90% in the presence of 50% concentration that when compared with control samples were significant. Aflatoxin was at its lowest level (>90% inhibition) when the concentration of neem extract was adjusted to 50%. In this study the interference of neem components in aflatoxin biosynthesis pathways is not ruled out [Allameh et al. 2001]. These results are inconsistent with previous reports on existence of a positive correlation between aflatoxin activity and glutathione S-transferase activity in toxigenic strains of *Aspergillus* [Saxena et al., 1989]. In this connection it has been reported that feeding high level of neem seed cake (>10%) has adverse effects on palatability and performance of poultry [Gowda et al., 2004]. Antifungal effects of neem leaf extract also reported from South America against Crinipellis perniciosa and Phytophthora species causing Witches broom and Pot Not of cocoa (Ramos et al., 2007). Azadirachtin, Azadiradione, nimonol and epoxy azadradione were yielded from the organic extract of seeds and leaves of neem. Nimonol (82%) is likely to be a major active component of neem organic extract. Inhibition of seed-born infection by neem leaf extract has been reported earlier [Massum et al., 2009]. According to Moslem & El-Kholie (2009), the extracts prepared from neem leave and seed are effective as antifungal against all tested fungi namely, *Alternaria solani, Fusarium oxysporum, Rhizoctonia solani,* and *sclerotinia sclerotiorum,* but *Fusarium oxysporum* and *Rhizoctonia solani* were the most sensitive fungi. The essential oil of Ocimum gratissimum may be recommended as a plant based safe (nontoxic) food additive in protecting the spices from deteriorating fungi as well as from aflatoxin contamination. Methyl cinnamate (48%) and γ-terpinene (26%) were recorded the major components of the oil through GC-MS analysis. The biological activity of an essential oil is related to the presence of bioactive compounds, the proportions in which they are present and due to the interactions between different compounds of the oil (Burt, 2004). The oils with antioxidant properties may be recommended in enhancing shelf life of products such as spices [Prakash et al., 2011].

Sinha et al. (1993) showed that cinnamon and clove essential oils were effective against aflatoxin formation in maize grain by *Aspergillus flavus* after 10 days under favorable conditions of mycotoxin production. It has also been shown that 500 μgg-1 of boldus, poleo, clove, anise, and mountain thyme were necessary to reduce growth rate and aflatoxin production in high pecentage (85–100%) in maize meal extract agar (MMEA) [Bluma et al., 2008b]. Also, the essential oils of Pimpinella anisum L. (anise), Pëumus boldus Mol (boldus), Hedeoma multiflora Benth (mountain thyme), Syzygium aromaticum L. (clove), and Lippia turbinate var. integrifolia (griseb) (poleo) have been shown to have significant inhibitory effect on lag phase, growth rate and aflatoxin B1 accumulation by *Aspergillus* section Flavi isolates in sterile maize grain at different water activity conditions. Only the highest concentration of the oils (3000 μgg-1) showed the ability to maintain antifungal activity [Bluma et al., 2008]. The effects of clove essential oil and its principal component, eugenol on growth and mycotoxin production by some toxigenic fungal genera such as *Aspergillus spp.*, *Penicillium spp.* and *Fusarium spp.* had been reported [Bullerman et al. 1977; Velluti et al., 2003, Velluti et al., 2004; Bluma et al., 2008].

The complete inhibition of mycelial growth of *Aspergillus flavus* and aflatoxin B production on rice grains can be assigned to eugenol extracted from clove which was effective at a concentration of 2.4 mg/g [Reddy et al., 2007]. Although clove oil is a good antifungal compound, cost is a major criterion for considering its inclusion in animal feeds [Gowda et al., 2004]. The seed extract of Ajowan (Trachyspermum ammi (L.) Sprague ex Turrill) showed the degradation of aflatoxin G1 (up to 65%). The dialyzed Trachyspermum ammi extract was more effective than the crude extract, capable of degrading >90% of the toxin. The aflatoxin detoxifying activity of the Trachyspermum ammi extract was drastically reduced upon boiling at 100 °C for 10 min. Significant levels of degradation of other aflatoxins viz., aflatoxin B1 (61%), aflatoxin B2 (54%) and aflatoxin G2 (46%) by the dialyzed Trachyspermum ammi extract was also observed. Time course study of aflatoxin G1 detoxification by dialyzed Trachyspermum ammi extract showed that more than 78% degradation occurred within 6 h and 91% degradation occurred 24 h after incubation [Velazhahan et al., 2010]. Other plant extracts namely, Syzigium aromaticum, Allium sativum, Curcuma longa, Ocimum sanctum, Annona squamosa, Azadirachta indica (Neem), Allium cepa, Eucalyptus terticornis, and Pongamia glaberima are among the list of the plant extracts in inhibiting both fungal growth and aflatoxin production by *Aspergillus* [Reddy et al., 2009]. In this conection, Haciseferogullary et al. (2005) reported the effect of garlic and onion extract on the mycoflora of pepper, cinnamon and rosemary and reported the effectiveness of garlic extract up to 0.25% (v/v) to inhibit the *Aspergillus flavus, Aspergillus fumigatus, Aspergillus niger, Aspergillus ochraceus, Aspergillus terreus, Penicillium chrysogenum, Penicillium puberulum, Penicillium citrinum, Penicillium corylophilum, Rhizopus stolonifer, Stachybotrys chartarum, Eurotium chevalieri and Emericella nidulans* growth. [Reddy et al., 2009]. Extract of garlic exhibited anti-fungal effects at all levels 0.1, 0.2, 0.5 and 1%. A maximum 84% reduction in toxin production occurred at the 1% level, but significant reductions in spore counts were recorded at all levels. The anti-fungal properties of garlic were also reported by Garcia and Garcia (1988) and Kshemkalyani et al. (1990).

Bilgrami et al. (1992) recorded up to a 60% reduction in aflatoxin production with onion extract supplementation by *Aspergillus flavus* in liquid SMKY medium and in maize grains. A lacrimatory factor (Thio propanol-S-oxide) in onion extract has a sporicidal effect on *Aspergillus parasiticus* [Sharma et al., 1981]. The anti-aflatoxigenic activity of Rosemary

(Rosmarinus officinalis L.) is probably due to borneol and other phenolics in the terpene fraction. In rosemary a group of terpenes (borneol, camphore, 1,8 cineole, α-pinene, camphone, verbenonone and bornyl acetate) were reported to be responsible [Davidson and Naidu; 2000; Rasooli et al., 2008; Jiang et al., 2011]. Antimicrobial activities of such extracts are mostly attributable to the presence of phenolic compounds such as thymol, and to hydrocarbons like γ-terpinene and p-Cymene with Limonenev being more active than p-Cymene [Dorman and Deans, 2000; Rasooli et al., 2007].

It has been reported that the chemical structures of the most abundant compounds in the essential oils is correlated with its antimicrobial activity. It seems possible that phenol components of essential oils may interfere with cell wall enzymes like chitin synthase/chitinase as well as with the α- and β-glucanases of the fungus [Adams et al., 1996]. Accordingly, the high content of phenol components may account for the high antifungal activity of oils [Adam et al., 1998]. Phenolics are secondary metabolites synthesized via phenylpropanoid biosynthetic pathway. These compounds are building blocks for cell wall structures, serving as defense against pathogens [Bluma et al., 2008a]. Also, the physical nature of essential oils, that is, low molecular weight combined with pronounced lipophilic tendencies allow them to penetrate cell membrane more quickly than other substances [Pawar & Thaker, 2007]. However, there is evidence that minor components have a critical part to play in antimicrobial activity, possibly by producing a synergic effect between other components [Burt, 2004]. The antimicrobial activity of essential oils or their constituents such as thymol, carvacrol and vanillin could act in different ways; (1) The result could be in the form of damage to the enzymatic cell system, including those associated with energy production and synthesis of structural compounds (2) denaturation of the enzymes responsible for spore germination or interference with the amino acid involved in germination [Nychas, 1995] and (3) irreversible damage in cell wall, cell membrane and cellular organelles. This was further confirmed when *Aspergillus parasiticus* and *Aspergillus flavus* were exposed to different essential oils (*Thymus eriocalyx* and *Thymus X-porlock*). The evidences presented here suggest that the essential oils could be safely used as preservative materials for certain food materials, particularly those which prevent fungal infections at relatively lower concentrations [Rasooli & Owlia, 2005]. The antiaflatoxigenic actions of essential oil may be related to inhibition of the ternary steps of aflatoxin biosynthesis involving lipid peroxidation and oxygenation [Alpsoy, 2010].It is clear that phenolic compounds inhibited one or more early rather than late steps in the aflatoxin B1 biosynthesis pathway. According to Farag et al. (1989) the presence of phenolic OH groups able to form hydrogen bonds with the active sites of target enzymes was thought to increase antimicrobial activity [Bluma et al., 2008a]. Natural products may regulate the cellular effects of aflatoxins and evidence suggests that aromatic organic compounds of spices can control the production of aflatoxins.

3.4 Biotechnological approaches for fighting aflatoxin-producing fungi

Molecular breeding of crops with an ability to degrade aflatoxins offers an alternative strategy for the management of aflatoxin contamination in agricultural commodities. Poppenberger et al., 2003 reported the isolation and characterization of a gene from Arabidopsis thaliana encoding a UDP-glycosyltransferase that is able to detoxify deoxynivalenol. Takahashi-Ando et al. [2004] isolated a zearalenone-detoxifying gene, zhd101, from Clonostachys rosea. These investigators further demonstrated that a recombinant Escherichia coli expressing zhd101 completely inactivated zearalenone and zearalenol within 1 h. It has been demonstrated that transgenic maize plants expressing the detoxification gene, zhd101 showed reduced contamination by the mycotoxin, zearalenone in maize kernels [Igawa et al., 2007]. Recently studies have been focused on identification of the aflatoxin detoxification genes from Trachyspermum ammi and to transfer them into crop plants in order to develop transgenic resistance to aflatoxin contamination [Velazhahan et al., 2010].

4. Conclusions

Several technologies have been tested to reduce mycotoxin risk. Field management practices that increase yields may also prevent aflatoxin. They include use of resistant varieties, timely planting, fertilizer application, weed control, insect control and avoiding drought and nutritional stress. Other options to control the toxin causing fungi Aspergillus flavus contamination in the field are use of non-toxigenic fungi to competitively displace toxigenic fungi, and timely harvest. Post-harvest interventions that reduce mycotoxins are rapid and proper drying, sorting, cleaning, drying, smoking, post harvest insect control, and the use of botanicals or synthetic pesticides as storage protectant. Another approach is to reduce the frequent consumption of 'high risk' foods (especially maize and groundnut) by consuming a more varied diet, and diversifying into less risky staples like sorghum and millet. Chemo-preventive measures that can reduce mycotoxin effect include daily consumption of chlorophyllin or oltipraz and by incorporating hydrated sodium calcium alumino-silicates into the diet. Detoxification of aflatoxins is often achieved physically, chemically and microbiologically by incorporating pro-biotics or lactic acid bacteria into the diet. There is need for efficient monitoring and surveillance with cost-effective sampling and analytical methods. Sustaining public education and awareness can help to reduce aflatoxin contamination. Phytochemicals may successfully replace physical and chemical agents and provide an alternative method to protect agricultural commodities of nutritional significance from toxigenic fungi such as Aspergillus flavus and aflatoxin production.

5. References

Adam, K.; Sivropoulou, A.; Kokkini, S.; Lanaras, T.; Arsenakis, M. (1998). Antifungal activities of Origanum vulgare subsp. hirtum, Mentha spicata, Lavandula angustifolia and Salvia fruticosa essential oils against human pathogenic fungi. J. Agric. Food Chem., 46, 1738-1745.

Adams, S.; Kunz, B.; Weidenbörner, M. (1996). Mycelial deformations of Cladosporium herbarum due to the application of Eugenol and Carvacrol. J. Essent. Oil Res., 8, 535-540.

Alberts, J.F., Gelderblom, W.C.A., Botha, A., & Van Zyl, W.H. (2009). Degradation of aflatoxin B1 by fungal laccase enzymes. International Journal of Food Microbiology 135, 47-52.

Allameh, A., Razzaghi-Abyaneh, M., Shams, M., Rezaei, M.B., & Jaimand, K., (2001). Effects of neem leaf extract on production of aflatoxins and fatty acid synthetase, citrate dehydrogenase and glutathione S-transferase in A. parasiticus. Mycopathologia, 154, 79–84.

Allameh, A., Safamehr, A., Mirhadi, S.A., Shivazad, M., Razzaghi-Abyaneh, M., Afshar-Naderi, A., (2005). Evaluation of biochemical nd production parametersof broiler chicks fed ammonia treated aflatoxin contaminated maize grain. Animal Food Science and Technology. 122, 289-301.

Alpsoy, L. (2010). Inhibitory effect of essential oil on aflatoxin activities. *African Journal of Biotechnology.* 9(17), 2474-2481.

Arora, D.S., Sharma, R.K., (2009). Ligninolytic fungal laccases and their biotechnological applications. Appl. Biochem. Biotechnol.,157(2), 174-209

Atanda, O.O., Akpan, I., Oluwafemi, F. (2007). The potential of some spice essen-tial oils in the control of A. parasiticus CFR 223 and aflatoxin production. *Food Control*, 18, 601–607.

Aziz, N.H., & Youssef, B. M. (2002). Inactivation of naturally occurring of mycotoxins in some Egyptian foods and agricultural commodities by gamma-irradiation. Egyptian Journal of Food Science, 30, 167–177.

Bankole, S. A., & Adebanjo, A. (2003). Mycotoxins in food in West Africa: Current situation and possibilities of controlling it. African Journal of Biotechnology, 2, 254–263.

Baptista, A. S., Horii, J., Calori-Domingues, M. A., da Gloria, E. M., Salgado, J. M., & Vizioli, M. R. (2004). The capacity of mannooligosaccharides thermolysed yeast and active yeast to attenuate aflatoxicosis.World Journal of Microbiology and Biotechnology, 20, 475–481.

Basilico, M.Z., Basilico, J.C., (1999). Inhibitory effect of some spice essential oils on Aspergillus ochraceus NRRL 3174 growth and ochratoxin production. Letters in Applied Microbiology 29 (4), 238–241.

Bata, A., & Lasztity, R. (1999). Detoxificatin of mycotoxin-contaminated food and feed by microorganisms. *Trends in Food Science and Technology.* 10, 223-228.

Beaver, R.W. (1991) Decontamination of mycotoxin-containing foods and feedstuffs. Trends in Food Sci.Technol. July, 170-173.

Beck, J., Ripka, S., Siegner, A., Schiltz, E., Schweizer, E., (1990). The multifunctional 6-methylsalicylic acid synthase gene of *Penicillium patulum. European Journal of Biochemistry* 192, 487–498.

Bennett, J.W., Klich, M., 2003. Mycotoxins. *Clinical Microbiology Reviews* 16, 497–516.

Bhatnagar D, Ehrlich KC, Cleveland TE (1992). Oxidation-reduction reactions in biosynthesis of secondary metabolites. In Biosynthesis of Secondary Metabolites Chapter 10: pp. 255-286. Edited by Town Publishers.

Bhatnagar, D., & McCormic, SP. (1988). The inhibitory effect of neem (*Azadirachta indica*) leaf extracts on aflatoxin synthesis in *Aspergillus parasiticus. J.Am. Oil Chem. Soc,* 65, 1166 –1168.

Bhatnagar, D., Yu, J., Ehrlich, K.C., (2002). Toxins of filamentous fungi. In: Breitenbach, M., Crameri, R., Lehrer, S.B. (Eds.), Fungal Allergy and Pathogenicity. *Chemical Immunology,* 81. Karger, Basel, pp. 167–206.

Bilgrami, K.S., Sinha, K.K., Sinha, A.K.. (1992). Inhibition of aflatoxin production and growth of *Aspergillus flavus* by eugenol and onion and garlic extracts. Indian J. Med. Res. Section-B Biomed. Res. 96, 171–175.

Bluma, R.V. & Etcheverry M.G. (2008). Application of essential oils in maize grain: Impact on *Aspergillus* section Flavi growth parameters and aflatoxin accumulation. *Food Microbiology*, 25, 324-334.

Bluma, R.V., Amaide´n, M.R., Etcheverry, M.G., (2008). Screening of Argentine plant extracts to check their impact on growth parameters and aflatoxin B1 accumulation by Aspergillus section Flavi. Int. J. Food Microbiol.29;122(1-2):114-25.

Bottone, E.J. and Peluso, R.W. (2003). Production by *Bacillus pumilus* (MSH) of an antifungal compound that is active against Mucoraceae and *Aspergillus* species: preliminary report. *Journal of Medical Microbiology*. 52, 69-74.

Brophy, J.J., Thubthimthed, S., Kitirattrakarn, T. and Anantachoke, C. (2002). Volatile leaf oil of *Melaleuca cajuputi*. In *Proceedings of the Forestry Conference*, pp.304-313.

Bullerman LB (1986). Mycotoxins and food safety. Food Technol. 40:59-66.

Bullerman, L.B., (1974). Inhibition of aflatoxin production by cinnamon. Journal of Food Science 39, 1163–1165.

Bullerman, L.B., Lieu, F.Y., Seier, S.A., (1977). Inhibition of growth and aflatoxin production by cinnamun and clove oils, cinnamic aldehyde and eugenol. J. Food Sci. 42, 637–646.

Burt, S. (2004). Essential oils: their antibacterial properties and potential applications in Foods- a review. International Journal of Food Microbiology, 94, 223−253.

Buser, M. D., & Abbas, H. K. (2002). Effects of extrusion temperature and dwell time on aflatoxin levels in cottonseed. Journal of Agricultural and Food Chemistry, 50, 2556–2559.

Cazzaniga, D., Basilico, J.C., Gonzalez, R.J., Torres, R.L., de Greef, D.M., (2001). Mycotoxins inactivation by extrusion cooking of corn flour. Lett. Appl. Microbiol. 33, 144–147.

Chang, P.K., & Hua S-S.T. (2007). Nonaflatoxigenic Aspergillus flavus TX9-8 competitively prevents aflatoxin accumulation by A.flavus isolates of large and small sclerotial morphotypes. *International Journal of Food Microbiology*. 114, 275-279.

Chiruvella, R.V., Jaluria, Y., Karwe,M.V., (1996). Numerical simulation of the extrusion process for food materials in a single-screw extruder. J. Food Eng. 30, 449–467.

Ciegler A. (1979). Control measures for aflatoxin contamination of agricultural commodities. Annals of the new York academy of Science. 329: 285-292.

Codifer L.P. Jr, Mann G.E, Dollear F.G. (1976). Aflatoxin inactivation: treatment of peanut meal with formaldehyde and calcium hidroxide. J Am Oil Chem Soc., 53:204-6.

Conner, D. E. (1993). Naturally occurring compounds, In *Antimicrobial in Foods*, pp.441- 468. Edited by Davidson, P. M. and Branen, A. L. Marcel Dekker Inc., New York

Coulombe Jr., R.A., (1993). Biological action of mycotoxins. J. Diary Sci.76, 880–891.

Daniels, J.M., Liu, L., Stewart, P.K., & Massey, T.E. (1990). Biotransformation of aflatoxin B1 in rabbit lung and liver microsomes. Carcinogenesis. 11(5), 823-827.

Davidson, P.M., Naidu, A.S., (2000). Phyto-phenols. In: Naidu, A.S. (Ed.), Natural Food Antimicrobial Systems. CRC Press, Boca Raton, FL, pp. 265–294.

Denissenko MF, Cahill J, Kondriakova TB, Gerber N, Pfeifer GP (1999). Quantitation and mapping of aflatoxin B1-induced DNA damage in genomic DNA using aflatoxin B1-8, 9-epoxide and microsomal activation systems. Mutat. Res. 425: 205-211.

Devegowda, G., Arvind, B. I. R., & Morton, M. G. (1996). Saccharomyces cerevisiae and mannanoligosaccharides to counteract aflatoxicosis in broilers. Proceedings of Australian poultry science symposium Sydney. pp. 103–106.

Dollear, F.G., Mann, G.E., Codifer, L.P., Gardner, H.K., Koltun, S.P., Vix, H.L.E., (1968). Elimination of aflatoxins from peanut meal. J. Am. Oil Chem. Soc. 45, 862.

Dorman, H.J.D., Deans, S.G., (2000). Antimicrobial agents from plants: antibacterial activity of plant volatile oils. Journal of Applied Microbiology 88, 308-316.

Dube, S., Upadhyay, P., Tripathi, S., (1989). Antifungal, physicochemical, and insect-repelling activity of the essential oil of Ocimum basilicum. Canadian Journal of Botany 67 (7), 2085-2087.

Dutton MF (1988). Enzymes and aflatoxin biosynthesis. Microbiological Reviews. 2: 274-295.

Dwividi, S.A., Dubey, B.L., 1993. Potentionial use of essential oil of the trachyepermum ammy against seed borne fungi of guar (Cyamopsis tetragonoloba L.). Mycopathologia 121 (2), 101-104.

Ela, M.A.A., El Shaer, N.S., Ghanem, N.B., (1996). Antimicrobial evaluation and chromatographic analysis of some essential and fixed oils. Pharmazie 51, 993-994.

El-Bazza, Z.E., Farrag, H.A., El-Fouly, M.Z., & El-Tablawy, S.Y. (2001). Inhibitory effect of gamma radiation and Nigella sativa seeds oil on growth spore germination and toxin production of fungi. *Radiation Physics and Chemistry*, 60, 181-189.

El-Bazza, Z.F., Mahmoud, M.T., Roushdy, H.M., Farrag, H.A. and El-Tablawy, S.Y. (1996). Fungal growth and mycotoxigenic production in certain medicinal herbs subjected to prolonged cold storage and possible control by gamma irradiation. *Egypt. J. Pharmaceutical Sci.*, 37: 85-95.

El-Maraghy, S.S.M., (1995). Effect of some spices as preservatives for storge of lentil (Lens esculenta L. seeds. Folia Microbiologica 40 (5), 490-492.

Fanelli, C., Fabbri, A.A., Finotti, E., Passi, S., (1983). Stimulation of aflatoxin biosynthesis by lipophilic epoxides. J. Gen. Microbiol. 129, 1721-1723.

Farag R.S., Daw Z.Y., & Abo-Raya S.H. (1989). Influence of some spice essential oils on *Aspergillus parasiticus* growth and production of aflatoxins in a synthetic medium. *J. Food Sci.*, 54, 74 -76.

Farag, R.S., Rashed, M.M., Abo-Hgger, A.A.A., (1996). Aflatoxin destruction by microwave heating. International Journal of FoodSciences and Nutrition 47, 197-208.

Frayssinet, C., LaFarge, C., (1970) Process for removing aflatoxins from peanut cake. French Patent 2.098,711, Recorded July 24, Accorded February 14, 1972.

Garcia, R.P., Garcia, M.L. (1988). Laboratory evaluation of plant extracts for the control of *Aspergillus* growth and aflatoxin production. In: Proceedings of the Japanese Association of Mycotoxicology. 1, 190-193.

Ghanem, I., Orfi, M., & Shamma, M. (2008). Effect of gamma radiation on the inactivation of aflatoxin B1 in food and feed crops. Brazilian Journal of Microbiology, 39, 787-791.

Ghannadi, A. (2002). Composition of the essential oil of Satureja hortensis L. seeds from Iran. Journal of Essential Oil Research, 14, 35-39.

Ghfir, B., Fonvieille, J.L., & Dargent, R. (1997). Effect of essential oil of Hyssopus officinalis on the chemical composition of the walls of *Aspergillus fumigatus*. *Mycopathologia*, 138, 7-12.

Gowda, N.K.S., Malathi, V., & Suganthi, R.U. (2004). Effect of some chemical and herbal compounds on growth of *Aspergillus parasiticus* and aflatoxin production. *Animal Feed Science and Technology*, 116, 281-291.

Gratz, S., Mykkanen, H., Ouwehand, A. C., Juvonen, R., Salminen, S., & El-Nezami, H. (2004). Intestinal mucus alters the ability of probiotic bacteria to bind aflatoxin B1 in vitro. Applied and Environmental Microbiology, 70, 6306-6308.

Griffin, G.S.; Markham, L.J.; Leach, N.D. (2000) An agar dilution method for the determination of the minimum inhibitory concentration of essential oils. *J. Essent. Oil Res.*, *12*, 149-255.

Haciseferogullary, H., Ozcan, M., Demir, F., & Caly´ sy´ r, S. (2005). Some nutritional and technological properties of garlic (Allium sativum L.). Journal of Food Engineering, 68, 463–469.

Hagler WM, Hutchins JE, Hamilton P.B (1982). Destruction of aflatoxin in corn Bi with sodiumbisulfite. J Food Prot., 45, 1287-91.

Hagler WM, Hutchins JE, Hamilton P.B (1983). Destruction of aflatoxin in corn B, with sodiumbisulfite: isolation of the major product aflatoxin BS. J. Food Prot., 46,295-300.

Halliwell B, Gutteridge JMC (1999). Oxidative stress: adaptation damage, repair and death, in: Halliwell B, Gutteridge JMC (Eds.), Free Radicals in Biology and Medicine, Oxford University Press, New York, pp. 246-350.

Haskard C.A., EL-Nezami H.S., Kankaanpaa P.E., Salminen S.,Ahokas J.T. (2001): Surface binding of aflatoxin B1 by lactic acid bacteria. Applied and Environmental Biology, 67, 3086-3091.

Heimbecher S.K., Jorgensen K.V, Price R.L., (1988) Interactive effects of duration of storage and addition of formaldheyde on levels of aflatoxin M, in milk. J Assoc Offic Anal Chem; 71:285-7.

Holmes RA, Boston RS, Payne GA (2008). Diverse inhibitors of aflatoxin biosynthesis. Appl. Microbiol. Biotechnol. 78: 559-572.

Hope, R., Aldred, D., Magan, N., (2005). Comparison of environmental profiles for growth and deoxynivalenol production by *Fusarium culmorum* and *F. graminearum* on wheat grain. *Letters in Applied Microbiology* 40, 295–300.

Hsieh, D.P.H. (1973). Inhibition of aflatoxin biosynthesis of dichlorvos. J. Agric. Food. Chem. 21(3), 468-470.

Hua S.T. (2004). Reduction of aflatoxinin pistachio through biological control of aspergillus flavus. Extension reports. California Pistachio Commision Production Research Reports. P 212- 220.

Huang, J.Q., Jiang, H.F., Zhou, Y.Q., Lei, Y., Wang, S.Y., & Liao B.S. (2009). Ethylene inhibited aflatoxin biosynthesis is due to oxidative stress alleviation and related to glutathione redox state changes in *Aspergillus flavus*. *International Journal of Food Microbiology*, 130, 17-21.

Hwang, J.H., & Lee, K.G. (2006). Reduction of aflatoxin B1 contamination in wheat by various cooking treatments. *Food Chemistry*, 98, 71-75.

IARC (1993). Some naturally occurring substances: Food items and constituents, heterocyclic aromatic amines and mycotoxins. IARC monographs on the evaluation of carcinogenic risks to humans. 56. International Agency for Research on Cancer (pp. 489–521).

Igawa T., Takahashi-Ando N., Ochiai N., Ohsato S., Shimizu T., Kudo T., Yamaguchi I., Kimura M. (2007). Reduced contamination by the Fusarium mycotoxin zearalenone in maize kernels through genetic modification with a detoxification gene. Appl Environ Microbiol. 73(5):1622-1629.

Inouye, S., Tsuruoka, T., Watanabe, M., Takeo, K., Akao, M., Nishiyama, Y., Yamaguchi, H., (2000). Inhibitory effect of essential oils on spical growth of Aspergillus fumigatus by vapour contact. Mycoses 43, 1–2, 17–23.

Inouye, S., Watanabe, M., Nishiyama, Y., Takeo, K., Akao, M., Yamaguchi, H., (1998). Antisporulating and respiration inhibitory effects of essential oils on filamentous fungi. Myoses 41(9–1 0), 403– 410.

Jalili, M., Jinap, S., & Noranizan, A. (2010). Effect of gamma radiation of mycotoxins in black pepper. Food Control, 21, 1388-1393.

Jayashree, T., Subramanyam, C., (1999). Antiaflatoxigenic activity of eugenol is due to inhibition of lipid peroxidation. Lett. Appl. Microbiol. 28, 179–183.

Jiang Y., Wu N., Fu N.J., Wang W., Luo M., Zhao C.J., Zu Y.J., Liu XL. (2011). Chemical composition and antimicrobial activity of the essential oil of Rosemary. Environ Toxicol Pharmacol. 32(1):63-8.

Joseph, G.S., Jayaprakasha, G.K., Selvi, A.T., Jena, B.S., Sakariah, K.K., (2005). Antiaflatoxigenic and antioxidant activities of Garcinia extracts. Intl. J. Food Microbiol. 101, 153–160

Karolewiez, A., Geisen, R., (2005). Cloning a part of the ochratoxin A biosynthetic gene cluster of Penicillium nordicum and characterization of the ochratoxin polyketide synthase gene. Systematic and Applied Microbiology 28, 588–595.

Kimura, M., Tokai, T., O'Donnell, K., Ward, T.J., Fujimura, M., Hamamoto, H., Shibata, T.,Yamagushi, I., (2003). The trichothecene biosynthesis gene cluster of Fusarium graminearum F15 contains a limited number of essential pathways genes and expressed non-essential genes. FEBS Letters 539, 105–110.

Kshemkalyani, S.B., Ragini, Telore, Madhavi, B., Patel, G.S., Telore, R., (1990). The effect of allicine and extracts of garlic on Aspergillus flavus and Aspergillus parasiticus. Indian J. Mycol. Plant Pathol. 20, 247–248.

Kuliman-Wahls, M.E.M., Vilar, M.S., De Nijs-Tjon, L., Maas, R.F.M., & Fink-Gremmels, J. (2002). Cyclopiazonic acid inhibits mutagenic action of aflatoxin B1. Environmental Toxicology and Pharmacology, 11, 207-212.

Kurita, N., Miyaji, M., Kurane, R., & Takahara, Y. (1981). Antifungal activity of essential oil components. Agricultural Biology and Chemistry, 45, 945–952.

Leonowicz, A., Matuszewska, A., Luterek, J., Ziegenhagen, D., Wojtas-Wasilewska, M., Cho, N.S., Hofrichter, M., Rogalski, J., (1999). Biodegradation of lignin by white rot fungi. Fungal Genet. Biol. 27, 175–185.

Lillehoj, E. B., Zuber, M.S., (1974). Aflatoxin problem in corn and possible solutions. In: Proc. Annu. Corn & Sorghum Res. Conf. 1974. Chicago IL. American Seed Trade Association, Alexandra, VA. pp 230-250.

Liu, D.L., Yao, D.S., Liang, R., Ma, L., Cheng, W.Q., & Gu, L.Q. (1998). Detoxification of aflatoxin B1 by enzymes isolated from Armillariella tabescens. Food and Chemical Toxicology, 36, 563-574.

Maeba, H., Takamoto, Y., Kamimura, M., Miura, T., (1988). Destruction and detoxification of aflatoxins with ozone. Journal of Food Science 53, 667–668.

Mahmoud AL., (1994) Antifungal action and antiaflatoxigenic properties of some essential oil constituents. Lett. Appl. Microbiol. 2, 110 - 3.

Mahoney, N., Molyneux, R.J., (2004). Phytochemical inhibition of aflatoxigenicity in Aspergillus flavus by constituents of walnut (Juglans regia). J. Agric. Food Chem. 52, 1882–1889.

Mann, G.E., Codifer, L.P., Garner, H.K., Koltun, S.P., Dollear, F.G., (1970). Chemical inactivation of aflatoxins in peanut and cottonseed meals. J. Am. Oil Chem. Soc. 47, 173.

Massum, M.M.I., Islam, S.M., and Fakir M.G.A. (2009). Effect of seed treatment practices in controlling of seed-born fungi in sorghum. Scient. Res. Essay. 4:22-27.

McKenzie, K.S., Sarr, A.B., Mayura, K., Bailey, R.H., Miller, D.R., Rogers, T.D., Norred, W.P., Voss, K.A., Plattner, R.D., Kubena, L.F., & Phillips, T.D. (1997). Oxidative degradation and detoxification of mycotoxins using a novel source of ozone. *Food and Chemical Toxicology*, 35, 807-820.

Mendez-Albores, A., Arambula-Villa, G., Loarca-Pina, M.G.F., Castano-Tostado, E., Moreno-Martinez, E., (2005). Safety and efficacy evaluation of aqueous citric acid to degrade B-aflatoxins in maize. Food Chem. Toxicol.43 (2), 233–238.

Mendez-Albores, A., Del Rio-Garcia, J.C., & Moreno-Martinez, E. (2007). Decontamination of aflatoxin duckling feed with aqueous citric acid treatment. *Animal Feed science and Technology*, 135, 249-262.

Mendez-Albores, A., Veles-Medina, J., Urbina-Alvarez, E., Martinez-Bustos, F., Moreno-Martinez, E. (2009). Effect of citric acid on aflatoxin degradation and on functional and textural properties of extruded sorghum. *Animal Feed Science and Technology*, 150, 316-329.

Miller, R.C., 1990. Unit operations and equipment IV. Extrusion and extruders. In: Fast, R.B., Galdwell, E.F. (Eds.), Breakfast Cereals and How They Are Made. American Association of Cereal Chemists, St Paul, MN, pp. 135–193.

Mishra, H.N., Das, C., 2003. A review on biological control and metabolism of aflatoxin. Crit. Rev. Food Sci. 43, 245–264.

Moerck KE, McElfresh P, Wohlman A, Hilton BW. Aflatoxin destruction in corn using sodium bisulfite, sodium hydroxide and aqueous ammonia. J Food Prot 1980; 43:571-4.

Monroe D.H. & Eaton D.L. (1987) Comparative effects of butylated hyroxyanisole on hepatic in vivo DNA binding and in vitro biotransformation of aflatoxin B1 in the rat and mouse. *Toxicol Appl Pharmacol* 90: 401-409.

Montes-Belmont, R., Carvajall, M., 1998. Control of Aspergillus flavus in maize with plant essential oils and their components. Journal of Food Protection 61 (5), 616–619.

Moore, G., 1994. Snack food extrusion. In: Frame, N.D. (Ed.), The Technology of Extrusion Cooking. American Association of Cereal Chemists, St Paul, MN, pp. 111–143.

Morozumi, S., 1978. Isolation, purficiation and antibiotic activity of o-methoxycinnamaaldyde from cinnamon. Applied Environmental Microbiology 36, 577–583.

Moslem, M.A., & El-Kholie, E.M. (2009). Effect of neem (Azadirachta indica A.Juss) seeds and leaves extract on some plant pathogenic fungi. *Pakistan Journal of Biological Sciences*, 12(14), 1045-1048.

Namazi, M., Allameh, A.A., Aminshahidi, M., Nohee, A., and Malekzadeh, F. (2002). Inhibitory effects of ammonia solution on growth and aflatoxin production by Aspergillus parasiticus NRRL. Acta Pol Toxicol. 10, 65-72.

Nychas G.J.E. (1995). Natural antimicrobial from plants. In New Methods of Food Preservations ed. Gould GW. pp. 58-89. Glasgow, UK: Blackie Academic and Professional.

O'Callaghan, J., Caddick, M.X., Dobson, A.D.W., 2003. A polyketide synthase gene required for ochratoxin A biosynthesis in Aspergillus ochraceus. *Microbiology* 149, 3485-3491.

Odamtten, G.T., Appiah, V., Langerak, D.I. (1986). Preliminary studies of the effects of heat and gamma irradiation on the production of aflatoxin B1 in static liquid culture, by *Aspergillus flavus* link NRRL 5906. *International Journal of Food Microbiology*, 3, 339-348.

Ogiehor, I.S., & Ikenebomeh, M.J. (2004). Antimicrobial effects of sodium benzoate on the growth, survival and aflatoxin production potential of some species of Aspergillus in garri during storage. *Pakistan Journal of Nutrition*, 3(5), 300-303.

Onyenekwe, P.C., Ogbadu, G.H., Hashimoto, S. (1997). The effect of gamma radiation on the microflora and essential oil of Ashanti pepper (Piper guineense) berries. *Postharvest Biology and Technology*, 10, 161-167.

Ozcan, M. (1998). Inhibitory effects of spice extracts on the growth of *Aspergillus parasiticus* NRRL2999 strain. *Z Lebensm Unters Forsch A*, 207, 253-255.

Park, B.J., Takatori, K., Sugita-Konishi, Y., Kim, I.H., Lee, M.H., Han, D.W., Chung, K.H., Hyun, S.O., Park, J.C. (2007). Degradation of mycotoxins using microwave-induced argon plasma at atmospheric pressure. *Surface and Coating Technology*, 201, 5733-5737.

Park, D.L. (1993) `Controlling aflatoxin in food and feed. Food Technology, 47, 92-96.

Paster, N., Menasherov, M., Ravid, U., & Juven, B. (1995). Antifungal activity of oregano and thyme essential oils applied as fumigants against fungi attacking stored grain. Journal of Food Protection, 58, 81-85.

Pawar VC, Thaker VS (2007) In vitro efficacy of 75 essential oils against Aspergillus niger. Mycoses, 49: 316-323.

Payne G.A., & Brown M.P. (1998). Genetics and physiology of aflatoxin biosynthesis. Annu. Rev. Phytopathol. 36, 329-362.

Penalva, M.A., Arst Jr., H.N., (2002). Regulation of gene expression by ambient pH in filamentous fungi and yeasts. Microbiology and Molecular Biology Reviews. 66, 426-446.

Phillips TD, Kubena LF, Harvey RB, Taylor DR, Heidelbaugh ND. (1988). Hydrated sodium calcium aluminosilicate: a high affinity sorbent for aflatoxin. Poult Sci 67:243-7.

Pier, A.C., (1992). Major biological consequence of aflatoxicosis in animal production. J. Anim. Sci. 70, (3964-3967).

Pietri A, Fusconi G, Blasi P, Moschini M. Azione sequestrante di una zeolite nei confronti di radionuclidi e aflatossina in diete per ruminanti e monogastrici. Proc X Congresso ASPA, 141-6, Italy, Bologna, 31 maggio-6 giugno 1993

Pinto E, Pina-Vaz C, Salgueiro L, Goncalves MJ, Costa-de-Oliverira S, Cavaleiro C, Palmeria A, Rodrigues A, & Martinez-de-Oliveria J. Antifungal activity of the essential oil of *Thymus pulegioides* on Candida, Aspergillus and dermatophyte species. *J. Med. Microbiol.* 2006; 55 (10), 1367 - 73

Piva, G., Galvano, F., Pietri, A., Piva, A. (1995). Detoxification methods of aflatoxins. A review. *Nutrition Research*, 15(5), 767-776.

Pomeranz, Y.I., Bechtel, D.B., Sauer, D.R. and Seitz, L.M. (1990) `Fusarium Headblight (scab) in Cereal Grain' in Advances in Cereal Science and Technology Vol.10., (Pomeranz, Y., ed), pp. 373±473, AACC, St. Paul

Poppenberger, B., Berthiller, F., Lucyshyn, D., Sieberer, T., Schuhmacher, R., Krska, R., et al. (2003). Detoxification of the Fusarium mycotoxin deoxynivalenol by a UDP-glucosyltransferase from Arabidopsis thaliana. Journal of Biological Chemistry, 278, 47905–47914.

Prado, G., Carvalho, E.P.D., Oliveira, M.S., Madeira, J.G.C., Morais, V.D., Correa, R.F., et al. (2003). Effect of gamma irradiation on the inactivation of aflatoxin B1 and fungal flora in peanut. Brazilian Journal of Microbiology, 34(Suppl. 1) (Sao Paulo).

Prakash, B., Shukla, R., Singh, P., Mishra, P.K., Dubey, N.K., Kharwar, R.N. (2011). Efficacy of chemically characterized Ocimum gratissmum L. essential oil as an antioxidant and a safe plant based antimicrobial against fungal and aflatoxin B1 contamination of spices. Food Research International, 44, 385-390.

Proctor, A.D., Ahmedna, M., Kumar, J.V., Goktepe, I., (2004). Degradation of aflatoxins in peanut kernels/flour by gaseous ozonation and mild heat treatment. Food Additives and Contaminants 21, 786–793.

Proctor, R.H., Brown, D.W., Plattner, R.D., Desjardins, A.E., (2003). Co-expression of 15 contiguous genes delineates a fumonisin biosynthetic gene cluster in Gibberella moniliformis. Fungal Genetics and Biology 38, 237–249.

Prudente Jr., A.D., King, J.M., (2002). Efficacy and safety evaluation of ozonation to degrade aflatoxin in corn. Journal of Food Science 67, 2866–2872.

Rahman, S., 1995. Food Properties Handbook. CRC Press, New York.

Ramos, A.R., Falco L.L, Barbosa G.S., Marcellino L.S., and Gander E.S. (2007). Neem (Azadirachta indica A. Crinipellis perniciosa and Phytophtora spp. Microbiol. Ress. 162: 238-243.

Rasooli, I., Allameh, A.A., Rezaee, M.B. (2005). Antimicrobial efficacy of thyme essential oils as food preservatives. In: Food Policy, Control and Research. Editor: Arthur P. Riley. Chapter one, pp. 1-33 Nova Science Publishers. Inc.

Rasooli, I., Fakoor, M.H., Allameh, A., Rezaee, M.B., & Owlia, P. (2009). Phytoprevention of aflatoxin production. Journal of Medicinal Plants, 8(5), 97-104.

Rasooli, I., Fakoor, M.H., Yadegarinia, D., Gachkar, L., Allameh, A.A., & Rezaei, M.B. (2007). Antimycotoxigenic characteristics of Rosmarinus officinalis and Trachyspermum copticum L. essential oils. Int. J. Food Microbiol., 29, 135-139.

Rasooli, I., Fakoor, M.H., Yadegarinia, D., Gachkar, L., Allameh, A., Rezaee, M.B. (2008). Antimycotoxigenic characteristics of Rosmarinus officinalis and Trachyspermum copticum L. essential oils. International Journal of Food microbiology. 122, 135-139.

Rasooli, I., Owlia, P. (2005). Chemoprevention by thyme oils of Aspergillus parasiticus growth and aflatoxin production. Phytochemistry, 66, 2852-2856.

Rasooli, I., Razzaghi-Abyaneh, M. (2004). Inhibitory effect of thyme oils on growth and aflatoxin production by Aspergillus parasiticus. Food Control, 15, 479-483.

Rasooli, I., Rezaei B., Allameh, A.A. (2006b). Ultrastructure studies on antimicrobial efficacy of thyme essential oils on Listeria monocytogenesis. International Journal of Infectious disease. 10, 236-241.

Rasooli, I., Rezaei, MB., & Allameh, A. (2006a). Growth inhibition and morphological alterations of *Aspergillus niger* by essential oils from *Thymus eriocalyx* and *Thymus x-porlock*. Food Control, 17, 359 –3 64.

Razzaghi-Abyaneh M., Allameh A., Al-Tiraihi T., and Shams M. (2005a). Studies on the mode of action of neem (Azadirachta indica) leaf and seed extracts on morphology and aflatoxin production ability of Aspergillus paraiticus. Acta Hort., 675, (123-127).

Razzaghi-Abyaneh, M., Allameh, A., & Shams, M. (2000). Screening of aflatoxin producing mould isolates based on fluorescence production on a specific medium under ultraviolet light. *Acta Med. Iranica,* 38, 67-73.

Razzaghi-Abyaneh, M., Allameh, A., Tiraihi, T., Shams-Ghahfarokhi, M., and Ghorbanian, M, (2005b). Morphological alterations in toxigenic *Aspergillus parasiticus* exposed to neem (Azadirachta indica) leaf and seed aqueous extracts. *Mycopathologia,* 159, 565–570.

Razzaghi-Abyaneh, M., Shams-Ghahfarokhi, M., Yoshinari, T., Rezaee, M B., Jaimand, K., Nagasawa, H., & Sakuda, S. (2008). Inhibitory effects of Satureja hortensis L. essential oil on growth and aflatoxin productionby *Aspergillus parasiticus*. *International Journal of Food Microbiology*, 12, 228-233.

Reddy, C. S., Reddy, K. R. N., Prameela, M., Mangala, U. N., & Muralidharan, K. (2007). Identification of antifungal component in clove that inhibits Aspergillus spp. colonizing rice grains. Journal of Mycology and Plant Pathology, 37(1), 87–94.

Reddy, K.R.N., Reddy, C.S. & Muralidharan, K. (2009). Potential of botanicals and biocontrol agents on growth and aflatoxin production by *Aspergillus flavus* infecting rice grains. *Food Control*, 20, 173-178.

Reichling, J., Schnitzler, P., Suschke, U., Saller, R. (2009). Essential oils of aromatic plants with antibacterial, antifungal, antiviral, and cytotoxic properties. an overview. *Forsch Komplementmed*, 16, 79-90.

Rokey, G.J., 1994. Petfood and fishfood extrusion. In: Frame, N.D. (Ed.), The Technology of Extrusion Cooking. American Association of Cereal Chemists, St Paul, MN, pp. 144–189.

Rustom, I. Y. S. (1997). Aflatoxin in food and feed: occurrence legislation and inactivation by physical methods. Food Chemistry, 59, 57–67.

Samarajeewa, U., Sen, A.C., Cohen, M.D., Wei, C.I., 1990. Detoxification of aflatoxins in foods and feeds by physical and chemical methods. Journal of Food Protection 53, 489–501.

Santin, E., Paulilo, A. C., Maiorka, A., Nakaghi, L. S. O., Macan, M., de Silva, A. V. F., et al. (2003). Evaluation of the efficiency of Saccharomyces cerevisiae cell wall to ameliorate the toxic effects of aflatoxin in broilers. International Journal of Poultry Science, 2, 241-344.

Santos, C. C. M., Lopes, M. R. V., & Kosseki, S. Y. (2001). Ocorrê^ncia de aflatoxinas em amendoim e produtos de amendoim comercializados na regia~o de Sa~o Jose´ de Rio Preto/SP. Revista do Instituto Adolfo Lutz, 60(2), 153–157.

Saxena, M., allameh A.A., mukerji, K.G., & Raj H.G. (1989). Studies on glutathione – transferase of Aspergillus flavus group in relation to aflatoxin production. J. Toxicology-Toxin Rev., 8; 319-328.

Schmidt-Heydt, M., & Geisen, R. (2007). A microarray for monitoring the production of mycotoxins in food. *International Journal of Food Microbiology*, 117, 131-140.

Šegvić Klarić, M., Kosalec, I., Mastelić, J., Piecková, E., Pepeljnack, S., 2006. Antifungal activity of thyme (Thymus vulgaris L.) essential oil and thymol against moulds from damp dwellings. Letters in Applied Microbiology 44, 36–42.

Sharma, A., Padwal Desai, S.R., Tewari, G.M., Bandyopadhyay, C., (1981). Factors affecting antifungal activity of onion extractives against aflatoxin producing fungi. J. Food Sci. 46, 741–744.

Shetty, P.H., & Jespersen, L. (2006). *Saccharomyces cerevisiae* and lactic acid bacteria as potential mycotoxin decontaminating agents. *Trends in Food science and Ttechnology*, 17, 48-55.

Singh, P., Srivastava, B., Kumar, A., & Dubey, N. K. (2008). Fungal contamination of raw materials of some herbal drugs and recommendation of Cinnamomum camphora oil as herbal fungitoxicant. Microbial Ecology, 56, 555−560.

Sinha, K.K., Sinha, A.K., Prasad, G., (1993). The effect of clove and cinnamon oils on growth of and aflatoxin production by Aspergillus flavus. Lett. Appl. Microbiol. 16, 114–117.

Sison JA. Sequestering mycotoxins in animal feeds. Feeds Cotnp 1992; 12:34-35.

Smiley, R.D., Draughon, F.A., 2000. Preliminary evidence that degradation of aflatoxin B1 by Flavobacteria aurantiacum is enzymatic. J. Food Prot. 63, 415–418.

Snyder, O.P. 1997 *Antimicrobial Effects of Spices and Herbs*. Hospitality Institute of Technology and Management, St Paul, Minnisota.

Sokovic M.D., Vukojevic J., Marin P.D., Brkic D.D., Vajs V., Van Griesnsven L.J.L.D (2009). Chemical composition of essential oils of Thymus and menthe species and their antifungal activities. Molecules 14:238-249.

Soliman, K.M., Badeaa, R.I. (2002). Effect of oil extracted from some medicinal plants on different mycotoxigenic fungi. *Food and Chemical Toxicology* , 40, 1669–1675.

Sommartya T, Jatumanusiri T, Konjing C, Maccormac C. Aspergillus flavus in peanut in Thailand with special reference to aflatoxin contamination, and detoxification. Proc. Jap. Assoc. Mycotox., 1988; Suppl. N" 1, 71-2.

Soni, K.B., Rajan, A., Kuttan, R., 1992. Reversal of aflatoxin induced liver damage by turmeric and curcumin. Cancer Lett. 66, 115–121.

Sova Z, Pohunkova H, Reisnerova H, Slamova A, Haisl K. Haematological and histological response to the diet containing aflatoxin B, and zeolite in broilers of domestic fowl. Acta Vet Brno 1991; 60:31-40.

Stankov M, Obradovic V, Obradovic J, Vukicevic 0. Effect of a micronized zeolite preparation on the health and production of weaned piglets. Veterinarsky Glasnik 1992; 46:91-6.

Strosnider H., Azziz-Baumgartner E., Banziger M., Bhat R.V., Breiman R., Brune M., et al. (2006). Workgroup report: public health strategies for reducing aflatoxin exposure in developing countries. Environ. Health Perspect, 114: 1989-1903.

Takahashi-Ando, N., Ohsato, S., Shibata, T., Hamamoto, H., Yamaguchi, I., & Kimura, M. (2004). Metabolism of zearalenone by genetically modified organisms expressing the detoxification gene from Clonostachys rosea. Applied and Environmental Microbiology, 70, 3239–3245.

Teniola, O.D., Addo, P.A., Brost, I.M., Farber, P., Jany, K.D., Alberts, J.F., Zyl, W.H.V., Steyn, P.S., & Holzapfel, W.H. (2005). Degradation of aflatoxin B1 by cell-free extracts of *Rhodococcus erythropolis* and *Mycobacterium fluoranthenivorans* sp. Nov. DSM44556. *International Journal of Food Microbiology*, 105, 111-117.

Thanaboripat, D., Suvathi, Y., Srilohasin, P., Sripakdee, S., Patthanawanitchai, O., Charoensettasilp, S. (2007). Inhibitory effect of essential oils on the growth of *Aspergillus flavus*. *KMITL Sci. Tech J*, 7:1, 1-7.

Thiesen, J., 1977. Detoxification of aflatoxins in groundnut meal. Anim. Feed Sci. Technol. 2, 67–75.

Trail F, Mahanti N, Linz J (1995). Molecular biology of aflatoxin biosynthesis. Microbiology, 141: 755-765.

Valko M, Rhodes CJ, Moncola J, Izakovic M, Mazura M (2006). Free radicals, metals and antioxidants in oxidative stress-induced cancer Chemico Biol. Interact. 160: 1-40.

Van Egmond, H. P. (1989). Current situation on the regulations for mycotoxins. Overview of tolerances and status of the standard methods of sampling and analysis. *Food Additives and Contaminants*, 6, 139–188.

Velazhahan, R., Vijayanandraj, S., Vijayasamundeeswari, A., Paranidharan, V., Samiyappan, R., Iwamoto T., Friebe B., & Muthukrishnan S. (2010). Detoxification of aflatoxins by seed extracts of the medicinal plant, Trachyspermum ammi (L.) Sprague ex Turrill- Structural analysis and biological toxicity of degradation product of aflatoxin G1. *Food Control*, 21, 719-725.

Veldman A. Effect of sorbentia on carry-over of aflatoxin from cow feed to milk (1992). Milchwissenschaft, 47:777-80.

Velluti, A., Sanchis, V., Ramos, A.J., Ergido, J.,Marin, S., 2003. Inhibitory effect of cinnamon, clove, lemongrass, oregano and palmarosa essential oils on growth and fumonisin B1 production by Fusarium proliferatum in maize grain. Int. J. Food Microbiol. 89, 145–154.

Velluti, A., Sanchis, V., Ramos, A.J., Marin, S., 2004. Effect of essential oil of cinnamon, clove, lemongrass, oregano and palmarosa on growth and fumonisin B1 production by Fusarium verticilloides in maize grain. J. Sci. Food Agric. 89, 145–154.

Wang JS, Groopman JD (1999). DNA damage by mycotoxins. Mutat.Res. 424: 167-181.

Winfree RA, Allred A. (1992). Bentonite reduces measurable aflatoxin B in fish feed. Progressive Fish-Culturist, 54: 157-62.

Wu F., Khlangwiset (2010). Health economic impacts and cost-effectiveness of aflatoxin reduction strategies in Africa: case studies in biocontrol and postharvest interventions. Food Additives and Contamination, 27 (4): 496-509.

Wu Q., Jezkova A., Yuan Z., Pavlikova L., Dohnal V., & Kamil Kuca K. (2009). Biological degradation of aflatoxins. *Drug Metabolism Reviews*, 41(1): 1–7

Xu, A., 1999. Use of ozone to improve the safety of fresh fruits andvegetables. Food Technology 53, 58–62.

Yabe K, Nakajima H (2004). Enzyme reactions and genes in aflatoxin biosynthesis. Appl. Microbiol. Biotechnol. 64: 745-755.

Yagen B, Hutchins JE, Hagler WM, Hamilton PB. Aflatoxin B,S: revised structure for the sodium sulfonate formed by destruction of aflatoxin B, with sodiumbisulfite. J Food Prot 1989; 52574-7.

Yin YN, Yan LY, Jiang JH, and Ma ZH (2008). Biological control of aflatoxin contamination of crops. Journal of Zhejiang University Science, 9(10): 787–792.

Yu, J., Bhatnagar, D., Cleveland, T.E., 2004a. Completed sequence of aflatoxin pathway gene cluster in Aspergillus parasiticus. FEBS Letters 564, 126–130.

Yu, J., Chang, P.-K., Ehrlich, K.C., Cary, J.W., Bhatnagar, D., Cleveland, T.E., Payne, G.A., Linz, J.E., Woloshuk, C.P., & Bennett, J.W. (2004b). Clustered pathway genes in aflatoxin biosynthesis. *Applied and Environmental Microbiology, 70*, 1253–1262.

Zaika, L.L. and Buchanan, R.L. (1987). Review of compounds affecting the biosynthesis or bioregulation of aflatoxins. *Journal of Food Protection*, 50(8), 691-708.

Zambonelli, A., Zechini D'Aurelio, A., Bianchi, A., & Albasini, A. (1996). Effects of essential oils on phytopathogenic fungi. *Journal of Phytopathology, 144*, 491–494.

Zhang, X. B., & Ohta, Y. (1991). Binding of mutagens by fractions of lactic acid bacteria on mutagens the cell wall skeleton. Journal of Dairy Science, 74, 1477–1481.

Ziglari, T., Allameh, A., Razzaghi-Abyaneh, M., Khosravi, A.R., & Yadegari M.H. (2008). Comparison of glutathione S-transferase activity and concentration in aflatoxin-producing and their no-toxigenic counterpart isolates. *Mycopathologia, 166*, 219-226.

15

Aflatoxin B1 - Prevention of Its Genetic Damage by Means of Chemical Agents

Eduardo Madrigal-Bujaidar[1], Osiris Madrigal-Santillán[2],
Isela Álvarez-González[1] and Jose Antonio Morales-González[2]
[1]Laboratorio de Genética, Escuela Nacional de Ciencias Biológicas, IPN,
[2]Instituto de Ciencias de la Salud, UAEH,
México

1. Introduction

Mycotoxins are structurally diverse groups largely composed of small molecular weight chemicals, which are generally produced by the mycelial structure of filamentous fungi. These toxins are secondary metabolites mainly synthesized during the end of the mould exponential phase of growth. They appear to have no biological significance with respect to their growth/development or competitiveness, but when ingested by higher vertebrates and other animals they can cause diseases called mycotoxicoses (Kabak et al., 2006; Madrigal-Santillán et al., 2010). Aflatoxin B1 (AFB1), in particular, is a tetrahydrofuran moiety fused to a coumarin ring and was chemically classified as cyclopenta[c]furo[3',2':4,5]furo[2,3h][1] benzopyran-1,11-dione,2,3,6a,9a-tetrahydro-4-methoxy-, (6aR,9aS) (Eaton et al., 1994; Hedayati et al., 2007) (Figure 1).

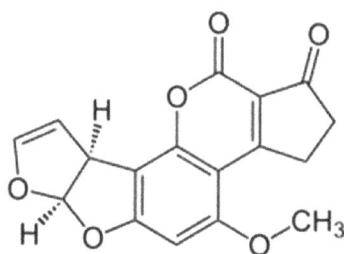

Fig. 1. Chemical structure of aflatoxin B1

The compound is a pale-white to yellow crystalline, odorless solid, soluble in water and in polar organic solvents, such as methanol, chloroform, acetone, acetonitrile, and dimethyl sulfoxide. It has a molecular weight of 312.3, a melting point between 268-269°C, and shows a blue fluorescence in the presence of ultraviolet light (Eaton et al., 1994; Hussein & Brasel, 2001). This secondary metabolite is produced by several strains of filamentous ascomycetes fungi, mainly *Aspergillus flavus* and *Aspergillus parasiticus* (Table 1) which are ubiquitous in the environment and highly resistant to heat and drying. They are saprophytic and frequently live in soil, vegetation, and feeds, acquiring nutrients from dead plants and animal matter.

Their spores are produced in large numbers and are spread widely by air currents. These molds grow within many commodities when temperatures are between 24-35 °C, and the moisture content exceeds 7% -10% (Kogbo et al., 1985; Williams et al., 2004).

The present chapter has the purpose of putting into perspective the worldwide relevance of the AFB1 contamination problem due to its effect on the aspects of economy and health, as well as to review the main strategies developed for coping with such contamination. In particular, we discuss the theoretical grounds and the practical approaches which have been carried out by using antimutagenesis and chemoprevention strategies. In these areas are included a description and discussion of the more relevant agents tested against the genotoxic and carcinogenic damage induced by AFB1.

2. The contamination problem

Aspergillus parasiticus often grows in oily products such as peanuts, walnuts, pistachios, pine nuts, pumpkin seeds and sunflower seeds, while *A. flavus* is commonly found contaminating agricultural fields of grains such as corn, sorghum, rice, barley, rye, and oats, as well as in spices (chili, pepper, mustard, and cloves). All of these commodities or products may be raw material for animal feed, which when ingested may pass into breast milk and can later be found in cheese, yogurt, cream, meat, and egg, constituting a source of secondary contamination for humans (Juan-López et al., 1995).

Fungal invasion and aflatoxin contamination often begin before harvest and can be promoted by production and harvest conditions, genotypes, drought, soil types, and insect activity, among other factors (Cole et al., 1995; Lynch & Wilson, 1991; Mehan et al., 1986, 1991). Therefore, timely harvest and rapid and adequate drying before storage are important factors to avoid or reduce post-harvest contamination, because even moisture generated by insect respiration and local condensation may develop local pockets favorable to aflatoxin growth (Mehan et al., 1986; Williams et al., 2004). This may partially explain differences in the range of contaminated products among countries. For example, in Japan aflatoxins were detected in about 50% of peanut butter and bitter chocolate samples, while their presence was not found in corn products; in contrast, a study in China reported contamination in 70% of corn products (Kumagai et al., 2008; Wang & Liu, 2007).

Aflatoxin contamination may be more severe in developing than in developed countries, yet this is a worldwide problem that could reach as much as 25% of the world's crops (Fink-Gremmels, 1999). In past years, a survey conducted in Midwestern states of the USA found 19.5% of corn samples contaminated with aflatoxin when assayed prior to any induced environmental stress, and 24.7% of them contaminated following stress induction (Russell et al., 1991). Also, Shane (1993) estimated losses in Southeastern USA for around 97 million dollars because of AFB1-contaminated corn with an additional 100 million dollars in production losses at hog farms feeding the contaminated grain.

Kingdom	Fungi
Phylum	Ascomycota
Class	Eurotiomycetes
Order	Eurotiales
Family	Trichocomaceae
Genus	*Aspergillus*

Table 1. Taxonomy of *Aspergillus*

There are diverse criteria for assessing the economic impact of aflatoxins. These include loss of human and animal life, health care and veterinary care costs, loss of livestock production, loss of forage crops and feeds, regulatory costs, and research cost focusing on relieving the impact and severity of the aflatoxin problem. However, most reports on the matter are on a single aspect of aflatoxin exposure or contamination.

With regard to the heavy impact of AFB1 contamination, India can be an example of the problem in emerging countries. A study in the Bihar region showed that nearly 51% of the 387 samples tested were contaminated with molds, and that from the 139 samples containing AFB1, 133 had levels above 0.02 mg/kg (Ranjan & Sinha, 1991). In other studies, authors found levels as high as 3.7 mg/kg of AFB1 in groundnut meal used for dairy cattle, as well as 0.05 to 0.4 mg/kg in 21 of 28 dairy feed samples from farms in and around Ludhiana and Punjab (Dhand et al., 1998; Phillips et al., 1996). Also, in raw peanut oil 65-70 % of AFB1 was found in the sediment and 30-35 % in the supernatant oil after centrifugation (Banu & Muthumary, 2010a). In this context, groundnut contamination was estimated to represent about a 10 million dollar loss in India's export within a decade (Hussein & Brasel, 2001; Vasanthi & Bhat, 1998). Regarding the extent of the problem in developing countries, Table 2 shows that a wide range of commodities are contaminated, even to a higher degree than usually allowed (Williams et al., 2004).

In Mexico the main contaminated crop is corn. This is a logical situation considering that the country has one of the highest rates of human consumption of this grain in the world (120 kg per capita per year) with a production of about 10.2 million tons for human consumption and 5 million tons for animal feed and other industries (Plasencia, 2004). One of the most significant episodes of aflatoxin contamination of maize was probably that which occurred in a northern state (Tamaulipas) in 1989, where levels of the toxin above 0.1 mg/kg were reported in practically all the plants harvested (García & Heredia, 2006). This represents a potential high health risk to the population, because corn is a basic food consumed as tortilla, with a consumption of 325 g per day (Anguiano-Ruvalcaba et al., 2005). However, this is not the only food susceptible to AFB1 that may pose a health risk, because a number of other maize-based foods are part of the Mexican diet. In regard to this contamination a few studies have been made. In kernelled corn for human consumption in the city of Monterrey, AFB1 was determined in 36 of the 41 samples tested, with concentrations ranging from 5 to 465 ng/g, with 59% of those samples above the Mexican legal limit of 0.02 mg/kg (Torres-Espinoza et al., 1995). Another study in 66 stored samples of maize and wheat in the state of Sonora showed 13 samples (20%) contaminated with AFB1, although the level was higher than 0.02 mg/kg in only one sample (Ochoa et al., 1989). Some general explanations for the contamination in the country are the following: 1- inadequate pre-harvest and storage management, as well as distribution procedures that may favor the development of *Aspergillus*; 2- corn growing under non-irrigation conditions in many places, predisposing plants to drought stress and mold infection; 3- limited possibilities of modern agricultural practices for low income farmers; 4- legal restriction for the use of transgenic maize manifesting insecticidal proteins or any other trait to reduce aflatoxin contamination; 5-infestation with the microleopterans *Carpophilus freemani*, the sap beetle, *Sitophilus zeamais*, the maize weevil, and *Cathartus quadricollis*, square-necked grain beetle, which may facilitate spore entry in the cobs; 6- growth of pollinated varieties which appear to be more prone to disease development and to the effect of environmental factors in comparison with maize hybrids (Figueroa, 1999; Plasencia, 2004; Zuber et al., 1983).

Country/ commodity/Number	Positive AFB1 samples (%)	Contamination rate (ppb)
Bangladesh		
Maize (95)	67	33.0 (mean)
Brazil		
Corn (96)	38.3	0.2-129.0
Peanut (97)	67	43.0-1099.0
China		
Corn (99)	76	>20.0
Costa Rica		
Maize (100)	80	>20.0
Egypt		
Peanut butter (101)	56.7	>10.0
Hazelnut (102)	90	25.0-175.0
Soybean (104)	35	5.0-35.0
Guatemala		
Incaparina (corn/ cottonseed flour) (106)	100	3.0-214.0
India		
Chilies (109)	18	>30.0
Maize (113)	26	>30.0
Korea		
Barley food (114)	12	26.0 (mean)
Corn food (114)	19	74.0
Malaysia		
Wheat (117)	1.2	>25.62
Mexico		
Corn (118)	87.8	5.0-465.0
Nigeria		
Corn (119)	45	25.0-770.0
Maize-based gruels (120)	25	0.002-19.716
Qatar		
Pistachio (121)	8.7 to 33	>20.0
Senegal		
Peanut oil (122)	85	40.0 (mean)

Table 2. Examples of market sample contamination, frequencies, and concentrations

Besides economical and educational actions that can be carried out to reduce the contamination problem in Mexico, other specific actions can be the following: 1- more research and breeding programs to identify varieties resistant to fungal infection and AFB1 contamination; 2- epidemiological data concerning liver cancer/AFB1 ingestion, as well as determination of AFB1 intake and its excretion in fluids, (particularly because cancer initiation may take about 6 years); 3- adoption of a standard method for measuring AFB1 content at both national and international levels, which must be sensitive, reliable, reproducible, and cost-effective.

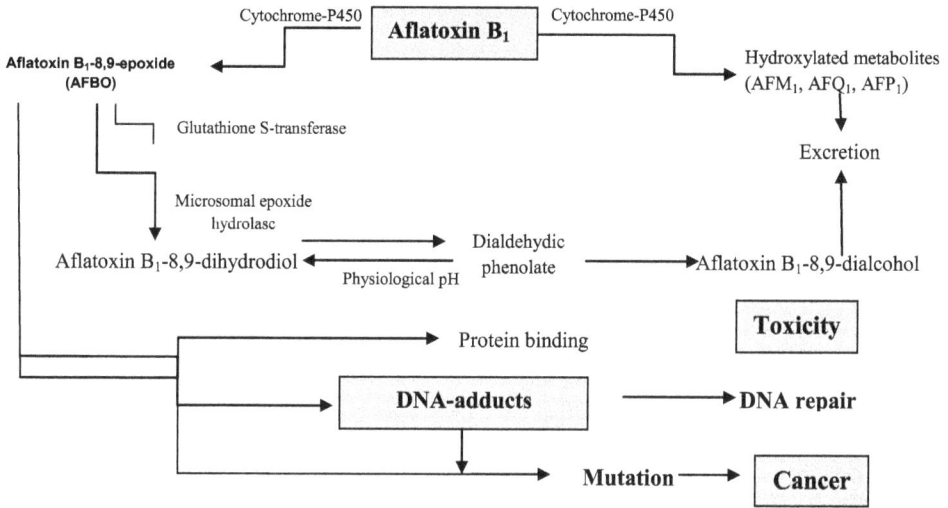

Fig. 2. Biotransformation pathways of aflatoxin B1

3. Toxicity and intervention strategies

AFB1 was first isolated some 40 years ago after outbreaks of disease and death occurred in turkeys and rainbow trout fed on contaminated peanut and cottonseed meals (Williams et al., 2004). From this time onwards a number of investigations have corroborated the strong toxicity of this mycotoxin in mammals, poultry, fish and other animals (Girish & Smith, 2008; Kensler et al., 2011; Madrigal-Santillán et al., 2010; Santacroce et al., 2008). Aflatoxicosis is the poisoning that results from ingesting AFB1, and two general forms of the affection have been identified. One is an acute, severe intoxication, which results in direct liver damage and subsequent illness or death, related with large doses; this type of aflatoxicosis includes symptoms such as hemorrhagic necrosis of the liver, bile duct proliferation, edema, lethargy, and liver cirrhosis The other, a chronic form of the disease, corresponds to a subsymptomatic exposure, which is related with nutritional and immunologic consequences, such as suppression of the cell-mediated immune responses; also, as dose exposure has a cumulative effect, there can be a significant risk increase of developing cancer (Steyn, 1995; Williams et al., 2004).

Studies on the matter have established a species-related susceptibility to health effect by AFB1, and a role of the dose and the duration of the exposure (Neiger et al., 1994; Pestka & Bondy, 1994; Silvotti et al., 1997); nevertheless, it has been clearly shown that AFB1 is a powerful carcinogen for humans and many animal species, including rodents, non-human primates, and fish (Kimura et al., 2004; Santacroce et al., 2008). The main target of the agent is the liver, although tumors may also develop in other organs, such as the lungs, kidney and colon (Wang & Groopman, 1999). Therefore since 1993, The International Agency for Research on Cancer (IARC) has classified it as a high potential carcinogenic agent (Class I) (IARC, 1993). Besides, a strong synergy between aflatoxin and the presence of hepatitis B and C viruses has also been determined, a combination that significantly increases the risk for having liver cancer, as shown in places like Gambia, and Qidong, China (Wang et al., 1996, 2001).

AFB1 is absorbed in the small intestine and distributed by the blood throughout the body. Examination of the physicochemical and biochemical characteristics of the AFB1 molecule has revealed two important sites for toxicological activity. One is the double bond in position C / C-8,9, of the furo-furan ring. The aflatoxin-DNA and protein interaction at this site can alter the functioning of these macromolecules leading to cellular deleterious effects. Another reactive group is the lactone ring in the coumarin moiety, which is easily hydrolyzed and therefore, vulnerable for degradation (Banu & Muthumary, 2010b). AFB1 is metabolically activated by cytochrome P450 enzymes to yield two chemically reactive epoxides: AFB1-8,9-*exo* and -8,9-*endo* epoxides (Figure 2). However, only the 8,9-*exo* isomer reacts readily with DNA, forming the N7-guanine and its derivative AFB1-formamidopyrimidine adduct (Johnson & Guengerich, 1997). These events constitute the basis of AFB1 genotoxicity, which includes promutagenic and mutagenic events that can result in the activation of protooncogenes and the inactivation or loss of tumor suppressor genes. The formed epoxide is very unstable in water but can be handled relatively easily in aprotic solvents. CYP enzymes, on the other hand, also oxidize AFB1 to deactivated products that are generally poor substrates for epoxidation, or to those which after that step do not interact with DNA, including AFM, AFQ, and the *endo*-epoxide (Johnson & Guengerich, 1997; Guengerich et al., 1998).

The genotoxic effects induced by AFB1 have been extensively documented. The chemical is known to inhibit DNA synthesis, as well as DNA-dependent RNA polymerase activity messenger RNA synthesis, and protein synthesis (McLean & Dutton, 1995; Wang & Groopman, 1999). Furthermore, its strong genotoxicity has been demonstrated in many endpoints and model systems which include HeLa cells, *Bacillus subtillis, Neurospora crassa, Salmonella typhimurium,* CHO cells, chromosomal aberrations, sister chromatid exchanges (SCE), micronucleus, unscheduled DNA synthesis, DNA strand breaks, and DNA adducts (Anwar et al., 1994; El-Zawahri et al., 1990; Le Hegarat et al., 2010; Miranda et al., 2007; Theumer et al., 2010).

The above mentioned genotoxicity is in line with the induction of cancer by aflatoxins. Hepatocellular carcinoma is one of the most common malignancies worldwide, and a major risk factor includes dietary exposure to AFB1. Genetic and epigenetic changes are involved in the pathogenesis of the disease, including G:C to T:A transversions at the third base of codon 249 of the tumor suppressor gene *p53*. Besides, chronic infection with hepatitis virus, and the generation of reactive oxygen/nitrogen species can also damage DNA and mutate cancer-related genes, such as *p53*. One of the functions of this gene is to regulate the transcription of protective antioxidant genes, however, when the DNA is damaged, *p53* regulates the transcription of protective antioxidant genes, but with extensive DNA damage it transactivates pro-oxidant genes that contribute to apoptosis. Also, genes from the hepatitis B virus can be integrated in the genome of hepatocellular carcinoma cells, and mutant proteins may still bind to p53 and attenuate DNA repair and apoptosis; thus, it is clear that viruses and chemicals may be involved in the etiology of mutations during the molecular pathogenesis of liver carcinoma (Hussain et al., 2007; Oyaqbemi et al., 2010).

The strong toxicity of AFB1, which may be reflected in financial and social problems, prompted countries to incorporate regulations concerning the levels of mycotoxins in food and feed. In the case of AFB1, the maximum tolerated level varies from 1 to 20 µg/kg. The limit of 4 µg/kg is usually applied in countries that follow the harmonized regulations of the European Free Trade Association (EFTA) and the European Union (EU), and the 20 µg/kg limit is mainly applied in Latin American countries, the United States, and Africa (Guzmán de la Peña & Peña Cabrales, 2005).

Actions to fulfill regulations or to correct possible failures can be taken at the phases of production, storage, and processing. At the initial steps, insect control can be performed, and improvements made in irrigation practices and storage structures as well as in the inoculation of non-aflatoxigenic strains; in the latter step, the actions can refer to the separation of the contaminated product, its dilution with grains lacking AFB1, or its decontamination through a number of physical and chemical methods which are designed to degrade, destruct, inactivate or remove the toxin. The ideal decontamination procedure should be easy to use and inexpensive, and it should not lead to the formation of compounds that are still toxic, or that may reverse to compounds that reform the parent mycotoxin or alter the nutritional and palatability properties of the grain or grain products. This has been a difficult task, and thus a number of methods have been proposed, showing variable results. Examples of these methods are presented in Table 3 (Madrigal-Santillán et al., 2010). The widespread contamination of AFB1, in addition to the complexity and danger of its toxicity, has suggested that not only one form of control and prevention can cope with the problem; this is a conviction that has promoted the development of different strategies. One of these refers to the application of antimutagenesis and chemoprevention procedures, as the basis to avoid or reduce DNA lesions, as well as other molecular and cellular alterations related with the process of cancer initiation. These studies can be carried out by inhibiting the formation of active AFB1 metabolites, avoiding the interaction with target macromolecules, or by accelerating the detoxication and repair processes, among other mechanisms. Comparison of antimutagenic or chemopreventive activities with biochemical and organic quantifications are relevant to confirm the efficacy of the prevention strategy.

Physical methods	Specific examples
Inactivation by heat	Vapor pressure Microwave treatment Nixtamalization
Inactivation by radiation	Ultraviolet light Gamma radiation
Elimination by adsorbents	Zeolites Bentonites Aluminosilicates
Chemical methods	
Extraction by organic solvents	Ethanol 95% Acetone 90%
Chemical destruction	Hexane-ethanol Hydrogen peroxide Ammonium hydroxyde Methylamine Sodium hypochlorite

Table 3. AFB1 decontamination procedures

4. Antimutagenesis and chemoprevention

The relationship between chemical exposure and cancer development was observed about 140 years ago when an increase was noted in the cancer mortality rate of workers managing coal tar. This effect was experimentally confirmed by Yamagiwa and Ichikawa (Weisburger, 2001) in exposed rabbits. Later, with the identification of the DNA structure and the development of new analytical methods, carcinogenesis was clearly shown to be related with alterations in this molecule. Such a relation was confirmed in the 60s, determining the effect exerted by metabolites of specific carcinogens on the structure and function of DNA and studying the activity of carcinogens on various cellular systems. In this context, the term genotoxic was then used for identifying carcinogens acting on DNA, in contrast with others whose effect was exerted through other routes (Weisburger, 2001). At this time, it was evident that the presence of numerous physical, chemical, and biological genotoxic agents could affect human health, and also that these genotoxicants may be present in a wide range of human activities, such as those related with work, food, health or personal habits. This produced a sort of genotoxic saturation, a condition which stimulated a search for knowledge about the agent's molecular, cellular, and metabolic peculiarities as well as detailed characteristics of the xenobiotic-DNA interaction; furthermore, the consequences of such an interaction were investigated, particularly with reference to human health.

A number of educational recommendations were then suggested in order to counteract the genotoxic effects, in addition to establishing regulatory measures related to permissible limits for specific substances; besides, efforts were made for substituting genotoxic drugs with less dangerous ones, including appropriate modification of their molecular structure. In the search for strategies to cope with the deleterious effect of mutagens, agents appeared which may reduce or eliminate such damage, the antimutagens. The basis for studying these substances is the knowledge that carcinogenesis is strongly related with mutagenesis, evidence supporting that carcinogenesis is highly due to the activity of environmental agents, and also, information that genotoxic inhibitors may frequently be found in plants and their products, as well as in other components of the diet; factors which favor their use under the appropriate conditions (Weisburger, 2001). In this context, it is pertinent to refer to Ames (1983) who suggested that a diet insufficient in fruits and vegetables may double the risk of acquiring cancer and cardiac diseases. This statement, as well as a number of reports on the matter increased the interest in determining the potency, toxicity, and mechanism of action of antimutagens as the necessary basis for incorporating the best candidates in preclinical and clinical chemoprevention trials. Moreover, studies on the matter have put into perspective the real beneficial action of this type of agents. Such studies have also considered various aspects that must be understood or solved, such as the fact that most antimutagens act on specific mutagens, and the possibility that the effect can be nullified or even reversed to mutagenicity in regard to the dose, time, and cell/organism tested, a possibility which is also complicated by the interactions that may occur between any compound and the complex human organism (De Flora & Ferguson, 2005; Ferguson, 2010).

The number of antimutagens and how they may act has been growing in recent years. Moreover, it is known that an agent may have more than one mechanism of action, and that two or more antimutagens could act synergistically. For more detailed information on the classification and the antimutagen's mechanism of action the reader may consult specific reports (De Flora et al., 2001; De Flora & Ferguson, 2005); however, for the sake of simplicity, in this revision antimutagens have been classified as desmutagens, impeding or limiting the effect of the mutagen before reaching the DNA molecule, such as the adsorbents that may interfere with the cellular absorption, or as those that avoid or reduce mutagen formation by blocking the biotransformation of premutagens through the inhibition of their activation at the cytochrome P-450. The other class of antimutagens corresponds to the bioantimutagens agents which may even reduce the level of mutations after the DNA has been damaged. To this group belong sequesters of mutagens and free radicals, agents that enhance the activity of phase II enzymes and the repair system, or those that reduce errors at the DNA replication level (Kada & Shimoi, 1987).

The term chemoprevention was coined by Sporn in the mid 70s, and during the following years it was defined as a procedure for the prevention, inhibition, delay, or reversal of carcinogenesis by means of a variety of agents which include different nutrients, extracts of plants or pharmacologic compounds, among others. The aim of the strategy refers to finding agents with several characteristics: 1-low cost, as related to cost-benefit analyses and to the size of the target population; 2- practicality of use, regarding availability, storage conditions and administration route, besides taking into account the need to be used for long periods of time; 3-efficacy; and 4-safety. The selected chemopreventive agents should protect target molecules, cells, general population and individuals at risk, against the initiation, promotion or progression phases of carcinogenesis (De Flora & Ferguson, 2005). The concept is based in that chronic diseases may have common pathogenic determinants, such as DNA damage, oxidative stress, and chronic inflammation, and that a number of agents have proved to be efficient in blocking such alterations and in improving the quality and span of human life. The more promising candidates are subjected to clinical trials, which should be designed and conducted properly, and should include well characterized agents, suitable cohorts, and reliable biomarkers for measuring efficacy, which can serve as surrogate endpoints for cancer incidence. Phase II chemoprevention trials test promising agents for biomarkers modulation in cohorts of 30 to 200 participants at greater than average risk of the cancer under study; in contrast, phase III trials test agents for their efficacy in cancer prevention in thousands of participants who are generally healthy or who may be at slightly elevated risk (Kelloff et al., 1995; Richmond & O'Mara, 2010).

5. Genotoxicity/antigenotoxicity tests

Genotoxicity/antigenotoxicity tests can be defined as in vitro and in vivo assays designed to detect both, compounds that induce genetic damage by various mechanisms, as well as those that prevent such damage. These tests enable hazard identification with respect to DNA damage and its fixation in the form of gene mutations, chromosomal aberrations or other alterations, all of which are considered essential for heritable effects and for the multi-step process of malignancy (Figure 3). In contrast, the same tests may provide information on the level of protection and the mechanism involved regarding the antigenotoxic agents (Food and Drug Administration [FDA], 2008). In this section we will briefly describe the more basic fundaments regarding some of the tests most used to evaluate the prevention of DNA damage induced by AFB1.

a. The Ames mutation assay, which was developed in *Salmonella typhimurium* in the mid 70s, is based in the use of strains with a mutation in the histidine locus which does not allow the bacteria to synthesize such aminoacid; thereby, reversion to the normal situation constitutes the mutagenic endpoint. The sensitivity of the test has been improved by incorporating mutations to the test organism, making it more permeable to chemicals and more resistant to DNA repair (Dearfield & Moore, 2005). Moreover, several strains which detect different base-pair substitutions have been constructed, thereby allowing the detection of oxidative damage or DNA cross-linking. Besides, limitation of the bacteria in regard to the absence of metabolic activation by enzymes common to the mammalian metabolism was overcome by adding rat liver microsome homogenate (S9 homogenate) to the bacterial cultures (Ames et al., 1973; Dearfield & Moore, 2005).

b. SCE represent the interchange of DNA replication products at apparently homologous chromosomal loci; such exchanges presumably involve DNA breakage and reunion (Latt & Schreck, 1980). The test can be made in vitro or in vivo. In the first case a number of cellular lines or primary cultures from different organisms can be used, with or without the addition of S9 for inducing metabolic activation. Also, a main step in making the test is to differentially stain the sister chromatids in such a way that they can be clearly visualized as a distinct chromatid in second division metaphases; this is essential for counting the number SCE per chromosome/cell, which is the evaluated genotoxic endpoint. The compound bromodeoxyuridine is usually added to the cultures or intraperitoneally injected to the test animal to visualize the sister chromatids; this compound is a thymidine analogue which is readily incorporated into the DNA chains and acts as a molecular marker during DNA replication, which is reflected as the differentially stained chromatids when colored with a Giemsa stain (Latt & Schreck, 1980).

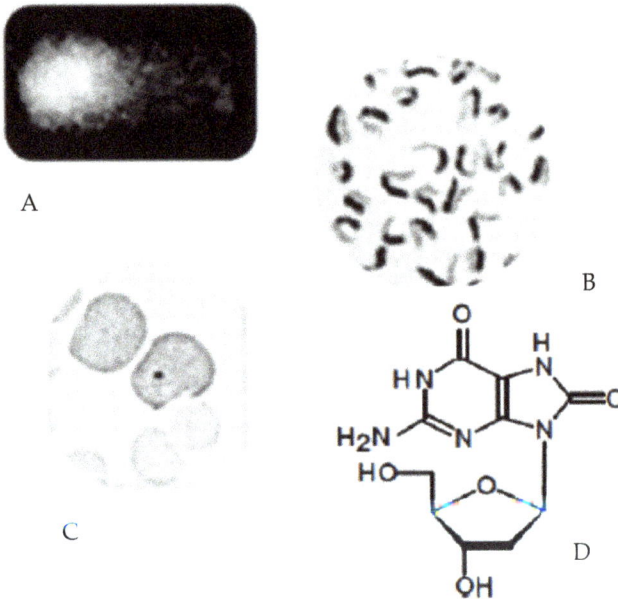

Fig. 3. Examples of genotoxic lesions. A) DNA damage observed with the comet assay, B) Sister cromatid exchanges in mouse bone marrow cells, C) micronuclei in mouse erythrocyte, D) DNA adduct (8-hydroxy-2´-deoxiguanosine)

c. Chromosomal aberrations are generally classified under two types: a numerical one and a structural one, and although both types are useful for the analysis of genotoxicity, the structural type of anomalies, is probably the more utilized. This type of aberrations is usually classified as chromosome and chromatid lesions, and more specifically, to deletions and fragments as well as various forms of chromosome intrachanges and interchanges. The studied cells and/or organisms can be selected based on their growth ability, stability of the karyotype, chromosome number, and spontaneous frequency of alterations (US Enviromental Protection Agency [EPA], 1998; Organisation for Economic Co-operation and Development [OECD], 2010). The in vivo or in vitro assay requires proliferating cells which are treated with the tested substance during an appropriate exposure time, and at the end, the test requires a hypotonic treatment, followed by the cell fixation and staining. Also, evaluating chromosomal aberrations in an initial round of cellular proliferation is important in order to have precise quantification of aberrations due to possible losses in subsequent divisions; therefore, the differential staining procedure applied to determine SCE can also be used to specifically identify cells in the first cellular division.

d. Micronuclei are genotoxic lesions that may originate from acentric chromosome fragments or whole chromosomes that are unable to migrate to the poles during the anaphase stage of cell divisions (Organisation for Economic Co-operation and Development [OECD], 2007). The test can be designed to detect the activity of chemicals with clastogenic and aneugenic potential, in cells that have undergone cell divisions during or after exposure to the test substance. Because of easier performance and microscopic detection, as well as high sensitivity for detecting mutagens/antimutagens, it has partially substituted the chromosomal aberration analysis. The assay can be made in vitro and in vivo, as well as in animals or plants. In cultured cells, the incorporation of cytochalasin, a cytokinesis blocker that allows micronuclei evaluation in synchronic binucleated cells, has become popular. In mouse in particular, micronuclei can be scored in both immature polychromatic erythrocytes or mature, normochromatic erythrocytes. Moreover, the test can be applied for examining exfoliated cells, and its sensitivity can be increased by means of flow cytometry which allows the analysis as many as 500000 events (Dertinger et al, 2011; Fenech et al., 2011).

e. The single cell gel electrophoresis assay, also known as comet assay, is a sensitive technique for detecting and analyzing DNA breakage in a variety of organs and various plant and animal cells. The basic principle resides in the migration of DNA in an agarose matrix under electrophoretic conditions, which depends on the level of breakage. When viewed under a microscope, a cell may have the appearance of a comet, with a head (the nuclear region) and a tail containing DNA fragments which have migrated in the direction of the anode. The length and the frequency of comets depend on the genotoxic potential of the tested agent (Oshida et al., 2008). The advantage of the comet assay is that it allows any viable eukariote cell to be analyzed for DNA damage, by detecting single or double strand breaks, alkali-labile sites that are expressed as single-strand breaks, and single-strand breaks associated with incomplete excision repair. Quantitative analysis for DNA damage has yielded several parameters, including tailed nuclei, tail length, percentage of DNA in the tail, and tail moment; besides, specific enzymes can be added to the test to analyze oxidative damage or the test can be integrated with the FISH assay for evaluating specific gene position/movement (Hartmann et al., 2003; Kumaravel et al., 2009).

f. An adduct corresponds to a stable complex formed when a chemical is covalently linked to a macromolecule, such as protein or DNA. The measurement of adducts in body fluids is highly sensitive and specific to determine the effect of the studied xenobiotic. DNA adducts have been clearly shown to be relevant to the disease process in prospective studies (Bonassi & Au, 2002). In relation to our present review, the adducts 8,9-dihydro-8-(N(7)-guanyl-)-9-hydroxyaflatoxin, as well as the AFB1-formamidopyrimidine compound, among others, are thought to be involved in the mutations caused by AFB1. The detection of these compounds can be made in various organs, as well as in serum and urine, by means of a variety of methods that include HPLC, ELISA, accelerator mass spectrometry, and liquid chromatography/electrospray ionization/mass spectrometry (Sharma & Farmer, 2004; Wang et al., 2008).

6. Chlorophyllin

Chlorophyll was detected as an antimutagenic agent since the early 80s. However, very soon afterwards most research on the matter was focused on the effect of chlorophyllin (CHL) (Figure 4), mostly because this chemical is a water-soluble, sodium and copper salt, chlorophyll derivative. In regard to its protective effect on AFB1 damage, one of the first studies with CHL was made by quantifying the number of revertants of specific base-pair mutants in *Neurospora crassa* (Whong et al., 1988). The study showed strong inhibition of the antimutagen on the AFB1-mutagenicity determined in growing cultures of the mold, a result which prompted the group to continue their studies but now in *Salmonella typhimuriun*, strain TA98, as a model (Whong et al., 1988). The authors used the plate-incorporation test in this bacteria and found a concentration-dependent inhibition of AFB1 mutagenicity when the cells were treated with the tested substances, concurrently; in their assay they observed that 860 nmol/plate of CHL completely abolished the mutagenicity of the toxin. However, when other approaches were tested, negative effects were detected; therefore, the authors suggested that CHL acted before the mutagen entered the bacterial cell by suppressing metabolic activation or by scavenging the mutagen.

The preventive capacity of CHL on the AFB1-DNA damage was confirmed by means of the arabinose-resistant *Salmonella* forward mutation assay (Warner et al., 1991); the results of the study were obtained in a preincubation test, and showed an inhibitory effect of CHL with and without the addition of S9 mix. With 2.5 mg/plate or less, the authors reported an almost complete inhibition of the aflatoxin mutagenicity.

In the following years, a number of studies were made, particularly in rainbow trout but also in microorganisms. Their purpose was to confirm the CHL antimutagenic potential by means of various models, as well as to investigate the involved mechanism(s) of action and to evaluate its cancer chemopreventive capacity (Breinholt et al., 1995a, 1995b; Dashwood et al., 1991). Results of these studies confirmed the in vitro efficiency of CHL when liver microsomes were added to the system, not only against the damage induced by AFB1, but also on the mutagenic effect induced by its precarcinogenic metabolite, 8,9-epoxide. In rainbow trout, the authors found an inhibitory effect of CHL on the precarcinogenic hepatic DNA adduction induced by AFB1, as well as a significant lowering in the number of liver tumors. In agreement with this finding, a study made in rat concluded that a concurrent administration of both compounds engendered an important reduction in the level of liver AFB1-DNA adducts (Kensler et al., 1998a), and another study showed a significant inhibitory effect of CHL (60%) on the morphological transformation of BALB/3T3 cells (Wu et al., 1994). In regard to the mechanism of action, a strong complex formation between mutagen and antimutagen was proposed.

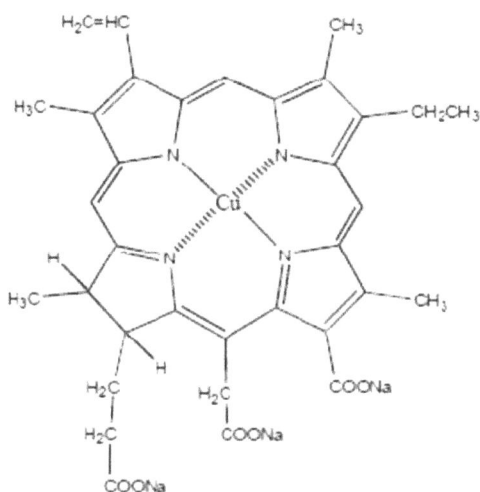

Fig. 4. Chemical structure of chlorophyllin

Based on the described positive experimental studies, efforts were initiated to determine the CHL chemopreventive capacity in humans. Qidong, People's Republic of China is a high risk region for hepatocellular carcinoma probably related with the consumption of AFB1 contaminated food; here, a randomized, double-blind, placebo-controlled trial was made to determine whether CHL administration altered the disposition of aflatoxin. CHL consumption at each meal led to an overall 55 % reduction in a median urinary level of the biomarker, aflatoxin-N[7]-guanine, compared with individuals taking placebo (Egner et al., 2001). The determined adduct biomarker derives from the carcinogenic metabolite, aflatoxin 8,9-epoxide; thus, the authors suggested that prophylactic interventions with CHL or supplementation of diets with chlorophyll rich foods may be useful to prevent the development of hepatocellular carcinoma or other environmentally induced cancers. This type of studies was supported by reports on the experimental effect of CHL in rats (Simonovich et al., 2007). The authors observed the inhibition of AFB1-albumin adducts and of AFB1-N[7]-guanine adducts, as well as the inhibition of AFB1 uptake when quantified in feces, besides a decrease in the number of colonic aberrant crypt foci induced by the aflatoxin. However, no modification in the activity of phase II enzymes was found.

In summary, a number of in vitro and in vivo studies have supported the antigenotoxic and chemopreventive capacity of CHL against the damage induced by AFB1, activities which can be related with the formation of a strong non-covalent complex, although additional mechanisms, such as its antioxidant potential, cannot be discarded. However, a word of caution about safety in using CHL is pertinent in light of the negative or controversial results that have been published in regard to the compound: for example, its effect as both inhibitor or promoter of genetic damage depending on the tested approach (Cruces et al.,

2003, 2009), its induction of embryo lethality in mouse (Garcia-Rodriguez et al., 2002), or its possible effect as a tumor promoter in rats (Nelson, 1992).

7. Oltipraz

Dithiolethiones are five-membered cyclic sulfur-containing compounds found in cruciferous vegetables, which have shown radioprotective, chemopreventive, chemotherapeutic, and antiviral activities (Kensler, 1997). In the context of our interest, a study made with lyophilized cabbage or cauliflower demonstrated a significant reduction in the rate of AFB1-induced carcinogenesis in rat (Boyd et al., 1982); subsequently, this chemopreventive effect has been thoroughly evaluated in regard to the drug oltipraz (OL), 5-(2-pirazinyl)-4-methyl-1,2-dithiole-3-thione (Figure 5). The mentioned dithiolethione was initially used because of its antischistosomal capacity in animals and humans, but later it was widely evaluated because of the detected potential to abolish or reduce the liver carcinogenesis induced by AFB1.

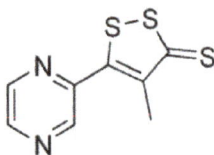

Fig. 5. Chemical structure of oltipraz

In an earlier study, rats were fed for 4 weeks with OL and gavaged with AFB1 during the second and third week of the assay (Kensler et al., 1987). In this report, the authors determined a significant reduction of focal areas with hepatocellular alterations achieved with the administration of OL, an effect shown by the staining of liver sections for gamma-glutamyl transpeptidase. The research concluded that dietary concentrations of OL as low as 0.01 % powerfully inhibited the formation of presumptive liver preneoplastic lesions. Similarly, the ameliorating effect of OL on the AFB1- toxicity induced in rats was shown by a decrease in the mortality rate, in the levels of serum alanine amino transaminase and sorbitol dehydrogenase, and also because of the normal growing rate propitiated by OL during the aflatoxin treatment (Liu et al., 1988). A few years later, another study made in rats, clearly confirmed the OL protective effect on AFB1-hepatocarcinogenesis (Roebuck et al., 1991). In this report, 11 % of hepatocellular carcinoma and 9 % hepatocellular adenomas were observed in AFB1-treated, diet-fed rats, in contrast with no tumor development in OL treated animals. Moreover, rats in the OL group had a significantly longer life span and an increased survival free of liver tumors in comparison with animals under aflatoxin treatment; besides, the authors found at least 65% reduction in the liver aflatoxin-N7-guanine adducts in the OL-fed animals, a finding which suggested that the protection against hepatocarcinogenesis might have resulted from the marked decrease in this type of hepatic DNA adducts. Studies about the chemopreventive efficacy of OL were also made in other animals, such as the marmoset monkey and the tree shrew. These studies gave rise to variable, but positive results (Bammler et al., 2000; Li et al., 2000).

With respect to humans, several studies have been made to examine the chemopreventive effect of OL. Phase IIa and phase IIb clinical trials were performed in a rural township in China (Kensler et al., 1998b; Wang et al., 1999). These trials were randomized, placebo-controlled, double-blind studies of people who had ingested AFB1 via their usual diet. For the rationale of the study the authors considered that the AFB1 epoxides are substrates for glutation S transferase (GST), an enzyme which catalyzes the conjugation of the epoxide with reduced glutation, thus mitigating the formation of DNA adducts. In fact, the results showed an increase in the activity of GST related with a sustained low dose of OL (125 mg/day), also yielding a high level of the AFB1 urinary metabolite, mercapturic acid; however, an intermittent, high dose of OL inhibited the activation of phase I enzymes, as reflected by a lowering in the excretion of the metabolite AFBm1. Nevertheless, the results of another study made in the same population suggested that prevention of oxidative DNA damage by OL was not a relevant mechanism to explain its effect against AFB1 (Glintborg et al., 2006).

The above-mentioned reports, as well as others, have suggested that the major preventive action of OL is through the activation of phase II enzymes, and secondarily on the inhibition of the carcinogen metabolism through the blocking of CYP enzymes. In addition to this, OL has been suggested to increase the nucleotide excision repair, which represents one of the major pathways for eliminating carcinogen DNA adducts; however, this effect has not been confirmed, as negative reports have also been published (O'Dwyer et al., 1997; Sparfel et al., 2002). With respect to GST, the mechanism by which OL enhances its level has been studied. Activation of such a cytoprotective enzyme seems to be related with a complex cellular signaling which includes the interaction of the Kelch ECH-associating protein (keap1) with the transcription factor NF-E2-related factor 2 (Nrf2), especially with the participation of the antioxidant response element-mediated regulation of Nrf2 (ARE) (Yates & Kensler, 2007).

At this time, other dithiolethiones seem even more promising than OL in preventing cancer. One such example is the compound 3H-1,2-dithiole-3-thione, which has shown a potent induction of phase II enzymes, a powerful inhibition of AFB1-induced hepatic toxicity including the formation of hepatic preneoplastic lesions, and inhibition of hepatic AFB1-DNA adducts; besides, the chemical was found to cause significant increases of GST and/or NQO1 in 12 tissues in addition to the liver, and it is probably not an inducer of CYP enzymes (Kensler et al., 1987; Roebuck et al., 2003; Zhang & Munday, 2008).

Future studies may include a structure-activity relationship among dithiolethiones, mainly to identify structural features that convey potent activation of Nrf2 and induction of phase II enzymes, the identification of mechanisms involved, as well as new biomarkers for evaluating their in vivo efficacy. Also, toxicity evaluations and clinical trials, especially with new dithiolethiones should be valuable.

8. Vitamins

These are organic compounds essential for the normal growth and development of a multicellular organism. A human fetus begins to develop from the nutrients it absorbs, including a certain amount of vitamins which facilitate the chemical reactions in different tissues. Vitamins have diverse biochemical functions such as the hormone-like regulation of mineral metabolism, regulation of cell and tissue growth as well as differentiation, besides acting as antioxidants, or precursors of enzyme cofactors. A number of efforts have been made to evaluate the useful impact of these compounds on the damaging effects induced by AFB1. Published studies have been made in in vitro and in vivo models, which include investigations in bacteria, cultured cells, experimental animals, or humans. The applied approaches go from the determination of their capacity to prevent various types of AFB1 genotoxic alterations, including adduction, to their modulatory effects on the AFB1 metabolism, or their participation as anticarcinogenic agents.

In regard to vitamin A, initial reports about its protective capacity were made by means of the Ames *Salmonella*/mammalian microsomes test. Such studies showed a relevant concentration-dependent decrease of the mutagenicity induced by AFB1. The effects were determined in strains TA98 and TA100, where the authors considered that the observed capacity of the vitamin could be related with the inhibition of AFB1 metabolism or with an increased breakdown of the active metabolite (Busk & Ahlborg, 1980; Raina & Gurtoo, 1985). Other strains (TA102 and TA1535) were also tested and revealed positive results (Qin & Huang, 1986). In Chinese hamster V79 cells, (Huang et al., 1982) found a similar effect. In this assay, dose and time dependent inhibition of AFB1-induced SCE, as well as correction of the cell cycle delay produced by the toxin was achieved by adding vitamin A to the cultures. Moreover, Qin et al. (1985) confirmed the indicated finding and extended it to determine a similar effect of the vitamin over the amount of chromosomal aberrations induced by AFB1. A few years later, S9 fractions obtained from mice with a high vitamin A liver level were found to be less potent in activating AFB1 than those with a low liver level; also, the first ones proved to be related with a stronger reduction of SCE in mice administered aflatoxin with respect to the effect in mice with a low vitamin A level, which therefore confirmed the role of such vitamin to ameliorate the genotoxic damage (Quin & Huang, 1985). In this period, Suphakarn et al. (1983) also determined an enhancement of liver and colon cancer in rats with a vitamin A deficient diet and exposed to AFB1. The authors evaluated factors such as liver morphology, enterohepatic recirculation, level of reduced glutathione in liver, and conjugating capacity to GST, and they suggested that their results may have been related with the influence of the vitamin on the binding of AFB1 to cellular macromolecules, partially influenced through enzymatic mechanisms. With the purpose of learning more about the preventive effect of vitamin A, Webster et al. (1996) applied the approach of modulating its ingested amount. They found that rats with a deficiency of vitamin showed a high level of DNA single strand breaks induced by AFB1, as well as a decrease in various repair enzymes subsequent to DNA damage, although correction of these two parameters was achieved with vitamin supplementation. In regard to the capacity of AFB1 for inducing DNA adducts, an in vitro assay using a microsome catalyzed reaction showed that the addition of vitamin A to the system produced a dose-dependent inhibition of the adduction (Firozi et al., 1987). Similar results were found studying woodchuck hepatocytes (Yu et al., 1994).

The information indicated above suggested to researchers that the main action of vitamin A (as well as of other vitamins) was on the initiating step of AFB1 carcinogenesis, yet there still remained studies to be done so as to clarify the issue on the preventive biochemical action of the vitamin (Bhattacharya et al., 1989; Decoudu et al., 1992).

The antigenotoxic and antitoxic potential of vitamin A was determined in experimental mice. In these animals, a decrease in the toxin-induced clastogenicity in both mitotic and meiotic chromosomes was reported, as well as inhibition in sperm abnormalities (Sinha & Darmshila, 1994). Besides, the antigenotoxic effect of vitamin A was also found in human lymphocytes (Alpsoy et al., 2009); in these cells the authors reported a significant, dose-dependent reduction of the SCE induced by 5 uM of AFB1, with the lowest protective concentration being 0.5 uM.

With respect to the anticarcinogenic potential, a report established such an effect in a 2-year follow up of AFB1-administered rats where it was observed that most animals fed a diet devoid of the vitamin developed liver cancer, contrary to few cases in rats which received it (Nyandieka & Wakhisi, 1993). However, a study aimed at determining the vitamin inhibitory effect on liver preneoplastic foci showed negative results, a finding which was probably related with an excess of the vitamin in the assay; this, nonetheless, helped to stress the relevance of selecting appropriate experimental conditions in chemoprevention studies (Gradelet et al., 1998).

Finally, although there is a deplorable scarcity of studies in humans, AFB1-albumin adducts were quantified in a high risk Ghanaian population, where a relationship was determined between a high mycotoxin level with decreased levels of the vitamins A and E, suggesting then, that such deficiency may significantly influence the incidence of adverse health effects (Tang et al., 2009).

Vitamin C and E are other compounds tested against the genotoxic damage induced by AFB1. With respect to these chemicals, a study made by means of the *Salmonella typhimurium* test (strains TA98 and TA100) showed that although both vitamins prevented the expression of AFB1-induced mutagenesis, vitamin E was more potent, and also that its effect was related with the metabolism of the mycotoxin, whereas vitamin C was involved in both the metabolic and post-metabolic levels of the AFB1 mutagenesis assay (Raina & Gurtoo, 1985). This result was congruent with the protective, dose-dependent effect determined for both vitamins against the SCE induced by AFB1 in cultured human lymphocytes (Alpsoy et al., 2009). In this study, the order of protective efficacy was vitamin C-vitamin E-vitamin A. In regard to vitamin E, however, the indicated positive results were contrary to those reported by Karekar et al. (2000) who applied two short term genotoxicity assays — the Ames test and the Drosophila wing spot test — and they found no antimutagenic response of the vitamin; moreover, woodchuck hepatocytes that were treated with four doses of [3H]AFB1 or with different combinations of the toxin and vitamins C and E for 6 h resulted in an effect of vitamin C for inhibiting AFB1-DNA binding; contrarily, an enhancement of covalent binding of AFB1 to DNA by vitamin E was observed (Yu et al., 1994). Also, negative results were found when evaluating the protection of such vitamin in SCE induced by AFB1 in V79 cells (Deng et al., 1988). These results clearly suggest the need for further research to understand the complex role of these vitamins in the mutagenesis and carcinogenesis of the aflatoxin. Such a complex response was also reported in rats fed on a variable diet of vitamin E (Cassand et al., 1993). Animals on a diet supplemented with a low amount of the vitamin (0.5 IU) increased P-450 IIB and IIIA enzyme activity, whereas a higher vitamin supplemented diet (5 IU) reduced these specific activities. However, lipid peroxidation was increased in the vitamin E free diet animals and strongly decreased in the supplemented group. Nevertheless, in a subsequent study (Karakilcik et al., 2004) a significant increase was found in the level of various liver enzymes in rabbits fed a diet with AFB1, while such activities were lower in the groups receiving the mycotoxin plus vitamins C or E, whether alone or combined. In spite of these controversial reports on antigenotoxicity, another study made in rats to determine the preventive capacity of vitamins C and E on the development of liver cancer gave strong positive results, because only few animals under vitamin treatment suffered the illness along the 24 months of the assay (Nyandieka & Wakhisi, 1993).

On the other hand, some studies aimed to test the effect of specific types of vitamin B have given inconclusive results. In the case of riboflavin, an earlier assay using the *Salmonella typhimurium,* strain T100, rat-liver microsome system, concluded that with lower concentrations of AFB1 the effect of the vitamin was very strong (Bhattacharya et al., 1987). However, another in vitro and in vivo study suggested a variable role of the compound with respect to the AFB1 metabolic activation, an effect which was related with the tested amount of the vitamin (Prabhu et al., 1989). Then, in a subsequent report made in rats under riboflavin supplementation, a clear, positive effect was determined on the DNA damage induced by AFB1 by quantifying the reversion of DNA single strand breaks (Webster et al., 1996). Folic acid has also been evaluated. In this case, a survey made in high risk Chinese individuals concluded that increased folate levels may be inversely associated with the development of liver damage and hepatocellular carcinoma (Welzel et al., 2007).

9. Probiotics and microbial cell wall components

In the context of the exposed theme, biological decontamination seems attractive because it works under mild, environmentally friendly conditions. The AFB1 detoxification potential of probiotics such as yeast and lactic acid bacteria, among other microorganisms, has been evaluated in light of their adsorbent capacity that prevents the transfer of aflatoxin to the intestinal tract of humans and animals (Wu et al., 2009). This effect has been reported for various species of *Lactobacillus,* including *L. casei, L. plantarum, L. fermentum, and L. rhamnosus*; moreover, the participation of teichoic acids has been suggested to play a key role in the binding ability of some species toward AFB1 (Fazeli et al., 2009; Gratz et al., 2007; Hernandez-Mendoza et al., 2009). Although the antimutagenic capacity of fermented foods and probiotics is known, few studies in respect to aflatoxin have been done with bacteria; an example is the report made in Caco-2cells treated with *Lactobacillus rhamnosus* strain GG, which showed protection against AFB1-induced reductions in transephitelial resistance, as well as reductions in DNA fragmentations assessed by extracting DNA and separating intact and damaged DNA by the use of gel electrophoresis (Grats et al., 2007). In regard to probiotics, there is an interesting study made on ninety healthy young men from Guangzhou, China whose diet was supplemented with a probiotic mixture that induced a reduction of the biologically effective dose of aflatoxin exposure, suggesting an effective dietary approach to decrease the risk of liver cancer. This conclusion was reached after quantifying the urinary excretion of AFB1-N7-guanine in the evaluated population (El-Nezami et al., 2006).

Research on the yeast *Saccharomyces cereviciae* (Sc) has confirmed its decontaminating ability through its binding with AFB1, which may depend on the used strain and other experimental conditions (Shetty et al., 2007). Besides, the potential of Sc to ameliorate the effects of aflatoxicosis was clearly established in broiler chicks or Japanese quail by evaluating a number of biochemical and organic parameters (Parlat et al., 2001; Stanley et al., 1993). The above-mentioned information is congruent with the antigenotoxic effect observed in mice fed with AFB1 contaminated corn (Madrigal-Santillán et al., 2006). In this study, the animals were experimentally fed with the tested chemical for six weeks; the results observed in the groups treated with the yeast showed a significant improvement in the weight loss induced by AFB1, and a decrease of more than 60 % in the level of micronuclei induced by the toxin in normochromatic erythrocytes, as well as a similar reduction in the level of SCE in mouse bone marrow cells, effects that were related with the adsorbing capacity of the yeast. Besides, the study revealed a recovery to normal parameters in about three weeks without the aflatoxin administration, which suggest the usefulness of periodical monitoring of commodities at risk.

The cell wall of yeast, as well as that of other microorganisms is composed of polyssacharides, mainly mannans, glucans, and glucomannans, some of which have been studied for their AFB1 protective effect. Mannan is a highly branched olygossacharide constituted by a main chain of α-(1,6)-D-mannoses linearly attached, and with α-(1,2) and α-(1,3)-D-mannose branches. In a first report mice were fed AFB1-contaminated corn, and AFB1 treated grain plus three doses of mannan (including the appropriate control groups). The assay lasted four weeks and the measurements included, weight, micronuclei, cytotoxicity index, and SCE (Madrigal-Santillán et al., 2007). Results showed that mice fed AFB1 had a significant weight decrease, as well as a significant increase in the rate of MNNE and SCE, while animals fed the combined regime presented a 25 % weight increase with respect to animals treated with AFB1 alone, as well as a reduction in the level of MNNE and SCE (about 70 % with the high two doses). In a subsequent report, the authors confirmed the protective effect of mannan in mouse hepatocytes which were analyzed with the comet assay at 4, 10 and 16 h of exposure (Madrigal-Santillán et al., 2009). In such study, the best preventive effect of mannan was found at 10 h with the high tested dose (700 mg/kg). Moreover, the authors proposed a supramolecular complex between mannan and the aflatoxin based on the melting points, and the UV spectra of the crystals from the independent compounds and a co-crystalization of both chemicals.

Glucans are a heterogeneous group of glucose polymers, consisting of a backbone of β(1,3)-linked β-D-glucopyranosyl units with β(1,6)-linked side chains of varying distribution and length (Akramiene et al., 2007). Besides its immunostimulant effect, the compound has been reported to have chemoprotective potential against a number of mutagenic agents (Akramiene et al., 2007; Mantovani et al., 2007); also, interaction of glucans with AFB1 including the participation of hydroxyl, ketone, and lactone groups was proposed as the basis for the formation of hydrogen bonds and van der Waals interactions (Yiannikouris et al., 2006). However, in spite of such information, very few studies have been made in regard to the antigenotoxic potential of glucans on the damage induced by AFB1. An investigation similar to the described above by Madrigal-Santillán et al. (2009) but testing the antigenotoxicity of glucan and glucomannan in mice hepatocytes showed a positive effect for the two agents (Madrigal-Santillán, 2004). DNA damage was quantified by means of the comet assay at 4, 10 at 16 h after the chemicals exposure. Glucan showed a protective effect with the two low doses tested (400, and 700 mg/kg), reaching about 40% as the highest reduction of the damage induced by AFB1; glucomannan, however, showed a significant response with all the three tested doses, reaching an inhibition as high as 80 % at 10 h of treatment.

10. Miscellaneous agents

Our purpose in this section is not to show an extensive list of the agents tested against the genotoxicity of AFB1, but rather the variability of such agents, which goes from single compounds to mixtures with different complexity. These investigations may be motivated by the mutagenic potency of AFB1, which make it a relevant candidate for demonstrating the capacity of antimutagens, as well as by the need for finding efficient agents to prevent the serious damage that such mutagen can provoke.

Ellagic acid and a phenolic extract obtained from the bean (*Phaseolus vulgaris*) are examples of phenolic compounds studied for their use in controlling the mutagenicity of AFB1. In both cases *Salmonella typhimurium* (strain TA98 and TA100) was used as the test model, and the obtained results showed a concentration-dependent antimutagenic effect, which was more clearly expressed when the compounds were tested at the same time. The authors suggested the formation of a chemical complex between the involved agents as an explanation for the protective effect (Loarca-Piña et al., 1998; Cardador-Martinez et al., 2002).

Green and black teas are known as efficient antimutagenic and anticarcinogenic agents. In the case of AFB1, tea polyphenols from both teas were tested against its mutagenicity by means of the *Salmonella typhimurium* assay (strain TA98). In the report the authors determined a sharp decrease toward the mutagenic effect of the aflatoxin (Weisburger et al., 1996). Besides, results obtained in rat bone marrow cells treated with AFB1 in vivo revealed that the administration of green tea 24 h before administering the mutagen produced a significant reduction in the number of structural chromosomal aberrations (Ito et al., 1989). A confirmation of the green tea protective effect was determined in 352 human blood and urine samples that corresponded to a 3-month trial of individuals under green tea consumption (Tang et al., 2008). The authors measured AFB1-albumin adducts, AFBM1, and AFB1-mercapturic acid, and concluded that green tea effectively modulates the metabolism and metabolic activation of AFB1.

A number of plant flavonoids were tested against the effect of AFB1 by means of the *Salmonella typhimurium* assay (TA98 and TA100 strains), and some of them showed an efficient antimutagenic capacity: kaempferol, morin, fisetin, biochanin A, and rutin (Francis et al., 1989). Also, it was reported that kolaviron, a flavonoid from the seeds of *Garcinia kola* was able to inhibit the amount of micronuclei and the hepatic oxidative damage induced by AFB1 in rats (Farombi et al., 2005).

The determination of ammonia as antimutagen is included in this review considering that the chemical has been used as one of the agents to decontaminate AFB1; therefore, confirmation of its utility through genetic endpoints seems interesting. In the described report, mice were fed for four weeks with AFB1 contaminated corn and concomitantly treated with ammonium hydroxide (Marquez-Marquez et al., 1993). The results showed a significant reduction in the rate of micronucleated normochromatic erythrocytes starting from the first week of the assay, and at the fourth week of treatment the inhibition reached 60 %; besides, at the last week of the test, the quantification of SCE showed an inhibition of 55 % in comparison with the level determined in the AFB1 treated group.

Coffee is a beverage of habitual consumption that has shown controversial results concerning its genotoxic/antigenotoxic potential; however, there is an interesting study by Abraham (1991) who evaluated the inhibitory effect of standard instant coffee on the number of mice bone marrow micronuclei. Mice were orally administered coffee 2 and 20 h before injecting the carcinogen, and observations made at 28 and 48 h showed a dose-dependent decrease in the rate of micronuclei, with a reduction of more than 60 % with the high tested dose (500 mg/kg).

In regard to constituents of apiaceous vegetables, such as carrots, parsnips, celery or parsley, Peterson et al. (2006) used a methoxyresorufin O-demethylase assay and a trp-recombination assay in *Saccharomyces cereviciae*, and found that 5-methoxypsoralen, and 8-methoxypsoralen reduced the CYP1A2-mediated mutagenesis of AFB1. In the same context, it was reported the hepatoprotective effect of ethanolic extract of *Phyllantus amarus* on AFB1-induced damage in mice, as well as the protective effect of soybean saponins against the aflatoxin in the *Salmonella typhimurium* assay, and a significant decrease of DNA-adduct formation in human liver hepatoma cells (Jun et al., 2002; Naaz et al., 2007).

11. Conclusions

In light of the serious effects that AFB1 contamination can originate, the authors agree that different socio-economical and toxicological approaches should be carried out for its elimination or control, including specific strategies regarding regulatory, supervisory, educational, scientific and technologic issues. Basic knowledge on the metabolism and the molecular and cellular fate of AFB1 is presently known, and various models have been used to test the effects of a number of chemopreventive agents, some of which have shown promising results, suggesting then, the pertinence of continuing with such strategy. However, it is reasonable to have a deeper knowledge on the chemical characteristics of each AFB1 metabolite, as well as on their interactions with macromolecules and cells, and to identify the more sensitive biomarkers for the assayed damage; this will be of help in designing more appropriate experimental projects, or clinical trials with the best candidates detected, in addition to preventing the selected genotoxic damage with more efficacy. At present, only a few agents have been tested in humans for evaluating their capacity of protection against AFB1 damage, although numerous chemicals have been evaluated in an almost isolated experimental form and have presented favorable results; therefore, extensive studies on these agents should be carried out so as to gain knowledge on their safety, efficacy, and mechanism of action, in order to select those more suitable for chemopreventive purposes.

12. References

Abraham, S. (1991). Inhibitory effects of coffee on the genotoxicity of carcinogens in mice. *Mutation Research.* Vol. 262, No. 2 (February 1991), pp. 109-114, ISSN 0027-5107

Akramienė, D., Kondrotas, A., Didžiapetrienė, J., & Kėvelaitis, E. (2007). Effects of B-glucans on the immune system. *Medicina (Kaunas),* Vol. 43, No. 8 (2007), pp. 597-605, ISSN 1010-660X

Alpsoy, L., Agar, G., & Ikbal, M. (2009). Protective role of vitamins A, C, and E against the genotoxic damage induced by aflatoxin B1 in cultured human lymphocytes. *Toxicology and Industrial Health,* Vol. 25, No. 3, (April 2009), pp. 183-188, ISSN 0748-2337

Ames, B., Durston, W., Yamasaki, E., & Lee, F. (1973) Carcinogens are Mutagens: A Simple Test System Combining Liver Homogenates for Activation and Bacteria for Detection. *Proceedings of the National Academy of Sciences of the United States of America,* Vol. 70, No. 8, (August 1973), pp. 2281-2285, ISSN 0027-8424

Ames, B. (1983) Dietary carcinogens and anticarcinogens. Oxygen radicals and degenerative diseases. *Science,* Vol. 221, No. 221, (September 1983), pp. 1256-1264, ISSN 0036-8075

Anguiano-Ruvalcaba, G., Vargas-Cortina, A., & Guzmán-De Peña, D. (2005). Inactivación de aflatoxina B1 y aflatoxicol por nixtamalización tradicional del maíz y su regeneración por acidificación de la masa. *Salud Pública de México.* Vol. 47, No. 5, (2005), pp. 369-375

Anwar, W., Khalil, M., & Wild, C. (1994). Micronuclei, chromosomal aberrations and aflatoxin albumin adducts in experimental animals after exposure to aflatoxin B1. *Mutation Research* Vol. 322, No. 1, (July 1994), pp. 61–67, ISSN 0027-5107.

Bammler, T., Slone, D., & Eaton, D. (2000). Effects of dietary oltipraz and ethoxyquin on aflatoxin B1 biotransformation in non-human primates. *Toxicological Sciences*, Vol. 54, No. 1, (March 2000), pp. 30-41, ISSN 1096-6080

Banu, M., & Muthumary J. (2010a). Aflatoxin B1 contamination in sunflower oil collected from sunflower oil refinery situated in Karnataka. Health. Vol. 2, No. 8, (2010), pp. 973-987, ISSN 1949- 4998

Banu, M., & Muthumary J. (2010b). Taxol as chemical detoxificant of aflatoxin produced by Aspergillus flower isolated from sunflower seeds *Health.*Vol. 2, No. 7, (2010), pp. 789-795, ISSN 1949-4998

Bhattacharya, R., Francis, A., & Shetty, T. (1987). Modifying role of dietary factors on the mutagenicity of aflatoxin B1: in vitro effect of vitamins. *Mutation Research*, Vol. 188, No. 2, (June 1987), pp. 121-128, ISSN 0027-5107

Bhattacharya, R., Prabhu, A., & Aboobaker, V. (1989). In vivo effect of dietary factors on the molecular action of aflatoxin B1: role of vitamin A on the catalytic activity of liver fractions. *Cancer Letters*, Vol. 44, No. 2, (February 1989), pp. 83-88, ISSN 0304-3835

Bonassi, S., & Au, W. (2002). Biomarkers in molecular epidemiology studies for health risk prediction. *Mutation Research*, Vol. 511, No. 1, (March 2002), pp. 73-86, ISSN 0027-5107

Boyd, J., Babish, J., & Stoewsand, G. (1982). Modification of beet cabbage diets of aflatoxin B1- induced rat plasma alpha-foetoprotein elevation, hepatic tumorigenesis and mutagenicity of urine. *Food and Cosmetics Toxicology*, Vol. 20, No. 1, (February 1982), pp. 47-52, ISSN 0015-6264

Breinholt, V., Hendricks, J., Pereira, C., Arbogast, D., & Bailey, G. (1995a). Dietary chlorophyllin is a potent inhibitor of aflatoxin B1 hepatocarcinogenesis in rainbow trout. *Cancer Research*, Vol. 55, No. 1, (January 1995), pp. 57-62, ISSN 0008-5472

Breinholt, V., Schimerlik, M., Dashwood, R. & Bailey, G. (1995b). Mechanisms of chlorophyllin anticarcinogenesis against aflatoxin B1: complex formation with the carcinogen. *Chemical Research in Toxicology*, Vol. 8, No. 4, (June 1995), pp. 506-514, ISSN 0893-228X

Busk, L., & Ahlborg, U. (1980). Retinol (vitamin A) as an inhibitor of the mutagenicity of aflatoxin B. *Toxicology Letters*, Vol. 6, No. 4-5, (September 1980), pp. 243-249, ISSN 0378-4274

Cardador-Martínez, A., Castaño-Tostado, E., & Loarca-Piña, G. (2002). Antimutagenic activity of natural phenolic compounds present in the common bean (*Phaseolus vulgaris*) against aflatoxin B1. *Food additives and contaminants*, Vol. 19, No. 1, (January 2002), pp. 62-69, ISSN 0265-203X

Cassand, P., Decoudu, S., Lévêque, F., Daubèze, M., & Narbonne, J. (1993). Effect of vitamin E dietary intake on in vitro activation of aflatoxin B1. *Mutation Research*, Vol. 319, No. 4, (December 1993), pp. 309-316, ISSN 0027-5107

Cole, R., Dorner, J., & Holbrook, C. (1995). Advances in mycotoxin elimination and resistance. In: *Advances in Peanut Science*. Pattee HE, Stalker HT, (Ed.), 456 –474. American Peanut Research and Education Society, Inc, USA.

Cruces, M., Pimentel, E., & Zimmering S. (2003) Evidence suggesting that chloropyllin (CHLN) may act as an inhibitor or a promoter of genetic damage induced by chromium(VI) oxide (CrO3) in somatic cells of Drosophila. *Mutation Research*, Vol. 536, No. 1-2, (April 2003), pp. 139-144, ISSN 0027-5107

Cruces, M., Pimentel, E., & Zimmering, S. (2009). Evidence that low concentrations of chlorophyllin (CHLN) increase the genetic damage induced by gamma rays in somatic cells of Drosophila. *Mutation Research*, Vol. 679, No. 1-2, (September-October 2009), pp. 84-86, ISSN 0027-5107

Dashwood, R., Breinholt, V., & Bailey, G. (1991). Chemopreventive properties of chlorophyllin: inhibition of aflatoxin B1 (AFB1)-DNA binding in vivo and anti-mutagenic activity against AFB1 and two heterocyclic amines in the Salmonella mutagenicity assay. *Carcinogenesis*, Vol. 12, No. 5, (May 1991), pp. 939-942, ISSN 0143-3334

De Flora, S., & Ferguson, L. (2005). Overview of mechanism of cancer chemopreventive agents, *Mutation Research*, Vol. 591, No. 1-2, (December 2005), pp. 8-15, ISSN 0027-5107

De Flora, S., Izzotti, A., D'Agostini, F., Balansky, R., Noonan, D., & Albini A. (2001). Multiple points of intervention in the prevention of cancer and other mutation-related diseases. *Mutation Research*. Vol. 480-481, (September 2001), pp. 9-22, ISSN 0027-5107

Dearfield, K., & Moore, M. (2005). Use of genetic toxicology information for risk assessment. *Enviromental and Molecular Mutagenesis*, Vol. 46, No. 4, (December 2005), pp. 236-245, ISSN 0893-6692

Decoudu, S., Cassand, P., Daubèze, M., Frayssinet, C., & Narbonne, J. (1992). Effect of vitamin A dietary intake on in vitro and in vivo activation of aflatoxin B1. *Mutation Research*, Vol. 269, No. 2, (October 1992), pp. 269-278, ISSN 0027-5107

Deng, D., Hu, G., & Luo, X. (1988). Effect of beta-carotene on sister chromatid exchanges induced by MNNG and aflatoxin B1 in V79 cells. *Zhonghua Zhong Liu Za Zhi [Chinese Journal of Oncology]*, Vol. 10, No. 2, (March 1988), pp. 89-91, ISSN 0253-3766

Dertinger, S., Torous, D., Hayashi, M., & MacGregor, J. (2011). Flow cytometric scoring of micronucleated erythrocytes: an efficient platform for assessing in vivo cytogenetic damage. *Mutagenesis*. Vol. 26, No. 1, (January 2011), pp. 139-145, ISSN 0267-8357

Dhand, N., Joshi, D., & Jand, S., (1998). Aflatoxins in dairy feeds/ingredients. *Indian Journal of Animal Nutrition*. Vol. 15 (1998), pp. 285–286, ISSN 0970-3209.

Eaton, D., Kallager E., & Groopman J. (1994). Mechanisms of aflatoxin carcinogenesis. *Annual Review of Pharmacology and Toxicology*. Vol. 34, (April 1994), pp. 135-172, ISSN 0362-1642

Egner, P., Wang, J., Zhu, Y., Zhang, B., Wu, Y., Zhang, Q., Qian, G., Kuang, S., Gange, S., Jacobson, L., Helzlsouer, K., Bailey, G., Groopman, J., & Kensler, T. (2001). Chlorophyllin intervention reduces aflatoxin-DNA adducts in individuals at high risk for liver cancer. *Proceedings of the National Academy of Sciences USA*. Vol. 98, No. 25, (December 2001), pp. 14601-14606, ISSN 0027-8424

El-Nezami, H., Polychronaki, N., Ma, J., Zhu, H., Ling, W., Salminen, E., Juvonen, R., Salminen, S., Poussa, T., & Mykkanen, M. (2006). Probiotic supplementation reduces a biomarker for increased risk of liver cancer in young men from Southern China. *The American Journal of Clinical Nutrition*. Vol. 20, No.83, (January 2006), pp. 1199-1203, ISSN 0002-9165

El-Zawahri, M., Morad, M., & Khishin, A. (1990). Mutagenic effect of aflatoxin G1 in comparison with B1. *Journal of Environmental Pathology, Toxicology & Oncology*. Vol. 10, No. 1-2, (January-April 1990), pp. 45–51, ISSN 0731-8898

Farombi, E., Adepoju, B., Ola-Davies, O., & Emerole, G. (2005). Chemoprevention of aflatoxin B1-induced genotoxicity and hepatic oxidative damage in rats by kolaviron, a natural bioflavonoid of *Garcinia kola* seeds. *European Journal of Cancer Prevention: the official journal of the European Cancer Prevention Organisation (ECP)*, Vol. 14, No. 3, (June 2005), pp. 207-214, ISSN 0959-8278

Fazeli, M., Hajimohammadali, M., Moshkani, A., Samadi, N., Jamalifar, H., Khoshayand, M., Vaghari, E., & Pouragahi, S. (2009). Aflatoxin B1 binding capacity of autochthonous strains of lactic acid bacteria. *Journal of Food Protection*. Vol. 72, No. 1, (January 2009), pp. 189-192, ISSN 0362-028X

Fenech, M., Holland, N., Zeiger, E., Chang, WP., Burgaz, S., Thomas, P., Bolognesi, C., Knasmueller, S., Kirsch-Volders, M., & Bonassi, S. (2011). The HUMN and HUMNxL international collaboration projects on human micronucleus assays in lymphocytes and buccal cells--past, present and future. *Mutagenesis*,Vol. 26, No. 1, (January 2011), pp. 239-245, ISSN 0267-8357

Ferguson, L. (2010). Dietary influences on mutagenesis--where is this field going?, *Enviromental and Molecular Mutagenesis*, Vol. 51, No. 8-9, (October-December 2010), pp. 909-918, ISSN 0893-6692

Figueroa, J. (1999). La tortilla vitaminada. *Avance y Perspectiva*, Vol. 18 (May-june 1999) pp. 149–158.

Fink-Gremmels, J. (1999). Mycotoxins: their implications for human and animal health. The *Veterinary Quarterly,*Vol.21, No. 4, (October 1999), pp. 115-20. ISSN 0165-2176

Firozi, P., Aboobaker, V., & Bhattacharya, R. (1987). Action of vitamin A on DNA adduct formation by aflatoxin B1 in a microsome catalyzed reaction. *Cancer Letters*, Vol. 34, No. 2, (February 1987), pp. 213-220, ISSN 0304-3835

Food and Drug Administration. (2008). S2(R1) Genotoxicity Testing and Data Interpretation for Pharmaceuticals Intended for Human Use. Retrieved from http://www.fda.gov/downloads/Drugs/GuidanceComplianceRegulatoryInform ation/Guidances/UCM074931.pdf

Francis, A., Shetty, T., & Bhattarcharya, R. (1989). Modifying role of dietary factors on the mutagenicity of aflatoxin B1: in vitro effect of plant flavonoids. *Mutation Research*. Vol. 222, No. 4, (April 1989), pp. 393-401, ISSN 0027-5107

Garcia, S., & Heredia, N. Mycotoxins in Mexico: Epidemiology, management, and control strategies. (2006). *Mycopathologia*, Vol. 162, No. 3, (September 2006), pp. 255–264. ISSN 0301-486X

Garcia-Rodriguez, M., Morales-Ramirez, P., & Altamirano-Lozano, M. (2002). Effects of chlorophyllin on mouse embryonic and fetal development in vivo. *Teratogenesis, carcinogenesis, and mutagenesis*, Vol. 22, No. 6, (2002), pp. 461-471, ISSN 0270-3211

Girish, C., & Smith, T. (2008). Impact of feed-borne mycotoxins on avian cell-mediated and humoral immune responses. *World Mycotoxin Journal*, Vol.1, No.2, (June 2008), pp. 105–121, ISSN 1875-0710

Glintborg, B., Weismann, A., Kensler, T., & Poulsen, H. (2006). Oltipraz chemoprevention trial in Qidong, People's Republic of China: unaltered oxidative biomarkers. *Free Radical Biology and Medicine*, Vol. 41, No. 6, (September 2006), pp. 1010-1014, ISSN 0891-5849

Gradelet, S., Le Bon, A., Bergès, R, Suschetet, M., & Astorg, P. (1998). Dietary carotenoids inhibit aflatoxin B1-induced liver preneoplastic foci and DNA damage in the rat: role of the modulation of aflatoxin B1 metabolism. *Carcinogenesis*, Vol. 19, No. 3, (March 1998), pp. 403-411, ISSN 0143-3334

Gratz, S., Wu, Q., El-Nezami, H., Juvonen, R., Mykkanen, H., & Turner, P. (2007). Lactobacillus rhamnosus strain GG reduces aflatoxin B1 transport, metabolism, and toxicity in Caco-2 cells. *Applied and Environmental Microbiology*, Vol. 73, No. 12,(June 2007), pp. 3958-3964, ISSN 0099-2240

Guengerich, F., Johnson, W., Shimada, T., Ueng, Y., Yamazaki, H., & Langouët, S. (1998). Activation and detoxication of aflatoxin B1. *Mutation Research.* Vol. 402, No. 1-2, (1998), pp. 121-128, ISSN 0027-5107

Guzmán de la Peña, D., &Peña-Cabrales, J. (2005). Regulatory consideration aflatoxin contamination of foods in Mexico. *Revista Latinoamericana de Microbiología.* Vol. 4, No. 1-2, (July 2005), pp. 6121-6125, ISSN 0034-9771

Hartmann, A., Agurell, E., Beevers, C., Brendler-Shwaab, S., Burlinson, B., Clay, P., Collins, A., Smith, A., Speit, G., Thybaud, V., & R.R. Tice. (2003). Recommendations for conducting the *in vivo* alkaline Comet assay. 4th International Comet Assay Workshop. *Mutagenesis,* Vol. 18 No. 1, (January 2003), pp. 45-51, ISSN 0267-8357

Hedayati, M., Pasqualotto, A., Warn, P., Bowyer, P., & Denning, D. (2007). Aspergillus flavus: human pathogen, allergen and mycotoxin producer. *Microbiology,* Vol. 153, No. Pt 6, (June 2007), pp. 1677–1692, ISSN 1350-0872

Hernandez-Mendoza A., Guzman de la Peña D., & Garcia H. (2009). Key role of teichoic acids on aflatoxin B binding by probiotic bacteria. *Journal of Applied Microbiology* Vol. 107, No. 2, (August 2009), pp. 395-403, ISSN 1364-5072

Huang, C., Hsueh, J., Chen, H., & Batt, T. (1982). Retinol (vitamin A) inhibits sister chromatid exchanges and cell cycle delay induced by cyclophosphamide and aflatoxin B1 in Chinese hamster V79 cells. *Carcinogenesis,* Vol. 3, No. 1, (1982), pp. 1-5, ISSN 0143-3334

Hussain, S., Schwank, J., Staib, F., Wang, X., & Harris, C. (2007). TP53 mutations and hepatocellular carcinoma: insights into the etiology and pathogenesis of liver cancer. *Oncogene,* Vol. 26 No. 15, (April 2007), pp. 2166-2176

Hussein., H., & Brasel, J. (2001). Toxicity, metabolism, and impact of mycotoxins on humans and animals. *Toxicology.* Vol. 167, No. 2, (October 2001), pp. 101–134, ISSN 0300-483X.

IARC. (1993). Evaluation of carcinogen risks to humans. Some naturally occurring substances: food, items and constituents, heterocyclic aromatic amines and mycotoxins. *IARC Monographs for Evaluation of Carcinogenic Risks in Humans.* Vol. 56 (1993) pp. 489–521.

Ito,Y., Ohnishi S., & Fujie K. (1989). Chromosome aberrations induced by aflatoxin B1 in rat bone marrow cells in vivo and their suppression by green tea. *Mutation Research.* Vol. 222, No. 3, (March 1989), pp. 253-261, ISSN 0027-5107

Johnson W., & Guengerich F.P (1997). Reaction of AFB1 exo-8,9-epoxide with DNA: Kinetic analysis of covalent binding and DNA-induced hydrolysis. *Proccedings of the National Academy of Sciences, USA.* Vol. 94, No. 12, (June 1997), pp. 6121-6125

Juan-López, M., Carvajal, M., & Ituarte, B. (1995). Supervising programme of aflatoxins in Mexican corn. *Food Additives and Contaminants,* Vol. 12, No. 3, (May-June 1995), pp. 297-312, ISSN 0265-203X

Jun, H., Kim, S., & Sung, M. (2002). Protective effect of soybean saponins and major antioxidants against aflatoxin B1-induced mutagenicity and DNA-adduct formation. *Journal of Medicinal Food.* Vol. 5, No. 4, (Winter 2002), pp. 235-240, ISSN 1096-620X

Kabak, B., Dobson, A., & Var, I. (2006). Strategies to prevent mycotoxin contamination of food and animal feed: a review. *Critical Reviews in Food Science and Nutrition,* Vol. 46, No. 8, (2006), pp. 593–619, ISSN 1040-8398

Kada, T. & Shimoi, K. (1987). Desmutagens and bio-antimutagens – their modes of action. *BioEssays,* Vol. 7, No. 3, (September 1987), pp. 113–116, ISSN 0265-9247

Karakilcik, A., Zerin, M., Arslan, O., Nazligul, Y., & Vural, H. (2004). Effects of vitamin C and E on liver enzymes and biochemical parameters of rabbits exposed to aflatoxin B1. *Veterinary and Human Toxicology*, Vol. 46, No. 4, (August 2004), pp. 190-192, ISSN 0145-6296

Karekar, V., Joshi, S., & Shinde, S. (2000). Antimutagenic profile of three antioxidants in the Ames assay and the Drosophila wing spot test. *Mutation Research*, Vol. 468, No. 2, (July 2000), pp. 183-194, ISSN 0027-5107

Kelloff, G., Boone, C., Crowell, J., Nayfieldm, S., Hawk, E., Steele, V., Lubet, R., & Sigman, C. (1995). Strategies for phase II cancer chemoprevention trials: cervix, endometrium, and ovary. *Journal of Cellular Biochemistry. Supplement*, Vol. 23, (1995), pp.1-9, ISSN 0733-1959

Kensler, T., Egner, P., Dolan, P., Groopman, J., & Roebuck, B. (1987). Mechanism of protection against aflatoxin tumorigenicity in rats fed 5-(2-pyrazinil)-4-methyl-1,2-dithiol-3-thione (oltipraz) and related 1,2-dithiol-3-thiones and 1,2-dithiol-3-ones. *Cancer Research*, Vol. 47, No. 16, (August 1987), pp. 4271-4277, ISSN 0008-5472

Kensler, T. (1997). Chemoprevention by inducers of carcinogen detoxication enzymes. *Environmental Health Perspectives*, Vol.105, Suppl. 4, (June 1997), pp. 965-970, ISSN 0091-6765

Kensler, T., Groopman, J., & Roebuck, B. (1998a). Use of aflatoxin adducts as intermediate endpoints to assess the efficacy of chemopreventive interventions in animals and man. *Mutation Research*, Vol. 402, No. 1-2, (June 1998), pp. 165-172, ISSN 0027-5107

Kensler, T., He, X., Otieno, M., Egner, P., Jacobson, L., Chen, B., Wang, J., Zhu, Y., Zhang, B., Wang, J., Wu, Y., Zhang, Q., Quian, G., Kuang, S., Fang, X., Li, F., Yu, L., Prochaska, H., Davidson, N., Gordon, G., Gorman, M., Zarba, A., Enger, C., Muñoz, A., Helzlsour KJ., & Groopman JD. (1998b). Oltipraz chemoprevention trial in Qidong, People's Republic of China: modulation of serum aflatoxin albumin adduct biomarkers. *Cancer Epidemiology Biomarkers & Prevention*, Vol. 7, No. 2, (February 1998), pp. 127-134, ISSN 1055-9965

Kensler, T., Roebuck, B., Wogan, G., & Groopman J. (2011). Aflatoxin: a 50 year odyssey of mechanistic and translational toxicology. *Toxicological Sciences*. Vol. 120 Suppl 1, (March 2011), pp. 528-548, ISSN 1096-6080

Kimura, M., Lehmann, K., Gopalan-Kriczky, P., & Lotlikar, P. (2004). Effect of diet on aflatoxin B1-DNA binding and aflatoxin B1-induced glutathione S-transferase placental form positive hepatic foci in the rat. *Experimental & Molecular Medicine*. Vol. 36, No. 4, (August 2004), pp. 351–357, ISSN 1226-3613

Kogbo, W., Lemarinier, S., & Boutibonnes, P. (1985). Morphological characteristics and physiological properties of aflatoxin B1 producing and non-producing Aspergillus flavus strains. *Mycopathologia*, Vol. 91, No. 3, (September 1985), pp. 181-186, ISSN 0301-486X

Kumagai, S., Nakajima, M., Tabata, S., Ishikuro, E., Tanaka, T., Norizuki, H., Itoh, Y., Aoyama, K., Fujita, K., Kai, S., Sato, T., Yoshiike, N., & Sugita-Konishi, Y. (2008). Aflatoxin and ochratoxin A contamination of retail foods and intake of these mycotoxins in Japan. *Food Additives & Contaminants. Part A*, Vol. 25, No. 9, (September 2008), pp. 1101–1106. ISSN 1944-0049

Kumaravel, T., Vilhar, B., Faux, S., & Jha, A. (2009). Comet Assay measurements: a perspective. *Cell Biology and Toxicology*, Vol. 25 No. 1, (February 2009), pp. 53-64, ISSN 0742-2091

Latt, S., & Schereck, R. (1980). Sister Chromatid Exchange Analysis. *American Journal of Human Genetics*, Vol. 32, No. 3, (May 1980), pp. 297-313, ISSN 1537-6605

Le Hegarat, L., Dumont, J., Josse, R., Huet, S., Lanceleur, R., Mourot, A., Poul, J., Guguen-Guillouzo, C., Guillouzo, A., & Fessard, V. (2010). Assessment of the genotoxic potential of indirect chemical mutagens in HepaRG cells by the comet and the cytokinesis-block micronucleus assays. *Mutagenesis*. Vol. 25, No. 6, (2010), pp. 555-560. ISSN 0267-8357

Li, Y., Su, J., Qin, L., Egner, P., Wang, J., Groopman, J., Kensler, T., & Roebuck, B. (2000). Reduction of aflatoxin B(1) adduct biomarkers by oltipraz in the tree shrew (Tupaia belangeri chinensis). *Cancer Letters*, Vol. 154, No. 1, (June 2000), pp. 79-83, ISSN 0304-3835

Liu, Y., Roebuck, B., Yager, J., Groopman, J., & Kensler, T. (1988). Protection by 5-(2-pyrazinil)-4-methyl-1,2-dithiol-3-thione (oltipraz) against the hepatotoxicity of aflatoxin B1 in the rat. *Toxicology and Applied Pharmacology*, Vol. 93, No. 3, (May 1988), pp. 442-451, ISSN 0041-008X

Loarca-Piña, G, Kuzmicky, P., de Mejía, E., & Kado, N. (1998). Inhibitory effects of ellagic acid on the direct-acting mutagenicity of aflatoxin B1 in the Salmonella microsuspension assay. *Mutation Research*. Vol. 398, No. 1-2, (February 1998), pp. 183-167, ISSN 0027-5107

Lynch, R., & Wilson, D. (1991). Enhanced infection of peanut, *Arachis hypogaea* L, seeds with *Aspergillus flavus* group fungi due to external scarification of peanut pods by the lesser cornstalk borer, *Elasmopalpus lignosellus* (Zeller). *Peanut Science*, Vol. 18, No. 2, (1991), pp.110–116, ISSN 0095-3679

Madrigal-Santillán, E. Ph. D. Thesis. Inhibición de la genotoxicidad inducida por la aflatoxina B1 en ratón mediante la administración de glucano, manano y glucomanano. Nacional Politechnic Institute, Mexico, 2004

Madrigal-Santillán, E., Madrigal-Bujaidar, E., Márquez-Márquez, R., & Reyes, A. (2006). Antigenotoxic effect of Saccharomyces serevisiae on the damage produced in mice fed aflatoxin B(1) contaminated corn. *Food and Chemical Toxicology*. Vol. 44, No. 12, (December 2006), pp. 2058-2063, ISSN 0278-6915

Madrigal-Santillán, E., Alvarez-González, I., Márquez-Márquez, R., Velázquez-Guadarrama, N., & Madrigal-Bujaidar, E. (2007). Inhibitory effect of mannan on the toxicity produced in mice fed aflatoxin B1 contaminated corn. *Archives of Environmental Contamination and Toxicology* Vol.53 No.3, (October 2007), pp. 466-472, ISSN 0090-4341

Madrigal-Santillán, E., Morales-González, J., Sánchez-Gutierrez, M., Reyes-Arellano, A., & Madrigal-Bujaidar, E. (2009). Investigation on the protective effect of alpha-mannan against the DNA damage induced by aflatoxin B(1) in mouse hepatocytes. *International Journal of Molecular Sciences*. Vol. 10, No. 2, (February 2009), pp. 395-406, ISSN 1422-0067

Madrigal-Santillán, E., Morales-González, J., Vargas-Mendoza, N., Reyes-Ramírez, P., Cruz-Jaime, S., Sumaya-Martínez, T., Pérez-Pastén, R., & Madrigal-Bujaidar, E. (2010). Antigenotoxic studies of different substances to reduce the DNA damage induced by aflatoxin B1 and ochratoxin A. *Toxins*, Vol. 2, No. 4, (2010), pp. 738-757.

Mantovani, M., Bellini, M., Angeli, J., Oliveira, R., Silva, A., & Ribeiro, L. (2007). B-glucans in promoting health: prevention against mutation and cancer. *Mutation Research*. Vol. 658, No. 3, (March-April 2007), pp. 154-165, ISSN 0027-5107

Marquez-Marquez, R., Madrigal-Bujaidar, E., & Tejada de Hernandez, I. (1993). Genotoxic evaluation of ammonium inactivated aflatoxin B1in mice fed with contaminated corn. *Mutation Research*. Vol. 299, No. 1, (March 1993), pp. 1-8, ISSN 0027-5107

McLean, M., & Dutton, M. (1995). Cellular interactions and metabolism of aflatoxin: an update. Pharmacology & Therapeutics. Vol. 65, No 2, (February 1995), pp. 163-192, ISSN 0163-7258

Mehan, V., McDonald, D., Ramakriahna, N., & Williams, J. (1986). Effects of genotype and date of harvest on infection of peanut seed by *Aspergillus flavus* and subsequent contamination with aflatoxin. *Peanut Science,* Vol. 13, (1986), pp. 46 –50 ISSN 0095-3679

Mehan, V., Mayee, C., Jayanthi, S., & McDonald, D. (1991). Preharvest seed infection by *Aspergillus flavus* group fungi and subsequent   aflatoxin contamination in groundnuts in relation to soil types. *Plant and Soil,* Vol. 136, No. 2, (March 1991), pp. 239–248, ISSN 0032-079X

Miranda, D., Arçari, D., Ladeira, M., Calori-Domingues, M., Romero, A., Salvadori, D., Gloria E., Pedrazzoli, J. Jr., & Ribeiro, M. (2007). Analysis of DNA damage induced by aflatoxin B1 in Dunkin-Hartley guinea pigs. Mycopathologia. Vol.163, No. 5, (2007), pp. 275-280, ISSN 0301-486X

Naaz, F., Javed, S., & Abdin, M. (2007). Hepatoprotective effect of ethanolic extract of Phyllantus amarus Schm. et Thonn. on aflatoxin B1-induced liver damage in mice. *Journal of Ethnopharmacology*. Vol. 113, No. 3, (September 2007), pp. 503-509, ISSN 0378-8741

Neiger, R., Hurley, D., Hurley, D., Higgins, K., Rottinghaus, G., & Starh, H. (1994). The short-term effect of low concentrations of dietary aflatoxin and R-2 toxin on mallard ducklings. *Avian Diseases*. Vol. 38, No. 4, (October 1994), pp. 738-743, ISSN 0005-2086

Nelson, R. (1992). Chlorophyllin, an antimutagen, acts as a tumor promoter in the rat-dimethylhydrazine colon carcinogenesis model. *Anticancer Research*, Vol. 12, No. 3, (May 1992), pp. 737-739, ISSN 0250-7005

Nyandieka, H., & Wakhisi, J. (1993). The impact of vitamins A, C, E, and selenium compound on prevention of liver cancer in rats. *East African Medical Journal*, Vol. 70, No. 3, (March 1993), pp. 151-153, ISSN 0012-835X

O'Dwyer, P., Johnson, S., Khater, C., Krueger, A., Matsumoto, Y., Hamilton, T., & Yao, K. (1997). The chemopreventive agent oltipraz stimulates repair of damaged DNA. *Cancer Research,* Vol. 57, No. 6, (March 1997), pp. 1050-1053, ISSN 0008-5472

Ochoa, M., Torres, Ch., Moreno, I., Yepiz, G., Alvarez Ch., Marroquin, J., Tequida, M., & Silveira, G. (1989). Incidencia de aflatoxina B1 y zearalenona en trigo y maíz almacenado en el estado de Sonora. *Revista de Ciencias Alimentarias.* Vol. 1, No. 1, (1989), pp. 16–20

OECD. (1997). Guideline for the Testing of Chemicals. Section 4: Health Effects. 475 Mammalian Bone Marrow Chromosome Aberration Test. Vol. 1, No. 4, (July 2010), pp.1-8

OECD. (2007) Guideline for the testing of chemicals draft proposal for a new guideline 487. In vitro Mammalian cell Micronucleus Test (MNvit). December 13, 2007 (Version 3)

Oshida, K., Iwanaga, E., Miyamoto-Kuramitsu, K., & Miyamoto, Y. (2008). An in vivo comet assay of multiple organs (liver, kidney and bone marrow) in mice treated with methyl methanesulfonate and acetaminophen accompanied by hematology and/or blood chemistry. *The Journal of Toxicological Sciences,* Vol. 33, No. 5, (December 2008), pp. 515-524, ISSN 0388-1350

Oyaqbemi, A., Azeez, O., & Saba, A. (2010). Hepatocellular carcinoma and the underlying mechanisms. *African Health Sciences.* Vol. 10 No. 1, (2010), pp. 93-98, ISSN 1680-6905

Parlat, S., Ozcan, M., & Oguz, H. (2001). Biological suppression of aflatoxicosis in Japanese quail (*Coturnix coturnix japonica*) by dietary addition of yeast (Saccharomyces cerevisiae). *Research in Veterinary Science,* Vol.71, No. 3, (December 2001), pp. 207-211, ISSN 0034-5288

Pestka, J., & Bondy, G. (1994). Mycotoxin-induced immunomodulation. In: *Immunotoxicology and Immunopharmacology.* Dean J., Luster M., Munson A., Kimber I. (Ed). 163-182. Raven Press, New York, ISSN 0892-3973

Peterson S., Lampe J., Bammler, T., Gross-Steimeyer, K., & Eaton, D. (2006). Apiaceous vegetable constituents inhibit human cytochrome P-450 1A2 (hCYP1A2) activity and hCYP1A2-mediated mutagenicity of aflatoxin B1. *Food and Chemical Toxicology.* Vol. 44, No. 9, (September 2006), pp. 1474-1484, ISSN 0278-6915

Phillips, S., Wareing, P., Ambika, D., Shantanu, P., & Medlock, V. (1996). The mycoflora and incidence of aflatoxin, zearalenone and sterigmatocystin in dairy feed and forage samples from Eastern India and Bangladesh. *Mycopathologia,* Vol. 133, (November 1995), pp. 15–21, ISSN 0301-486X

Plasencia, J. (2004). Aflatoxins in maize: a Mexican perspective. *Toxin Reviews,* Vol. 23, No. 2-3, (January 2004), pp. 155–177, ISSN 1556-9543

Prabhu, A., Aboobaker, V., & Bhattacharya, R. (1989). In vivo effect of dietary factors on the molecular action of aflatoxin B1: role of riboflavin on the catalytic activity of liver fractions. *Cancer Letters,* Vol. 48, No. 2, (November 1989), pp. 89-94, ISSN 0304-3835

Qin, S., Batt, T., & Huang, C. (1985). Influence of retinol on carcinogen-induced sister chromatid exchanges and chromosome aberrations in V79 cells. *Environmental Mutagenesis,* Vol. 7, No. 2, (1985), pp. 137-148, ISSN 0192-2521

Qin, S., & Huang, C. (1985). Effect of retinoids on carcinogen-induced mutagenesis in Salmonella tester strains. *Mutation Research,* Vol. 142, No. 3, (March 1985), pp. 115-120, ISSN 0027-5107

Qin, S., & Huang, C. (1986). Influence of mouse liver stored vitamin A on the induction of mutations (Ames tests) and SCE of bone marrow cells by aflatoxin B1, benzo(a)pyrene, or cyclophosphamide. *Environmental Mutagenesis,* Vol. 8, No. 6, (1986), pp. 839-847, ISSN 0192-2521

Raina, V., & Gurtoo, H. (1985). Effects of vitamins A, C, and E on aflatoxin B1-induced mutagenesis in Salmonella typhimurium TA-98 and TA-100. *Teratogenesis, Carcinogenesis, and Mutagenesis,* Vol. 5, No. 1, (1985), pp. 29-40, ISSN 0270-3211

Ranjan, K., & Sinha, A. (1991). Occurrence of mycotoxigenic fungi and mycotoxins in animal feed from Bihar, India. *Journal of the Science of Food and Agriculture,* Vol. 56, No. 1, (September 1991), pp. 39-47, ISSN 0022-5142

Richmond, E., & O´Mara, A. (2010). Conducting chemoprevention clinical trials: challenges and solutions. *Seminars in Oncology,* Vol. 37, No. 4, (August 2010), pp. 402-406, ISSN 0093-7754

Roebuck, B., Liu, Y., Rogers, A., Groopman, J., & Kensler, T. (1991). Protection against aflatoxin B1-induced hepatocarcinogenesis in F344 rats by 5-(2-pyrazinil)-4-methyl-1,2-dithiole-3-thione (oltipraz): predictive role for short-term molecular dosimetry. Cancer Research, Vol. 51, No. 20, (October 1991), pp. 5501-5506, ISSN 0008-5472

Roebuck, B., Curphey, T., Li, Y., Baumgartner, K., Bodreddigari, S., Yan, J., Gange, S., Kensler, T., & Sutter, T. (2003). Evaluation of the cancer chemopreventive potency of dithiolethione analogs of oltipraz. *Carcinogenesis*, Vol. 24, No. 12, (December 2003), pp. 1919-1928, ISSN 0143-3334

Russell, L., Cox, D., Larsen, G., Bodwell, K., & Nelson, C. (1991). Incidence of molds and mycotoxins in commercial animal feed mills in seven Midwestern states, 1988–1989. *Journal of Animal Science*, Vol. 69, No. 1, (January 1991), pp. 5-12, ISSN 0021-8812

Santacroce, M., Conversano, M., Casalino, E., Lai, O., Zizzadoro, C., Centoducati, G., & Crescenzo, G. (2008). Aflatoxins in aquatic species: metabolism, toxicity and perspectives. *Reviews in Fish Biology and Fisheries*. Vol. 18 No. 1, (June 2008), pp. 99–130.

Shane, S. (1993). Economic issues associated with aflatoxins. In: *The Toxicology of Aflatoxins: Human Health, Veterinary, and Agricultural Significance*, Eaton DL, Groopman JD, (Ed.), 513-527, Academic Press, London UK.

Sharma, R., & Farmer, P. (2004). Biological relevance of adduct detection to the chemoprevention of cancer. *Clinical Cancer Research: An Official Journal of the American Association for Cancer Research*, Vol. 10, No. 15, (August 2004), pp. 4901-4912, ISSN 1078-0432

Shetty, P., Hald, B., & Jespersen, L. (2007). Surface binding of aflatoxin B1 by Saccharomyces cereviciae strains with potential decontaminating abilities in indigenous fermented foods. *International Journal of Food and Microbiology*, Vol. 113, No. 1, (January 2007), pp. 41-46, ISSN 0168-1605

Silvotti, L., Petterino, C., Bonomi, A., & Cabassi, E. (1997). Immunotoxicological effects on piglets of feeding sows diets containing aflatoxins. *The Veterinary Record*, Vol.141, No. 18, (November 1997), pp. 469 –472, ISSN 0042-4900

Simonovich, M., Egner, P., Roebuck, B., Orner, G., Jubert, C., Pereira, C., Groopman, J., Kensler, T., Dashwood, R., Williams, D., & Bailey, G. (2007). Natural chlorophyll inhibits aflatoxin B1-induced multi-organ carcinogenesis in the rat. *Carcinogenesis*, Vol. 28, No. 6, (June 2007), pp. 1294-1302, ISSN 0143-3334

Sinha, S., & Dharmshila, K. (1994). Vitamin A ameliorates the genotoxicity in mice of aflatoxin B1-containing Aspergillus flavus infested food. *Cytobios*, Vol. 79, No. 317, (1994), pp. 85-95, ISSN 0011-4529

Sparfel, L., Langouët, S., Fautrel, A., Salles, B., & Guillouzo, A. (2002). Investigations on the effects of oltipraz on the nucleotide excision repair in the liver. *Biochemical Pharmacology*, Vol. 63, No. 4, (February 2002), pp. 745-749, ISSN 0006-2952

Stanley, V., Ojo, R., Woldesenbet, S., Hutchinson, D., & Kubena, L. (1993). The use of Saccharomyces cerevisiae to suppress the effects of aflatoxicosis in broiler chicks. *Poultry Science*, Vol. 72, No. 10, (October 1993), pp. 1867-1872, ISSN 0032-5791

Steyn, P.S. (1995). Mycotoxins, general view, chemistry and structure. *Toxicology Letters*, Vol. 82-83, (December 1995), pp. 843-851, ISSN 0378-4274

Suphakarn, V., Newberne, P., & Goldman, M. (1983). Vitamin A and aflatoxin: effect on liver and colon cancer. *Nutrition Cancer*, Vol. 5, No. 1, (1983), pp. 41-50, ISSN 0163-5581

Tang, L., Tang, M., Xu, L., Luo, H., Huang, T., Yu, J., Zhang, L., Gao, W., Cox, S., & Wang, J. (2008). Modulation of aflatoxin biomarkers in human blood and urine by green tea polyphenols intervention. *Carcinogenesis*, Vol. 29, No. 2, (February 2008), pp. 411-417, ISSN 0143-3334

Tang, L., Xu, L., Afriyie-Gyagu, E., Liu, W., Wang, P., Tang, Y., Wang, Z., Huebner, H., Ankrah, N., Ofori-Adjei, D., Williams, J., Wang, J., & Phillips, T. (2009). Aflatoxin-albumin adducts and correlation with decreased serum levels of vitamins A and E in an adult Ghanaian population. *Food Additives & contaminants. Part A*, Vol. 26, No. 1, (January 2009), pp. 109-118, ISSN 1944-0049

Theumer, M., Cánepa, M., López, A., Mary, V., Dambolena, J., & Rubinstein HR. (2010). Subchronic mycotoxicoses in Wistar rats: assessment of the in vivo and in vitro genotoxicity induced by fumonisins and aflatoxin B(1), and oxidative stress biomarkers status. *Toxicology*, Vol. 268, No. 1-2, (2010), pp. 104-110, ISSN 0300-483X

Torres-Espinoza, E., Acuña-Askar, K., Naccha-Torres, L., & Castellon-Santa Ana, J. (1995). Quantification of aflatoxins in corn distributed in the city of Monterrey, Mexico. *Food Additives and Contaminants*, Vol. 12, No. 3, (May 1995), pp. 383-386, ISSN 0265-203X

United States Enviromental Protection Agency. (1998). Health Effects test Guidelines. OPPTS 870.5375. In vitro Mammalian Chromosome Aberration Test. (7101). EPA 712-C-98-223 August 1998. Retrieved from www.regulations.gov/fdmspublic/ContentViewer?objectId...pdf

Vasanthi, S., & Bhat, R. (1998). Mycotoxins in foods-occurrence, health & economic significance & food control measures. *The Indian Journal of Medical Research*, Vol. 108, (November 1998), pp. 212–224, ISSN 0971-5916

Wang, J., Qian, G., Zarba, A., He, X., Zhu, Y., Zhang, B., Jacobson, L., Gauge, S., Muñoz, A., Kensler, T., & Groopman, J. (1996). Temporal patterns of aflatoxin albumin adducts in hepatitis B surface antigen-positive and antigennegative residents of Daxin, Qidong county, People's Republic of China. *Cancer Epidemiology Biomarkers & Prevention*, Vol. 5, No 4, (April 1996), pp. 253– 561, ISSN 1055-9965

Wang, J., Shen, X., He, X., Zhu,Y., Zhang, B., Wang, J., Quian, G., Kuang, S., Zarba, A., Egner, P., Jacobson, L., Muñoz, A., Helzlsouer, K., Groopman, J., & Kensler, T. (1999). Protective alterations in phase 1 and 2 metabolism of aflatoxin B1 by Oltipraz in Residents of Qidong, People's Republic China. *Journal of the National Cancer Institute*, Vol. 91, No. 4, (February 1999), pp. 347-354, ISSN 1052-6773

Wang, J., & Groopman, J. (1999). DNA damage by mycotoxins. *Mutation Research*, Vol. 424, No. 1-2, (July 1999), pp. 167–181, ISSN 0027-5107

Wang, J., Huang, T., Su, J., Liang, Y., Luo, H., Kuang, S., Quiang, G., Sun, G., He, X., Kensler, T., & Groopman, J. (2001). Hepatocellular carcinoma and aflatoxin exposure in Zhuqing Village, Fusui County, People's Republic of China. *Cancer Epidemiology Biomarkers & Prevention*, Vol. 10, No 2, (February 2001), pp. 143– 146, ISSN 1055-9965

Wang, J., & Liu, X. (2007). Contamination of aflatoxins in different kinds of foods in China. *Biomedical and Environmental Sciences*, Vol. 20, No. 6, (December 2007), pp. 483–487, ISSN 0895-3988

Wang, P., Afriyie-Gyawu, E., Tang, Y., Johnson, N., Xu, L., Tang, L., Huebner, H., Ankrah, N., Ofori-Adjei, D., Ellis, W., Jolly, P., Williams, J., Wang, J., & Phillips, T. (2008). NovaSil clay intervention in Ghanaians at high risk for aflatoxicosis: II. Reduction in biomarkers of aflatoxin exposure in blood and urine. *Food additives & contaminant. Part A, Chemistry, analysis, control, exposure & risk assessment*, Vol. 25, No. 5, (May 2008), pp. 622-634, ISSN 1944-0049

Warner, J., Nath, J., & Ong, T. (1991). Antimutagenicity studies of chlorophyllin using the Salmonella arabinose-resistant assay system. *Mutation Research*, Vol. 262, No. 1, (January 1991), pp. 25-30, ISSN 0027-5107

Webster, R., Gawde, M., & Bhattacharya, R. (1996). Modulation of carcinogen-induced DNA damage and repair enzyme activity by dietary riboflavin. *Cancer Letters*, Vol. 98, No. 2, (January 1996), pp. 129-135, ISSN 0304-3835

Weisburger, J., Hara, Y., Dolan, L., Luo, F., Pittman, B., & Zang, E. (1996). Tea polyphenols as inhibitors of mutagenicity of major clases of carcinogens. *Mutation Research*, Vol. 371, No. 1-2, (November 1996), pp. 53-67, ISSN 0027-5107

Weisburger, J. (2001). Antimutagenesis and anticarcinogenesis, from the past to the future. *Mutation Research*, Vol. 480-481, (September 2001), pp. 23-35, ISSN 0027-5107

Welzel, T., Katki, H., Sakoda, L., Evans, A., London, W., Chen, G., O'broin, S., Shen, F., Lin, W., & McGlynn, K. (2007). Blood folate levels and risk of liver damage and hepatocellular carcinoma in a prospective high-risk cohort. *Cancer Epidemiology, Biomarkers & Prevention*, Vol. 16, No. 6, (June 2007), pp. 1279-1282, ISSN 1055-9965

Whong, W., Stewart, J., Brockman, H., & Ong, T. (1988). Comparative antimutagenicity of chlorophyllin and five other agents against aflatoxin B1-induced reversion in Salmonella typhimurium strain TA98. *Teratogenesis, Carcinogenesis, and Mutagenesis*, Vol. 8, No. 4, (1988), pp. 215-224, ISSN 0270-3211

Williams, J., Phillips, D., Jolly, P., Stiles, J., Jolly, C., & Aggarwal, D. (2004). Human aflatoxicosis in developing countries: a review of toxicology, exposure, potential health consequences, and interventions. *The American Journal of Clinical Nutrition*, Vol. 80, No. 5, (November 2004), pp. 1106-1122, ISSN 0002-9165

Wu, Z., Chen, J., Ong, T., Brockman, H., & Whong, W. (1994). Antitransforming activity of chlorophyllin against selected carcinogens and complex mixtures. *Teratogenesis, Carcinogenesis, and Mutagenesis*, Vol. 14, No. 2, (1994), pp. 75-81, ISSN 0270-3211

Wu, Q., Jezkova, A., Yuan, Z., Pavlikova, L., Dohnal, V., & Kuca, K. (2009). Biological degradation of aflatoxins. *Drug Metabolism Reviews*, Vol. 41, No. 1, (2009), pp. 1-7, ISSN 0360-2532

Yates, M., & Kensler, T. (2007). Keap1 eye on the target: chemoprevention of liver cancer. *Acta Pharmacologica Sinica*, Vol. 28, No. 9, (September 2007), pp. 1331-1342, ISSN 1671-4083

Yiannikouris, A., André, G., Poughon, L., Francois, J., Dussap, C., Jeminet, G., Bertin, G., & Jouany, J. (2006). Chemical and conformational study on the interactions involved in mycotoxin complexation with beta-D-glucans. *Biomacromolecules*. Vol. 7, No. 4, (April 2006), pp. 1147-1155, ISSN 1525-7797

Yu, M., Zhang, Y., Blaner, W., & Santella, R. (1994). Influence of vitamins A, C, and E and beta-carotene on aflatoxin B1 binding to DNA in woodchuck hepatocytes. *Cancer*, Vol. 73, No. 3, (February 1994), pp. 596-604, ISSN 0008-543X

Zhang, Y., & Munday, R. (2008). Dithiolethiones for cancer chemoprevention: where do we stand?. *Molecular Cancer Therapeutics*, Vol. 7, No. 11, (November 2008), pp. 3470-3479, ISSN 1535-7163

Zuber, M., Darrah, L., Lillehoj, E., Josephson, L., Manwiller, A., Scott, G., Gudauskas, R., Horner, E., Widstrom, N., Thomposn, D., Bockholt, A., & Brewbaker, J. (1983). Comparison of open-pollinated maize varieties and hybrids for preharvest aflatoxin contamination in the Southern United States. *Plant Disease*, Vol. 67, (February 1983), pp. 185–187, ISSN 0191-2917

16

Aflatoxins Biochemistry and Molecular Biology - Biotechnological Approaches for Control in Crops

Laura Mejía-Teniente,
Angel María Chapa-Oliver, Moises Alejandro Vazquez-Cruz,
Irineo Torres-Pacheco and Ramón Gerardo Guevara-González
Facultad de Ingeniería, CA Ingeniería de Biosistemas,
Universidad Autónoma de Querétaro,
Centro Universitario Cerro de las Campanas s/n, Querétaro, Qro,
México

1. Introduction

Fungi play a very important, but yet mostly unexplored role. Their widespread occurrence on land and in marine life makes them a challenge and a risk for humans (Bräse *et al.*, 2009). Fungi are ingenious producers of complex natural products which show a broad range of biological activities (Bohnert *et al.*, 2010). However, a specific characteristic is the production of toxins. Mycotoxins (from "*myco*" fungus and toxin), are nonvolatile, relatively low-molecular weight, fungal secondary metabolic products (Bräse *et al.*, 2009). The most agriculturally important micotoxins are aflatoxins (AF) which are a group of highly toxic metabolites, studied primarily because of their negative effects on human health. Aflatoxins belong to a group of difuranocumarinic derivatives structurally related, and are produced meanly by fungi of genus *Aspergillus* spp. Its production depends on many factors such as substrate, temperature, pH, relative humidity and the presence of other fungi. It has been identified 18 types of aflatoxins; the most frequent in foods are B_1, B_2, G_1, G_2, M_1, and M_2 (Bhatnagar *et al.*, 2002). These secondary metabolites contaminate a number of oilseed crops during growth of the fungus and this can result in severe negative economic and health impacts (Cary *et al.*, 2009). The higher levels of aflatoxins have been found in cotton and maize seeds, peanuts, and nuts. In grains like wheat, rice, rye or barley the presence of aflatoxins is less frequent. Mycotoxins may also occur in conjugated form, either soluble (masked mycotoxins) or incorporated into/ associated with/attached to macromolecules (bound mycotoxins). These conjugated mycotoxins can emerge after metabolization by living plants, fungi and mammals or after food processing. Awareness of such altered forms of mycotoxins is increasing, but reliable analytical methods, measurement standards, occurrence, and toxicity data are still lacking (Berthiller *et al.*, 2009). A variety of studies has been conducted in order to understand the process of crop contamination by aflatoxins. Mycotoxins are dangerous metabolites that are often carcinogenic, and they represent a serious threat to both animal and human health (Reverberi *et al.*, 2010). Mycotoxins are considered secondary metabolites because they are not necessary for fungal growth and are simply a product of primary metabolic processes. The functions of mycotoxins have not clearly established, but they are believed to play a role in eliminating other microorganisms competing in the same environment (Bräse *et al.*, 2009). The biosynthesis and regulation of these toxins represent one of the most studied areas of all the fungal secondary metabolites. Much of the information obtained on the AF biosynthetic genes and regulation of AF biosynthesis was obtained through studies using *A. flavus* and *A. parasiticus* and also the model fungus *Aspergillus nidulans* that produces sterigmatocystin (ST), the penultimate precursor to AF. Further studies in *A. nidulans* and *A. flavus* and also of the fungus-host plant interaction have identified a number of genetic factors that link secondary metabolism and morphological differentiation processes in *A. flavus* as well as filamentous fungi in general (Cary *et al.*, 2009). Recent investigations of the molecular mechanism of AF biosynthesis showed that the genes required for biosynthesis are in a 70 kb gene cluster. These genes encode for the proteins required in the oxidative and regulatory steps in the aflatoxins byosinthesis. A positive regulatory gene, *aflR*, coding for a sequence-specific, zincfinger DNA-binding protein is located in the cluster and is required for transcriptional activation of most, if not all, of the aflatoxin structural genes. Some of the genes in the cluster also encode other enzymes such as cytochrome P450-type monooxygenases, dehydrogenases, methyltransferases, and polyketide and fatty acid synthases (Bhatnagar *et al.*, 2003). The application of genomic DNA sequencing and functional genomics, powerful technologies that allow scientists to study a whole set of genes in an organism, is one of the most exciting developments in aflatoxin research (Yu *et al.*, 2004; Bennett *et al.*, 2007). Moreover, the rapid development of high throughput sequencing made it possible in genetic research to advance from single gene cloning to whole genome sequencing. Tremendous advances have also been made in understanding the genetics of four non-aflatoxigenic *Aspergillus* species, *A. oryzae, A. sojae, A. niger and A. fumiga*tus. Currently, the whole genome sequencing and/or Expressed Sequence Tag (EST) projects for *A. flavus* have been completed (Bhatnagar *et al.*, 2006). The characterization of genes involved in aflatoxin formation affords the opportunity to examine the mechanism of molecular regulation of the aflatoxin biosynthetic pathway, particularly during the interaction between aflatoxin-producing fungi and plants (Bhatnagar *et al.*, 2003).Aflatoxin contamination in crops is a worldwide food safety concern due that are compound carcinogenic highly and mutagenic in animals and human (Yin *et al.*, 2008). Therefore their management in agricultural (pre-harvest, harvest and post-harvest) is of importance vital, so quantity in food and feed is closely monitored and regulated in most countries for example, in the European Union has a maximum level of 2 ng/g for B1 and 4 ng/g for total aflatoxins in crops (van Egmond and Jonker, 2004).

2. Occurrence of mycotoxins

Mycotoxins occur in many varieties of fungi. Several mycotoxins are unique to one species, but most mycotoxins are produced by more than one species. The most important mycotoxins are aflatoxins, ochratoxins, deoxynivalenol (DON), searalenone, fumonisin, T-2 toxin, and T-2 like toxins. However, food borne mycotoxins likely to be of greatest significance in tropical developing countries are the fumonisins and aflatoxins (Kumar *et al.*, 2008; Muthomi *et al.*, 2009). Aflatoxins are carcinogenic secondary metabolites produced by several species of *Aspergillus* section *Flavi*, including *Aspergillus flavus* Link,

Aspergillus parasiticus Speare, and *Aspergillius nominus* Kutzman, Horn, and Hesseltime. The fungus forms sclerotia which allow it to survive in soil for extended periods of time (Schneiddeger & Payne, 2003). Conditions such as high temperatures and moisture, unseasonal rains during harvest and flash floods lead to fungal proliferation and production of mycotoxins (Bhat & Vasanthi, 2003). About 4.5 billion people in developing countries are chronically exposed to aflatoxin and the CODEX recommended sanitary and phytosanitary standards set for aflatoxins adversely affect grain trade in developing countries (Gebrehiwet *et al.*, 2007). Concerns for human and livestock health have led several countries to constantly monitor and regulate aflatoxin contamination of agricultural commodities (Wang & Tang, 2005). Since the discovery of aflatoxins in the early 1960s, many studies have been conducted to assess the occurrence and to describe the ecology of aflatoxin-producing fungi in natural and agricultural environments. *Aspergillus flavus* is the most abundant aflatoxin-producing species associated with corn (Abbas *et al.*, 2004*a*). While aflatoxins occur mostly in maize and groundnuts, the prevalence of fumonisins in maize is 100% (Wagacha & Muthomi, 2008). Mycotoxins have negative impact on human health, animal productivity and trade (Wagacha & Muthomi, 2008; Wu, 2006). Aflatoxin B_1 is the most toxic and is associated with liver cancer and immune suppression (Sheppard, 2008). Exposure to large doses (> 6000 mg) of aflatoxin may cause acute toxicity with lethal effect, whereas exposure to small doses for prolonged periods is carcinogenic (Groopmann & Kensler, 1999). There may be an interaction between chronic mycotoxins exposure and malnutrition, immune-suppression, impaired growth, and diseases such as malaria and HIV/AIDS (Williams *et al.*, 2004). Mycotoxin poisoning may be compounded by the co-ocurrence of aflatoxins with other mycotoxins such as fumonisins, zearalenone and deoxynivalenol (Kimanya *et al.*, 2008; Pietri *et al.*, 2009).

However, the presence of mycotoxins in food is often overlooked due to public ignorance about their existence, lack of regulatory mechanisms, dumping of food products and the introduction of contaminated commodities into the human food chain during chronic food shortage due to drought, wars, political, and economic instability. The largest mycotoxin-poisoning epidemic in the last decade occurred in Kenya in 2004. Aflatoxin poisoning was associated with eating home grown maize stored under damp conditions (Lewis et al., 2005). Acute aflatoxin poisoning has continued to occur severally in Eastern and Central provinces of Kenya (CDC, 2004). In the 2004 aflatoxin-poisoning outbreak, the concentrations of aflatoxin B1 in maize was high as 4,400 ppb, which is 220 times greater than the 20 ppb regulatory limit. The outbreak covered more than seven districts and resulted in 317 case-patients and 125 deaths (Lewis *et al.*, 2005). The association of mycotoxins with human and animal health is not a recent phenomenon; for example, in the past, ergotism was suspected of being a toxicosis resulting from these toxic fungal metabolites. Nowadays, more is known regarding this family of compounds. Mycotoxins were considered as a storage phenomenon whereby grains becoming moldy during storage allowed for the production of these secondary metabolites proven to be toxic when consumed by man and other animals. Subsequently, aflatoxins and mycotoxins of several kinds were found to be formed during development of crop plants in the field. The determination of which of the many known mycotoxins are significant can be based upon their frequency of occurrence and/or the severity of the disease that they produce, especially if they are known to be carcinogenic. The diseases (mycotoxicoses) caused by these mycotoxins are quite varied and involve a wide range of susceptible animal species including humans. Most of these diseases occur after consumption of mycotoxin contaminated grain or products made from such grains but other routes of exposure exist. The diagnosis of mycotoxicoses may prove to be difficult because of the similarity of signs of disease to those caused by other agents. Therefore, diagnosis of a mycotoxicoses is dependent upon adequate testing for mycotoxins involving sampling, sample preparation and analysis (Richard, 2007).

2.1 Toxicology of mycotoxins

Mycotoxins primarily occur in the mycelium of the toxigenic moulds and may also be found in the spores of these organisms and cause a toxic response, termed a mycotoxicoses, when ingested by higher vertebrates and other animals (Bennett & Klich, 2003). These secondary metabolites are synthesized during the end of the exponential phase of growth and appear to have no biological significance with respect to mould growth/development or competitiveness. All moulds are not toxigenic and while some mycotoxins are produced by only a limited number of species, others may be produced a relatively large range from several genera (Hussein & Brasel, 2001). The toxic effect of mycotoxin ingestion in both humans and animals depends on a number of factors including intake levels, duration of exposure, toxin species, mechanisms of action, metabolism, and defense mechanisms (Galvano et al., 2001). Consumption of mycotoxin-contaminated food or feed does however lead to the induction of teratogenic, carcinogenic, oestrogenic, neurotoxic, and immunosuppressive effect in humans and/or animals (Atroshi et al., 2002). The mycotoxins of most significance from both a public health and agronomic perspective include the aflatoxins, trichotecenes, fumonisins, ocharotoxin A (OTA), patulin, tremorgenic toxins, and ergot alkaloids (Papp et al., 2002).

3. Aflatoxins

Aflatoxin was initially identified as toxic after investigations of the death of 100,000 turkeys in the United Kingdom in 1960 (Blout, 1961). This prompted a major revolution in mycotoxin research resulting in intensive testing of mycotoxins in any moldy products. Since then several Aspergilli have been identified as capable of producing aflatoxins. The two most agriculturally important species are Aspergillus flavus and A. parasiticus, which are found throughout the world, being present in both the soil and the air (Abbas et al., 2005). When conidia (spores) encounter a suitable nutrient source and favorable environmental conditions (hot and dry conditions) the fungus rapidly colonizes and produces aflatoxin (Payne, 1992). Contamination of agricultural commodities by aflatoxin is a serious problem due to the substantial health effect it has on humans and animals. The use of agrochemicals (fungicides), timely irrigation, and alternate cropping systems have independently shown limited success in preventing aflatoxin contamination. Integration of these tactics will be required to manage such a difficult problem (Cleveland et al., 2003). A more recent and promising technology is the use of non toxigenic strains of Aspergillus as biocontrol agents. However, to maximize this methodology and to prevent the colonization of multiple crops by A. flavus and related species (A. parasiticus and A. nominus), it is critical that a complete understanding of the ecology of these unique fungi be developed (Abbas et al., 2009). Aflatoxins are toxic compounds chemically related to bisfuranocoumarin that are produced by A. flavus and A. parasiticus strains (Abbas et al.,

2004b). These two aflatoxigenic species have been frequently studied due to their impact on agricultural commodities and their devastating effects on livestock. The name aflatoxin comes from the genus *Aspergillus* which is where the letter "a" in aflatoxin is derived and "fla" from the species name *flavus*. In agricultural grains the fungi *A. flavus* and *A. parasiticus* are capable of producing four major aflatoxins (AfB1, AfB2, AfG1, and AfG2). *A. flavus* tipically produces only the B toxins (Abbas *et al.*, 2004b). Corn and cottonseed are typically contaminated with the aflatoxin B1, produced after colonization by *A. flavus* (Klich, 1986). *A. parasiticus* is more prevalent in peanuts than any other crop; however, it is typically outcompeted by *A. flavus* when the two fungi are both present (Horn *et al.*, 1995). These fungi are ubiquitous in the environment, being readily isolated from plants, air, soil, and insects (Wicklow *et al.*, 2003). Soil populations of *A. flavus* in soils under maize cultivation can range from 200 to >300,000 colony-forming units (CFU) g^{-1}soil (Zablotowicz *et al.*, 2007) and can constitute from ≤0.2% to ≤8% of the culturable soil fungi population. The major soil property associated with maintaining soil populations of *A. flavus* is soil organic matter. Higher populations of A. flavus are maintained in the soil surface of no-till compared to conventional-till soils (Zablotowicz *et al.*, 2007). The presence of Aspergillus species in dust can compromise individuals with elevated allergies to the fungus or its products (Benndorf et al., 2008). Of more concern is the colonization of certain food and feed crops (corn, cottonseed, peanuts, and some tree nuts) by the fungus, where it may produce a high concentration of these chemical compounds, specifically aflatoxin, to cause them to be considered contaminated and unfit for their intended use (Abbas *et al.*, 2009). When suitable environmental conditions arise, sclerotia and conidia germinate into mycelia that produce numerous conidiophores and release conidia into the air that can be available for colonizing plants. Although *A. flavus* colonizes a plant structure, it doesn´t necessarily produce aflatoxin to excessive levels. In this manner, *A. flavus* is an opportunistic pathogen in a similar context to the opportunistic human pathogens *Pseudomonas fluorescens* and *Burkholderia capacia*. These bacteria may colonize in low levels in compromised individuals, such as burn patients or the immunocompromised, and become pathogenic. In the same context, healthy plant tissues are less prone to be extensively colonized by *A. flavus*. However, under heat stress and moisture deficit, corn reproductive structures are readily susceptible to high levels of aflatoxin contamination, (O'Brian *et al.*, 2007). Therefore, inoculum potential modified by life cycle of the fungus is as critical as the environment and the host. The *A. flavus* life cycle can be divided into two major phases: the colonization of plant residues in the soil, and the infection of crop tissues, including grain and seeds of actively growing plant tissues. At the beginning of the growing season, usually in spring and sometimes at the end of winter, when sclerotia are exposed to the soil surface, they quickly germinate and form new conidial inoculum. This new inoculum will be vectored by insects and carried by the wind to begin the colonization and infection of the freshly planted crops (Horn, 2007). During the growing season, infected plant tissues can serve as sources of secondary conidial inoculum, which colonize new non-infected plant tissues (Fig.1). Despite our understanding of how the initial and secondary inocula occur for plant infection, little information is available about the saprotrophic activities of these fungi in soil. Recently, Accinelli *et al.* (2008) confirmed the presence of A. flavus in the soil actively synthesizing aflatoxins. However, not all *A. flavus* and *A. parasiticus* isolates produce aflatoxins (Abbas *et al.*, 2004b).

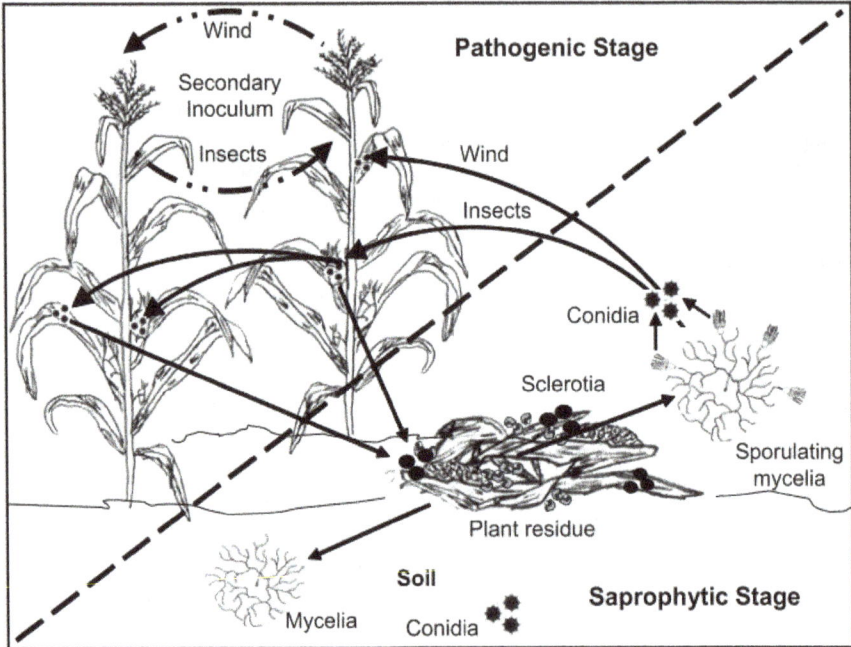

Fig. 1. Life cycle of *A. flavus* in a corn cropping system (Abbas *et al.*, 2009)

Fungi are classified as nonaflatoxigenic if they do not produce aflatoxins but produce other toxins. If fungi produce no toxins at all, they are classified as nontoxigenic. Generally, in any environment, the frequency of aflatoxigenic isolates can range from 50% to 80% (Abbas *et al.*, 2004a). The relative distribution of aflatoxigenic versus nonaflatoxigenic isolates is modulated by many factors including plant species present, soil composition, cropping history, crop management, and environment conditions, including rain fall and temperature (Abbas *et al.*, 2004b). Each of these factors can reduce the levels of *A. flavus*, for example, noncultivated fields near cultivated land are observed to have very low populations of *A. flavus* (Horn, 2007). Similarly, the frequency of drought is a factor in populations of fungi, with significant drops in soil populations of *A. flavus* after several years without drought. The conidia remain dormant in soil and only germinate when nutrient sources are present (Zablotowicz *et al.*, 2007). The behavior of Aspergilli structures in soil needs to be investigated and evaluated thoroughly, especially in agricultural soils, due to the fungal structures serving as the primary inoculum resulting in aflatoxin contamination in agricultural commodities (Abbas *et al.*, 2009).

3.1 Biosynthesis

Aflatoxins the most carcinogenic substances known to date have gained much interest among organic chemists since the elucidation of their structure by Buchi and co-workers in 1963. Even though numerous syntheses of racemic aflatoxins were reported in the following years , it took 40 years for the first enantioselective total synthesis of (-)-aflatoxin B_1 and B_2 to be published by Trost *et al.* (2003), their approach resembles in part (construction of the DE ring system) the first total synthesis of (±)-aflatoxin by Buchi *et al.* (1967). The biosynthetic pathway in *A. flavus* consists of approximately 23 enzymatic reactions and at least 15 intermediates (reviewed in Bhatnagar *et al.*, 2006; Bräse *et al.*, 2009) encoded by 25 identified genes clustered within a 70-kb DNA region on chromosome III (Bhatnagar *et al.*, 2006a; Cary & Ehrlich, 2006; Smith *et al.*, 2007; Cary & Calvo, 2008). The initial substrate acetate is used to generate polyketides with the first stable pathway intermediate being anthraquinone norsolorinic acid (NOR) (Bennett *et al.*, 1997). This is followed by anthraquinones, xanthones, and ultimately aflatoxins synthesis (Yu *et al.*, 2004). Few regulators of this process have been identified (Cary & Calvo, 2008), and a general model based on Aspergillus has recently been reviewed by Georgianna & Payne (2009) (Fig. 2). In addition to pathway-specific regulators, production of aflatoxins is also under the control of a number of global regulatory networks that respond to environmental and nutritional cues. These include responses to nutritional factors such as carbon and nitrogen sources and environmental factors such as pH, light, oxidative stress, and temperature. Nitrogen source plays an important role in aflatoxin biosynthesis (Bhatnagar *et al.*, 1986). In general, nitrate inhibits aflatoxin production, while ammonium salts are conducive (Cary & Calvo, 2008). Ammonium acetate does not have any significant impact on the level of OTA-related pks (the gene encoding for a polyketide synthase) expression. Nevertheless, this compound does lead to an increase in OTA production (Abbas *et al.*, 2009). Some aminoacids as proline, asparagines, and tryptophane significantly increase the biosynthesis of aflatoxins B1 and G1 in *A. parasiticus* (Payne and Hagler, 1983). Tryptophane acts by up-regulating aflatoxin gene expression in *A. parasiticus* and down-regulating it in *A. flavus*. Some nitrogen sources can also be non-conducive for OTA production in A. ochraceus, and their inhibitory effect is probably exerted at the transcriptional level (O'Callaghan *et al.*, 2006). The influence of carbon sources on aflatoxins and OTA biosynthesis has been studied for decades and it has produce contradictory results (Abbas *et al.*, 2009). Aflatoxin biosynthesis is induced by simple sugars such as glucose and sucrose that are present or generated by fungal hydrolytic enzymes during invasion of seed tissues (Cary & Calvo, 2008). A key factor determining whether a carbon source can support aflatoxin production and fungal growth is its availability to both hexose monophosphate and glycolitic pathways. This finding was confirmed by the identification of a set of genes including *enoA* and *pbcA* genes, both these genes are up-regulated in response to sucrose supplementation (Price *et al.*, 2006). The addition of different simple sugars may have opposite effects on OTA synthesis depending on the culture media used. Nevertheless, lactose exhibited a significant enhancing effect on OTA biosynthesis both in restrictive and conducive media, whilst glucose can show a repressive effect on OTA synthesis (Abbas *et al.*, 2009). This negative effect may be partially explained by the involvement of *CreA*, the regulator of the carbon repression system which also acts as a controller of the secondary metabolism in many fungal species (Roze *et al.*, 2004). Other environmental factors, such as temperature, water activity and pH, strongly influence mycotoxin biosynthesis. Some examples have been provided for OTA and aflatoxins biosynthesis (Ramirez *et al.*, 2006; Ribeiro *et al.*, 2006). The optimal temperature for production of aflatoxins is approximately 30°C (Boller & Schroeder, 1974). The establishment of temperature as an important component of infection by A. flavus and subsequent aflatoxin contamination has been clearly demonstrated under controlled greenhouse conditions (Payne *et al.*, 1988). Some efforts to illustrate a relationship between temperature and aflatoxin contamination were unsuccesfull (Stoloff and Lillehoj, 1981). The reason for this phenomenon can be traced to the finding that a detectable relationship exists only during years when amounts of contamination are high (McMillian *et al.*, 1985). Conclusions of this work were that high temperatures do significantly contribute to the contamination process and the ultimate amount of aflatoxin which is produced. Naturally, nothing can be done to control ambient temperatures, but it is possible to avoid their full impact during the later stages of kernel filling by early planting (Abbas *et al.*, 2009). Relative humidity above 86% also promotes colonization and aflatoxin production in the field (Plasencia, 2004).

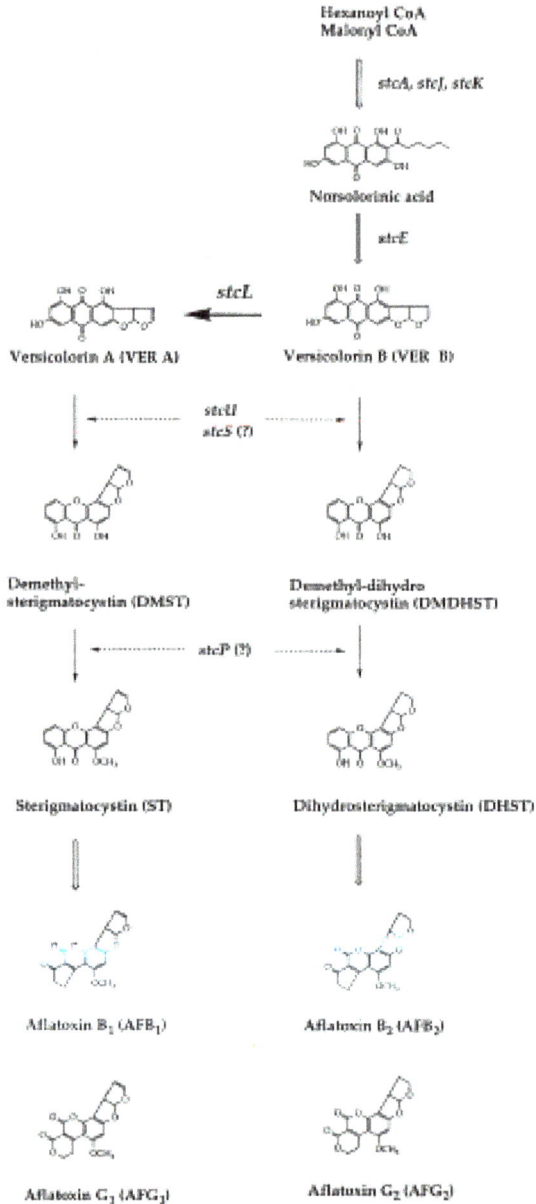

Hexanoyl CoA
Malonyl CoA

stcA, stcJ, stcK

Norsolorinic acid

stcE

Versicolorin A (VER A)   ← *stcL*   Versicolorin B (VER B)

stcU
stcS (?)

Demethyl-
sterigmatocystin (DMST)   Demethyl-dihydro
                              sterigmatocystin (DMDHST)

stcP (?)

Sterigmatocystin (ST)   Dihydrosterigmatocystin (DHST)

Aflatoxin B₁ (AFB₁)   Aflatoxin B₂ (AFB₂)

Aflatoxin G₁ (AFG₁)   Aflatoxin G₂ (AFG₂)

Fig. 2. AF/ST biosynthetic pathway in *Aspergillus spp.* (Kelkar *et al.*, 1997)

Aflatoxin production, in general, is greatest in acidic medium and tends to decrease as the pH of the medium increases (Keller et al., 1997). Response to changes in pH is regulated by the globally acting transcription factor PacC, which is posttranslationally modified by a pH-sensing protease (Tillburn et al., 1995). PacC binding sites indentified in the promoters of aflatoxins biosynthetic genes could be involved in negative regulation of aflatoxins biosynthesis during growth at alkaline pH (Ehrlich et al., 2002). Fungal development also appears to respond to changes in pH as sclerotial production was found to be reduced by 50% at pH 4.0 or less while aflatoxins production was at its maximal (Cotty, 1988). According to Georgianna and Payne (2009), only temperature has a greater influence on aflatoxin biosynthesis than pH. pH values lower than 4.0 are needed for aflatoxin production, and generally, the lower the pH value, the higher is the toxin synthesis (Klich, 2007).

In addition to temperature, water activity, and pH, the application of suboptimal concentrations of fungicides can boost mycotoxin biosynthesis (Schmidt-Heydt et al., 2007; D'Mello et al., 1998). A more appropriate general strategy is therefore to investigate natural products within the crop which confer resistance to *Aspergillus* colonization and growth, and/or aflatoxin biosynthesis. Two classes of protective natural factors exist in nature: phytoalexins, inducible metabolites, formed after invasion *de novo*, e.g. by activation of latent enzyme systems; phytoanticipins, constitutive metabolites, present *in situ*, either in the active form or easily generated from a precursor. Since phytoalexins are produced only in response to fungal attack, it is obvious that their presence would lag behind the infection and levels capable of suppressing aflatoxin would be difficult to regulate. In contrast, phytoanticipins are always present and such factors offer the potential for enhancement through breeding and selection of more resistant cultivar, or even genetic manipulation to introduce or enhance their levels. Once such compounds have been identified, it is only necessary to ensure that they are present in large enough quantities and in tissues from which fungal growth and aflatoxin deposition must be excluded (Campbell et al., 2003). Currently available methods of removing aflatoxins from tree nuts after contamination are impractical and expensive (Scott, 1998). There is a need to design new and environmentally safe methods of reducing infection by aflatoxigenic aspergilla and to inhibit aflatoxin biosynthesis.

3.2 Genetics of aflatoxin biosynthesis

Cloning of genes involved in aflatoxin biosynthesis is the key to understanding the molecular biology of the pathway (Trail et al., 1995). There are 21 enzymatic steps required for aflatoxin biosynthesis and the genes for these enzymes have been cloned (Bhatnagar et al., 2003). Molecular research has targeted the genetics, biosynthesis, and regulation of aflatoxin formation in *A. flavus and A. parasiticus*. Aflatoxins are biosynthesized by a type II polyketide synthase and it has been known for a long time that the first stable step in the biosynthetic pathway is the norsolorinic acid, an anthraquinone (Bennett et al., 1997). A complex series of post-polyketide synthase steps follow, yielding a series of increasingly toxigenic anthraquinone and difurocoumarin metabolites (Trail et al., 1995). Sterigmatocystin (ST) is a late metabolite in the aflatoxin pathway and is also produced as a final biosynthetic product by a number of species. It is now known that ST and aflatoxins share almost identical biochemical pathways (Bhatnager et al., 2003). Aflatoxin (AF) was one of the first fungal secondary metabolites shown to have all its biosynthetic genes organized within a DNA cluster (Fig. 3). These genes, along with the pathway specific regulatory genes *aflR* and *aflS*, reside within a 70 kb DNA cluster near the telomere of chromosome 3 (Sweeney et al., 1999; Georgianna and Payne, 2009). Research on *A. flavus*, *A. parasiticus and A. nidulans* has led to our current understanding of the enzymatic steps in the AF biosynthetic pathway, as well as the genetic organization of the biosynthetic cluster. *A. nidulans* does not produce AF but has all of the genes and enzymatic steps preceding the production of ST. The AF and ST pathways appear to have a common biosynthetic scheme up to the formation of ST, and thus information gained from both pathways has been used to study AF regulation (Georgianna & Payne, 2009). The biosynthetic and regulatory genes required for ST production in *A. nidulans* are homologous to those required for aflatoxin production in *A. flavus* and *A. parasiticus* and they also are clustered. The physical order of the genes in the cluster largely coincides with the sequential enzymatic steps of the pathway and both gene organization and structure are conserved within *A. favus* and *A. parasiticus* (Sweeney *et al.*, 1999; Bhatnagar *et al.*, 2006). Of the 25 genes identified in the pathway, only four (*norA*, *norB*, *aflT*, and *ordB*) have yet to have the function of their protein product determined experimentally. Only one of these genes, *aflR*, appears to encode a transcription factor (Bhatnagar *et al.*, 2006, 2003). The expression of the structural genes in both aflatoxin and ST biosynthesis is regulated by a regulatory gene, *aflR*, which encodes a GAL4-type C6 zinc binuclear DNA-binding protein (Bhatnagar *et al.*, 2003). This gene is located in the cluster and is required for transcriptional activation of most, if not all, of the aflatoxin structural genes. Adjacent to and divergently transcribed from the *aflR* gene is *aflJ*. This gene is also involved in the regulation of the aflatoxin gene cluster because no aflatoxin pathway intermediates are produced when it is disrupted. The gene product of *aflJ* has no sequence homology with any other genes or proteins present in databases. It interacts with *aflR* but not with the structural genes of the pathway. It has been speculated that *aflJ* is an *aflR* coactivator (Yu *et al.*, 2002; Bennett *et al.*, 2007). The function of most of the aflatoxin gene products has been deduced either by genetic or biochemical means (Bhatnagar *et al.*, 2006). Two of the genes of the ST gene cluster in *A. nidulans*, *stcJ* and *stcK*, encode the K- and L-subunit of a fatty acid synthase (FAS) which is specific for the formation of the hexanoate starter of ST. Disrupted *stcJ/stcK* mutants do not synthesise ST, but retain the ability to do it when provided with hexanoic acid (Sweeney *et al.*, 1999). The protein set requested for ST/AF transduction regulatory pathways includes: FlbA, an RGS (regulator of G-protein signaling) protein; FluG, an early acting development regulator; FadA, the alpha subunit of a heterotrimeric G-protein; and PkaA, encoding the catalytic subunit of protein kinase A. When FadA is activated following the signal "perception" both directly and indirectly it is able to inhibit AflR activity. FlbA whose activation is dependent on FluG, suppresses FadA and triggers AflR activation (Reverberi *et al.*, 2010)

3.2.1 The pathway specific regulator gene

Two genes, *aflR* and *aflS*, located divergently adjacent to each other within the AF cluster are involved in the regulation of AF/ST gene expression. The gene *aflR* encodes a sequence-specific DNA-binding binuclear zinc cluster (Zn(II)2Cys6) protein, required for transcriptional activation of most, if not all, of the structural genes (Georgianna and Payne, 2009). It was first cloned from an A. flavus cosmid library by showing that it could restore aflatoxin-producing ability to a mutant blocked in all steps of aflatoxin biosynthesis. An increase in the copy number of *aflR* somehow altered normal regulation of aflatoxin biosynthesis (Bhatnagar *et al.*, 2003). The aflR locus has been compared among isolates of AF producers such as *A. parasiticus* and *A. flavus*. These comparisons revealed differences in

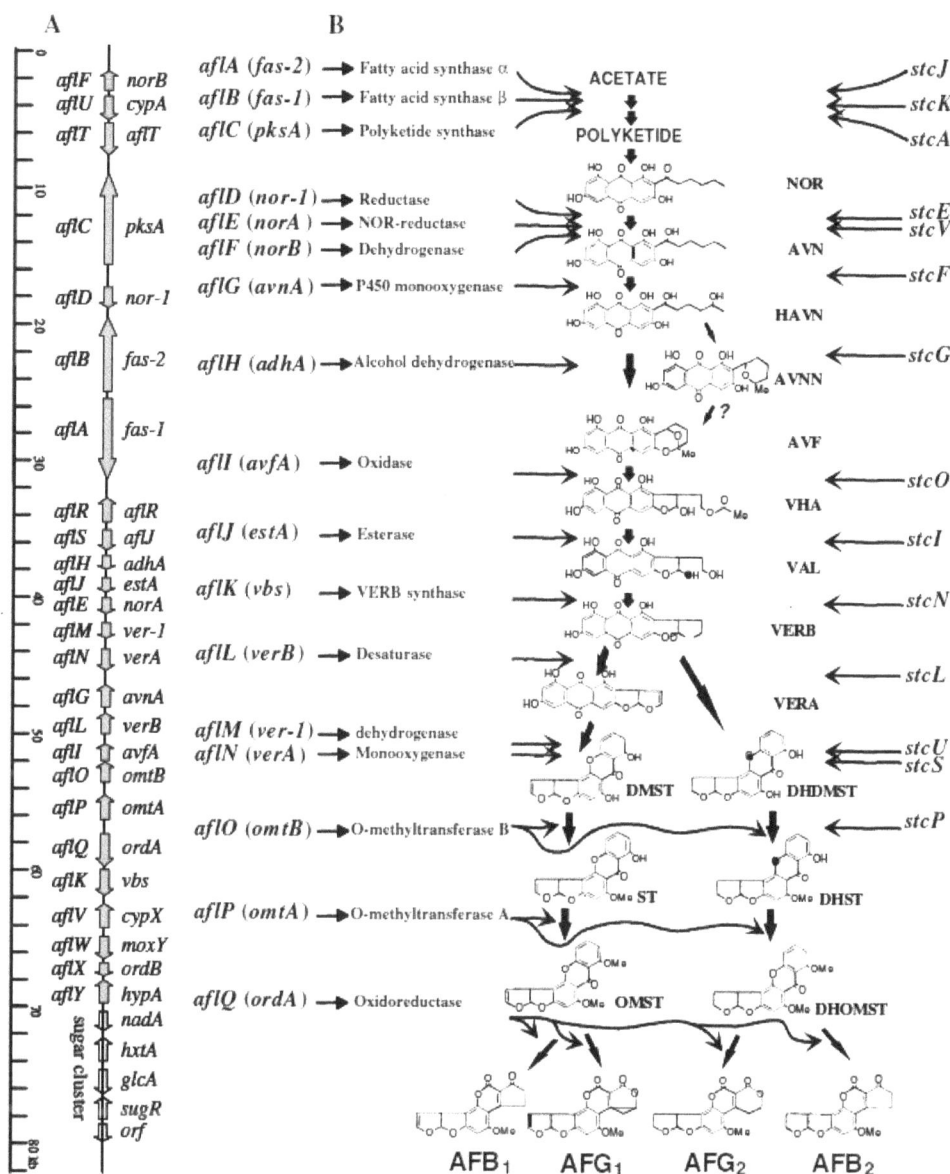

Fig. 3. The gene cluster responsible for aflatoxins biosynthesis in *A. flavus* and *A. parasiticus*. A) Clustered genes (arrows indicate the direction of gene transcription) and B) the AF biosynthetic pathway (Bhatnagar *et al.*, 2006).

many promoter regulatory elements such as PacC and AreA binding sites. The *aflR* gene is also found in *A. nidulans* and *A. fumigatus*. Despite clear differences in the sequence of AflR between A. nidulans and A. flavus, function is conserved. AflR from *A. flavus* is able to drive expression of the ST cluster in an A. nidulans *aflR* deletion strain (Carbone *et al.*, 2007; Georgianna and Payne, 2009). AflR binds to the palindromic motif 5'-TCGN5CGA-3' (also called AflR binding motif) in the promoter region of aflatoxin structural genes in *A. parasiticus*, *A. flavus*, and *A. nidulans*. The promoter regions of the majority of aflatoxin genes have at least one 5'-TCGN5CGA-3' binding site within 200 bp of the translation start site, though some putative binding sites have been identified further upstream. AflR probably binds to its recognition site as a dimer. The gene, *aflR* may be self-regulated, as well as, under the influence of negative regulators. Upstream elements may be involved in negative regulation of *aflR* promoter activity. When *aflR* is disrupted, no structural gene transcript can be detected. Introduction of an additional copy leads to overproduction of aflatoxin biosynthetic pathway intermediates (Fernandes et al., 1998; Bennett et al., 2007).

Electrophoretic mobility shift assays (EMSA) have been used to thoroughly examine promoters for AflR binding in 11 different genes from the AF cluster, with three of these genes having sites that deviate from the predicted AflR binding motif, and an additional three AF genes for which AflR binding sites could not be demonstrated. Among these genes are *aflE*, *aflC*, *aflJ*, *aflM*, *aflK*, *aflQ*, *aflP*, *aflR*, and *aflG*. All of these genes have predicted sites and demonstrate some degree of AflR binding in EMSA assays. Moreover, they were differentially expressed between WT and the *DaflR* mutant, suggesting that AflR is required to activate their expression (Price et al., 2006; Georginna and Payne, 2009). Aflatoxins biosynthesis is also regulated by *aflS* (formerly *aflJ*), a gene that resides next to *aflR*. The genes *aflS* and *aflR* are divergently transcribed, but have independent promoters. The intergenic region between them, however, is short and it is possible that they share binding sites for transcription factors or other regulatory elements (Ehrlich and Cotty, 2002; Georgianna and Payne, 2009). The roles of AflR and AflS were examined by studying the expression of pathway genes in transformants of *A. flavus* strain 649-1 that received the respective genes individually. Strain 649-1 lacks the entire AF biosynthetic cluster but has the necessary upstream regulatory elements to drive the transcription of *aflR* (Du et al., 2007). These studies showed that AflR is sufficient to initiate gene transcription of early, mid, and late genes in the pathway, and that AflS enhances the transcription of early and mid aflatoxin pathway genes. Moreover, the induced expression of *A. flavus aflR* in *A. nidulans*, under conditions in which ST biosynthesis is normally suppressed, resulted in activation of genes in the ST biosynthetic pathway. These studies demonstrated that *aflR* function is conserved in widely different *Aspergillus* spp (Bhatnagar et al., 2003). Roles for AflS have been suggested to be as diverse as aiding in transport of pathway intermediates to the interaction of AflS with AflR for altered AF pathway transcription. The observation that AflS binds to AflR argues that AflS modulates aflatoxin expression through its interaction with AflR (Chang, 2003; Georgianna and Payne, 2009). Metabolite feeding studies showed that a functional *aflR* allele is required for accumulation of NOR, the first stable intermediate in the aflatoxin biosynthetic pathway. When this gene was disrupted, the fungi were incapable of aflatoxin metabolite production or transcription of nor-1, but otherwise grew normally (Bhatnagar et al., 2003). In addition to the binding sites for AflR, there are binding sites within the cluster for other transcriptional factors that may play important roles in transcriptional regulation of the AF cluster. A novel cAMP-response element, CRE1, site has been studied specifically in the *aflD* (nor-1) promoter of *A. parasiticus* (Georgianna and Payne, 2009).

3.2.2 Aflatoxins and fungal development

The association between fungal morphological development and secondary metabolism, including aflatoxin production, has been observed for many years (Calvo et al., 2002). The environmental conditions required for secondary metabolism and sporulation are similar, and both processes occur at about the same time (Reiss, 1982; Bennett et al., 2007). A number of studies have identified a genetic connection between aflatoxin/sterigmatocystin biosynthesis and fungal development. In *Aspergillus*, several observations linked a fluffy phenotype to loss of AF/ST production. The available well characterized fluffy mutants in *A. nidulans* were instrumental in the discovery of a signal transduction pathway regulating both conidiation and ST/AF biosynthesis. These mutants are deficient in ST formation (Weiser et al., 1994). Proteins identified as belonging to this signal transduction pathway include FlbA, an RGS (Regulator of G-protein Signaling) protein, FluG, an early acting development regulator, FadA, the alpha subunit of a heterotrimeric G-protein and PkaA, encoding the catalytic subunit of protein kinase A (Gerogianna and Payne, 2009). Furthermore, a possible transcription regulatory gene, *veA*, has been identified in *A. nidulans* and *A. parasiticus* and this gene controls both toxin production and sexual development. Both *A. nidulans* and *A. parasiticus veA* mutants fail to produce ST or aflatoxin. Moreover, *A. nidulans* and *A. parasiticus* do not produce cleistothecia (sexual fruiting bodies harboring ascospores) and sclerotia (asexual overwintering structures) respectively. Finally, a number of genetic loci were identified in *A. nidulans* mutants that resulted in loss of ST production but had normal developmental processes. Complementation studies with one of these mutants identified a gene called *laeA*. This gene encodes an enzyme with sequence similarity to methyltransferases and appears to be required for expression of ST. LaeA homologs have been found in a number of filamentous fungi and in all species examined, disruption of *laeA* resulted in loss of secondary metabolite production while overexpression of *laeA* results in hyperproduction of the secondary metabolite (Bhatnagar et al., 2006; Reverberi et al., 2010).

3.3 Economic impact of aflatoxins

Aspergillus spp. is a fungal that grows and produces aflatoxins in climes ubiquitous but is commonly found in warm and humid climates (Dohlman, 2003). Hence most commodities from tropical countries, especially peanut and maize, are likely to be easily contaminated with aflatoxins (Bley, 2009). Aflatoxin contamination of human and animal feeds poses serious health and economic risks worldwide (Bley, 2009). The economic impact of aflatoxin contamination is difficult to measure, but the following losses have been documented. In United States (US) from 1990 to 1996, litigation costs of $34 million from aflatoxin contamination occurred. In 1998, corn farmers lost $40 million as a result of aflatoxin contaminated grain (AMCE, 2010). The FAO estimates that 25% of the world food crops are affected by mycotoxins each year and constitute a loss at post-harvest (FAO, 1997). According to Cardwell et al (2004) aflatoxin contamination of agricultural crops causes annual losses of more than $750 million in Africa. Dohlman (2003) defined mycotoxin as toxic by-products of mould infestations affecting about one-quarter of global food and feed crop output. Newly in the US, it was reported that income losses due to AF contamination cost an average of more than US$100 million per year to US producers (Coulibaly et al., 2008). As of this date, the average direct loss to the US is estimated at $200 million annually for corn. Indirect losses because of contaminated byproducts, such as distillers' grain, compound these losses. Ultimately, all contribute to increased costs to consumers (AMCE, 2010). Jolly et al. (2009) also reveal that post-harvest losses of crops are greater than the improvements made in primary production. In other hand, Otsuki et al (2001) has calculated that the European Union (EU) regulation on aflatoxins costs Africa $670 million each year in exports of cereals, dried fruit and nuts. But another study (World Bank, 2005) indicated that Otsuki et al. had overestimated the impact of the EU aflatoxin standard on Africa, and that the largest losses were incurred by Turkey, Brazil, and Iran. However, several studies have indicated that these costs may increase not only for Africa but for other countries that are suppliers of grains of the EU (Otsuki *et al.*, 2001; Wu, 2004). This due to that the regulation on aflatoxins is among the strictest in the world, at 4 ng/g total aflatoxins for all foods except peanuts (15 ng/g). The EU regulation standards on aflatoxins are base in the ALARA principle (As Low As Reasonably Achievable) which has a strong potential impact on nations attempting to export foods that are susceptible to aflatoxins contamination into the EU (Wu, 2008). In the study of 2004, Wu estimated a $450 million annual loss to the U.S., China, Argentina, and sub-Saharan African peanut markets if the EU aflatoxin standard were adopted worldwide. Nevertheless, in other study realized in 2008, Wu also mentions that under certain conditions, export markets may actually benefit from the strict EU standard. These conditions include a consistently high-quality product, and a global scene that allows market shifts. Even lower-quality export markets can benefit from the strict EU standard, primarily by technology forcing. Nevertheless, if the above conditions are not met, export markets suffer from the strict EU standard. Recent studies have linked aflatoxins production in foods to environmental conditions, poor processing and lack of proper storage facilities in developing countries (Farombi, 2006; Hell et al., 2000; Kaaya and Kyamuhangire, 2006).

3.4 Control of aflatoxin contamination in crops

Mycotoxin contamination often is an additive process, beginning in the field and increasing during harvest, drying, and storage (Wilson and Abramson, 1992). Environmental conditions are extremely important in pre-harvest mycotoxin contamination of grain and oilseed crops. Aflatoxin generation is favored in years with above average temperature and below average rainfall (Wilson and Abramson, 1992). Fungal contamination both at pre-harvest and post-harvest is determined by a range of factors which can be classified into four main groups including: intrinsic nutritional factors, extrinsic factors, processing factors and implicit microbial factors (Sinha, 1995). The Fig. 4 summarises the factors which affect fungal colonization of stored grain (Megan and Aldred, 2008). Strategies to address the food safety and economic issues employ both pre-harvest and post harvest measures to reduce the risk of mycotoxin contamination in food and feed (Dorner, 2004). Pre-harvest control includes good cultural practices, biocontrol and development of resistant varieties of crops through new biotechnologies. The good cultural practices consist in planting adapted varieties, proper fertilization, weed control, and necessary irrigation as well as crop rotation, cropping pattern, and use of biopesticides as protective actions that reduce mycotoxin contamination of field crops. Among the strategies of biotechnology in the pre-harvest control is the development of transgenic plants resistant to fungal infection as well as crops capable of catabolism/interference with toxin production. Pre-harvest prevention especially through host resistance is probably the best and widely explored strategy for control of mycotoxins (Kumar and Kumari, 2010; Bhatnagar, 2010). Post-harvest control is based mainly eliminate or inactivate mycotoxins in grains and other commodities. Among the methods used in this control, are physical separation, detoxification, biological inactivation, chemical inactivation, and decreasing the bioavailability of mycotoxins to the host animal

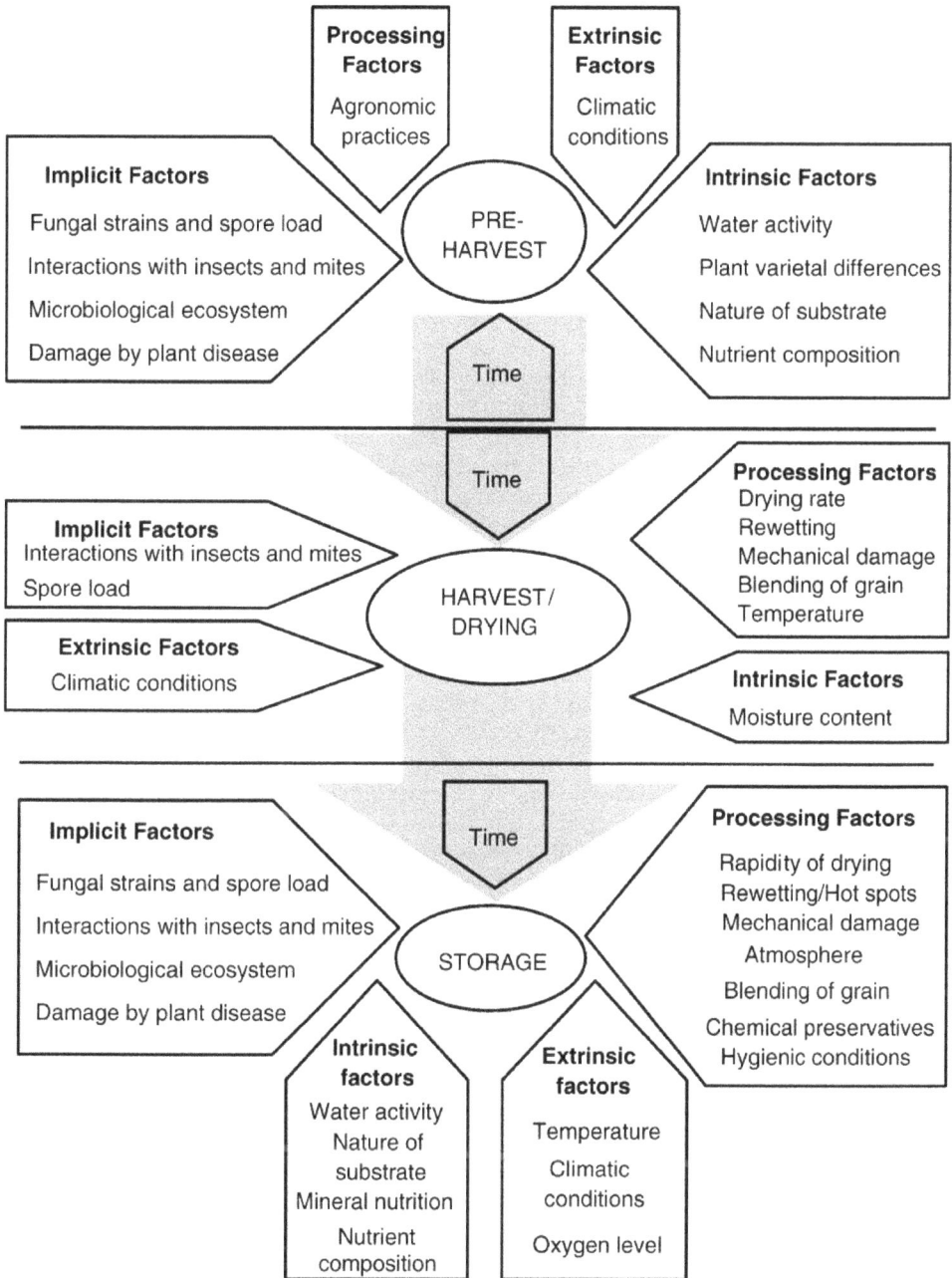

Fig. 4. Interaction between intrinsic and extrinsic factors in the food chain which influences mould spoilage and mycotoxin production in stored commodities (Magan *et al.*, 2004)

(Richard, J.L. et al., 2003). Because of the detrimental effects of mycotoxins, a number of strategies have been developed to help prevent the growth of mycotoxigenic fungi as well as to decontaminate and/or detoxify mycotoxin contaminated foods and animal feeds (Rustom, 1997). These strategies include: the prevention of mycotoxin contamination, detoxification of mycotoxins present in food and feed, as well as the inhibition of mycotoxin absorption in the gastrointestinal tract. Mycotoxin contamination may occur in the field before harvest, during harvesting, or during storage and processing. Thus methods can conveniently be divided into pre-harvest, harvesting and postharvest strategies (Heathcote & Hibbert, 1978). Whereas certain treatments have been found to reduce specific mycotoxin formation in different commodities, the complete elimination of mycotoxin contaminated commodities is currently not realistically achievable. Several codes of practice have been developed by Codex Alimentarius for the prevention and reduction of mycotoxins in cereals, peanuts, apple products, and raw materials. The elaboration and acceptance of a General Code of Practice by codex will provide uniform guidance for all countries to consider in attempting to control and manage contamination by various mycotoxins. In order for this practice to be effective, it will be necessary for the producers in each country to consider the general principles given in the Code, taking into account their local crops, climate, and agronomic practices, before attempting to implement provisions in the Code. The recommendations for the reduction of various mycotoxins in cereals are divided into two parts: recommended practices based on Good Agricultural Practice (GAP) and Good Manufacturing Practice (GMP); a complementary management system to consider in the future is the Hazard Analysis Critical Control Point (HACCP) (Codex Alimentarius Commission, 2002). Mycotoxins are secondary metabolites which are produced by several fungi mainly belonging to the genera: *Aspergillus*, *Penicillium*, *Fusarium*, and *Alternaria*. While Aspergillus and Penicillium species are generally found as contaminants in food during dry and storage, Fusarium and Alternaria spp. can produce mycotoxins before or immediately after harvesting (Sweeney & Dobson, 1999). Up until now, approximately 400 secondary metabolites with toxigenic potential produced by more than 100 moulds, have been reported, with the Food and Agricultural Organization (FAO) estimating that as much as 25% of the world´s agricultural commodities are contaminated with mycotoxins leading to significant economic losses (Kabak *et al.*, 2006).

Although *A. flavus* is readily isolated from diverse environmental samples, soil and plant tissues or residues are considered the natural habitat of this fungus (Jaime-Garcia & Cotty, 2004). Soil serves as a reservoir for primary inoculums for the infection of susceptible crops. Information concerning the soil ecology of *A. flavus* is consequently considered a prerequisite for developing effective measures to prevent and to control aflatoxin contamination of crops (Zablotowics *et al.*, 2007). Soil and crop management practices and a number of environmental factors can influence the population size and spatial distribution of *A. flavus* in cultivated soils (Abbas *et al.*, 2004b). The population size of A. flavus has been correlated with soil organic matter and nutritional status, with the most fertile soils containing the greatest concentration of aspergilli (Zablotowics *et al.*, 2007). Subsequently, as more soils are managed under no tillage systems, a higher inoculums of this fungus may result, which could contribute to increased pre-harvest aflatoxin contamination of susceptible crops. It should be noted that the post-harvest control is a corrective method, and in this Chapter be addressed essentially biotechnological approaches that serve as preventive methods from emergency and development of *Aspergillus flavus* and consequently; inhibition synthesis of aflatoxins –a pre-harvest level–. Such approaches include of biologic control methods and use of elicitors.

3.4.1 Pre-harvest control strategies

It is well established that mycotoxin contamination of agricultural product can occur in the field as well as during storage (Wilkinson, 1999). Since phytopathogenic fungi such as *Fusarium* and *Alternaria* spp can produce mycotoxins before or immediately post harvesting several strategies have been developed including biological and cultural control practices to help mycotoxin contamination occurring in this way.

3.4.1.1 Prevention strategies in cereals

The main mycotoxin hazards associated with wheat pre-harvest in Europe are the toxins that are produced by fungi belonging to the genus *Fusarium* in the growing crop. Mycotoxins produced by these fungi include zearalenone (ZEN) as well as trichothecenes and include nivalenol (NIV), deoxynivalenol (DON) and T-2 toxin. *Fusarium* species are also responsible for a serious disease called *Fusarium* Head Blight (FHB), which can result in significant losses in both crop yield and quality. It is important to note that although *Fusarium* infection is generally considered to be a pre-harvest problem, it is possible for poor drying practices to lead to an increased susceptibility for storage mycotoxin contamination (Aldred & Magan, 2004).

3.4.1.2 Resistant varieties and transgenics

Research has demonstrated that insecticides cannot be applied economically to control corn insects well enough to reduce aflatoxin to acceptable levels. The most successful approach has been the use of corn resistant to ear-feeding insects. Several authors have shown that *Bacillus thuringensis* (Bt)-transformed corn hybrids, which are resistant to ear-feeding insects, reduce aflatoxin contamination of the grain. The adoption of *Bt* corn hybrids has given producers crop with increased insect resistance, however these hybrids may only reduce aflatoxin contamination under certain circumstances. However, commercial production of these genetically modified hybrids is not allowed in some nations. Several sources of natural resistance to insects have been identified, and crosses between insect- and aflatoxin-resistant lines have shown potential to increase resistance to both insect damage and aflatoxin contamination (Williams *et al.*, 2002). Ideally, management of aflatoxin contamination should begin with the employment of resistant genotypes as has been demonstrated by several U.S. breeding programs. In Mexico the wide genetic diversity of maize has not been fully exploited to identify resistance to aflatoxin contamination in breeding programs, thus impeding the reduction of aflatoxin levels in the field. Additional complications come from the fact that transgenic maize expressing insecticidal protein or any other trait to reduce aflatoxin is not viable in Mexico due to a government prohibition on the use of genetically modified maize (Plasencia, 2004). Four major genetically controlled components for which variability exist appear to be involved in determining the fate of *A. flavus*-grain interaction: 1) resistance to the infection process, 2) resistance to toxin production, 3) plant resistance to insect damage, and 4) tolerance to environmental stress (Widstrom, 1987). The latter two components have an indirect influence since their effects only reduce aflatoxin contamination but do not prevent it. Although differences among genotypes have been found, heritability of the trait appears to be low, and the genotype/environment interaction may often mask true differences among genotypes (Plasencia, 2004). There are many new and exciting pre-harvest prevention strategies being explored that involve new biotechnologies. These new approaches involve the design and production of plants that reduce the incidence of fungal infection, restrict the growth of toxigenic fungi, or prevent toxic accumulation. Biocontrols using non-toxigenic biocompetitive agents is also a potentially useful strategy in corn. However, the possibility of recombination with toxigenic strains is a concern (Abdel-Wahhab & Kholif, 2008). The differences between crop species appear to differ between countries. This is probably due to the differences in genetic pool within each country's breeding program and the different environmental and agronomic conditions in which crops are cultivated (Edwards, 2004). With respect to genetic resistance to Aspergillus infection and subsequent aflatoxin production, since the early 1970s, much work has been done to identify genetically resistant crop genotypes in both laboratory and field based experiments to help control of aflatoxigenic mould growth and aflatoxin and aflatoxin biosynthesis (D'Mello *et al.*, 1998). This has led to the identification of a number of well-characterized sources of both resistance of *Aspergillus flavus* infection and to aflatoxin production. These include kernel proteins such as a 14-kDa trypsin-inhibiting protein and others including globulin 1 and 2 and a 22-kDa zeamatin protein (Chen *et al.*, 2001). Although the role of insects in fostering *Aspergillus* colonization of maize kernels is well documented, there is little evidence that transgenic corn expressing insecticidal proteins has a significant effect on reducing aflatoxin contamination. In contrast, several studies have reported a protective effect of Cry-type proteins in maize to *Fusarium* kernel rot and fumonisin accumulation (Dowd, 2003). Cry-type proteins constitute a family of insecticidal proteins from *Bacillus thuringensis*, whose genes have been incorporated into several crops to confer protection against insect pests. In corn, several hybrids expressing distinct Cry-type proteins have been developed and widely used in the U.S., Canada, Argentina, and other maize-producing countries (Plasencia, 2004). The distribution of aflatoxin in agricultural commodities has been fairly well characterized because of its importance to food supply. However, little is known on the occurrence and fate of aflatoxin in soil. Radiological assays conducted to assess the fate of aflatoxin B1 (AFB1) in soil indicated that a low level of mineralization of AFB1 to CO_2 was observed, with less than 1-8% mineralized in 120 days (Angle, 1986). Not surprisingly, several microorganisms have the potential to degrade aflatoxins, especially bacteria, e.g., *Flavobacterium* and *Mycobacterium* (Hormisch *et al.*, 2004). In addition, *A. flavus* also is capable of degrading aflatoxins during later stages of mycelial growth in pure culture (Huyhn & Lloyd, 1984). In recent years, molecular techniques have increased the possibilities to characterize soil microbial ecology. While molecular methods have been extensively used for studying soil bacteria, these techniques have been applied to studying soil fungi, such as the biological control agents Colletotrichum coccodes (Dauch *et al.*, 2003), Trichoderma (Weaver *et al.*, 2005), and mycorrizal fungi (Ma *et al.*, 2005). Amplification of specific DNA fragments using polymerase chain reaction (PCR) and specific gene probes is extremely sensitive and has the potential to detect the presence of *A. flavus* in agricultural commodities (Manonmani *et al.*, 2005). Since all of the genes involved in the aflatoxin biosynthesis pathway have been identified and cloned (Yu *et al.*, 2004a, 2004b), and the entire genome of *A. flavus* sequenced (Payne *et al.*, 2006), molecular methods for the detection of Aspergillus should be fairly readily adapted by using biosynthetic pathway genes as probes, as evidenced by the recent work differentiating toxigenic and atoxigenic *A. flavus*-utilizing aflatoxin gene expression using the reverse transcription-polymerase chain reaction (RT-PCR) (Degola *et al.*, 2007). Application of these molecular techniques to *A. flavus* soil ecology should greatly enhance our understanding of this fungus. *Aspergillus flavus* is commonly considered a saprophytic fungus; however, its ability to colonize growing crops and inflict economic damage clearly shows that it can and does function as an opportunistic pathogen. Despite the elucidation of many aspects influencing *A. flavus* ability to colonize crops and accumulate aflatoxins, its activity and potential to produce aflatoxins in soil and in crop residues has remained unexplored (Accinelli *et al.*, 2008). One interesting approach is the engineering of cereal plants to catabolize fumonisins *in situ*. Typically, these approaches require considerable research and development but have the potential of ultimately producing low cost and effective solutions to the mycotoxin problem in corn and other cereals. Thus this level of prevention is the most important and effective plan for reducing fungal growth and mycotoxin production.

3.4.1.3 Field management

Appropriate field management practices including crop rotation, soil cultivation, irrigation, and fertilization approaches are known to influence mycotoxin formation in the field. Crop rotation is important and focuses on breaking the chain production of infectious material, for example by using wheat/legume rotations. The use of maize in a rotation is to be avoided however, as maize is also susceptible to *Fusarium* infection and can lead to carry-over onto wheat via stubble/crop residues (Nicholson *et al.*, 2003). Dill-Macky and Jones (2000) observed that FHB disease severity and DON contamination of grain was significantly different when the previous crop was maize, wheat, or soya bean; with the highest levels following maize and the lowest levels following soya bean. Soil cultivation can be divided into ploughing, where the top 10-30 cm of soil is inverted; minimum tillage, where the crop debris is mixed with the top 10-20 cm of soil; and no till, where seeds are directly drilled into the previous crop stubble with minimum disturbance to the soil structure. Ay crop husbandry that results in the removal, destruction, or burial of infected crop residues is likely to reduce the *Fusarium* inoculum for the following crop. Dill-Mackey and Jones (2000) reported that no till (direct drilling) after wheat or maize significantly increase DON contamination of the following wheat crop compared to ploughing, but no till had no effect when the previous crop was soya bean. Irrigation is also a valuable method of reducing plant stress in some growing situations. It is first necessary that all plants in the field have an adequate supply of water if irrigation is used. It is known that excess precipitation during anthesis makes conditions favorable for dissemination and infection by Fusarium spp., so irrigation during anthesis and during ripening of the crops, specifically wheat, barley, and rye, should be avoided. The soil must be tested to determine if there is need to apply fertilizer and soil conditioners to assure adequate soil pH and plant nutrition to avoid plant stress, especially during seed development. Fertilizer regimes may affect FHB incidence and severity either by altering the rate of residue decomposition, by creating a physiological stress on the host plant or by altering the crop canopy structure. Martin et al., (1991) observed the increasing N from 70 to 170 kg/ha significantly increased the incidence of *Fusarium* infection grain in wheat, barley, and triticale. Recent work by Lemmens *et al.*, (2004) has shown that a significant increase in FHB intensity and DON contamination in the grain was observed with increasing a mineral N fertilizer from 0 to 80 kg/ha. This group concluded that in practical crop husbandry, FHB cannot be sufficiently controlled by only manipulating the N input.

3.4.1.4 Environmental conditions

Environmental conditions such as relative humidity and temperature are known to have an important effect on the onset of FHB. For example, it has been shown that moisture conditions at anthesis are critical in Fusarium infection of the ears (Aldred & Magan, 2004); while Lacey et al. (1999) have shown that Fusarium infection in the UK is exacerbated by wet periods at a critical time in early flowering in the summer, which is the optimum window for susceptibility. Equally, there is evidence that droughed-damaged plants are more susceptible to infection, so crop planting should be timed to avoid both high temperature and drought stress during the period of seed development and maturation. On the other hand, the planning of harvesting grain at low moisture content and fully maturity may be an important control point in the preventing of mycotoxin contamination, unless allowing the crop to continue to full maturity would subject it to extreme heat, rainfall or drought conditions. Delayed harvest of the grain already infected by Fusarium species is known to cause a significant increase in the mycotoxin content of the crop.

3.4.2 Biotechnological approaches

3.4.2.1 Biological control of aflatoxins

The first approach which we will discuss is the biological control, which is focuses in the use of living organisms to control pests (insects, weeds, diseases and disease vectors) in agriculture. The objective of the biologic control is to stimulate the colonization of antagonist organism on plant surfaces to reduce the inoculum of the pathogens (FAO, 2004). Different organisms, including bacteria, yeasts and nontoxigenic *Aspergillus* fungi, have been tested for their ability in the control of aflatoxin contamination (Yin et al., 2008). According to reported by Palumbo et al., (2006) several bacterial species as *Bacillus* spp., *Lactobacilli* spp., *Pseudomonas* spp., *Ralstonia* spp. and *Burkholderia* spp., have shown the ability to inhibit fungal growth and production of aflatoxins by *Aspergillus* spp. in laboratory experiments (Yin et al., 2008), the same effect was observed in strains of *B. subtilis* and *P. solanacearum* isolated from the non-rhizophere of maize soil were also able to inhibit aflatoxin accumulation (Nesci *et al.*, 2005). In other experiments, is showed that *Bacillus subtilis* prevented aflatoxin contamination in corn in field tests when ears were inoculated with the bacterium 48 hours before inoculation with A. *flavus* (Cuero et al., 1991). However, no reduction in aflatoxin occurred when bacteria were inoculated 48 hours after inoculation with A. flavus. Bacillus subtilis (NK-330) did not inhibit aflatoxin contamination in peanuts when it was applied to pods prior to warehouse storage for 56 days (Smith et al., 1990). Saprophytic yeasts isolated from fruits of almond, pistachio, and walnut trees inhibited aflatoxin production by A. *flavus* in vitro (Hua et al., 1999; Masoud and Kaltoft, 2006). A strain of *Candida krusei* and a strain of *Pichia anomala* reduced aflatoxins production by 96% and 99%, respectively, in a Petri dish assay. Efforts are underway to apply these yeasts to almond and pistachio orchards to determine their potential for aflatoxin reduction under crop production conditions (Hua, 2002). Although they were considered to be potential biocontrol agents for management of aflatoxins, further field experiments are necessary to test their efficacies in reducing aflatoxins contamination under field conditions (Yin et al., 2008). Alternatively, a limited number of biocompetitive microorganisms have been shown for the management of *Fusarium* infections. Antagonistic bacteria and yeasts may also lead to reductions in pre-harvest mycotoxin contamination. For instance, *Bacillus subtilis* has been shown to reduce mycotoxin contamination by F. *verticilloides* during the endophytic growth phase. Similarly antagonistic yeasts such as *Cryptococcus nodaensis* have also been shown to inhibit various *Fusarium* species (Cleveland *et al.*, 2003). Recent glasshouse studies by Diamond and Coke (2003) involving the pre-inoculation of wheat ears at anthesis, with the two non-host pathogens, *Phoma betae* and *Pytium ultimum* showed a reduction in disease development and severity caused by F. *culmorum*, F. *avenaceum*, F. *poae*, and M. *nivale*. A. *flavus* is not considered to be an aggressive invader of pre-harvest corn ear tissue. However, developing grain when damaged is easily contaminated by the pathogen (Diener *et al.*,

1987). The association between insect damage and fungal infection of corn ears was first recognized by Riley (1882) reported molds appearing on corn-ear tips soon after being infested with insect larvae. Garman and Jewett (1914) reported that in years with high insect populations, the incidence of moldy ears in field corn increased. Efforts to determine the specific role of insects in the *A. flavus* infection process increased dramatically when aflatoxin was recognized as a health concern, leading to recognition that ear feeding insects (e.g., corn earworm, *Helicoverpa zea*; European corn borer, *Ostrinia nubilalis*; fall armyworm, *Spodoptera frugiperda*; western bean cutworm, *Striacosta albicosta*; and southwestern corn borer, *Diatraea grandiosella*) can increase aflatoxin levels in pre-harvest corn (Catangui & Berg, 2006). The difficulty in establishing the relationship between insect damage and aflatoxin incidence is in part due to A. flavus ability to colonize silks, infect kernels, and produce aflatoxins in developing ears under insect-free conditions (Jones et al., 1980), and in part due to unknown factors that result in conflicting information (Abbas *et al.*, 2009). Because the relationship between insect damage to corn ears and aflatoxin is heavily influenced by environmental conditions, success in managing aflatoxin contamination via insect control has been highly variable. The greatest success to date regarding biological control of aflatoxins contamination in the field has been achieved through competitive exclusion by applying on aflatoxigenic strains of *Aspergillus flavus* and *Aspergillus parasiticus* to soil of developing crops. These strains are typically referred to as atoxigenic or nontoxigenic, but those designations are often used with reference to production of aflatoxins only (Dorner, 2004). According to Yin *et al.*, (2008) the use of non-toxigenic *Aspergillus* strains is a strategy based on the application of nontoxigenic strains to competitively exclude naturally toxigenic strains in the same niche and compete for crop substrates. Thus, for competitive exclusion to be effective, the biocontrol nontoxigenic strains must be predominant in the agricultural environments when the crops are susceptible to be infected by the toxigenic strains (Cole and Cotty, 1990; Cotty, 1994; Dorner, 2004). For this to work, the applied strains must occupy the same niche as the naturally occurring toxigenic strains and compete for crop substrates (Dorner, 2004). Two primary factors exist that determine the effectiveness of this strategy. First, the applied strain(s) must be truly competitive and dominant relative to the toxigenic strains that are already present. Second, the formulation used to apply the competing strain(s) must be effective in delivering the necessary quantity of conidia to achieve a competitive advantage. In addition, the timing of that application is crucial for ensuring that the necessary competitive level is present when the threat of crop infection is greatest (Cotty, 1989; Dorner, 2004). Should be noted, that not only species of *Aspergillus* used for biological control are capable of producing aflatoxins, but also a variety of other toxins and toxic precursors to aflatoxins including cyclopiazonic acid, sterigmatocystin and related compounds, and the versicolorins (Cole and Cox, 1981). In the research realized by Cotty (1990) in greenhouse, demonstrated the ability of seven non-aflatoxigenic strains of *A. flavus* to reduce aflatoxins contamination of cottonseed when were co-inoculated with toxigenic strains. Six of these strains show significantly reduced the amount of aflatoxins produced in cottonseed by the toxigenic strain. Strain 36 (AF36) produced the largest reduction in aflatoxin under these conditions and it was Biological Control of Aflatoxin Contamination of Crops 429 subsequently shown to reduce aflatoxin contamination of cottonseed in the field when applied on colonized wheat seed (Cotty, 1994). This strain has been registered on cotton for control of aflatoxin contamination of cottonseed in Arizona, USA. It is also on a schedule for registration on pistachio in California. Additionally, this biocontrol agent was also tested for control of aflatoxin in corn (Cotty, 1996). When corn ears were either co-inoculated with AF36 and a toxigenic strain of *A. flavus* or inoculated with AF36 at 24 h prior to inoculation with the toxigenic strain, subsequent aflatoxins concentrations were significantly reduced, compared to inoculation with the toxigenic strain alone (Brown *et al.*, 1991). Also have been demonstrated that other strains of *A. flavus* and *A. parasiticus* are capable of reduce aflatoxin contamination in crops; as is case of *A. flavus* NRRL 21882, a naturally occurring strain isolated from a peanut in Georgia in 1991, that has been used in diverse studies where has been verified its efficacy for reducing contamination in the field. This strain is the active ingredient in an EPA-registered biopesticide called afla-guard1. A color mutant of this strain, NRRL 21368, was used in several early studies and also found to be effective when used in conjunction with a color mutant of *A. parasiticus* (NRRL 21369) (Dorner et al., 1998, 1999b). Atoxigenic strain technology based provides an opportunity to reduce the overall risk of contamina-tion during all phases of aflatoxin contamination including in the field during crop development, in storage or at any other time after harvest until the mature crop is eventually utilized. Atoxigenic strains are but one example of how improved knowledge of both the contamination process and the etiologic agents can result in improved methods for limiting human exposure to aflatoxins.

3.4.2.2 Chemical agents and use of elicitors to aflatoxin inhibition

Another factor which is known to increase the susceptibility of agricultural commodities to toxigenic mould is injury due to insect, bird, or rodent damage (Smith *et al.*, 1994). Insect damage and fungal infection must be controlled in the vicinity of the crop by proper use of registered insecticides, fungicides, and other appropriate practices within an integrated pest management control. Part of the integrated control of FHB in wheat production involves the use of fungicides, but this introduces a complication as far as trichothecenes are concerned as there is evidence that under certain conditions, fungicide use may actually stimulate toxin production. This raises particular concerns, since circumstances may arise where the obvious manifestations of FHB are reduced or even eliminated and yet high levels of mycotoxins may be present. Clearly grain affected in this way cannot be identified by visual inspection for signs of FHB (e.g., pink grains) and, in fact, cannot be identified until a specific mycotoxin analysis is carried out (Simpson *et al.*, 2001). Early investigation in vitro indicates that the fungicide chlobenthiazone is highly effective in inhibiting aflatoxin biosynthesis by cultures of *A. flavus*; however, aflatoxin synthesis by *A. parasiticus* was, in fact, stimulated by the fungicide (Wheeler, 1991). Various surfactants, including some used in pesticide formulations, reduced aflatoxin biosynthesis by >96% (Rodriguez & Mahoney, 1994). Use of natural oils from thyme (Kumar *et al.*, 2008), and other herbs has also been studied and shown to repress aflatoxin in certain crops in Asia. The herbicide glufosinate has been reported as having antifungal activity against certain phytopathogenic fungi in vitro (Uchimiya *et al.*, 1993) and has shown activity in reducing infection of corn kernels in vitro (Tubajika & Damann, 2002). Higher levels of aflatoxin were observed in glyphosate-resistant corn compared with traditional corn hybrids. Thus, effects of glyphosate on in vitro growth of *A. flavus* in pure culture and on native soil populations were examined, finding that high levels of glyphosate (> 5mM) were required for inhibition. In addition, application of greater amounts was found to have no effect on *A. flavus* populations. Interestingly, *A. flavus* when grown on glyphosate water agar media, produced 20% of aflatoxin produced on water agar without glyphosate (Abbas *et al.*, 2009). Research carried out on fungicide use in terms of FHB and mycotoxin development has produced very interesting results. In particular, fungicides in common use have been shown to have differential effects against toxin-forming Fusarium species and related non-toxing-forming pathogens such as *Microdochium nivale* on ears (Simpson *et al.*, 2001). The outcome of the use of fungicides seems to depend on the fungal species present, and the effect that the particular fungicide has on these species. For example, in recent work commissioned by the Home Grown Cereal Authority, in an experimental situation where *Fusarium culmorum* and *M. nivale* where both present, the use of azoxystrobin showed a significant reduction in disease levels while increasing the levels of DON present in grain. This was believed to be the result of selective inhibition of *M. nivale* by azoxystrobin. *M. nivale* is a natural competitor of toxin-forming *Fusarium* species, particularly *F. culmorum*. Removal of *M. nivale* by the fungicide probably allowed development of the toxigenic species in its place with concomitant increase in toxin formation. This is an important finding as it indicates that the impact of the fungicide is not directly related to mycotoxin production. It follows from these findings that where FHB is caused by *Fusarium* species in the absence of *Microdochium*, disease development is associated with higher levels of toxin (Magan *et al.*, 2002). Ioos *et al.* (2005) also carried out a screen on the efficacy of fungicides, azeoxystrobin, metconazole, and tebuconazole at anthesis against *Fusarium* spp., *M. nivale* and on years on naturally infected fields of soft wheat, durum wheat, and barley. The infection levels of *F. graminearum*, *F. culmorum*, and *M, nivale* were significantly reduced by the application *Fusarium* mycotoxin concentration over three of fungicides, with tebuconazole and metconazole effectively controlling the Fusarium spp., but they had little effect on M. nivale. Although this conclusion concurs with Simpson *et al.* (2001) for tebuconazole, their benefits were apparently seasonal-with tebuconazole controlling these fungi in 2001, while having little effect in 200 and 2002. The second approach involves the application of elicitors in crops susceptible to *A. flavus*, with the aim of protecting the plant of subsequent aflatoxins contamination. This because that the elicitors are capable molecules from activating multiple reactions defense that are induced and agrouped both histological level of physical barrier as a biochemist with the de novo synthesis of proteins associated with pathogenicity (PR), in the absence of the pathogen. Besides serves as aguide of intracellular events that end in activation of signal transduction cascades and hormonal pathways, triggering the induced resistance (IR) and consequently activation of plant immunity to ivironmental stresses (Riveros, 2001; Odjacova and Hadjiivanova, 2001; Garcia-Brugger et al., 2006; Bent and Mackey, 2007; Holopainen et al., 2009; Mejía-Teniente et al., 2011). Between the elicitors that have been investigated for more control of aflatoxin contamination in crops of commercial interest is the jasmonic acid (JA) and related compounds, as well as ethylene (ET). One factor influencing the production of aflatoxin is the presence of high levels of oxidized fatty acids such as fatty acid hydroperoxides, which can form in plant material either preharvest under stress or postharvest under improper storage conditions, correlates with high levels of aflatoxin production (Goodrich-Tanrikulu et al., 1995). Fatty acid hydroperoxides can be formed by autooxidation, or enzymically by lipoxygenases acting on a-linoleic and a-linolenic acids (Vick, 1993). These hydroperoxides stimulate the formation of aflatoxins by *A. flavus* and *A. parasiticus* (Fabbri et al., 1983; Fanelli and Fabbri, 1989). Degradation of the hydroperoxides by later steps in the plant lipoxygenase pathway leads to multiple byproducts, depending on the polyunsaturated fatty acid substrate, the positional specificity of the lipoxygenase, and the activities of enzymes catalysing the subsequent steps. The jasmonic acid (JA) is α-linolenic acid metabolite, via lipoxygenase and hydroperoxide dehydratase, is jasmonic acid (JA) (Vick, 1993). JA and closely related compounds, such as its methyl ester, MeJA, are endogenous plant growth regulators both higher and lower plants (Staswick, 1992; Sembdner and Parthier, 1993). JA and MeJ are two well-characterized plant growth regulators that exert a vast variety of biological activities in plants as the activation of defense responses (for review see Sembdener and Parthier, 1993). Among the diverse plant defense mechanisms, recent findings have demonstrated that low-concentrations of JA or MeJ induce protein inhibitors (Farmer and Ryan, 1992), thionin (Andresen et al., 1992; Epple et al., 1995) and several plant defense enzymes such as PAL (Gundlach et al., 1992), LOX (Bell and Mullet, 1991) and chalcone synthase (Creelman et al., 1992). MeJ is volatile suggesting its action could be exerted in gaseous form, similar to the plant hormone, ethylene. Goodrich-Tanrikulu et al., (1995) reporting the effect of MeJA on aflatoxins production and growth of *Aspergillus flavus* in vitro. They Found that at concentrations MeJA of 10-3-10-8 M in the growth medium was inhibited aflatoxin production, by as much as 96%. Besides that when cultures were exposed to MeJA vapour similarly was inhibited aflatoxin production, observing that the amount of aflatoxin produced depended on the timing of the exposure. MeJA treatment also delayed spore germination and was inhibited the production of a mycelial pigment. These fungal responses resemble plant jasmonate responses. In other hand, Zeringue (2002) carried out a series of experiments where artificially wounded 22–27-day old developing cotton bolls were initially inoculated with, (1) a cell-free, hot water-soluble mycelial extract (CFME) of an atoxigenic strain of Aspergillus flavus or with, (2) chitosan lactate (CHL) or with, (3) CFME or CHL and then exposed to gaseous methyl jasmonate (MJ) or, (4) exposed to MJ alone. The results indicated a two- or three-fold increase in the production of the phytoalexins when gaseous MJ was added in combination to the CFME or the CHL elicitors. While the effects of aflatoxin B1 production after the developing cotton bolls pretreated with CFME, CHL or with CFME–MJ, CHL–MJ or only with MJ, showed a lower aflatoxin (Table 2, taken of Zeringue, 2002). All pretreatments resulted in some degree of aflatoxin B1 inhibition in the seeds underlying the treatment. CFME pretreatment resulted in a 88% inhibition of aflatoxin B1 and CHL resulted in a 64% inhibition (Table 2). CFME–MJ boll treatment resulted in the maximum aflatoxin B1 inhibition (95%) compared to CHL–MJ (75%). These series of experiments demonstrate a correlation between increased phytoalexin induction with a decreased aflatoxin B1 formation under the influence of volatile MJ in combination with selected elicitors. Phytoalexins are synthesized and accumulated at the site of microbial infection or as shown in this study localized at the site of the placement of elicitors (carpel discs). Besides, these results demonstrate an added inducement of phytoalexins and aflatoxin B1 inhibition produced by MJ treatment in combination with elicitors. This inducement is perhaps produced by an added signal/signals that activates other secondary pathways that either enhance the concentrations of the demonstrated phytoalexins or inhibit aflatoxin B1 biosynthesis or both. These results further demonstrate the innate, natural defense responses of the cotton plant and its ability to defend itself upon microbial attack, with the possibility to extrapolate to other seeds (Zeringue, 2002).

3.4.3 Harvest management
For cereals, harvest is the first stage in the production chain where moisture management becomes the dominant control measure in the prevention of mycotoxin development. Since the moisture content may vary considerably within the same field, the control of moisture in several spots of each load of the harvested grain during the harvesting operation is very important. Another equally important control measure is an effective assessment of the crop

for the presence of disease such as FHB. This should be accompanied by an efficient strategy for separation of the diseased material from healthy grain. There is evidence that fungal infection can be minimized by avoiding the mechanical damage to the grain and by avoiding contact with soil at this stage.

3.4.4 Post-harvest management

Post-harvest strategies are important in the prevention of mycotoxin contamination and include improved drying and storage conditions, together with the use of natural and chemical agents, as well as irradiation.

3.4.4.1 Improving of drying and storage conditions

In cereals, mycotoxigenic fungal growth can arise in storage as a result of moisture variability within the grain itself or as a result of moisture migration results from the cooling of grains located near the interface with the wall of the storage container/silo (Topal *et al.*, 1999). Thus control of adequate aeration and periodical monitoring of the moisture content of silos plays an important role in the restriction of mycotoxin contamination during the storage period (Heathcote & Hibbert, 1978). The moisture level in stored crops is one of the most critical factors in the growth of mycotoxigenic moulds and in mycotoxin production (Abramson, 1998), and is one of the main reasons for mycotoxin problems in grain produced in developing countries. Cereal grains are particularly susceptible to grow by Aspergilli in storage environments. The main toxigenic species are *A. flavus* and *A. parasiticus* for aflatoxins, and *Penicillium verrucosum* is the main producers in cereals for OTA (Lund & Frisvad, 2003), while *A. ochraceus* is tipically associated with coffee, grapes, and species, aflatoxins can be produced at a_w values ranging from 0.95 to 0.99 with a minimum a_w value of 0.82 for *A. flavus*, while the minimum a_w for OTA production is 0.80 (Sweeney & Dobson, 1998). It has been reported that *A. flavus* will not invade grain and oilseeds when their moisture contents are in equilibrium with a relative humidity of 70% or less. The moisture content of wheat at this relative humidity is about 15%, and around 14% for maize, but it is lower for seeds containing more oil, approximately 7% and 10% for peanuts and cottonseeds, respectively (Heathcote & Hibbert, 1978), while *A. parasiticus* has been reported to produce aflatoxins at 14% moisture content in wheat grains after 3 months of storage (Atalla *et al.*, 2003). The second critical factor influencing post-harvest mould growth and mycotoxin production is temperature. Both the main aflatoxin producing Aspergillus strains *A. flavus* and *A. parasiticus* can grow in the temperature range from 10-12°C to 42-43°C, with an optimum in the 32 to 33°C range, with several studies highlighting the relatively high incidence of mycotoxins such as aflatoxins and ochratoxins in foods and feeds in tropical and subtropicals regions (Soufleros *et al.*, 2003). The control of temperature of the stored grain at several fixed time intervals during storage may be important in determining mould growth. A temperature rise of 2-3°C may indicate mould growth or insect infestation. Until recently, little if any work has been carried out on monitoring how spoilage fungi interact with each other in the stored grain ecosystem, and the effect that this has on mycotoxin production. Magan *et al.* (2003), have shown that the system is in a state of dynamic flux with niche overlap altering in direct response to temperature and a_w levels. It appears that the fungi present tended to occupy separate niches, based on resources utilization, and this tendency increased with drier conditions. Initially, *A. flavus* and other *Aspergillus* spp. were considered exclusively storage fungi, and aflatoxin contamination was believed to be primarily a storage problem. This is very severe in many rural areas that lack of infrastructure for drying and other appropriate storage conditions. Usually, corncobs are harvested at moisture contents that vary between 25-30% and are dried under the sunlight to reach 12-14% moisture content. Research has been conducted to determine the optimum temperature and moisture content of grains during storage to prevent Aspergillus spp. growth and aflatoxin production. In maize inoculated with *A. flavus* and stored at 27°C for 30 days with varying moisture contents, an association between moisture content and aflatoxin levels was established. At 16% moisture, aflatoxin levels reach 116 µg/kg while a 22% moisture 2166 µg/kg aflatoxin levels were obtained (Moreno-Martínez *et al.*, 2000). In this same study, the authors tested the protective effects of propionic acid salts (6.5-12.5 L/t) on fungal growth and aflatoxin production. All grains treated with ammonium, calcium, or sodium propionates yielded very low *Aspergillus flavus* growth and aflatoxin levels (2 - 5.6 µg/kg) at all moisture contents. It is well established that rapid crop drying may be useful in controlling aflatoxin contamination in storage and that in addition that crops containing different moisture values are not stored together. It is also well established that mould invasion is facilitated as a result of increased moisture levels of stored commodities. Moisture abuse can even occur in crops with very low moisture content. Another factor to bear in mind is the fact that if fungal growth does occur in storage, moisture will be released during metabolism, which will be released during metabolism, leading to the growth of other fungal species and to the production of mycotoxins such as OTA.

4. Detection of mycotoxins in food

Aflatoxigenic fungi can contaminate food commodities, including cereals, peanuts, spices and figs. Foods and feeds are especially susceptible to colonization by aflatoxigenic *Aspergillus* species in warm climates where they may produce aflatoxins at several stages in the food chain, i.e. either at pre-harvest, processing, transportation or storage (Ellis *et al.*, 1991). The level of mold infestation and identification of the governing species are important indicators of raw material quality and predictors of the potential risk of mycotoxin occurrence (Shapira *et al.*, 1996). Traditional methods for the identification and detection of these fungi in foods include culture in different media and morphological studies. This approach, however, is tim-consuming, laborious and requires special facilities and mycological expertise (Edwards et al., 2002). Moreover, these methods have a low degree of sensitivity and do not allow the specification of mycotoxigenic species (Zhao *et al.*, 2001). PCR-based methods that target DNA are considered a good alternative for rapid diagnosis due to their high specificity and sensitivity, and have been used for the detection of aflatoxigenic strains of *A. flavus* and *A. parasiticus* (Somashekar *et al.*, 2004). However, as yet, none of these methods can reliably differentiate *A. flavus* from other species of the *A. fluvus* group. In particular, *A. flavus* and *A. parasiticus* have different toxigenic profiles, *A. flavus* produces aflatoxin B1 (M1), B2, cyclopiazonic acid, aflatrem, 3-nitropropionic acid, sterigmatocystin, verdsicolorin A and aspetoxin, whereas *A. parasiticus* produced aflatoxin B1 (M1), B2, G1, G2 and versicolorin A. Another important fact is that *A. flavus* and *A. fumigatus* are responsible for 90% of the aspergillosis in human beings (González-Salgado *et al.*, 2008). It is evident that one fundamental solution to the problem of mycotoxins in food would be to ensure that no contamination of edible crops occurred during harvesting and storage. It is equally clear, however, that such a solution is virtually unattainable, and hence that the presence of mycotoxins in food will have to be accommodated. Three approaches to the problem are most widely encountered; one involves physico-chemical methods of analysis, other relies on biological assays, and another one is microscopic examination. The former approach has found most widespread acceptance for routine purposes, but some authorities feel that a chemical diagnosis should be supported with some form of demonstration that the detected material is, in fact biologically toxic. The validity of this requirement is open to debate, but, for specific legal purposes, it may well become obligatory (Robinson, 1975).

5. References

Abbas A, Vales H, Dobson ADW. 2009. Analysis of the effect of nutritional factors on OTA and OTB biosynthesis and polychetide syntase gene expression in Aspergillus ochraceus. Int. J. Food Microbiol, 135:22-27

Abbas HK, Weaver MA, Zablotowics RM, Horn BW, Shier WT. 2005. Relationships between aflatoxin production, sclerotia formation and source among Mississippi Delta *Aspergillus* isolates. Eur J Plant Pathol, 112:283-287

Abbas HK, Wilkinson JR, Zablotowics RM, Accinelli C, Abel CA, Bruns HA, Weaver Ma. 2009. Ecology of *Aspergillus flavus*, regulation of aflatoxin production, and management strategies to reduce aflatoxin contamination of corn. Toxin Reviews, 28:142-153

Abbas HK, Zablotowics RM, Locke MA. 2004*a*. Spatial variability of Aspergillus flavus soil populations under different crops and corn grain colonization and aflatoxins. Botany, 82:1768-1775.

Abbas HK, Zablotowics RM, Weaver MA, Horn BW, Xie W, Shier WT. 2004*b*. Comparison of cultural and analytical methods for determination of aflatoxin production by Mississippi Delta *Aspergillus* isolates. Can. J. Microbiol. 50:193-199

Abdel-Wahhab MA, Kholif AM. 2008. Mycotoxins in animal feeds and prevention strategies: A review. Asian Journal Of Animal Sciences, 2(1):7-25

Abramson D. 1998. Mycotoxin formation and environmental factors. In: Sinha, K.K., and Bhatnagar D., Eds., Mycotoxins in Agriculture and Food Safety. Marcel Dekker, Inc, New York, 255-277

Accinelli C, Abbas HK, Zablotowicz RM, Wilkinson JR. 2008. Aspergillus flavus aflatoxin occurrence and expression of aflatoxin biosynthesis genes in soil. Can. J. Microbiol. 54:371-379

Aflatoxin Mitigation Center of Excellence (AMCE). 2010. Preventing Health Hazards and Economic Losses from Aflatoxin. Texas Corn Producers.

Aldred D, Magan N. 2004. Prevention strategies for trichothecenes. Toxicol. Lett., 153:165-171.

Andresen I, Becker W, Schluter K, Burges J, Parthier B. 1992. The identification of leaf thionin as one of the main jasmonate-induced proteins of barley (*Hordeum Vulgare*). Plant Mol. Biol. 19:193–204.

Angle JS. 1986. Aflatoxin decomposition in various soils. J. Environ. Sci. Health B. 21:277-288

Atalla MM, Hassanein NM, El-Beith AA, Youssef YA. 2003. Mycotoxin production in wheat grains by different Aspergilli in relation to different relative humidities and storage periods. Nahrung, 47:6-10

Atroshi F, Rizzo A, Wastermack T, Ali-Vehmas T. 2002. Antioxidant nutrients and mycotoxins. Toxicol., 180:151-167

Bell E, Mullet JE. 1991. Lipoxygenase gene expression is modulated in plants by water deficit, wounding, and methyl jasmonate. Mol. Gen. Genet. 230:456–462.

Benndorf D Müller A, Bock K, Manuwald O, Herbarth O, Van Bergen M. 2008. Identification of spore allergens from the indoor mold Aspergillus versicolor. Allergy, 63:454-460

Bennett JW, Chang PK, and Bhatnagar D. 1997. One gene to whole pathway: the role of norsolorinic acid in aflatoxin research. Adv. Appl. Microbiol. 45:1–15.

Bennett JW, Kale S, Yu J. 2007. Aflatoxins: Backround, Toxicology and Molecular Biology. From Infectious Disease: Foodborne Diseases Edited by S.Simjee. Human Press Inc., Totowa, NJ. 355-374.

Bennett JW, Klich M. 2003. Mycotoxins. Clin. Microbiol. Rev., 16:497-516

Bent AF, Mackey D (2007). Elicitors, effectors, and R genes: the new paradigm and a lifetime supply of questions. Annu. Rev. Phytopathol. 45: 399-436

Berthiller F, Schumacher R, Adam G, Krska R. 2009. Formation, determination, and significance of masked and other conjugated mycotoxins. Anal Bioanal Chem, 395:1243-1252

Bhat RV, Vasanthi S. 2003. Mycotoxin food safety risks in developing countries, food safety in food security and food trade. Vision 2020, Agriculture and Environment, Focus 10, pp: 1-2

Bhatnagar D, Ehrlich KC, Yu J, Cleveland TE. 2003. Molecular genetic analysis and regulation of aflatoxin biosynthesis. Appl Microbiol Biotechnol. 61:83-93.

Bhatnagar D, Cary JW, Ehrlich KC, Yu J, Cleveland TE (2006). Understanding the genetics of regulation of aflatoxin production and Aspergillus flavus development. Mycopathologia 162:155-166

Bhatnagar D, Proctor R, Payne GA, Wilkinson J, Yu J, Cleveland TE, Nierman WC. 2006. Genomics of mycotoxigenic fungi. In: Barug D, Bhatnagar D, van Egmond HP, van der Kamp JW, van Osenbruggen WA, Visconti A, eds. The mycotoxigenic factbook (Food & Feed Topics). Wageningen, The Netherlands: Wageningen Academic Publishers, pp.157-178

Bhatnagar D, Yu J, Ehrlich KC. 2002. Toxins in filamentus fungi. In: Breitenbach M, Crameri R, Lehrer SB (Eds.), Fungal Allergy and Pathogenicity. Chem. Immunol. Basel, Karger 81:167-206.

Bhatnagar RK, Ahmad SK, Mukerji G. 1986. Nitrogen metabolism in Aspergillus parasiticus NRRL 3240 and A. flavus NRRL 3537 in relation to aflatoxin production. J. Appl. Bacteriol., 60:203-211

Bhatnagar, D. 2010. Elimination of postharvest and preharvest aflatoxins contamination; 10th International working conference on stored product protection, Section: Microbiology, mycotoxins and food safety: 425.

Bley NC. 2009. Economic Risks of Aflatoxin Contamination in the Production and Marketing of Peanut in Benin. Thesis Submitted to the Graduate Faculty of Auburn University in Partial Fulfillment of the Requirements for the Degree of Master of Science.

Blout WP. 1961. Turkey "X" disease. Turkeys, 52:55-58

Bohnert M, Wackler B, Hoffmeister D. 2010. Spotlights on advances in mycotoxin research. Appl Microbiol. Biotechnol. 81:1-7

Bräse S, Encinas A, Keck J, Nising CF. 2009. Chemistry and biology of mycotoxins and related fungal metabolites. Chem Rev, 109:3903-3990

Brown RL, Cotty PJ, Cleveland TE. 1991. Reduction in aflatoxin content of maize by atoxigenic strains of *Aspergillus flavus. J. Food Prot.*, 54(8):623-626.

Buchi G, Foulkes DM Kurono M, Mitchell GF, Schneider RS. 1967. The total synthesis of racemic aflatoxin B1. Journal of the American Chemical Society, 89:6745-6753

Calvo AM, Wilson RA, Bok JW, Keller NP (2002). Relationship between secondary metabolism and fungal development. Microb. Mol. Biol. Rev. 66:447–459.

Campbell BC, Molyneux RJ, Schatzki TF. 2003. Current research on reducing pre- and post-harvest aflatoxin contamination of U.S. almond, pistachio, and walnut. Journal Of Toxicology, Toxin Reviews, 22:225-266.

Carbone I, Ramirez-Prado JH, Jakobek JL, Horn BW (2007). Gene duplication, modularity and adaptation in the evolution of the aflatoxin gene cluster. BMC Evol. Biol. 7: 111.

Cardwell, K.F., D. Desjardins, S. H. Henry, et al. 2004. The Cost of Achieving Food Security and Food Quality. http://www.apsnet.org/online/ festure/mycotoxin/ top.html.

Cary JW, Calvo AM. 2008. Regulation of Aspergillus mycotoxin biosynthesis. Toxin Reviews, 27:347-370

Cary JW, Ehrlich K. 2006. Aflatoxigenicity in Aspergillus: molecular genetics, phylogenetic relationships and evolutionary implications. Mycopathologia,162:167-177

Cary JW, Szerszen L, Calvo AM. 2009. Regulation of *Aspergillus flavus* aflatoxin biosynthesis and development. American Chemical Society, 13:183-203

Catangui MA, Berg RK. 2006. Western beat cutworm, *Striacosta albicosta* (Smith) (Lepidoptera:Noctuidae), as a potential pest of transgenic Cry1Ab *Bacillus thuringensis* corn hybrids in South Dakota. Environ Entomol. 35:1439-1452

Center for Disease Control and Prevention (CDC). 2004. Outbreak of aflatoxin poisoning-Eastern and Central provinces. Kenya, January-July, 2004.

Chang PK. 2003. The *Aspergillus parasiticus* protein AFLJ interacts with the aflatoxin pathway-specific regulator AFLR. Mol. Genet. Genomics 268: 711–719.

Chen ZY, Brown RL, Cleveland TE, Damann KE, Russin JS. 2001. Comparison of constitutive and inducible maize kernel proteins of genotypes resistant or susceptible to aflatoxin production. J. Food Prot., 64:1785-1792

Cleveland T, Dowd PF, Desjardins AE, Bhatnagar D, Cotty PJ. 2003. United States Department of Agriculture-Agricultural research service research on preharvest prevention of mycotoxins and mycotoxigenic fungi in US crops. Pest Manag. Sci., 59:629-642

Cleveland TE, Yu J, Fedorova N, Bhatnagar D, Payne GA, Nierman WC, Bennett JW. 2009. Potential of *Aspergillus flavus* genomics for applications in biotechnology. Trends in Biotechnology 27:151-157.

Codex Alimantarius Commission. 2002. Proposed draft code of practice for the prevention (reduction) of mycotoxin contamination in cereals, including annexes on ochratoxin A, zearalenone, fumonisins, and trichothecens, CX/FAC 02/21, Joint FAO/WHO Food Standards Programme, Rotterdam, the Netherlands.

Cole RJ, Cox RH. 1981. Handbook of Toxic Fungal Metabolites. New York: Academic Press, 937 pp.

Cole RJ, Cotty PJ. 1990. Biocontrol of aflatoxin production by using biocompetitive agents. In Robens, J., Huff, W. and Richard, J. (eds.) *A Perspective on Aflatoxin in Field Crops and Animal Food Products in the United States: A Symposium; ARS-83*. U.S. Department of Agricul-ture, Agricultural Research Service, Washington, D.C., pp. 62-66.

Cotty P. 1988. Aflatoxin and sclerotial production by *Aspergillus flavus*: influence of pH. Phytopathology, 78:1250-1253

Cotty, P.J. 1989. Virulence and cultural characteristics of two *Aspergillus flavus* strains pathogenic on cotton. *Phytopathology* 79, 808-814.

Cotty PJ. 1994. Influence of field application of an atoxigenic strain of *Aspergillus flavus* on the population of *A. flavus* infecting cotton bolls and on the aflatoxin content of cottonseed. *Phyto-pathology* 84, 1270-1277.

Cotty PJ. 1996. Aflatoxin contamination of commercial cottonseed caused by the S strain of *Asper-gillus flavus*. *Phytopathology* 86, S71.

Cotty PJ, Probst C, Jaime-Garcia R. 2008. Etiology and Management of Aflatoxin from Contamination.Mycotoxins: detection methods, management, public health and agricultural trade. ISBN: 978-1-84593-082-0. DOI: 10.1079/9781845930820.0287

Coulibaly O, Hell K, Bandyopadhyay R, Hounkponou S, Leslie JF. 2008. "Mycotoxins: Detection Methods, Management, Public Health and Agricultural Trade", Published by CAB International, ISBN 9781845930820.

Creelman RA, Tierney ML, Mullet JE. 1992. Jasmonic acid/methyl jasmonate accumulate in wounded soybean hypocotyls and modulate wound gene expression. Proc. Natl. Acad. Sci. U.S.A. 89, 4938–4941.

Cuero RG, Duffus E, Osuji G, Pettit R. 1991. Aflatoxin control in preharvest maize: effects of chitosan and two microbial agents. J. Agr. Sci. 117:165–169.

D'Mello JPF, McDonald AMC, Postel D, Dijksma WTP, Dujardin A, Placinta CM. 1998. Pesticide use and mycotoxin production in *Fusarium* and *Aspergillus phytopathogenes*. Eur. J. Plant Pathol., 104:741:751

Dauch AL, Watson AK, Jabaji-Hare SH. 2003. Detection of the biological control agent Colletotrichum coccodes (183088) from the target weed velvetleaf and soil by strain specific PCR markers. J. Microbiol. Methods, 55:51-64

Degola F, Berni E, Dall'Asta C, Spotti E, Marchelli R, Ferrero I, Restivo FM. 2007. A multiplex RT-PCR approach to detect aflatoxigenic strains of *Aspergillus flavus*. J. Appl. Microbiol. 103:409-417

Diamond H, Cooke BM. 2003. Preliminary studies on biological control of the Fusarium ear blight complex of wheat. Crop Prot., 22:99-107

Diener UL, Cole RJ, Sanders TH, Payne GA, Lee LS, Klich MA (1987). Epidemiology of aflatoxin formation by *Aspergillus flavus*. Annu Rev Phypathol. 25:249-270

Dill-Macky R, Jones RK. 2000. The effect of previous crop residues and tillage on Fusarium head blight of wheat. Plant Dis., 84:71-76

Dohlman, E. 2003. "Mycotoxin Hazards and Regulations: Impacts on Food and Animal Feed Crop Trade." International Trade and Food Safety: Economic Theory and Case Studies, Jean Buzby (editor), Agricultural Economic Report 828. USDA, ERS.

Dorner, J. W., Cole, R. J., Blankenship, P. D. (1998). Effect of inoculum rate of biological control agents on preharvest aflatoxin contamination of peanuts. Biol. Control 12:171–176.

Dorner, J. W., Cole, R. J., Wicklow, D. T. (1999). Aflatoxin reduction in corn through field application of competitive fungi. J. Food Prot. 62:650–656.

Dorner, J.W. 2004. Biological Control of Aflatoxin Contamination of Crops. Journal of Toxicology-Toxin Reviews. Vol. 23, Nos. 2 & 3, pp. 425–450, 2004.

Dowd PF. 2003. Insect management to facilitate preharvest mycotoxin management. J. Toxicol. Toxin Rev. 22(2):327-350

Du W, O'brian GR, Payne GA (2007). Function and regulation of *aflJ* in the accumulation of aflatoxin early pathway intermediate in Aspergillus flavus. Food Addit. Contam. 24: 1043–1050.

Edwards SG. 2004. Influence of agricultural practices on Fusarium infection of cereals and subsequent contamination of grain by trichothecene mycotoxins. Toxicol. Lett., 153:29-35

Ehrlich KC, Cotty PJ (2002). Variability in nitrogen regulation of aflatoxin production by Aspergillus flavus strains. Appl. Microbiol. Biotechnol. 60: 174– 178.

Ehrlich KC, Montalbano BG, Cary JW, Cotty PJ. 2002. Promoter elements in the aflatoxin pathway polyketide synthase gene. Biochim. Biophys. Acta 1576:171-175

Ellis WO, Smith JP, Simpson BK. 1991. Aflatoxin in food: Occurrence, biosynthesis, effects on organisms, detection, and methods of control.

Epple P, Apel K, Bohlmann H. 1995. An *Arabidopsis thaliana* thionin gene is inducible via a signal transduction pathway different from that for pathogenesis-related proteins. Plant Physiol. 109:813–820.

Fabbri AA, Fanelli C, Panfili G, Passi S, Fasella P. 1983. Lipoperoxidation and aflatoxin biosynthesis by *Aspergillus parasiticus* and *A. flavus. Gen Microbiol,* 29: 3447-3452.

Fanelli C, Fabbri AA. 1989. Relationship between lipids and aflatoxin biosynthesis. *Mycopathologia* 107:115-120.

Farmer EE, Ryan CA. 1992. Octadecanoid precursors of jasmonic acid activate the synthesis of wound inducible proteinase inhibitors. Plant Cell 4:129–134.

Fernandes M, Keller NP, Adams TH. 1998. Sequence-specific binding by Aspergillus nidulans AflR, a C6 zinc cluster protein regulating mycotoxin biosynthesis. Mol. Microbiol. 28: 1355-1365.

Food and Agriculture Organization (FAO). 1997. Worldwide Regulations for Mycotoxins 1995: A compendium. FAO Food and Nutrition Paper. No. 64. Rome, Italy.

Food and Agriculture Organization FAO. 2004. Manual Técnico: Manejo Integrado de Enfermedades en Cultivos Hidropónicos. Oficina Regional para América Latina y el Caribe.

Galvano F, Piva A, Ritieni A, Galvano G. 2001. Dietary strategies to counteract the effects of mycotoxins: A review. J. Food Prot., 64:120-131.

Garcia-Brugger AG, Lamotte O, Vandelle E, Bourque S, Lecourieux D, Poinssot B, Wendehenne D, Pugin A (2006). Early signaling events induced by elicitors of plant defenses. MPMI, 19(7): 711-724.

Garman H, Jewett HH.1914. The life-history and habits of the corn-ear worm (*Chloridae obsoleta*). Kentucky Agricultural Experimental Station Bulletin. 187:388-392

Gebrehiwet Y, Ngqangweni S, Kirsten JF. 2007. Quantifying the trade effect of sanitary and phytosanitary regulations of OECD countries on South African foods exports. Agrekon, 46:23-38

Georgianna DR, Payne GA. 2009. Genetic regulation of aflatoxin biosynthesis: from gene to genome. Fungal Genet Biol., 46:113-125

Goodrich-Tanrikulu, M., Mahoney, N. E. and Rodriguez, S.B. 1995. The plant growth regulator methyl jasmonate inhibits af latoxin production by *Aspergillus flavus.* Microbiology. 141: 2831-2837.

Groopman JD, Kensler TW. 1999. The light at the end of the tunnel for chemical-specific biomarkers: Daylight or headlight?, Carcinogenesis, 20:1-11

Gundlach, H., Muller, M.J., Kutchan, T.M., Zenk, M.H., 1992. Jasmonic acid is a signal transducer in elicitor-induced plant cell structures. Proc. Natl. Acad. Sci. U.S.A. 89, 2389–2393.

Heathcote JG, Hibbert JR. 1978. Aflatoxin chemical and biological aspects. Elsevier Scientific Publishing Company, Amsterdam.

Holopainen JK, Heijari J, Nerg AM, Vuorinen M, Kainulainen P (2009). Potential for the use of exogenous chemical elicitors in disease and insect pest management of conifer seedling production. Open. For. Sci. J. 2: 17-24.

Hormisch D, Brost I, Kohring GW, Gifthorn F, Krooppenstedt E, Farber P, Holzapfel WH. 2004. *Mycobacterium fluoranthenivorans* sp. nov., a fluoranthene and aflatoxin B1 degrading bacterium from contaminated soil of a former coal gas plant. Syst. Appl. Microbiol. 27:653-660

Horn BW, Greene RL, Dorner JW. 1995. Effect of corn and peanut cultivation on soil populations of *Aspergillus flavus* and *A. parasiticus* in southwestern Georgia. Appl Environ Microbiol, 61:2472-2475

Horn BW. 2007. Biodiversity of Aspergillus section Flavi in the United States: a review. Food Addit. Contam. 24:1088-1101.

Hua SS. 2002. Biological Control of Aflatoxin in Almond and Pistachio by Preharvest Yeast Application in Orchards. In: Special Issue: Aflatoxin/Fumonisin Elimination and Fungal Genomics Workshops. Phoenix, Arizona, October 23–26, 2001. Mycopathologia, 65.

Hua SS, Baker T, Flores-Espiritu M. 1999. Interactions of saprophytic yeasts with a nor mutant of *Aspergillus flavus*. Appl. Environ. Microbiol. 65:2738–2740.

Hussein HS, Brasel JM. 2001. Toxicity, metabolism, and impact of mycotoxins on humans and animals. Toxicol., 167:101-134

Huyn VL, Lloyd AB. 1984. Synthesis and degradation of aflatoxins by *Aspergillius parasiticus*. I. Synthesis of aflatoxin B1 by young mycelium and its subsequent degradation in aging mycelium. Aust. J. Biol. Sci. 37:37-43

Ioos R, Belhadj A, Menez M, Faure A. 2005. The effects of fungicideson *Fusarium* spp. and *Microdochium nivale* and their associated trichothecene mycotoxins in French naturally-infected cereal grains. Crop Prot., 24:894-902

Jaime-Garcia R, Cotty PJ. 2004. *Aspergillus flavus* in soils and corncobs in South Texas: implications for management of aflatoxins in corn-cotton rotations. Plant Dis. 88:1366-1371.

Jolly CM, Bayard B, Awuah RT, Fialor SC, Williams JT. 2009. "Examining the Structure of Awareness and Perceptions of Groundnut Aflatoxin among Ghanaian Health and Agricultural Professionals and its influence on their Actions" The Journal of Socio-Economics, 38:280-287.

Jones RK, Duncan HE, Payne GA, Leonard KJ. 1980. Factors influencing infection by *Aspergillus flavus* in silk-inoculated corn. Plant. Dis. 64:859-863

Kaaya AN, Warren HL. 2005. A Review of Past and Present Research on Aflatoxin in Uganda. African Journal of Food Agriculture Nutrition and Development (AJFAND) 5(1):1-18.

Kabak B, Dobson ADW, Var I. 2006. Strategies to prevent mycotoxin contamination of food and animal feed: A review. Critical Reviews in Food Science and Nutrition, 46:593-619

Kelkar HS, Skloss TW, Haw JF, Keller NP, Adams TH. 1997. Aspergillus nidulans stcL encodes a putative cytochrome P-450 monooxygenase required for bisfuran desaturation during aflatoxin/sterigmatocystine biosynthesis. Journal of Biological Chemistry, 272(3): 1589-1594

Keller NP, Nesbitt C, Sarr B, Phillips TD, Burow GB. 1997. pH regulation of sterigmatocystin and aflatoxin biosynthesis in Aspergillus spp. Phytopathol. 87:643-648

Kimanya ME, De Meulenaer B, Tiisekwa B, Ndomondo-Sigonda M, Devlieghere F. 2008. Co-ocurrence of fumonisins with aflatoxins in home-stored maize for human consumption in rural villages of Tanzania. Food Additives and Contaminants, 25:1353-1364

Klich MA. 1986. Mycroflora of cotton seed from the southern United States: a three year study of distribution and frequency. Mycologia, 94:21-27

Kumar A, Shukla R, Singh P, Prasad CS, Dubey NK. 2008. Assessment of Thymus vulgaris L. essential oil as a safe botanical preservative against post-harvest fungal infestation of food commodities. Innovative Food Science and Emerging Technologies, 9:575-580

Kumar V, Basu MS, Rajendran TP. 2008. Mycotoxin research and mycoflora in some commercially important agricultural commodities. Crop Prot, 27:891-905

Lacey J, Bateman GL, Mirocha CL. 1999. Effects of infection time and moisture on the development of ear blight and deoxynivalenol production by Fusarium spp. in wheat. Ann. Appl. Biol., 134:277-283

Lacey J. 1989. Prevention of mould growth and mycotoxin production through control of environmental factors. In: Natori S, Hashimoto K, and Ueno Y. Eds., Mycotoxins and Phycotoxins 1988. Elsevier, Amsterdam, 161-168

Lemmens M, Haim K, Lew H, Ruckenbauer P. 2004. The effect of nitrogen fertilization on Fusarium head blight development and deoxynivalenol contamination in wheat. J. Phytopathol., 152:1-8

Lewis L, Onsongon M, Njapau H, Schurz-Rogers H, Luber G. 2005. Aflatoxin contamination of commercial maize products during an outbreak of acute aflatoxicosis in Eastern and Central Kenya. Environ. Health Perspect., 113:1763-1767

Lund F, Frisvad JC. 2003. Penicillium verrucosum in wheat and barley indicates presence of ochratoxin A. J. Appl. Microbiol., 95:1117-1123

Ma WK, Sicilliano SD, Germida JJ. 2005. A PCR-DGGE method for detecting arbuscular mycorrizal fungi in cultivated soil. Soil Biol. Biochem, 37:1589-1597.

Magan N, Hope R, Cairns V, Aldred D. 2003. Post-harvest fungal ecology: impact of fungal growth and mycotoxin accumulation in stored grain. Eur. J. Plant Pathol., 109:723-730

Magan N, Hope R, Colleate A, Baxter ES. 2002. Relationship between growth and mycotoxin production by Fusarium species, biocides and environment. Eur. J. Plant Pathol., 108:685-690

Magan N, Aldred D. 2008. Post-harvest control strategies: Minimizing mycotoxins in the food chain. International Journal of Food Microbiology 119 (2007) 131–139.

Magan, N., Sanchis, V., Aldred, D., 2004. Role of spoilage fungi in seed deterioration. In: Aurora, D.K. (Ed.), Fungal Biotechnology in Agricultural, Food and Environmental Applications. Marcell Dekker, pp. 311–323. Chapter 28.

Manonmani HK, Anand S, Chandrashekar A, Rati ER. 2005. Detection of atoxigenic fungi in selected food commodities by PCR. Process Biochem., 40:2859-2864

Martin RA, MacLeod JA, Caldwell C. 1991. Influences of production inputs on incidence of infection by Fusarium species on cereal seed. Plant Dis., 84:71-76

Masoud W, Kaltoft CH. 2006. The effects of yeasts involved in the fermentation of coffea arabica in East Africa on growth and ochratoxin A (OTA) production by *Aspergillus ochraceus*. *Int. J. Food Microbiol.*, 106(2): 229-234.

McMillian WW, Wilson DM, Widstrom NW. 1985. Aflatoxin contamination of preharvest corn in Georgia: a six-year study of insect damage and visible Aspergillus flavus. J. Environ. Qual. 14:200-202

Mejía-Teniente L, Torres-Pacheco I, González-Chavira MM, Ocampo-Velazquez RV, Herrera-Ruiz G, Chapa-Oliver AM and Guevara-González RG. Use of elicitors as an approach for sustainable agriculture. African Journal of Biotechnology. 9 (54): 9155-9162.

Moreno-Martínez E, Vázquez-Badillo M, Facio-Parra F. 2000. Use of propionic acid salts to inhibit aflatoxin production in stored grains of maize. Agrociencia, 34(2):477-484

Muthomi JW, Njenga LN, Gathumbi JK, Chemining'wa GN. 2009. The occurrence of aflatoxins in maize and distribution of mycotoxin-producing fungi in Eastern Kenya. Plant Pathology Journal, 8(3):113-119

Nesci AV, Bluma RV, Etcheverry MG. 2005. In vitro selection of maize rhizobacteria to study potential biological control of *Aspergillus* section *Flavi* and aflatoxins production. *Eur. J. Plant Pathol.*, 113(2):159-171.

Nicholson P, Turner JA, Jenkinson P, Jennings P, Stonehouse J, Nuttall M, Dring D, Weston G, Thomsett M. 2003. Maximising control with fungicides of *Fusarium* ear blight (FEB) in roder to reduce toxin contamination of wheat. Project report No. 297, HGCA, London.

O'Brian GR, Georgianna DR, Wilkinson JR, Abbas HK, Wu J, Bhatnagar D, Cleveland TE, Nierman W, Payne GA. 2007. The effect of elevated temperature on gene expression and aflatoxin biosynthesis. Mycologia, 90:232-239

O'Callaghan J, Stapleton PC, Dobson ADW. 2006. Ochratoxin A biosynthetic genes in Aspergillus ochraceus are diferentially regulated by pH and nutritional stimuli. Fungal Genet Biol, 43:213-221

Odjacova M, Hadjiivanova C (2001). The complexity of pathogen defense in plants. Bulg. J Plant. Physiol. 27: 101-109.

Otsuki T, Wilson JS, Sewadeh M. 2001. What price precaution? European harmonization of aflatoxin regulations and African groundnut exports. European Review of Agricultural Economics 28: 263-283.

Papp E, H-Otta G, Zaray G, Mincsovics E. 2002. Liquid chromatographic determination of aflatoxins. Microchemical J., 73:39-46

Payne GA, Hagler WM. 1983. Effect of specific aminoacids on growth and aflatoxin by *Aspergillus parasiticus* and *A. flavus* in defined media. Appl Environ Microbiol, 171(3): 1539-1545

Payne GA, Nierman WC, Wortman JR, Pritchard BL, Brown D, Dean RA. 2006. Whole genome comparison of *Aspergillus flavus* and *A. oryzae*. Med. Mycol. 44:9-11

Payne GA. 1992. Aflatoxin in maize. CRC Crit Rev Plant Sci. 10:423-440

Pietri A, Zanetti M, Bertuzzi T. 2009. Distribution of aflatoxins and fumonisins in dry-milled maize fractions. Food Additives and contaminants, 26:372-380

Plasencia J. 2004. Aflatoxins in maize: A Mexican perspective. Journal of Toxicology, 23:155-177

Price MS, Yu J, Nierman WC, Kim HS, Pritchard B. 2006. The aflatoxin pathway regulator AflR induces gene transcription inside and outside of the aflatoxin biosynthetic cluster. FEMS Microbiol Lett, 255:275-279

Ramirez ML, Chulze S, Magan N. 2006. Temperature and water activity effects on growth and temporal deoxynivalenol production by two Argentinean strains of Fusarium graminearum on irradiated wheat grain. Int J. Food Microbiol, 106:291-296

Reiss J (1982). Development of *Aspergillus parasiticus* and formation of aflatoxin B1 under the influence of conidiogenesis affecting compounds. Arch. Microbiol. 133: 236–238.

Reverberi M, Ricelli A, Zjalic S, Fabbri AA, Fanelli C (2010). Natural functions of mycotoxins and control of their biosynthesis in fungi. Appl Microbiol Biotechnol 87:899–911.

Reverberi M, Ricelli A, Zjalic S, Fabbri AA, Fanelli C. 2010. Natural functions of mycotoxins and control of their biosynthesis in fungi. Appl. Microbiol. Biotechnol, 87:899-911

Ribeiro JMM, Cavaglieri LR, Fraga ME, Direito GM, Dalcero AM, Rosa CAR. 2006. Influence of water activity, temperature and time on mycotoxins production on barley rootlets. Lett Appl Microbiol, 42:179-184

Richard JL. 2007. Some major mycotoxins and their mycotoxicoses: An overview. International Journal of Food Microbiology, 119(2):3-10

Richard JL. 2003. Mycotoxins: Risks in Plant, Animal, and Human Systems. Task Force Report. Council for Agricultural Science and Technology, Ames, Iowa, USA Printed in the United States of America. ISSN 0194-4088; No.139.

Riley CV. 1882. The boll-worm alias corn-worm (*Heliothis armigera* Hubn.) order Lepidoptera; family Noctuidae. In: Report of the Commissioner of Agriculture for the years 1881-1882. Washington Printing Office, pp.145-152

Riveros AS. 2001. Moléculas activadoras de la resistencia inducida, incorporadas en programas de Agricultura Sostenible. Revista Manejo Integrado de Plagas (Costa Rica) 61: 4-11.

Robinson RK. 1975. The detection of mycotoxins in food. Intern. J. Environmental Studies, 8:199-202

Rodriguez SB, Mahoney NE. 1994. Inhibition of aflatoxin production by surfactants. Appl Environ Microbiol. 60:106-110

Roze LV, Miller MJ, Rarick M, Mahanti N, Linz J. 2004. A novel cAMP-response element, CRE1, modulates expression of *nor-I* in *Aspergillus parasiticus*. J. Biol. Chem, 279(26):27428-27439

Rustom IYS. 1997. Aflatoxin in food and feed: occurrence, legislation, and inactivation by physical methods. Food Chem., 59:57-67

Schmidt-Heydt M, Magan N, Geisen R. 2008. Stress induction of mycotoxin biosynthesis genes by abiotic factors. FEMS Microbiol Lett, 284:142-149

Scott PM. 1998. Industrial and farm detoxification processes for mycotoxins. Rev. Med. Vet. 149:543-548

Sembdener, G., Parthier, B., 1993. The biochemistry and the physiological and molecular actions of jasmonates. Annu. Rev. Plant Physiol. 44, 569–589.

Shapira R, Paster N, Eyal O, Menasherov M, Mett A, Salomon R. 1996. Detection of aflatoxigenic molds in grains by PCR. Appl. Environ. Microbiol. 62:3270-3273

Sheppard GS. 2008. Impact of mycotoxins on human health in developing countries. Food Additives and contaminants, 25:146-151

Simpson DR, Weston GE, Turner JA, Jennings P, Nicholson P. 2001. Differential control of head blight pathogens of wheat by fungicides and consequences for mycotoxin contamination in grain. Eur. J. Plant Pathol., 107:421-431

Sinha RN. 1995. The stored grain ecosystems. In: Jayas, D.S., White, N.D.G., Muir, W.E. (Eds.), Stored Grain Ecosystems. Marcell Dekker, New York, pp. 1–32.

Smith CA, Woloshuk CP, Robertson D, Payne GA. 2007. Silencing of the aflatoxin gene cluster in a diploid strain of Aspergillus flavus is suppressed by ectopic aflR expression. Genetics, 176:2077-2086

Smith JE, Lewis CW, Anderson JG, Solomons GL.1994. Mycotoxins in human health, Report EUR 16048 EN, European Commision, directorate –General XII, Brussels

Smith JS, Dorner JW, Cole RJ. 1990. Testing Bacillus subtilis as a possible aflatoxin inhibitor in stored farmers stock peanuts. Proc. Am. Peanut Res. Educ. Soc. 22:35.

Stoloff L, Lillehoj EB. 1981. Effect of genotype (open pollinated vs hybrid) and environment on preharvest aflatoxin contamination of maize grown in Southeastern United States. J. Am. Oil Chem. Soc. 58:976A-980A

Sweeney MJ, Dobson ADW (1999). Molecular biology of micotoxin biosynthesis. FEMS Microbiology Letters 175:149-163.

Sweeney MJ, Dobson ADW. 1998. Mycotoxin production by Aspergillus, Fusarium and Penicillium species. Int. J. Food Microbiol., 43:141-158

Sweeney MJ, Dobson ADW. 1999. Molecular biology of mycotoxin biosynthesis. FEMS Microbiol. Lett., 175:149-163

Tilburn J, Sarkar S, Widdick DA, Espeso EA, Orejas M, Mungroo J, Penalva MA, Arst HN. 1995. The Aspergillus PacC zinc finger transcription factor mediates regulation of both acid and alkaline expressed genes by ambient pH. EMBO. J. 14:779-790

Topal S, Aran N, Pembezi C. 1999. Turkiye'nin tarmsal mikroflorasinin mikotoksin profilleri. Gida Dergisi, 24:129-137

Trail F, Mahanti N, Linz J. 1995. Molecular biology of aflatoxins biosynthesis. Microbiology 141:755-765.

Trost BM, Toste FD. 2003. Palladium catalyzed kinetic and dynamic kinetic asymmetric transformation of γ-acyloxybutenolides. Enantioselective total synthesis of (+)-aflatoxin B1 and B2a, J. Am. Chem. Soc., 125:3090-3100

Tubajika KM, Damann KE. 2002. Glufosinate-ammonium reduces growth and aflatoxin B1 production by Aspergillus flavus. J. Food Prot., 65:1483-1487

Uchimiya H, Iwate M, Nojiri C, Samarajeewa PK, Takamatsu S, Ooba S, Anzai H, Christensen AH, Quail PH, Toki S. 1993. Bialaphos treatment of transgenic rice plants expressing a bar gene prevents infection by the sheath blight pathogen (Rhizoctonia solani). Biotechnology, 11:190-197

van Egmond, H.P., Jonker, M.A., 2004. Worldwide Regulations on Aflatoxins — The Situation in 2002. J. Toxicol. Toxin Rev., 23(2&3):273-293.

Vick BA. 1993. Oxygenated fatty acids of the lipoxygenase pathway. In *Lipid Metabolism in Plants*, pp. 167-191. Edited by T. S. Moore, Jr. Boca Raton, FL: CRC Press.

Wagacha JM, Muthomi JW. 2008. Mycotoxin problem in Africa current status implications to food safety and health and possible management strategies. Int. J. Food Microbiol., 124:1-12

Wang JS, Tang L. 2005. Epidemiology of aflatoxin exposure and human liver cancer. In aflatoxin and food safety. Edited by H.K. Abbas. Taylor & Francis, Boca Raton, Fla. pp.195-211

Weaver MA, Vadenyapina E, Kenerley CM. 2005. Fitness, persistence, and responsiveness of an engineered strain Thrichoderma virens in soil mesocosms. Appl. Soil Ecol., 29:125-134

Wheeler MH. 1991. Effects of chlobenthiazone on aflatoxin biosynthesis in *Aspergillus parasiticus* and *A. flavus*. Pesticide Biochemistry and Phisiology, 41:190-197

Wicklow DT, Wilson DM, Nelsen TE. 1993. Survival of Aspergillus flavus sclerotia and conidia buried in soil in Illinois and Georgia. Phytopathology, 83:1141-1147

Widstrom NW. 1987. Breeding strategies to control aflatoxin contamination of maize through host plant resistance. In: Zuber, MS, Lillehoj EB, Renfro BL, eds. Aflatoxin in Maize: A proceedings of the workshop. Mexico: CIMMYT, 212-220

Wieser J, Lee BN, Fondon JW, Adams TH (1994). Genetic requirement for initiating asexual development in *Aspergillus nidulans*. Curr. Genet. 27: 62–69.

Wilkinson JM. 1999. Silage and animal health. Nat. Toxins., 7:221-232

Williams J, Phillips TD, Jolly PE, Stiles JK, Jolly CM, Aggarwal D. 2004. Human aflatoxicosis in developing countries: A review of toxicology, exposure, potential health consequences and interventions. Am. J. Clin. Nutr., 80:1106-1122

Williams WP, Buckley PM, Windham GL. 2002. Southwestern corn borer (Lepidoptera: Crambidae) damage and aflatoxin accumulation in maize. J. Econ Entomol. 95:1049-1053

World Bank, 2005. Food safety and agricultural health standards. Challenges and opportunities for developing country exports. Report No. 31207, Washington, D.C., USA.

Wu F. 2006. Mycotoxin reduction in Bt Corn: Potential economic, health and regulatory impacts. ISB News Report, September 2006.

Wu, F. 2008. A tale of two commodities: how EU mycotoxin regulations have affected u.s. tree nut industries. World Mycotoxin Journal. 1(1): 95-102.

Wu, F., 2004. Mycotoxin risk assessment for the purpose of setting international standards. Environmental Science & Technology 38:4049-4055.

Yabe K (2002). Pathway and genes of aflatoxin biosynthesis. In: Microbial Secondary Metabolites: Biosynthesis, Genetics and Regulation, Research Signpost (Fierro, F. and Martin, J. F., eds.), Kerala, India, pp. 227–251.

Yin Y, Lou T, Jiang J, Yan L, Michailides TJ, Ma Z. 2008. Molecular characterization of toxigenic and atoxigenic *Aspergillus flavus* isolates collected from soil in various agroecosystems in China. *Food Microbiol.*, manuscript sumbitted for publication.

Yin Y, Yan L, Jiang J. 2008. Biological Control of Aflatoxin Contamination of Crops. Journal of Zhejiang University Science B. 9(10):787-792.

Yu J, Bhatnagar D, Cleveland TE. 2004a. Completed sequence of aflatoxin pathway gene cluster in *Aspergillus parasiticus*. FEBS Lett. 564:126-130

Yu J, Bhatnagar D, Ehrlich KC. 2002. Aflatoxin biosynthesis. Rev. Iberoam. Microbiol. 19: 191–200.

Yu J, Chang PK, Ehrlich KC, Cary JW, Bhatnagar D, Cleveland TE. 2004*b*. Clustered pathway genes in aflatoxin biosynthesis. Appl. Environ. Microbiol. 70:1253-1262.

Yu J, Proctor RH, Brown DW (2004). Genomics of economically significant Aspergillus and Fusarium species. In: Applied Mycology and Biotechnology (Arora, K. D. and Khachatourians, G. G., eds.), Vol. 3, Elsevier, Amsterdam, pp. 249–283.

Zablotowics RM, Abbas HK, Locke MA. 2008. Population ecology of Aspergillus flavus associated with Mississippi Delta soils. Food Additives Contam. 24:1102-1108.

Zablotowicz RM, Abbas HK, Locke MA. 2007. Population ecology of Aspergillus flavus associated with Mississippi Delta soils. Food Addit Contam, 24:1102-1108

Zeringue H.J. Jr. 2002. Effects of methyl jasmonate on phytoalexin production and aflatoxin control in the developing cotton boll. Biochemical Systematics and Ecology 30: 497–503.

17

Selenium and Aflatoxins in Brazil Nuts

Ariane Mendonça Pacheco[1] and Vildes Maria Scussel[2]

[1]*Faculty of Pharmaceutical Sciences, Federal University of Amazonas, Manaus-AM,*
[2]*Food Science and Technology Department, Centre of Agricultural Sciences,*
Federal University of Santa Catarina, Florianopolis
Brazil

1. Introduction

Brazil nuts (*Bertholletia excelsa* H.B.K.) are considered a high nutritious food. Apart from carbohydrates, lipids, sulphur proteins and minerals, Brazil nuts are known to be rich in selenium - Se (Barclay et al., 1995; Coutinho et al., 2002, Souza et al., 2004, Pacheco and Scussel, 2007). Despite of that, when their shell are cracked either when pods fall on the ground, or during pod opening for nut extraction (done by an axe) and exposed to high moisture and temperature of the tropical forest, fungi may grow, leading to nut spoilage. If fungi are toxigenic they may produce aflatoxins (AFLs).

Se has been reported to be an antioxidant and studies have reported differences on its levels in Brazil nuts from the two Amazon regions being the Eastern nuts richer in Se than the Westerns (Chang et al., 1995; Pacheco and Scussel, 2007). Its content may vary when grown in different soils of the Amazon basin. The aflatoxigenic *Aspergillus* species of *A. flavus* and *parasiticus* are intimately related to agricultural crops, including tree nuts and their growth are influenced by environmental conditions. Although Brazil nuts have tested positive for *A. flavus*, less is known about its populations on Brazil nuts, how they grow and vary among the two Amazon regions, in the different stages of nut collection prior reaching the factory and how the processing affect them (Castrillon and Purchio, 1988; Freire et al., 2000; Candlish et al, 2001, Caldas et al., 2002). However, AFLs have been reported contaminating Brazil nuts. Their pods are harvested after they fall onto the forest soil. They stay directly in contact with the soil for several days or weeks prior to collection. It is during that time that pods may get contaminated with *Aspergillus* sp. and so with AFLs. Post harvesting operations are expected to have major influence on further contamination of the nuts (Bayman, 2002; Campos/Pas, 2004). For fungi growth and for their normal maintenance, a number of metals are required in different amounts. Many microorganisms are known to be able to use Se (i.e., selenite, selenate of other forms) in their metabolism (Roux et al., 2001; Fleet-Stalder et al., 2000). Se has been added to the media as sodium selenite (Na_2SeO_3), to understand its effect in different concentrations on fungi behaviour. The inhibitory action of Se on the growth rate of various fungi such as *Aspergillus*, *Penicillum* and *Fusarium* was reported by several authors (Ramadan et al., 1983, Ragab, et al., 1986, Zohri et al., 1997, Li et al, 2003). When it was used in Czapek Dox agar medium to evaluate the *A. parasiticus* behavior concerning morphological growth and toxin production, it was observed that fungus growth decreased by the increasing of the Se concentration (Zohri et al., 1997). In an experiment carried out by Li et al. (2003) utilizing Na_2SeO_3, the authors observed from stimulating to toxic effects on organisms depending on the levels applied in the media and concluded that the presence of Se in high concentration in the growth environment can lead to morphological distortion of the fungi characteristics. Se compounds have been reported also been acting as antagonists to the mutagenic and carcinogenic activities of some agents such as UV light and AFLs (Martin et al., 1984, Gregory, 1984, Overvad et al., 1985). AFLs are produced by *Aspergilllus flavus* in the Brazil nuts that have Se in their composition and the Amazon forest has optimal conditions for fungi growth such as high temperatures (> 25 °C) and relative humidity (RH) > 80 % (Bayman et al., 2002; Freire et al, 2000; Scussel, 2004; Arrus et al., 2005a; Arrus et al, 2005b; Pacheco and Scussel, 2006). Considering that Brazil nuts are rich in Se, its concentration may vary among regions, AFL contamination has been reported on Brazil nuts and that Na_2SeO_3 has either stimulation or toxic effects on organisms depending on its levels in media, a work was carried out in order to evaluate the effect of Brazil nut Se content from different Amazon regions and Na_2SeO_3 on the aflatoxigenic *A. flavus* FC1087 in terms of (a) fungal growth behavior, (b) colonies characteristics and (c) AFL production.

2. Material and methods

2.1 Material

a. *Brazil nuts:* two batches of raw, medium size nuts, with 18.4 and 43.5 mg/kg Se in their composition. Those nuts were from the Western and Eastern regions of the Brazilian Amazon basin, respectively. Nuts were previously tested for AFL. No AFL was detected up to the method limit of detection (LOQ) of 1.95 µg/kg.

b. *Toxigenic Aspergillus strain:* aflatoxigenic *A. flavus* FC1087 isolated from the Amazon forest and supplied by the Oswaldo Cruz Foundation (Fiocruz), Manaus, Amazon State, Brazil. AFLs total $(AFB_1+AFB_2+AFG_1+AFG_2)$ and AFB_1: 90.3 and 109.2 µg/kg, respectively.

c. *Culture media*: aqueous tween 80 and potato dextrose agar (PDA) from Merck.

d. *Se standard:* Na_2SeO_3, analytical grade, Baker. *(d.1) for ICP analysis:* acidified aqueous solution with nitric acid at 10 % (certificate N° *SRM 3149*), NIST and *(d.2) for mycological study:* acidified aqueous stock solution (100 µg/mL) with sulfuric acid at 10% - sterile.

e. *Aflatoxin standards:* AFB_1, AFB_2, FG_1 and AFG_2 from Sigma.

f. *Chemicals:* acetonitrile and methanol (HPLC grade), Baker. Ammonium acetate, ammonium sulfate, hydrochloric acid, nitric acid and anhydrous sodium sulfate (analytical grade) also from Baker. Ultrapure water (Mili-Q), Milipore.

g. *Equipment* for *(g.1) Se analysis:* atomic emission spectrophotometer with inductively coupled plasma (ICP) -optical emission spectrometry (OES), Model Otima 2000, Perkin Elmer and for *(g.2) Aflatoxins analysis:* ultra violet cabinet (365 nm), Tecnal; spectrophotometer, Hitachi; (g.3) *Mycology:* bacteriological oven and autoclave, Fanem, colony counter, Marconi and microscope stereoscope, lenses 40X, model Q714TZ-1, Quimis. Thin layer chromatographic aluminum sheets (20 x 20) with G60 silica gel from Merck and industrial Brazil nut cracker, CIEX.

h. *Other materials:* sterilized stainless steel blenders, trays (400 x 250 mm) and scissors. Sterilized polyethene bags. Round and straight platinum wires, Petri dishes - plates (90x15 mm, volume: 15 ml) and Neubauer counting chamber, Optik.

2.2 Methods

a. *Brazil nut preparation:* 1 kg of in-shell Brazil nuts, of each Se content, after de-shelling was finely grounded (particle size <100 μm), homogenized and three portions of 250 g were separated for Se, AFLs and the mycology tests.

b. *Se analysis:* the method used was that reported by US EPA (1996) using atomic emission spectrometry-ICP-OES. The limit of detection (LOD) was 2.00 mg/kg and the limit of quantification (LOQ) was 3.50 mg/kg. LOQ was defined as the lowest point of the calibration curve with high repeatability-axial view. Three replicates each (n=3).

c. *AFLs analysis:* by AOAC (2005) with a LOD and LOQ for total AFLs of 0.97 and 1.95 μg/kg, respectively. Two replicates each (n=3).

d. *Se culture media preparation:* Media were divided into three groups containing different Se origin and concentrations. They were prepared by adding Se in PDA as follows. **Group I**: ground Brazil nuts from Western (18.4 mg/kg) and Eastern (43.5 mg/kg) Amazon basin (portions of each Brazil nut batch were added to PDA to obtain final Se concentration of 0.018 and 0.044 mg/kg in the plates, respectively). **Group II**: Se as Na_2SeO_3 into six different Se concentrations (volumes of the Na_2SeO_3 solution was added to PDA accordingly to get the following increasing concentrations: 0.01; 0.02; 0.1; 0.2; 0.4 and 0.6 mg/kg in the plates). Poured plates were gently shaken to allow proper homogenization of the metal into the medium. **Group III:** no Se was added to the media - as a Control - to evaluate the strain normal behaviour. The media were sterilized by autoclavation for 15 min at 121°C. See details of the media on the Groups and Se concentration in Table 1.

e. *Spore suspension preparation:* the suspension concentration of the *A. flavus* spores was set using a Neubauer counting chamber and diluting the reference strain original suspension in 0.2% aqueous tween 80 to get a final concentration of 3.0 x 10 spores/mL.

f. *Fungal total count:* the suspension was spread on the media plates (Group I; Group II and Group III) previously prepared and incubated for 5 days at 28 °C. After that period the total colonies were counted utilizing the colony counter. Note: to observe further colony behavior and development, the plates were kept incubated up to 14 days at the same temperature. Four replicates each (n=4).

g. *The effect of Se on A. flavus FC1087:(g.1) growth and colony features:* the fungi spore suspension was also inoculated on another set of the three Groups media, by means of a straight wire to give a single point inoculum on the plate centre, to evaluate the colony radius growth (by a fine measuring scale) and other features behaviour such as shape and colour. After fungi inoculation, dishes were incubated for 14 days at 28 °C. Fungi growth rate related to Se content was checked every day by measuring their radius, examining their morphological changes and their verse and reverse colour. At the end of the incubation period, the morphological colonies features (size, colour, shape, reverse colour) were registered (Table 1). Colony colour was defined according to the Methuen Handbook of Colour (Kornerup and Wanscher, 1989).

g. *(2) AFL production*: The strain AFL production related to Se concentration was evaluated in each media by examining the AFL characteristic fluorescence (white/bluish) development. Each media fluorescence positive had their AFLs extracted (Moss and Badii, 1982) and analyzed as in (c). Four replicates each (n=4).

h. *Statistical analysis:* the data were statistically analyzed using analysis of variance (ANOVA).

3. Results and discussion

The influence of Se on fungi growth, colony diameter, verse/reverse colour, fluorescence development and AFL production are shown in Table 1. The data obtained shows that Se affected the *A. flavus* F1087 strain proliferation and AFL production when inoculated in media added either of ground Brazil nuts or Na₂SeO₃ at different concentrations.

3.1 Effect of Se concentration on the total fungi growth

A. flavus FC1087 was able to grow in presence of both Brazil nut Se concentrations (18.4 and 43.5 mg/kg: Western and Eastern region nuts, respectively) – Group I. It was observed that the highest total *A. flavus* colonies count and growth rate was obtained in the media containing the Western Brazil nuts (Se: 0.02 mg/kg in the plate with 2.3 x 10 cfu) and so for the Control (2.8 x 10 cfu) - Group III. The Eastern nuts media with higher Se content (0.04 mg/kg) presented lower number of colonies than the Western's (1.2 x 10 cfu). Similar to the Eastern nut total colonies count, was observed in the tested Na₂SeO₃ media concentrations (0.01 to 0.20 mg/kg) – Group II, with total counts of 1.5, 1.3, 1.3 and 1.1 x 10 cfu, respectively. Growth decreased from Na₂SeO₃ level of 0.20 to 0.4 mg/kg reaching 0.9 x 10 cfu. At the highest Na₂SeO₃ content (0.60 mg/kg) fungi failed completely to grow, probably because of its toxic level to the strain. These findings are corroborated by previous authors (Badii et al, 1986; Aboul-Dahad 1991; Zohri et al, 1997) that have also found reduction on fungi growth as the concentration of Na₂SeO₃ increase in the media. The toxic Se effect at higher doses reduces microorganism growth due to the detoxification process that takes place. That involves fungi transformation of the inorganic Se (more toxic) to organics (less toxic). It also involves the reduction of the Se oxyanions, Na₂SeO₃ and selenate to inert elemental Se within the mycelium. The toxic action of Se in fungi is believed to be due to its incorporation into the protein amino acids instead of sulfur which can lead to alteration of the tertiary structure thus to dysfunction of the fungi proteins and enzymes (Gharieb and Gadd, 1998). Figure 1 shows the *A. flavus* FC1087 growth from Day 5 to Day 14 of incubation in the media containing Se Western and Eastern Brazil nuts Se.

3.2 Effect of Se concentration on *A. flavus* FC1087 colony features

When *A. flavus* strain was inoculated in a single point on the media that contained either the two Brazil nut Se concentrations (Group I) or the six increasing Na₂SeO₃ concentrations (Group II), it was possible to observe their diameter, colours (verse/reverse) and fluorescence variations. Their diameters increased (5±2 to 19±4 mm) with the reduction of Na₂SeO₃ concentration (0.4 to 0.01 mg/kg). The highest colony diameter was obtained in the Western Brazil nut media with 31±5 mm. Eastern nuts colonies diameter was smaller with only 18±2 mm. Most of the colonies grown in Na₂SeO₃ media presented a characteristic green colour with the exception of the ones grown in the lowest Na₂SeO₃ concentration (0.01 mg/kg) that had only the centre green with a white, broad, non sporulating margin. That non sporulating margin was also observed in the Brazil nut both regions media colonies. On the other hand, the colonies reverse colour changed as the selenite concentration increased in the media from brown (lower Na₂SeO₃ concentration = 0.01 and 0.05 mg/kg) to orange-red (higher Na₂SeO₃ concentration = 0.2 and 0.4 mg/kg). Some authors have studied and reported these colour changes. McCready et al. (1966) reported that orange intracellular granules in bacteria (*Salmonella Heidelberg*), when grown in the presence of Na₂SeO₃, as amorphous red elemental Se. Badii et al in 1986, also observed a deep orange pigment on the undersurface of the *A. parasiticus* colonies in presence of Na_2SeO_3. Aboul-Dahab (1991) found that these colours were due to the reduction of Na_2SeO_3 with deposition of elemental Se within the fungal cells, as well as, in the growth media. Presumably, the fungus metabolizes Se to produce strong reducing and oxidizing agents into the media environment. Moss et al. (1987) reported that *A. parasiticus* is able to reduce also sodium bis selenite with the deposition of granules of elemental Se within the mycelium, being that red-orange colour a biological reduction of Se compounds into elemental colloidal Se. As far as the fluorescence of possible AFL presence is concerned, all media, containing Se despite of its origin or concentration, presented the characteristic white/bluish fluorescence and so the Control plates (except at the selenite highest concentration). That was an indication of the AFL presence, thus they were submitted to AFL analysis. However, it was observed that the fluorescence intensity reduced as the Na_2SeO_3 level increased in the media, reaching the highest concentration (0.60 mg/kg) with no fluorescence at all. In fact, *A. flavus* FC1087 failed to grow at that concentration caused probably by the toxic effect of the Na_2SeO_3 (Zohri, et al., 1997, Li et al., 2003). Figure 2 shows the *A. flavus* FC1087 growth in selenite at the 0.01 and 0.4 mg/kg with the development of orange to red pigment.

Group	PDA[a] Se content[a] (mg/kg)	Total fungi count (cfu/ml)	Colony features Diameter[b] (mm)	Colour Verse	Colour Reverse	Fluorescence[c]	Aflatoxin (µg/kg) Total[d]	AFB$_1$[e]
I	Brazil nuts							
	Western[f] 0.02	2.3 x 10	31±5	Green[g]	Colourless	+	ND	ND
	Eastern[h] 0.04	1.2 x 10	18±2	Green[g]	Orange Red	+	50.2±0.6	28.5±0.6
II	Na_2SeO_3[i]							
	0.01	1.5 x 10	19±4[j]	Green[g]	Brown	+	98.7±0.3	75.0±0.5
	0.05	1.3 x 10	13±4	Green	Brown	+	102.6±0.3	70.4±0.2
	0.10	1.3 x 10	10±3	Green	Light brown	+	90.3±0.5	67.9±0.3
	0.20	1.1 x 10	9±2	Gray	Orange Red	+	70.3± 0.5	49.2±0.5
	0.40	0.9 x 10	5±2	Gray	Orange Red	+	30.6±0.5	19.2±0.5
	0.60	None	NG[j]	NA[k]	NA	-	NA	NA
III	Control							
	NS[l]	2.8 x 10	30.5±5	Green	Colourless	+	109.2 ±0.2	90.3±0.5

[a] Potato dextrose Agar and Se concentration in each Petri dish
[b] The values of colony diameters are means based on four replicates (n = 4)
[c] Under ultra violet light at 365 nm
[d] $\sum$ total AFL LOD: 0.390 µg/kg
[e] AFB$_1$ LOD: 0.04 µg/kg
[f] PDA+ Brazil nuts from the Western Amazon region with Se content of 18.4 mg/kg
[g] With white broad nonsporulating margins
[h] PDA+ Brazil nuts from the Eastern Amazon region with Se content of 43.5 mg/kg
[i] In Se standard acidic solution (sodium selenite)
[j] No fungi growth
[k] Not applicable
[l] No Se added

Table 1. Effect of Brazil nuts natural Se content and Se as Na_2SeO_3 on the behaviour aflatoxigenic *A. flavus* FC1167

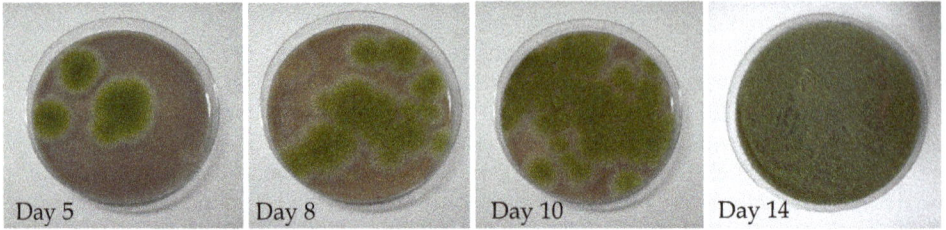

(a) PDA media with Western region Brazil nuts: 0.02 mg/kg - green colour with non-sporulating margin

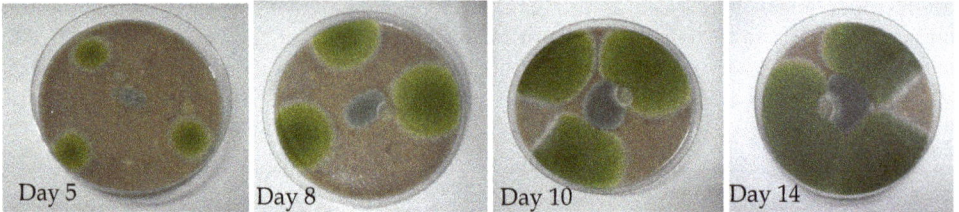

(b) PDA media with Eastern region Brazil nuts: 0.04 mg/kg- green colour with non -sporulating margin

Fig. 1. Toxigenic *Aspergillus flavus* FC1087 strain growth during 14 days of incubation at 28°C in PDA media containing natural Se content of ground Brazil Nut from (a) Western and (b) Eastern Amazon regions (*spreading technique*); Se content in the nuts: 18.4 and 43.5 mg/kg, respectively.

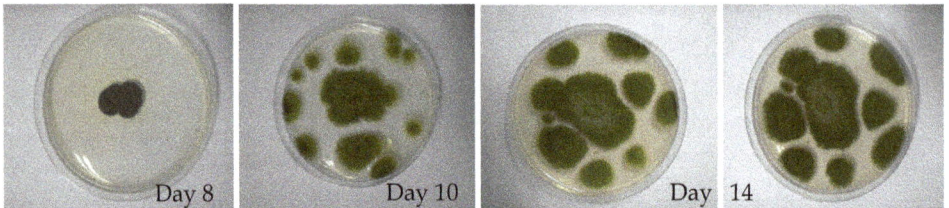

(a) PDA media with Se as Na_2SeO_3: 0.01 mg/kg - green colour

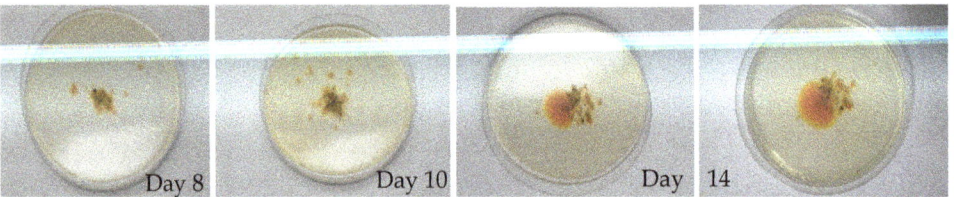

(b) PDA media with Se as Na_2SeO_3: 0.40 mg/kg – orange to red colour

Fig. 2. Toxigenic *Aspergillus flavus* FC1087 strain growth during 14 days of incubation at 28°C in PDA media containing Se as Na_2SeO_3 solution (*single point technique*); Se content (a) 0.01 and (b) 0.40 mg/kg.

3.3 Effect of Se concentration on *AFls* production

AFls were detected in the media, containing Eastern Brazil nuts and Na_2SeO_3 media, with the exception of the 0.60 mg/kg Na_2SeO_3 media, as no *A. flavus* was able to grow. Only in the Brazil nut medium with 0.04 mg/kg of Se, corresponding to nuts from the Eastern Amazon region, the strain produced AFL at a level of 50.2±0.6 µg/kg for total AFL and 28.5±0.6 µg/kg for AFB_1. No AFLs were detected in the nuts from Western region, however fungi growth were much abundant in that media, and larger the colonies diameter (31±5 mm). *A. flavus* FC1087 was able to produce in the Control media 109.2 ±0.2 µg/kg of AFL. It was observed that, the strain was Na_2SeO_3 doses-dependent either for its AFL synthesis and growth, as an AFL decrease with the increased of Na_2SeO_3 concentration occurred, concomitant to the fungi growth (Table 1). Both, the total AFLs ($AFB_1+AFB_2+AFG_1+AFG_2$) and AFB_1 production decreased as the Na_2SeO_3 levels increased. Similar to the Control plate, tested fungus was able to produce high amounts of AFls total, reaching at the two lowest Se concentrations (0.01 and 0.05 mg/Kg) a total AFLs of 98.7±0.3 and 102.6±0.3 µg/kg, respectively. In contrary, at the highest concentrations of 0.2 and 0.4 mg/kg, to where fungi were still able to grow, toxin production reduced from 70.3±0.5 to 30.6±0.5 µg/kg. The higher Se content in the nuts, the less fungi proliferation and rate in the plate. That could be caused by the Se toxic effect on the strain leading to fungi stress thus activating second metabolism of AFL formation. That could be explained for the toxicity of high amount of Se in organism, maybe causing oxidative stress in the strain (Letavayova et al., 2006, Valko et al., 2006). The oxidative stress is a prerequisite for AFL production by *A. parasiticus* (Jayashree and Subramanyam, 2000). Thus, a Se amount could activate the mechanisms of AFL production and contaminating Brazil nuts. Bronzetti *et al* (2001) demonstrated Se compounds in yeasts exerted both mutagenic and anti-mutagenic effect at different concentrations. On the other hand, other factors, such as the interaction or the competition with different strains of *Aspergillus* seem to affect the increasing of AFL production in some substrates (Martins, Martins, Bernardo, 2000).

3.4 Brazil nut composition, Se content versus fungi and AFls production

As far as Brazil nuts Se concentration is concerned, it is important to emphasize that there are two approached to take into account related to its benefits: First is the Se content in the nuts for (a) human consumption against diseases and the other is for (b) fungi proliferation and AFLs production in the nuts substrate. (a) Se is important to health as it has been reported being an excellent antioxidant for reducing toxic effects such as the carcinogenicity of some compounds. Its antioxidant protective effect is primarily associated with the presence of glutathione peroxides that protect DNA and other cellular components from damage by oxygen radicals. Se is an essential component of glutathione peroxides (Agar and Alpsoy, 2005). It inclusive, in some concentrations (such as 8 mg/kg), can inhibit the AFB_1 and AFG_1 mutagenic and carcinogenic effects in human blood cell culture (Geyikoglu and Turkez, 2006). (b) Regarding fungi and AFLs, in our study, it was observed that the *A. flavus* FC1087 on the Western Brazil nuts media presented similar behavior as the Control with 2.3 x 10 total count / 31±0.5 diameter / green colour / colourless reverse colonies however without AFLs production. On the other hand, the Eastern's media presented lower total *A. Flavus* FC1087 count, growth rate, and diameter, however with AFL production. Data suggests that: the higher Se content in the nuts, the less fungi proliferation and rate in the plate and high AFLs. That could be caused by the Se toxic effect on the strain leading to fungi stress thus activating the second metabolism of AFL formation.

4. Conclusion

The higher Se content present in the Brazil nuts lead to less fungi proliferation, growth rate in the plate and AFL production. That could be caused by its toxic effect to the fungus provoking stress and activating AFL production. Despite of data obtained, further studies need to be carried out utilizing Brazil nuts media with a more wide Se concentration range to find out the Se role on fungi and AFL production. It would be also necessary to investigated possible interactions and/or competition of different strains of *Aspergillus* on AFL production in Brazil nuts.

5. References

Aboul-Dahab, N.F. (1991). Responses of certain fungal communities towards environmental pollutants. *MSc thesis* Botany Depart Fac. Sci., Al-Azhar Univ. Cairo, Egypt.

Agar, G.& Alpsoy, L. (2005). Antagonistic effect of selenium against aflatoxin G_1 toxicity induced chromosomal aberrations and metabolic activities of two crop plants. *Bot. Bull. Acad. Sin.*, 46: 301-305

AOAC - Association of Official Analytical Chemists (2005). Official Methods of Analysis of AOAC International 18th, Horwitz, W. and Latimer, G. W. Jr. eds. Gaithersburg, Maryland, USA.

Arrus, K., Blank, G., Clear, R. & Holley, R.A. (2005a). Aflatoxin production by *Aspergillus flavus* in Brazil nuts. *J. of Stored Products Research*, 41, 513-527.

Arrus, K., Blank, G., Clear, R., Holley, R.A & Abramson, D. (2005b). Microbiological and aflatoxin evaluation of Brazil nut pods and the effects of unit processing operations. *J. of Food Protection*, 68(5), 1060-1065.

Badii, F., Moss, M.O. & Wilson, K. (1986). The effect of sodium selenite on the growth and aflatoxin production of *Aspergillus parasiticus* and the growth of other aspergilli. *Lett. appl. Microbiol.* 2:61-64.

Bayman, P., Baker, J & Mahoney, N. (2002). *Aspergillus* on tree nuts: incidence and associations. *Mycopathologia* 155:161-169.

Barclay, M.N.I., McPherson, A. & Dixon, J. (1995). Selenium content of a range of UK foods. *J. of Food Composition and Analysis.* 8, 307-318.

Bronzetti, G., Cini, M., Andreolli, E., Caltavuturo, L., Panunzio, M. & Croce, C.D. (2001). Protective effects of vitamins and selenium compounds in yeast. *Mutation research.* 496: 105-115.

Caldas, E. Silva, S. C. & Oliveira, J.N. (2002). Aflatoxins and ochratoxin A in food and the risks to human health. *Rev. S. Publica.* 36:319-323.

Candlish, A.A.G., Perason, S.M., Aidoo, K.E., Smith, J.E., Kelly, B. & Irvine, H. (2001). A survey of ethic foods for microbial quality and aflatoxin content. *Food addit. and Contaminants*, v18, nº2, p.129-136.

Castrillon, A.L. & Purchio, A. (1988). Aflatoxin occurrence in Brazil nuts. *Acta Amazonica*, 18, 49-56.

Chang, J.C., Gutenmann, W.H., Reid, C.M. & Lisk, D.J. (1995). Selenium content of Brazil nuts from two geographic locations in Brazil. *Chemosphere.* V. 30, 801-802, 1995.

Coutinho, V.F., Bittencourt, V.B. & Cozzolino, S.M.F. (2002). Effects of supplementation with Brazil nuts (CP, *Bertholletia excelsa* H.B.K.) in capoeira players on selenium (Se) concentration and glutathione peroxidases activity (GSH-PX, E.C.1.11.1.9). In: trace elements in man and animal. Springer ed., Part II.

Campo/PAS. (2003). Manual of quality and food safety for the Brazil nut. *Projeto PAS Campo*, Brasília: Embrapa/Sede., 69p.

Fleet-Stalder, V.V., Chasteen, T.G., Pickering, I.J., George, G.N. & Prince, R.C. (2000). Fate of selenate and selenite metabolized by *Rhodobacter sphaeroides*. *Appl. Environ. Microbiol.* 66:4849-4853.

Freire, F.C.O. & Offord, L. (2000). Mycoflora and mycotoxins in Brazilian black pepper, white pepper and Brazil nuts. *Mycopathologia*, 149, 13-19.

Geyikoglu F. & Turkez, H.   (2006). Protective effect of sodium selenite against the genotoxicity of aflatoxin B_1 in human whole blood cultures. *Braz. Arch. Biol. Technol.*, 49:393-398.

Gharieb, M. & Gadd. G. (1998). Role of glutathione in detoxification of metal(loid)s by Saccharomyces cerevisiae. *Bio-Metals.* 2: 183:188.

Gregory, J.F.& Edds G.T. (1984) Effect of dietary selenium on the metabolism of aflatoxin B1 in turkeys. *Food and Chemistry Toxicol.* 22:637-642.

Jayashree, T. & Subramanyam, C. (2000). Oxidative stress as a prerequisite for aflatoxin production by *Aspergillus parasiticus. Free Radical Biology & Medicine*, v.29, N.10, 981-985.

Kornerup, A. and Wanscher, J.H. (1989). Methuen Handbook of Colour 3rd ed. London: Eyre Methuen.

Letavayová, L., VIcková, V. & Brozmanová, J. (2006). Selenium: From cancer prevention to DNA damage. *Toxicology*, 227, 1-14.

Li. Z., Guo, S. & Li, L. (2003). Bioeffects of selenite on the growth of *Spirulina platensis* and its biotransformation. *Bioresource Technology*, 89: 171–176

Martin, S.E. & Schillaci, M.(1984). Inibitory effects of selenium on mutagenicity. *J. Agricult. Food Chemistry.* 32:425-433.

Martins, H.M., Martins, M.L. & Bernardo, F.A. (2000). Interaction of strains of non-toxigenic *Aspergillus flavus*, with *Aspergillus parasiticus* of Aflatoxin production. *Braz. J. vet. Res. Anim Sci*, v.37.n.6.

McCready, R.G.L. Campbell, J.N. & Payne, J.L. (1966). Selenite reduction by *Salmonella heidelberg. Canadian J. Microbiol.* 12:703-714.

Moss, M.O. & Badii, F. (1982). Increased production of aflatoxin by *A. parasiticus* Spear in the presence of rubratoxin B. *Appl. Environ. Microbiol.* 43:895-898.

Moss, M.O., Badii, F. & Gibbs, G. (1987). Reduction of biselenite to elemental selenium by *Aspergillus parasiticus*. Trans. Br. Mycol. Soc. 89(4):578-580.

Overvad, K., Thorling, E.B., Bjerring, P. & Ebbesen, P. (1985). Selenium inhibits UV light-induced skin carcinogenesis in hairless mice. *Cancer Lett.* 27:163-170.

Pacheco, A.M. & Scussel, V.M. (2007). Selenium and aflatoxin levels in raw Brazil nuts from the Amazon Basin. *J. of Agric.and Food Chemistry.* In press.

Pacheco, A.M. & Scussel, V.M. (2006). *Brazil Nut: From Tropical Forest to Consumer*. Editograf. 173pp.

Ragab, A.M., Ramadan, S.E., Razak, A.a. & Ghonamy, E.A. (1986). Selenium sportion by some seleno-tolerant fungi. In DI Alani and Moo-young (eds), *Perspective in Biotechnology and Applied Microbiology* 343-353.

Ramadan, S.E., Haroum, B.M. & Rarak, A.A. (1983). Morphological distortions of some fungi by the action of Cadmium, selenium and tellurium. *Proc. V. Conf. Microbiol.* 1983, May, 1(1):159-170, Ciro, Egypt.

Roux, M., Sarret, G., Pignot-Paintrand, I., Frontecave, M. & Coves, J. (2001). Mobilization of selenite by *Ralstonia metallidurans* CH34. *Appl. Environ. Microbiol.* 67:769-773.

Scussel, V.M. (2004). Aflatoxin and food safety: Recent South American Perspectives. *J. of Toxicology. Toxin Reviews*, 23(2 & 3):179-216.

Souza, M. L. & Menezes, H. C. (2004). Processing of Brazil nut and meal cassava flour: quality parameters. *Cienc. Tecnol. Aliment.*, v. 24, n° 01, 120-128.

US EPA- United States Environmental Protection Agency (EPA). (1996). "Method 6010B: Inductively Coupled plasma-atomic emission spectrometry." *SW-846 Test Methods for Evaluating Solid Wastes, Physical-chemical Methods.*

Valko,M., Rhodes, B.C, Moncol, J., Izakovic, M. & Mazur, M. (2006). Free radicals, metals and antioxidants in oxidative stress-induced cancer. *Chemico-Biological Interactions*, 160: 1–40.

Zohri, A.A., Saber, S.M. & Mostafa, E. (1997). Effect of selenite and tellurite on the morphological growth and toxin production of *Aspergillus parasiticus* var. *globosus* IMI120920. *Mycopathologia*, 139:51-57.

Permissions

The contributors of this book come from diverse backgrounds, making this book a truly international effort. This book will bring forth new frontiers with its revolutionizing research information and detailed analysis of the nascent developments around the world.

We would like to thank Dr. Irineo Torres Pacheco, for lending his expertise to make the book truly unique. He has played a crucial role in the development of this book. Without his invaluable contribution this book wouldn't have been possible. He has made vital efforts to compile up to date information on the varied aspects of this subject to make this book a valuable addition to the collection of many professionals and students.

This book was conceptualized with the vision of imparting up-to-date information and advanced data in this field. To ensure the same, a matchless editorial board was set up. Every individual on the board went through rigorous rounds of assessment to prove their worth. After which they invested a large part of their time researching and compiling the most relevant data for our readers. Conferences and sessions were held from time to time between the editorial board and the contributing authors to present the data in the most comprehensible form. The editorial team has worked tirelessly to provide valuable and valid information to help people across the globe.

Every chapter published in this book has been scrutinized by our experts. Their significance has been extensively debated. The topics covered herein carry significant findings which will fuel the growth of the discipline. They may even be implemented as practical applications or may be referred to as a beginning point for another development. Chapters in this book were first published by InTech; hereby published with permission under the Creative Commons Attribution License or equivalent.

The editorial board has been involved in producing this book since its inception. They have spent rigorous hours researching and exploring the diverse topics which have resulted in the successful publishing of this book. They have passed on their knowledge of decades through this book. To expedite this challenging task, the publisher supported the team at every step. A small team of assistant editors was also appointed to further simplify the editing procedure and attain best results for the readers.

Our editorial team has been hand-picked from every corner of the world. Their multi-ethnicity adds dynamic inputs to the discussions which result in innovative outcomes. These outcomes are then further discussed with the researchers and contributors who give their valuable feedback and opinion regarding the same. The feedback is then collaborated with the researches and they are edited in a comprehensive manner to aid the understanding of the subject.

Apart from the editorial board, the designing team has also invested a significant amount of their time in understanding the subject and creating the most relevant covers. They scrutinized every image to scout for the most suitable representation of the subject and create an appropriate cover for the book.

The publishing team has been involved in this book since its early stages. They were actively engaged in every process, be it collecting the data, connecting with the contributors or procuring relevant information. The team has been an ardent support to the editorial, designing and production team. Their endless efforts to recruit the best for this project, has resulted in the accomplishment of this book. They are a veteran in the field of academics and their pool of knowledge is as vast as their experience in printing. Their expertise and guidance has proved useful at every step. Their uncompromising quality standards have made this book an exceptional effort. Their encouragement from time to time has been an inspiration for everyone.

The publisher and the editorial board hope that this book will prove to be a valuable piece of knowledge for researchers, students, practitioners and scholars across the globe.

List of Contributors

Huili Zhang and Xianjun Meng
Academy of Food Science Shenyang Agricultural University, Shenyang, China

Huili Zhang and Jianwei He
School of Life Science, Liaoning University, Shenyang, China

Bing Li, Hui Xiong and Wenjie Xu
College of Light Industry, Liaoning University, Shenyang, China

Laura Anfossi, Claudio Baggiani, Cristina Giovannoli and Gianfranco Giraudi
Department of Analytical Chemistry, University of Turin, Italy

Alonso V.A., Dalcero A.M., Chiacchiera S.M. and Cavaglieri L.R.
Departamento de Microbiología e Inmunología, Facultad de Ciencias Exactas, Físico-Quími-cas y Naturales, Universidad Nacional de Río Cuarto, Cuarto, Córdoba, Argentina

Alonso V.A., González Pereyra M.L., Armando M.R. and Dogi C.A.
Consejo Nacional de Investigaciones Científicas y Técnicas (CONICET), Argentina

Chiacchiera S.M.
Departamento de Química, Facultad de Ciencias Exactas, Físico-Químicas y Naturales, Uni-versidad Nacional de Río Cuarto, Cuarto, Córdoba, Argentina

Dalcero A.M., Chiacchiera S.M. and Cavaglieri L.R.
Consejo Nacional de Investigaciones Científicas y Técnicas (CONICET), Argentina

Rosa C.A.R.
Departamento de Microbiologia e Imunología Veterinaria, Universidade Federal Rural do Rio de Janeiro, Instituto de Veterinaria, Rio de Janeiro, Brazil

Rosa C.A.R.
Conselho Nacional de Pesquisas Científicas (CNPq), Brazil

Eduardo Micotti da Gloria
University of Sao Paulo – ESALQ, Brazil

Simone Aquino
Universidade Nove de Julho/ UNINOVE, São Paulo, Brazil

Benedito Corrêa
Instituto de Ciências Biomédicas/ USP, São Paulo, Brazil

Anna Chiara Manetta
Department of Food and Feed Science, University of Teramo, Italy

Imtiaz Hussain
Govt. Post Graduate College Samundri, Faisalabad, Pakistan

Lucia Mosiello and Ilaria Lamberti
ENEA, Italian National Agency for New, Technologies, Energy and the Environment, Rome, Italy

Alejandro Espinosa-Calderón, Luis Miguel Contreras-Medina, Rafael Francisco Muñoz-Huerta, Jesús Roberto Millán-Almaraz, Ramón Gerardo Guevara González and Irineo Torres-Pacheco
C.A. Ingeniería de Biosistemas, División de Estudios de Posgrado, Facultad de Ingeniería, Universidad Autónoma de Querétaro, Querétaro, Qro, México

Da-Ling Liu, Hui-Yong Tan, Jun-Hua Chen, Ada Hang-Heng Wong, Chun-Fang Xie, Shi-Chuan Li, Hong Cao and Dong-Sheng Yao
Institute of Microbial Biotechnology, Jinan University, Guangzhou, P.R.China

Chun-Fang Xie and Dong-Sheng Yao
National Engineering Research Center of Genetic Medicine, Guangzhou, P.R.China

Da-Ling Liu
Guangdong Provincial Key Laboratory of Bioengineering Medicine, Guangzhou, P.R.China

Jun-Hua Chen
Institutes of Biomedicine and Health, Chinese Academy of Sciences, Guangzhou, P.R.China

Meng-Ieng Fong
IACM Laboratory, Macau, P.R.China

Peiwu Li and Qi Zhang
Key Laboratory of Biotoxin Analysis of Ministry of Agriculture, Key Laboratory of Oil Crops Biology of the Ministry of Agriculture, Oil Crops Research Institute, Chinese Academy of Agricultural Sciences, Wuhan, China

Daohong Zhang, Di Guan, Xiaoxia, Ding Xuefen Liu, Sufang Fang, Xiupin Wang and Wen Zhang
Key Laboratory of Biotoxin Analysis of Ministry of Agriculture, Key Laboratory of Oil Crops Biology of the
Ministry of Agriculture, Oil Crops Research Institute, Chinese Academy of Agricultural Sciences, Wuhan, China

Maria Pia Santacroce, Valentina Zacchino and Gerardo Centoducati
Department of Public Health and Animal Science, Division of Aquaculture, Faculty of Veterinary Medicine, University of Bari,

Marcella Narracci, Maria Immacolata Acquaviva and Rosa Anna Cavallo
Institute for Coastal Marine Environment, National Research Council of Italy, Taranto, Italy

T. Mahmood and T.N. Pasha
Department of Food and Nutrition, University of Veterinary and Animal Sciences, Lahore, Pakistan

F.M. Khattak
Avian Science Research Centre, SAC, Edinburgh, UK

Abdolamir Allameh
Department of Biochemistry, Faculty of Medical Sciences, Tarbiat Modares University, Tehran, I. R. Iran

Tahereh Ziglari
Faculty of Medicine, Islamic Azad University, Qeshm International Branch, Qeshm Island, I. R. Iran

Iraj Rasooli
Faculty of Basic Sciences, Shahed University, Tehran, I. R. Iran

Eduardo Madrigal-Bujaidar and Isela Álvarez-González
Laboratorio de Genética, Escuela Nacional de Ciencias Biológicas, IPN, México

Osiris Madrigal-Santillán and Jose Antonio Morales-González
Instituto de Ciencias de la Salud, UAEH, México

Laura Mejía-Teniente, Angel María Chapa-Oliver, Moises Alejandro Vazquez-Cruz, Irineo Torres-Pacheco and Ramón Gerardo Guevara-González
Facultad de Ingeniería, CA Ingeniería de Biosistemas, Universidad Autónoma de Querétaro, Centro Universitario Cerro de las Campanas s/n, Querétaro, Qro, México

Ariane Mendonça Pacheco
Faculty of Pharmaceutical Sciences, Federal University of Amazonas, Manaus-AM, Brazil

Vildes Maria Scussel
Food Science and Technology Department, Centre of Agricultural Sciences, Federal University of Santa Catarina, Florianopolis, Brazil

www.ingramcontent.com/pod-product-compliance
Lightning Source LLC
Chambersburg PA
CBHW070715190326
41458CB00004B/989